T0211108

Lecture Notes in Computer Science   11880

Founding Editors

Gerhard Goos, Germany
Juris Hartmanis, USA

FoLLI Publications on Logic, Language and Information

Subline of Lectures Notes in Computer Science

More information about this series at http://www.springer.com/series/7407

Ronald de Haan

Parameterized Complexity in the Polynomial Hierarchy

Extending Parameterized Complexity Theory
to Higher Levels of the Hierarchy

 Springer

Author
Ronald de Haan
Institute for Logic, Language
and Computation
University of Amsterdam
Amsterdam, The Netherlands

ISSN 0302-9743   ISSN 1611-3349   (electronic)
Lecture Notes in Computer Science
ISBN 978-3-662-60669-8   ISBN 978-3-662-60670-4   (eBook)
https://doi.org/10.1007/978-3-662-60670-4

LNCS Sublibrary: SL1 – Theoretical Computer Science and General Issues

"Parameterized Complexity in the Polynomial Hierarchy" was co-recipient of the E. W. Beth Dissertation Prize 2017 for outstanding dissertations in the fields of logic, language, and information. This work extends the theory of parameterized complexity to higher levels of the Polynomial Hierarchy (PH). For problems at higher levels of the PH, a promising solving approach is to develop fixed-parameter tractable reductions to SAT, and to subsequently use a SAT solving algorithm to solve the problem. In this dissertation, a theoretical toolbox is developed that can be used to classify in which cases this is possible. The use of this toolbox is illustrated by applying it to analyze a wide range of problems from various areas of computer science and artificial intelligence.

This Springer imprint is published by the registered company Springer-Verlag GmbH, DE
part of Springer Nature
The registered company address is: Heidelberger Platz 3, 14197 Berlin, Germany

Preface

This book came about during my PhD studies at the Technische Universität Wien, from 2012 until 2016.

It originally appeared as my PhD dissertation in 2016, for which I was awarded the E. W. Beth Dissertation Prize for outstanding contributions in the domains of logic, language, and information by the Association for Logic, Language, and Information (FoLLI) in 2017.

As a result, this book is now being published in the FoLLI Publications on Logic, Language, and Information, in Springer's *Lecture Notes in Computer Science* series.

The theme of this book is "The Extension of Parameterized Complexity Theory to Higher Levels of the Polynomial Hierarchy."

This theme allows for theoretical explorations that are motivated by the aim to develop theoretical tools that can be used to classify whether computational problems can be solved by the approach of first reducing them to SAT in fixed-parameter tractable time, and then solving them by invoking a SAT solving algorithm.

This toolbox is relevant for computational problems from all kinds of domains, in fact, many of the theoretical developments and applications in this book are driven by concrete, natural problems from a variety of areas of computer science and artificial intelligence.

I hope that this book will serve as a starting point for future theoretical investigations on this topic.

I am grateful to all the people that supported me throughout my PhD studies; helping me directly or indirectly in writing this book.

In particular, I owe many thanks to my PhD supervisor Stefan Szeider, for showing me how to do theoretical research, and for giving me the freedom to discover, as well as become an expert in an area of research that fits my interests.

September 2019   Ronald de Haan

Contents

Chapter 1
Introduction

Perhaps the single most profound fact that has been uncovered by computer science so far is the ubiquity of computational intractability.

— Rod Downey and Mike Fellows,
Parameterized Complexity [67]

Life in modern societies is full of hard problems. What is the best way to organize the economy? Or can you finally make the decision to quit that job that pays well but makes you unhappy? The hard problems that are pertinent to this thesis, however, are of a more computational nature, and consist of search problems for which even the fanciest, most modern supercomputers often fail to find a solution within a reasonable amount of time—by any standard—due to the combinatorial explosion in the search space. Such problems show up in a myriad of settings—they play a role in abstract, scientific domains such as computer science, physics and bioinformatics, but also lie at the basis of worldly tasks like finding optimal routes or schedules. A striking example of such a *computationally intractable* problem is the problem of finding the shortest total route for a traveller that wants to visit every city on a given map. The relevance of such problems for copious areas of human activity is hard to overestimate.

Research on computer science and engineering has led to a multitude of ways of dealing with computational intractability. Two of the most productive of these approaches are (1) the use of *fixed-parameter tractable* algorithms to efficiently solve problem inputs that exhibit certain types of structure, and (2) solving intractable problems by *encoding problem inputs into* the language of *key problems* for which powerful algorithms are available that work well in many cases (e.g., the problem of propositional satisfiability, also called SAT)—and subsequently using these algorithms to solve the problem. Recently, a potentially more powerful method has been put forward that combines these two approaches: using fixed-parameter tractable algorithms to encode problem inputs as inputs for problems such as SAT.

© Springer-Verlag GmbH Germany, part of Springer Nature 2019
R. de Haan: Parameterized Complexity in the Polynomial Hierarchy, LNCS 11880,
https://doi.org/10.1007/978-3-662-60670-4_1

In this thesis, we perform a theoretical analysis of the possibilities and limits of this novel approach. We use the framework of parameterized complexity theory to make precise the general concept of using fixed-parameter tractable algorithms to encode problems into SAT, and we distinguish several different formal interpretations of this concept. We develop a new completeness theory that allows us to adequately characterize the computational complexity of problems for which this new approach might be applied successfully, and we show how these new theoretical tools can be used—in combination with existing tools—to identify the limits of the method of fixed-parameter tractable SAT encodings. Moreover, we initiate a structured investigation of the possibilities of this method by using the developed theoretical tools for a complexity analysis of a wide range of problems from numerous domains in computer science and artificial intelligence.

1.1   Context: Intractability and a New Way of Coping

In order to properly describe the problem addressed in this thesis, we firstly give a brief overview of existing research on computational intractability, and the two methods for coping with intractability that we mentioned above.

One of the most productive ways of analyzing the running time of an algorithm is by studying its worst-case complexity—that is, identifying the maximum number $z(n)$ of steps that the algorithm takes on any input of size n. This allows researchers, for instance, to make the crucial distinction between exponential-time algorithms—with running times such as 2^n—and polynomial-time algorithms—with running times such as n^2. Problems that are solvable using polynomial-time algorithms are often described as *tractable*, as these algorithms can generally be employed successfully in practice. Similarly, problems that (suspectedly) do not admit algorithms with such favorable running times are commonly described as *intractable*.

Unfortunately, intractable problems are ubiquitous in a wide range of areas. The seminal work of Garey and Johnson [86], for instance, lists hundreds of relevant intractable problems. Therefore, intractability cannot simply be ignored. Research over the last several decades has led to many ways of dealing with intractability (see, e.g., [59, 63, 68]).

For the purposes of this thesis, we are interested in two algorithmic methods that have led to the development of many algorithms that can be used to effectively tackle intractable problems in practice. The first of these methods involves extending the notion of tractability to running times that are exponential (or worse), but where each non-polynomial contribution to the running time depends only on a limited part of the problem input that can be assumed to be small in practice (the *parameter*). This extended notion of tractability is called *fixed-parameter tractability*. Examining the running time of algorithms in terms of such a parameter—in addition to the input size—allows for the design of algorithms that are tractable because they exploit structure that is present in the input. The research area that investigates fixed-parameter tractable algorithms is called *parameterized complexity*, and has been very productive over the last few decades (see, e.g., [27, 57, 66, 67, 85, 166]). Even though fixed-parameter tractable algorithms are not tractable in the sense of polynomial-time computation, they perform extremely well in many settings.

The second method consists of encoding problem inputs (in polynomial time) as inputs for one of a number of key problems, for which algorithms have been engineered that work surprisingly well in many cases (but still take exponential time in the worst case). Perhaps the most prominent target problem for such encodings is the problem of propositional satisfiability (SAT). This method lies at the basis of some of the most successful algorithmic techniques for important problems in hardware and software design, operations research, artificial intelligence, and many other areas (see, e.g., [22]).

Both of these methods have been very successful in many important settings in computer science and artificial intelligence. However, both methods also have their limits. For fixed-parameter tractable algorithms to be useful in practice, the parameter needs to be small. This often severely restricts the set of problem inputs for which fixed-parameter tractable algorithms can be employed effectively. The use of polynomial-time SAT encodings is restricted to a class of problems that is known as NP. Many important problems fall outside this class, and thus these problems cannot be encoded into SAT efficiently (that is, in polynomial time).

The motivation behind the work in this thesis comes from the idea of combining the above two techniques for developing effective algorithms, which was recently put forward (see, e.g., [82, 173]). In this combination, problem inputs are encoded into instances of SAT by means of an algorithm that runs in fixed-parameter tractable time, rather than in polynomial time. This way, the benefit of using parameterized complexity to exploit structure that is present in the input can be combined with the great performance of modern SAT solving algorithms. The use of such *fpt-reductions to SAT* (short for: fixed-parameter tractable reductions to SAT) offers great potential for increasing the range of problems for which SAT solving algorithms can be applied.

1.2   Structured Complexity Investigation

The concept of employing fixed-parameter tractable algorithms to encode inputs of intractable problems as propositional formulas and subsequently calling a SAT solving algorithm to decide the satisfiability of the propositional formula is straightforward to implement. One identifies a suitable parameter, and uses the intuition behind this parameter to construct an algorithm to perform the encoding that runs in fixed-parameter tractable time. Many algorithmic techniques to develop fixed-parameter tractable algorithms are known, and these can be readily applied to develop fixed-parameter tractable SAT encodings.

Unsurprisingly, not in all cases the problem input can be encoded as a propositional formula in fixed-parameter tractable time. There are problems for which it is evident that an fpt-reduction to SAT is not possible. For instance, if a problem falls outside the class NP already for a single constant value of the parameter, then we cannot hope to construct a fixed-parameter tractable SAT encoding. In such cases, we can use known tools from parameterized and classical complexity theory to rule out fpt-reductions to SAT. There are, however, also problems for which the situation is less obvious. For many problems, there exists a SAT encoding that runs in polynomial-time for each separate value of the parameter, where the order of the polynomial depends on the parameter value. Such SAT encodings do not run in fixed-parameter tractable time, because the polynomial running time—when the contribution of the parameter value on the running time is disregarded—must be completely independent of any contribution of the parameter value to the running time. In these more subtle cases, known parameterized complexity theory does not provide adequate tools to rule out the possibility of fpt-reductions to SAT.

There are also cases where a problem input can be encoded in fixed-parameter tractable time into two propositional formulas and can subsequently be solved by determining the satisfiability of these formulas, but cannot be solved by an fixed-parameter tractable encoding into a single propositional formula. More generally, the notion of fpt-reductions to SAT can be extended to encodings into multiple propositional formulas, and every increase in the number of propositional formulas leads to higher solving power. Known parameterized complexity tools are not sufficient to explore the power of such *fixed-parameter tractable Turing reductions to SAT*. Both for identifying when it is possible to construct an fpt-reduction to a certain number of propositional formulas, and for identifying when this is not possible, the known theory is lacking.

In short, in order to get a good understanding of how far exactly the potential of the various notions of fpt-reductions to SAT reaches, a structured complexity investigation needs to be performed—and for this an adequate complexity-theoretic framework needs to be developed.

1.2.1 Problem Statement

In this thesis, we set out to enable and start a structured complexity-theoretic investigation of the possibilities and limits of the method of using fpt-reductions to SAT to solve intractable problems originating in computer science and artificial intelligence. We do so, by and large, by addressing the following four gaps in the literature on parameterized complexity theory and the application of parameterized complexity methods to investigate problems arising in computer science and artificial intelligence.

1. *The palette of the theoretical possibilities of encoding problem inputs in fixed-parameter tractable time into various numbers of propositional formulas remains largely unexplored.* That is, only for the case of encodings into a single propositional formula, the corresponding parameterized complexity class has been considered in the literature— this class is known as para-NP. Moreover, even for this

case, membership in the class para-NP has only been interpreted as a positive result in a handful of cases.

2. *The theoretical techniques that are available to rule out the existence of fpt-reductions to SAT are insufficient.* Hardness results for several known parameterized complexity classes (for instance, hardness for para-Σ_2^p) can be used to provide evidence that fpt-reductions to SAT are not possible. However, hardness for these classes can be established only in radical cases, where the impossibility of fpt-reductions to SAT is obvious.

3. *More generally, an appropriate fine-grained parameterized complexity toolbox is lacking, to characterize the complexity of parameterized problems that lie at various levels between the extremes of the spectrum*—that is, between para-NP and para-Σ_2^p. Such a toolbox is necessary to obtain matching lower and upper bounds for the complexity of many interesting parameterized problems that originate in numerous areas of computer science and artificial intelligence.

4. *The concept of employing fpt-reductions to SAT to deal with computational intractability has been applied only for a very limited number of problems.* For many relevant and interesting problems that arise in various domains of computer science and artificial intelligence, practically useful algorithms could potentially be developed using the method of fpt-reductions to SAT. Therefore, there is need for a structured parameterized complexity investigation for such problems, that is focused on the possibilities and limits of fpt-reductions to SAT.

1.3 Contributions

In this section, we describe the contributions of this thesis, addressing the four shortcomings of existing research about fpt-reductions to SAT that we pointed out above.

We investigate the different possibilities of solving parameterized problems by means of fixed-parameter tractable many-to-one or Turing reductions to SAT, and the additional power that is provided by more queries to a SAT oracle (for the case of Turing reductions).

- *We provide the first structured parameterized complexity investigation where membership in* para-NP *and in* para-co-NP *is the foremost target for positive results.* The known parameterized complexity class para-NP consists of all problems that admit a many-to-one fpt-reduction to SAT. Similarly, its dual class para-co-NP consists of all problems that are many-to-one fpt-reducible to UN-SAT, the co-problem of SAT. Even though these classes are known from the literature, they have been used as a target for positive complexity results only in a few cases.

- *We consider several parameterized complexity classes that characterize parameterized problems that can be solved by a fixed-parameter tractable Turing reduction to SAT*—that is, by an fpt-algorithm that can make a certain number of queries to a SAT oracle. In addition to several parameterized complexity classes that are a result of a generic scheme known from the literature for constructing parameterized complexity classes, we consider the parameterized complexity class $FPT^{NP}[few]$ that consists of those parameterized problems that can be solved by a fixed-parameter tractable Turing reduction to SAT that uses at most $f(k)$ oracle queries, where k denotes the parameter value and f is some computable function.

- *We develop theoretical tools for providing lower bounds on the number of oracle queries for any fixed-parameter tractable algorithm to solve certain problems.* These tools allow us to distinguish between parameterized problems that admit a many-to-one fpt-reduction to SAT, on the one hand, and problems that can only be solved in fixed-parameter tractable time with multiple queries to a SAT oracle, on the other hand.

To enable an investigation of the limits of fpt-reductions to SAT, we develop parameterized complexity tools for showing that in certain cases fpt-reductions to SAT are not possible.

- *We show that hardness for the known parameterized complexity class A[2] can be used to argue that a parameterized problem does not admit a many-to-one fpt-reduction to SAT.* This argument is based on a complexity-theoretic assumption that is related to the (non)existence of some subexponential-time algorithms. In particular, we prove that if any A[2]-hard parameterized problem is many-to-one fpt-reducible to SAT, then there exists a subexponential-time reduction from a canonical problem at the second level of the Polynomial Hierarchy—$QSAT_2(3DNF)$—to SAT. Assuming that such a subexponential-time reduction does not exist, we can then rule out many-to-one fpt-reducibility to SAT by showing that a problem is A[2]-hard. (This line of reasoning can also be applied to rule out many-to-one fpt-reductions to UNSAT.)

- *We show that hardness for A[2] can additionally be used to rule out fixed-parameter tractable Turing reductions to SAT.* Concretely, we prove that if any A[2]-hard parameterized problem is solvable by a fixed-parameter tractable algorithm with access to a SAT oracle, then there exists a subexponential-time Turing reduction from $QSAT_2(3DNF)$ to SAT.

We develop new parameterized complexity classes to accurately distinguish the subtly different levels of complexity for parameterized variants of problems at higher levels of the Polynomial Hierarchy.

- *We provide formal evidence for the claim that the known parameterized complexity classes are insufficient to adequately characterize the complexity of various parameterized variants of problems at the second level of the Polynomial Hierarchy.* We do so by considering the consistency problem for disjunctive answer set programming as an example. We show that various parameterizations of this example cannot be complete for any of the known classes (under various complexity-theoretic assumptions).

- *We develop novel parameterized complexity classes that map out the parameterized complexity landscape between the first and the second level of the Polynomial Hierarchy.* We denote these parameterized complexity classes by the names $\Sigma_2^p[k*]$ and $\Sigma_2^p[*k, t]$, as they are based on various weighted parameterized variants of the quantified satisfiability problem QSAT_2 that is canonical for the class Σ_2^p at the second level of the Polynomial Hierarchy. They can be considered as generalizations of the parameterized complexity classes of the well-known Weft hierarchy.

- *We strengthen the intuition behind the type of non-determinism that plays a role in the problems in these parameterized complexity classes by establishing alternative characterizations of several of these classes.* The new parameterized complexity classes are based on various variants of a quantified satisfiability problem for Boolean circuits. We give alternative characterizations that are based on (i) variants of a quantified satisfiability problem for propositional formulas, (ii) the model checking problem for a particular class of first-order logic formulas, and (iii) a particular class of alternating Turing machines. The third characterization can be seen as an analogue of the Cook-Levin Theorem for the new parameterized complexity classes.

- *We show that the new parameterized complexity classes give rise to a completeness theory that can be used to characterize the complexity of interesting parameterized problems, which is not possible using previously known parameterized complexity classes.* We establish this point by looking again at the example of the consistency problem for disjunctive answer set programming.

- *We further substantiate the new completeness theory by showing that many natural parameterized variants of interesting problems at the second level of the Polynomial Hierarchy are complete for (one of) the novel parameterized complexity classes.* These problems originate from a variety of domains from computer science and artificial intelligence.

- *We demonstrate that the new parameterized complexity classes can be used to strengthen the toolbox for providing evidence that fpt-reductions to SAT are not possible in certain cases.* In particular, we show that any problem that is hard for several of the new parameterized complexity classes does not admit an fpt-reduction to SAT, unless there exists a subexponential-time reduction from QSAT_2 to SAT.

- *We draw connections between the newly developed parameterized complexity classes and other areas of (parameterized) complexity theory.* Concretely, in addition to establishing relations to the (non)existence of various subexponential-time algorithms, we relate these classes to the area of non-uniform parameterized complexity, resulting in several parameterized analogues of the Karp-Lipton Theorem.

- *We generalize the newly developed parameterized complexity classes to arbitrary levels of the Polynomial Hierarchy.* These additional classes have the capability of providing a very fine-grained analysis of the complexity of parameterized problems at every level of the Polynomial Hierarchy.

- *We develop another novel parameterized complexity class that arises when investigating natural parameterized variants of* PSPACE-*complete problems.* This class adds to the richness of the parameterized complexity toolbox for classifying parameterized variants of problems at various levels of the Polynomial Hierarchy and problems in PSPACE.

We initiate a structured parameterized complexity investigation for problems from various domains of computer science and artificial intelligence that is focused on identifying settings where fpt-reductions to SAT are possible.

- *We demonstrate the potential of fpt-reductions to SAT for obtaining positive results that could lead to useful algorithms by pointing out several results from the literature that can be seen as fpt-reductions to SAT.* Several of these existing results had not previously been identified as fpt-reductions to SAT. For one of the cases that we consider, the algorithmic technique that gives rise to the fpt-reduction to SAT underlies one of the most competitive algorithmic approaches available to solve this problem— namely, the method of bounded model checking for (a fragment of) linear-time temporal logic, which has important applications in many areas of computer science and engineering.
- *We show productive techniques from parameterized complexity theory can be used to develop fixed-parameter tractable reductions to SAT.* In particular, we show how the concepts of treewidth and backdoors can be used to identify settings where fpt-reductions to SAT are possible.
- *We use the developed theoretical machinery to investigate whether fpt-reductions to SAT are possible for many natural parameterizations of a wide range of problems from various areas of computer science and artificial intelligence.* For several cases we construct fpt-reductions to SAT, and for many other cases we establish completeness results for various classes in the parameterized complexity landscape at higher levels of the Polynomial Hierarchy.
- *We provide an overview of this parameterized complexity investigation focused on the possibilities and limits of fpt-reductions to SAT in the form of a compendium.* This compendium provides a list of all parameterized problems that we consider in this thesis, together with the complexity results that we establish for these problems. These parameterized problems are based on problems whose complexity lies at the second level of the Polynomial Hierarchy or higher.

1.3.1   Main Contributions

The most important contributions of this thesis can be summarized as follows.

- *We pave the way for future research that investigates whether useful algorithms can be developed using the method of fpt-reductions to SAT for concrete problems from computer science, artificial intelligence, and other domains.*

- We develop a theoretical parameterized complexity framework that contains both (1) parameterized complexity classes for the various possible incarnations of the general scheme of fpt-reductions to SAT, and (2) parameterized complexity classes for problems that do not admit fpt-reductions to SAT.
- We show how this framework can be used for many relevant parameterized problems to determine whether an fpt-reduction to SAT is possible, and if so, what kind of fpt-reduction to SAT is needed to solve the problem.
- We initiate a structured investigation of the parameterized complexity of problems at higher levels of the Polynomial Hierarchy by considering many natural parameterized variants of such problems, and analyzing their complexity using the new theoretical framework.

1.3.2 Overview of the Parameterized Complexity Framework

We provide a graphical overview of the most prominent parameterized complexity classes that feature in the theoretical framework that we develop in this thesis in Fig. 1.1 on p. 10. This figure shows how the newly developed parameterized complexity classes relate to previously known parameterized complexity classes, by indicating where they fit in the parameterized complexity landscape. The classes that are most relevant for the results in this thesis are located between the classes para-NP and para-co-NP, on the one hand, and the classes para-Σ_2^p and para-Π_2^p, on the other hand.

1.3.3 Research Impact

Even though it might seem at first sight that the results in this thesis only affect a highly specialized and technical subfield of theoretical computer science, the impact of our work extends to a broader range of research and applications in the area of computer science and engineering. For a description of the impact of our results that is aimed at an audience without specialized training in theoretical computer science or (parameterized) complexity theory, we refer to Sect. 17.2.

1.4 Roadmap

The remainder of this thesis, after this introductory chapter, is divided into six parts. In the first of these parts (**Part I: Foundations**), we provide an overview of relevant concepts and results from the areas of traditional computational complexity theory and parameterized complexity theory, in Chaps. 2 and 3, respectively. In particular, in

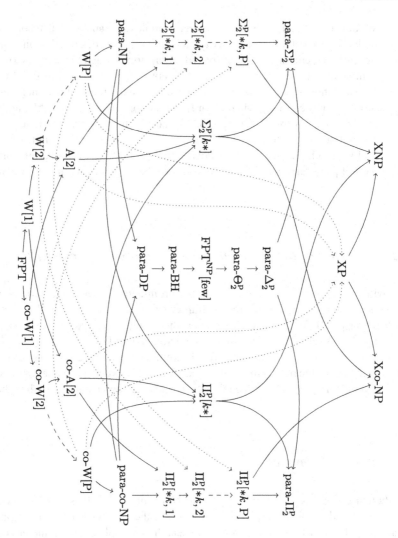

Fig. 1.1 The most prominent parameterized complexity classes that feature in this thesis.

Chap. 2 we put particular focus on complexity classes beyond NP, such as the classes of the Polynomial Hierarchy. In *Chap.* 3, after introducing and discussing the notion of fixed-parameter tractability and the corresponding parameterized complexity class FPT, we consider several other relevant parameterized complexity classes, such as the classes of the Weft hierarchy and the classes para-NP, para-co-NP, para-Σ_2^p and para-Π_2^p.

Then, in the next part (**Part II: Beyond para-NP**), we motivate and develop the parameterized complexity framework that can be used to investigate the possibilities and limits of fpt-reductions to SAT. We begin in *Chap.* 4 by surveying to what extent the existing parameterized complexity literature can be used to investigate the power of fpt-reductions to SAT. In particular, we consider several algorithmic results that can be seen as fpt-reductions to SAT, and we illustrate how productive parameterized complexity techniques can be used to obtain fpt-reductions to SAT. We also consider the first examples of parameterized problems that are unlikely to admit an fpt-reduction to SAT— claims that we support by showing hardness for the parameterized complexity classes A[2] and para-Σ_2^p.

Then, in *Chap.* 5, we argue that new parameterized complexity classes are needed to adequately characterize the complexity of many relevant parameterized variants of problems at higher levels of the Polynomial Hierarchy. In particular, we introduce the consistency problem for disjunctive answer set programming as a running example— this problem is complete for the second level of the Polynomial Hierarchy—and we consider various parameterized variants of this problem. We prove that several of these parameterized variants cannot be complete for any of the known parameterized complexity classes (under various complexity-theoretic assumptions).

In *Chap.* 6, we introduce and develop the parameterized complexity classes that underlie our framework. These classes are parameterized variants of the classical complexity class Σ_2^p that are based on various weighted variants of the quantified Boolean satisfiability problem that is canonical for the second level of the Polynomial Hierarchy, and are denoted by $\Sigma_2^p[k*]$ and $\Sigma_2^p[*k, t]$. We also make some observations about the relation between these new classes and known parameterized complexity classes, and we provide alternative characterizations of the new classes based on first-order logic model checking and alternating Turing machines. Moreover, we show completeness for several of the new parameterized complexity classes for those parameterized variants of the running example whose complexity we could not characterize adequately using known parameterized complexity classes.

Finally, in *Chap.* 7, we investigate several parameterized complexity classes that capture the possibilities of more powerful incarnations of the general scheme of fpt-reductions to SAT—namely fixed-parameter tractable Turing reductions to SAT. Most prominently, we introduce the parameterized complexity class FPT^{NP}[few] consisting of those parameterized problems that can be solved in fixed-parameter tractable time by an algorithm that can query a SAT oracle up to $f(k)$ times, where f is some computable function and where k denotes the parameter value. We illustrate this complexity class by giving several examples of problems that are complete for this class. Moreover, we develop some theoretical tools to establish lower bounds on the number of oracle queries that need to be made by fpt-algorithms to solve certain problems.

In the next part (**Part III: Applying the Theory**), we demonstrate how the newly developed theory can be used to analyze whether concrete problems from various areas of computer science and artificial intelligence admit fpt-reductions to SAT. In *Chap.* 8, we perform such an analysis for various problems from the area of Knowledge Representation and Reasoning, including abductive reasoning and a variant of the constraint satisfaction problem. In *Chap.* 9, we turn to the model checking problem for various fragments of temporal logics, where the Kripke structures are represented symbolically. The parameterized complexity investigation of this problem leads us to consider the parameterized complexity class PH(level), which is a new parameterized variant of the classical complexity class PSPACE. Then, in *Chap.* 10, we investigate various problems related to propositional satisfiability—such as minimizing DNF formulas and implicants of DNF formulas, and repairing inconsistent knowledge bases. In *Chap.* 11, we look at several parameterized variants of two problems that arise in the area of judgment aggregation—which is a subdomain of the research area of computational social choice. In *Chap.* 12, we study the parameterized complexity of a number of parameterized variants of the problem of propositional planning, before we finish this part in *Chap.* 13 by analyzing a number of graph problems, such as the problem of extending 3-colorings of the leaves of a graph to complete proper 3-colorings of the entire graph.

In the fourth part (**Part IV: Relation to Other Topics in Complexity Theory**), we investigate the relation between the parameterized complexity classes $\Sigma_2^p[k*]$ and $\Sigma_2^p[*k, t]$ that we developed, and concepts from other areas of computational complexity theory. Most notably, in *Chap.* 14, we extend a known result that relates parameterized complexity with subexponential-time complexity to the newly developed parameterized complexity classes in our framework. This known result relates the conjecture that FPT $\neq$ W[1] to the hypothesis that 3SAT cannot be solved in subexponential time. In particular, we show that parameterized problems that are hard for A[2], $\Sigma_2^p[k*]$ or $\Sigma_2^p[*k, t]$ do not admit fpt-reductions to SAT unless there exists one of various types of subexponential-time reductions from one of several quantified Boolean satisfiability problems that are canonical for the second level of the Polynomial Hierarchy to SAT or UNSAT.

Then, in *Chap.* 15, we study how the new parameterized complexity classes in our framework relate to various non-uniform parameterized complexity classes. This culminates in several parameterized analogues of the Karp-Lipton Theorem, that establish a close connection between non-uniform parameterized complexity and different parameterized variants of the classes of the Polynomial Hierarchy. For instance, one of these results states that if all problems in W[1] are solvable by circuits of fixed-parameter tractable size, then $\Pi_2^p[*k, 1] \subseteq$ para-Σ_2^p. In addition, to further motivate the investigation of non-uniform parameterized complexity classes, we show how several non-uniform parameterized complexity classes can be used in the setting of parameterized compilability.

We wrap up our investigation in the last regular part (**Part V: Conclusions**). In *Chap.* 16, we discuss open questions that remain and directions for future research. For instance, we elaborate on the possibility of gaining additional solving power by considering fixed-parameter tractable algorithms that have access to SAT oracles that can return satisfying assignments for propositional formulas that are satisfiable. Also, we suggest a way of distinguishing between various types of many-to-one fpt-reductions to SAT, which is based on a generalization of the concept of kernelization. Then, in *Chap.* 17, we summarize the results obtained in this thesis, and the impact of these results within the area of computer science and engineering.

In the final part of the thesis (**Appendices**), we provide some additional material. In *Appendix* A, we give an overview of the parameterized complexity results that we obtained—for all problems from the different areas of computer science and artificial intelligence that we considered—in the form of a compendium. Finally, in *Appendix* B, we generalize the parameterized complexity classes $\Sigma_2^p[k*]$ and $\Sigma_2^p[*k, t]$ to higher levels of the Polynomial Hierarchy.

We provide an overview of all parameterized problems that we consider in this thesis (grouped by their computational complexity), in the Index of Parameterized Problems on p. 397.

To help the reader find their way through the many chapters of this thesis, we provide a visual overview of the dependencies between the different chapters of this thesis in Fig. 1.2 on p. 14, together with several suggestions for what chapters to read and what chapters to skip. These suggestions allow the reader to focus on one of several different themes.

1.5 Reflecting on the Theoretical Paradigm

We conclude this introductory chapter by briefly reflecting on several aspects of the paradigm of parameterized complexity that we adopt and on several properties of the theoretical framework that we develop to investigate the limits and possibilities of fpt-reductions to SAT. Many aspects of the parameterized complexity paradigm that we discuss are not unique to the work in this thesis. Nevertheless, this discussion does help to shed light on the power and limitations of the framework that we develop.

1.5.1 Black Box Algorithms for NP-complete Problems

The parameterized complexity framework that we develop is based on a worst-case asymptotic complexity perspective. However, the exceptional performance of SAT solving algorithms cannot (yet) be satisfactorily explained in such a worst-case framework—in the worst case, all known SAT solving algorithms take exponential time, for instance. Therefore, the worst-case complexity paradigm alone cannot completely analyze the potential of the method of (1) first encoding problem inputs in fixed-parameter tractable time into one or more propositional formulas, and (2) subsequently using a SAT solving algorithm as a black box to decide the satisfiability of these formulas, and thereby solving the original problem input. Nevertheless, we use the worst-case complexity approach, for the following two reasons.

Firstly, the worst-case asymptotic complexity perspective turns out to be mathematically most productive. By adopting this perspective, we can use many techniques and tools from previous research on the topic of (parameterized) complexity theory that is also based on this perspective. Moreover, this way, we can relate our framework and the results that we establish to existing results in parameterized complexity theory. In short, using the worst-case complexity approach gives us access to a large and well-developed toolbox of mathematical methods.

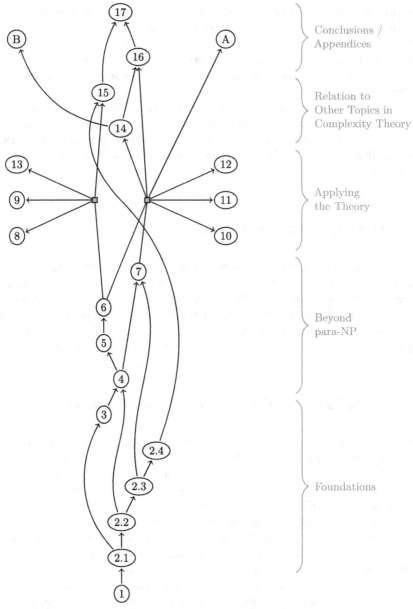

Theme	Chapters to read (in order)
Development of the classes $\Sigma_2^P[k*]$ and $\Sigma_2^P[*k, t]$	1–2.2, 3–6
Development of the class $\mathrm{FPT}^{\mathrm{NP}}[\mathrm{few}]$	1–2.3, 3, 4, 7
Structural parameterized complexity	1–7, 14, 15, B
Application to problems in CS & AI	1–13, A

Fig. 1.2 Dependencies among the chapters (and sections) of this thesis. An arrow from x to y indicates that x should be read before y.

Secondly, by using the worst-case complexity viewpoint, we can give performance guarantees on the running time of the algorithms that encode problem inputs as inputs for SAT. This way, we can ensure that the computational bottleneck for the entire solving method is not in the phase where the problem input is encoded as a propositional formula. After all, these encodings are intended to be used to obtain an effective problem solving approach.

Moreover, so far, in the discussion of NP-complete problems for which there exist highly efficient algorithms that work well in many settings in practice, we restricted our attention to the propositional satisfiability problem (SAT). For instance, we speak of "fpt-reductions to SAT." However, as the parameterized complexity framework that we develop is based on a worst-case asymptotic complexity perspective, one could use any other NP-complete problem instead. Take the problem of *integer linear programming (ILP)* as an example. This problem is NP-complete, and so for any parameterized problem there exists an fpt-reduction to SAT if and only if there exists an fpt-reduction to ILP. For this reason, all NP-complete problems can be used interchangeably in our theoretical framework. For the sake of convenience, in the remainder of the thesis, we will continue to speak of fpt-reductions to SAT.

In practice, it can of course make a huge difference for the efficiency of solving a particular problem whether you use a SAT solving algorithm or an algorithm for ILP. However, the aim of the work in this thesis is to develop a general theoretic framework, that is not restricted to specific problems and that can be used to investigate the theoretical possibilities and limits of the method of combining fpt-reductions and effective algorithms for NP-complete problems, rather than differentiating between the advantages of algorithms for one NP-complete problem over algorithms for another.

1.5.2 Focus on the Second Level of the Polynomial Hierarchy

In the development of our parameterized complexity framework, we focus mainly on parameterized variants of the complexity classes of the second level of the Polynomial Hierarchy. One reason for this is that the second level of the Polynomial Hierarchy is populated with many natural interesting problems that arise in many areas of computer science and artificial intelligence [179], whereas this is less the case for higher levels of the Polynomial Hierarchy. Another reason is that focusing on the second level results in a relatively tame setting, where we can conveniently investigate the various phenomena that play a role. Then, after having understood these phenomena and the way in which they interact in this setting, we can generalize our findings to higher levels of the Polynomial Hierarchy.

1.5.3   Complexity-Theoretic Assumptions

As is typically done in research on computational complexity, several lower bound
results that we establish in this thesis are based on various complexity-theoretic assumptions. Because it is not known for certain whether the complexity classes P
and NP are different or not, the use of such assumptions is necessary to establish
suitable lower bounds. Namely, if P = NP, then all problems in the Polynomial Hierarchy can trivially be reduced to SAT—in fact, in this case, they can all be solved in
polynomial time. Therefore, it is reasonable to use various complexity-theoretic assumptions to establish complexity results, especially if these assumptions are widely
believed and related to many important topics in complexity theory. An example
of such a widely-believed assumption states that the Polynomial Hierarchy does
not collapse—this is conjectured, for instance, in several textbooks on the topic of
computational complexity theory [7, 170].

However, for a number of the results that we develop in this thesis, we use
complexity-theoretic assumptions that are less standard—consequently, there is a
much less wide belief that these assumptions hold. For instance, our result that
A[2]-hard parameterized problems do not admit an fpt-reduction to SAT is based
on the assumption that there is no subexponential-time reduction from the problem $QSAT_2(3DNF)$ to SAT. Nonetheless, the use of such atypical assumptions to
establish complexity results is part of a productive methodology, for the following
reason. Suppose, on the one hand, that the assumption turns out to be true. Then
the complexity results that have been developed on top of this assumption are valid,
and thus we have made valuable discoveries. On the other hand, if the assumption
turns out to be false, we also gain useful knowledge about the computational complexity of important problems. For instance, if we were to learn that there in fact
do exist subexponential-time reductions from $QSAT_2(3DNF)$ to SAT, this would be
an enormous breakthrough in our understanding of the Polynomial Hierarchy. In
other words, the use of complexity-theoretic assumptions (related to important topics in complexity theory) to establish lower bound results can act as a double-edged
sword—either way it cuts, we gain useful knowledge.

Nevertheless, we urge the reader to be skeptical about the truth (or falsity) of
the complexity-theoretic assumptions that feature in the results in this thesis. Even
if a complexity-theoretic assumption serves as a useful working hypothesis in the
development of theoretical results that relate different concepts in complexity theory
to each other, it need not be true. Accordingly, it is probably wise to exercise a
certain amount of care when communicating results that involve atypical complexity-
theoretic assumptions [192].

1.5.4 Worst-Case Behavior of Fpt-Reductions

In our framework, we use the standard requirement on the running time of an algo-rithm (or reduction) in order for it to qualify as fixed-parameter tractable—namely, that its running time must be bounded by $f(k)n^c$, where f is an arbitrary computable function, where c is an arbitrary constant, where n denotes the input size, and where k denotes the parameter value. One can easily think of examples where such a run-ning time can be very effective—for instance, when $f(k) = 2^k$ and $c = 2$. However, this definition also allows for examples where the function f is much wilder. For instance, an algorithm whose running time is bounded by $A(k, k)n^2$—where A de-notes the Ackermann function—also qualifies as fixed-parameter tractable, since the Ackermann function is computable, even though it grows faster than any primitive recursive function. Already for very small values of k (e.g., $k \geq 4$), this running time can become very impractical. Nevertheless, there are several reasons why the admissive definition of fixed-parameter tractability is useful.

First of all, allowing arbitrary computable functions f leads to a robust and well-behaved mathematical theory. For instance, when composing two fpt-reductions, the upper bound on the running time of the resulting reduction contains a factor $f_1(f_2(k))$, where f_1 and f_2 are computable functions. Because we admit any computable func-tion in the upper bound of fixed-parameter tractable algorithms, we know from the fact that the composed function $(f_1 \circ f_2)$ is computable that the composed reduction is also an fpt-reduction. In short, allowing arbitrary computable functions f in the definition of fixed-parameter tractability results in useful mathematical properties of the framework.

Secondly, by allowing arbitrary computable functions f, the lower bound results that we develop are stronger. Namely, when we establish that an fpt-reduction to SAT is not possible in certain cases, for no choice of computable function f, we also rule out fpt-reductions to SAT whose running time bound includes reasonable functions f such as $f(k) = 2^k$. However, conversely, ruling out the existence of a reduction to SAT that runs in time $2^k n^c$, for some constant c, leaves open the possi-bility of a reduction that runs in time $2^{2^k} n^c$, for instance. In other words, the weaker our restrictions on the notion of fixed-parameter tractability, the more powerful our negative results can be.

Interestingly, it turns out that for many problems that are classified as fixed-parameter tractable, the running time of algorithms to solve these problems can be bounded using reasonably tame functions f. Often, once a problem is proved to be fixed-parameter tractable, insight is gained that can be used to construct efficient algorithms, and further algorithms for the problem are developed whose running time bound is based on a reasonable function f. We expect that a similar phenomenon occurs in the discovery and development of fpt-reductions to SAT.

We point out that a similar objection can be made for the admissive definition of polynomial-time solvability—namely that the corresponding requirement on the running time of $O(n^c)$, for some arbitrary constant c, also allows for unreasonable running times. For instance, a running time of n^{1000} is considered polynomial, but is highly impractical already when $n = 2$. Arguments similar to the ones described above can be made to justify the choice of this liberal definition of polynomial-time solvability.

Notes

The results in this thesis appeared in conference papers in the proceedings of COM-SOC 2014 [77], KR 2014 [116], SAT 2014 [112], SOFSEM 2015 [113], AA-MAS 2015 [78], IJCAI 2015 [109], KR 2016 [115], COMSOC 2016 [106], and ECAI 2016 [107], in an article appearing in the Journal of Computer and System Sciences [114], in an article appearing in the ACM Transactions on Computational Logic [108], in technical reports [105, 111, 117] and in an unpublished manuscript [110].

Foundations

Chapter 2
Complexity Theory
and Non-determinism

*Pretty well everybody outside the area of computer science
thinks that if your program is running too slowly, what you need
is a faster machine.*

— Rod Downey and Mike Fellows,
Fundamentals of Parameterized Complexity [66]

In order to establish some common ground on various concepts from computational complexity theory that will play a role in this thesis, we begin with giving a brief overview of these concepts. This will also allow us to clarify some notation that we will use throughout the thesis.

We begin by reviewing the overall framework that is used to analyze the complexity of computational problems, and by reviewing arguably the most commonly used complexity classes: P and NP. Then we move to the classes of the Polynomial Hierarchy and the class PSPACE. We discuss several classes of problems that can be solved with a bounded number of queries to an NP oracle, and we finish by mentioning some central concepts and results related to a branch of complexity theory known as non-uniform complexity.

Readers that are familiar with complexity theory can safely cherry-pick what parts of this chapter to read. For a more detailed treatment of these topics, we refer to textbooks on complexity theory [7, 69, 93, 94, 136, 170, 183].

2.1 Basics of Complexity Theory: P, NP

With the aim of abstracting away from immaterial details, in complexity theory we often restrict our attention to *decision problems*. In such problems, one is given an input $x \in \Sigma^*$ represented as a string over some finite alphabet Σ. (We fix an arbitrary but fixed finite alphabet Σ.) The problem is then to decide whether the string x satisfies a certain property Q. Formally, this property can be expressed as

© Springer-Verlag GmbH Germany, part of Springer Nature 2019
R. de Haan: Parameterized Complexity in the Polynomial Hierarchy, LNCS 11880,
https://doi.org/10.1007/978-3-662-60670-4_2

a formal language $Q \subseteq \Sigma^*$ of strings satisfying the property. We also equate the decision problem with this set Q. We say that an algorithm solves the problem Q if for all inputs $x \in \Sigma^*$, the algorithm correctly decides whether $x \in Q$. (We use the words *input* and *instance* interchangeably.)

Complexity theory studies the number of steps that an optimal algorithm needs to solve such problems. Here, the number of steps is measured in terms of the input size $|x| = n$. In most cases, a *worst-case perspective* is taken. This means that for each input size n, we measure the maximum number of steps that the algorithm takes on any input of size n. When expressing the running time of an algorithm, we focus on *upper bound* guarantees. For instance, we say that an algorithm runs in time n^2 if for each input $x \in \Sigma^*$, the algorithm takes at most (but possibly less) than $|x|^2$ steps.

Moreover, when we express the running time of algorithms, we often do not explicitly give an exact function that upper bounds the running time, but we say what the *order* of this function is. In order to explain this more precisely, we introduce the concept of "big-O". We also consider several related notions.

Let $f, g : \mathbb{N} \rightarrow \mathbb{N}$ be arbitrary functions. Then we say that f is *order of g*—written f is $O(g)$, or $f(n)$ is $O(g(n))$—if there is some $c \in \mathbb{N}$ and some $n_0 \in \mathbb{N}$ such that for all $n \geq n_0$ it holds that $f(n) \leq cg(n)$. Intuitively, if $f(n)$ is $O(g(n))$, it means that $f(n)$ *grows asymptotically at most as fast as* $g(n)$.

A counterpart of big-O is little-o, that expresses that a function $f(n)$ grows asymptotically strictly slower than the function $g(n)$. There are various ways of formally expressing this. We will use the following definition of little-o, as used by Flum and Grohe [85, Definition 3.22]. Let f, g be computable functions with the positive integers as domain and range. We say that f is $o(g)$ if there is a computable function h such that for all $\ell \geq 1$ and $n \geq h(\ell)$, we have that $f(n) \leq g(n)/\ell$. Equivalently, f is $o(g)$ if and only if there exists a positive integer n_0 and a nondecreasing and unbounded computable function ι such that for all $n \geq n_0$ we have that $f(n) \leq g(n)/\iota(n)$ [85, Lemma 3.23].

There are also counterparts of big-O and little-o that express lower bounds, rather than upper bounds. We write that $f(n)$ is $\Omega(g(n))$ if there is some $c \in \mathbb{N}$ and some $n_0 \in \mathbb{N}$ such that for all $n \geq n_0$ it holds that $f(n) \geq cg(n)$. Moreover, we write that $f(n)$ is $\omega(g(n))$ if for all $c \in \mathbb{N}$ there is some $n_0 \in \mathbb{N}$ such that for all $n \geq n_0$ it holds that $f(n) > cg(n)$.

To illustrate how the concept of big-O is usually used to express the important aspect of the running of an algorithm, consider an algorithm whose running time can be upper bounded by the function $2n + 1.5n^2$. Unless we want to be really precise, we will often say that this algorithm runs in time $O(n^2)$, as $2n + 1.5n^2$ is $O(n^2)$. A function $p : \mathbb{N} \rightarrow \mathbb{N}$ is called a *polynomial* if it is $O(n^c)$, for some constant $c \in \mathbb{N}$.

2.1.1 Traditional Tractability: P

Traditionally, research in computer science has focused on the complexity class P as the embodiment of the notion of *tractable problems*. This complexity class is defined as the class of all decision problems for which there exists an algorithm that solves the problem in *polynomial time*—that is, in time $O(n^c)$, for some constant $c \in \mathbb{N}$.

So far, we have been a little imprecise about how we count the amount of time (or the number of steps) than an algorithm takes. A mathematically formal way of defining this is provided by the notion of *Turing machines*, which were introduced in the foundational work of Alan Turing [189].

Formally, a *(deterministic) Turing machine* is a tuple $\mathbb{M} = (S, \Sigma, \Delta, s_0, F)$, where:

- S is the finite, non-empty set of *states*;
- Σ is the finite, non-empty *alphabet*;
- $s_0 \in S$ is the *initial state*;
- $F \subseteq S$ is the set of *accepting states*; and
- $\Delta : S \times (\Sigma \cup \{\$, \square\}) \to S \times (\Sigma \cup \{\$\}) \times \{\mathbf{L}, \mathbf{R}, \mathbf{S}\}$ is the *transition function*.

Here $\$, \square \notin \Sigma$ are special symbols. A Turing machine operates on an infinite, one-dimensional *tape*, containing a *cell* for each $n \in \mathbb{N}$. (In this section, we only consider Turing machines with a single tape. Turing machines with multiple tapes are also commonly considered.) Intuitively, this tape is bounded to the left and unbounded to the right. The leftmost cell, the 0-th cell, of each tape carries a "$", and initially, all other tape cells carry the blank symbol. The input is written on the first tape, starting with the first cell, the cell immediately to the right of the "$".

A *configuration* is a tuple $C = (s, x, p)$, where $s \in S$, $x \in \Sigma^*$, and $p \in [0, |x| + 1]$. Intuitively, $\$x\square\square \ldots$ is the sequence of symbols in the cells of the tape, and the head scans the p-th cell of the tape. The *initial configuration* for an input $x \in \Sigma^*$ is $C_0(x) = (s, x, 1)$.

A *computation step* of $\mathbb{M}$ is a pair (C, C') of configurations such that the transformation from C to C' obeys the transition function. Here, the symbol $\mathbf{L}$ indicates that the head moves one step to the left, the symbol $\mathbf{R}$ indicates that the head moves one step to the right, and the symbol $\mathbf{S}$ indicates that the head does not move. We omit the formal details. If (C, C') is a computation step of $\mathbb{M}$. we call C' a *successor configuration* of C. A *halting configuration* is a configuration that has no successor configuration. A halting configuration is *accepting* if its state is in F. We say that the machine $\mathbb{M}$ *accepts* an input x if the initial configuration $C_0(x)$ leads to an accepting halting configuration.

We use Turing machines to formally model algorithms. We say that an algorithm, given in the form of a Turing machine $\mathbb{M}$, solves a problem Q if for each $x \in \Sigma^*$ the machine $\mathbb{M}$ accepts x if and only if $x \in Q$. We then formally define the amount of time that the algorithm takes on input x as the number of computation steps between the initial configuration $C_0(x)$ and the halting configuration.

Turing machines can also be used to model algorithms that give an output. We say that a Turing machine $\mathbb{M}$ outputs the string $y \in \Sigma^*$ on some input $x \in \Sigma^*$ if in the accepting configuration leading from the initial configuration $C_0(x)$, the tape contains the string y.

2.1.2   Traditional Intractability: NP and NP-Completeness

Using the notion of tractability, is it easy to define intractability: a problem is intractable if it is not tractable. However, it turned out that this naive notion of intractability is unproductive for a large class of problems from many areas of computer science. For these problems, nobody has been able to find a polynomial-time algorithm, and nobody has been able to prove that no such algorithm exists. For this reason, the concept of *NP-completeness* was introduced.

The complexity class NP consists of all problems that are solvable using a *non-deterministic Turing machine* (NTM). Non-deterministic Turing machines are defined similarly to deterministic Turing machines, with the only difference that instead of a transition function, a *transition relation* $\Delta \subseteq S \times (\Sigma \cup \{\$, \square\}) \times S \times (\Sigma \cup \{\$\}) \times \{\mathbf{L}, \mathbf{R}, \mathbf{S}\}$ is used. This way, for each configuration, multiple successor configurations are possible. Then, a non-deterministic Turing machine $\mathbb{M}$ solves a problem Q if for each input $x \in \Sigma^*$ it holds that $x \in Q$ if and only if there exists some computation path from the initial configuration $C_0(x)$ to some accepting halting configuration. The running time of $\mathbb{M}$ is defined to be the length of the longest path from $C_0(x)$ to any halting configuration.

There are problems in the class NP for which the best known algorithms run in time $2^{\Omega(n)}$. To substantiate the suspicion that a certain problem is not polynomial-time solvable, one can relate this problem to other problems in NP using the concept of reductions.

Let Q_1 and Q_2 be two decision problems. A *polynomial-time reduction from Q_1 to Q_2* is a polynomial-time algorithm that for each input $x_1 \in \Sigma^*$ produces an output x_2 such that $x_1 \in Q_1$ if and only if $x_2 \in Q_2$. (Such reductions are also called *many-to-one reductions*, or *Karp reductions*.)

We then say that a problem Q is NP-*hard* if for each problem $Q' \in$ NP, there is a polynomial-time reduction from Q' to Q. Intuitively, an NP-hard problem Q is as hard as any problem in NP, because if Q were polynomial-time solvable, then each problem in NP would be polynomial-time solvable. A problem Q is NP-*complete* if it is both in NP and NP-hard.

A practically useful way of proving NP-completeness is offered by the Cook-Levin Theorem [54, 144]. This seminal result identified a first NP-complete problem: SAT. As a result, subsequent NP-hardness proofs only need to provide a polynomial-time reduction from this single problem, rather than providing a reduction from arbitrary problems in NP.

SAT is the satisfiability problem of propositional logic. In propositional logic, formulas are built from a countably infinite set of propositional variables $x_1, x_2, \ldots,$ the Boolean constants 0 and 1, and the Boolean operators $\wedge, \vee, \neg, \rightarrow,$ and $\leftrightarrow$. For any propositional formula, we let $\text{Var}(\varphi)$ denote the set of propositional variables occurring in φ. Truth of such propositional formulas is defined in the usual way. That is, a truth assignment $\alpha : \text{Var}(\varphi) \rightarrow \mathbb{B}$ satisfies a formula φ if the formula evaluates to the truth value 1 when the assignment α is applied to the variables occuring in φ. (By a slight abuse of notation, we use 0 and 1 to denote both truth values and syntactical constants.) If a truth assignment α satisfies a formula φ, we write $\alpha \models \varphi$. A formula φ is *satisfiable* (or *consistent*) if there exists a truth assignment α such that $\alpha \models \varphi$. Similarly, we say that a set Φ of formulas is satisfiable (or consistent) if there is a truth assignment α that simultaneously satisfies all formulas in Φ. If every truth assignment that satisfies a formula φ_1 also satisfies a formula φ_2, we write $\varphi_1 \models \varphi_2$.

We briefly introduce some notation for propositional formulas that we will use throughout the thesis. If $\gamma = \{x_1 \mapsto d_1, \ldots, x_n \mapsto d_n\}$ is a function that maps some variables of a formula φ to other variables or to truth values, then we let $\varphi[\gamma]$ denote the application of γ to the formula φ—here we simplify the resulting formula as much as possible, after the application of γ, e.g., $1 \wedge (x_1 \vee x_2)$ becomes $(x_1 \vee x_2)$. We also write $\varphi[x_1 \mapsto d_1, \ldots, x_n \mapsto d_n]$ to denote $\varphi[\gamma]$. A *literal* is a propositional variable x or a negated variable $\neg x$. The *complement* $\overline{x}$ of a positive literal x is $\neg x$, and the complement $\overline{\neg x}$ of a negative literal $\neg x$ is x. For literals $l \in \{x, \neg x\}$, we let $\text{Var}(l) = x$ denote the variable occurring in l. A *clause* is a finite set of literals, not containing a complementary pair $x, \neg x$, and is interpreted as the disjunction of these literals. A *term* is a finite set of literals, not containing a complementary pair $x, \neg x$, and is interpreted as the conjunction of these literals. A formula in *conjunctive normal form (CNF)* is a finite set of clauses, interpreted as the conjunction of these clauses. A formula in *disjunctive normal form (DNF)* is a finite set of terms, interpreted as the disjunction of these terms. Let $r \geq 2$. A formula is in rCNF if it consists of a conjunction of clauses that each contain at most r literals. Similarly, a formula is in rDNF if it consists of a disjunction of terms that each contain at most r literals. We define the *size* of a propositional formula φ to be the number of occurrences of Boolean operators in φ plus the number of occurrences of propositional variables in φ. (Note that the size of a propositional formula differs from the bitsize—i.e., the number of bits needed to represent the formula in binary—which is larger by a logarithmic factor.) The *size* of a CNF formula φ is linear in $\sum_{c \in \varphi} |c|$. The number of clauses of a CNF formula φ is denoted by $|\varphi|$. Similarly, the size of a DNF formula φ is linear in $\sum_{t \in \varphi} |t|$. The number of terms of a DNF formula φ is denoted by $|\varphi|$.

The Cook-Levin Theorem [54, 144] states that the satisfiability problem of propositional logic—which is defined as the set SAT containing all strings $x \in \Sigma^*$ that encode a satisfiable propositional formula—is NP-complete. In fact, it states that this problem is NP-hard even when restricted to propositional formulas in 3CNF. This restriction of the problem is denoted by 3SAT.

It is widely believed that the classes P and NP are different (see, e.g., [87]), but no formal proof of this statement is known. In fact, this is arguably the most famous and most important open problem in theoretical computer science. Nevertheless, the concept of NP-hardness is traditionally used in computer science to capture intractability, and has been used to indicate the absence of polynomial-time algorithms for many problems. For instance, the opus of Garey and Johnson [86] lists hundreds of relevant problems that are NP-complete.

A complexity class that is also often considered is the dual co-NP of NP, that consists of all problems Q for which the problem co-$Q = \{x \in \Sigma^* : x \notin Q\}$ is in NP. It is believed that NP $=$ co-NP, but no formal proof of this statement is known. The class co-NP bears a similar relation to P as NP, that is, P = NP if and only if P = co-NP. An important example of a co-NP-complete problem is the problem UNSAT, which consists of all strings $x \in \Sigma^*$ that encode an unsatisfiable propositional formula.

2.2   The Polynomial Hierarchy and Polynomial Space

There are also many natural decision problems that are not contained in the classical complexity classes P, NP, and co-NP. In this section, we consider several complexity classes that can be used to characterize the complexity of many such problems.

2.2.1   Polynomial Hierarchy

The *Polynomial Hierarchy (PH)* [163, 170, 185, 198] contains a hierarchy of complexity classes Σ_i^p and Π_i^p, for all $i \geq 0$. These classes Σ_i^p and Π_i^p are defined by means of non-deterministic Turing machines with an oracle. Let O be a decision problem, e.g., $O = \text{SAT}$. A Turing machine $\mathbb{M}$ with access to an O *oracle* is a Turing machine with a dedicated *oracle tape* and dedicated states q_{query}, q_{yes} and q_{no}. Whenever $\mathbb{M}$ is in the state q_{query}, it does not proceed according to the transition relation, but instead it transitions into the state q_{yes} if the oracle tape contains a string x that is a yes-instance for the problem O, i.e., if $x \in O$, and it transitions into the state q_{no} if $x \notin O$.

For any complexity class C, we let NP^C be the set of decision problems that is decided in polynomial time by a non-deterministic Turing machine with an oracle for a problem that is complete for the class C. Then, the classes Σ_i^p and Π_i^p, for $i \geq 0$, are defined by letting:

$$\Sigma_0^p = \Pi_0^p = \text{P},$$

and for each $i \geq 1$:

$$\Sigma_i^p = \text{NP}^{\Sigma_{i-1}^p}, \text{ and}$$
$$\Pi_i^p = \text{co-NP}^{\Sigma_{i-1}^p}.$$

In particular, the class Σ_1^p coincides with the class NP. It is believed that the PH is strict—that is, that for each $i \in \mathbb{N}$ it holds that $\Sigma_i^p \neq \Pi_i^p$ and thus that $\Sigma_i^p \neq \Sigma_{i+1}^p$—but no formal proof of this statement is known.

We give an alternative characterization of the classes Σ_i^p using the satisfiability problem of various classes of quantified Boolean formulas. A *(prenex) quantified Boolean formula (QBF)* is a formula of the form $Q_1 X_1 Q_2 X_2 \ldots Q_m X_m . \psi$, where each Q_i is a quantifier in $\{\exists, \forall\}$, where the X_i are disjoint sets of propositional variables, and ψ is a Boolean formula over the variables in $\bigcup_{i=1}^{m} X_i$. We call ψ the *matrix* of the formula. Truth of such formulas is defined in the usual way. That is, a formula of the form $\exists X.\varphi$ is true if there exists a truth assignment $\alpha : X \to \mathbb{B}$ such that $\varphi[\alpha]$ is true, and a formula of the form $\forall X.\varphi$ is true if for all truth assignment $\alpha : X \to \mathbb{B}$ it holds that $\varphi[\alpha]$ is true—here $\varphi[\alpha]$ denotes the formula obtained from φ by applying the truth assignment α to the matrix of φ. We say that a QBF is *in QDNF* if the matrix is in DNF, and we say that a QBF is *in QCNF* if the matrix is in CNF.

Alternatively, the semantics of quantified Boolean formulas can be defined using QBF models [178]. Let $\varphi = Q_1 x_1 \ldots Q_n x_n.\psi$ be a quantified Boolean formula. A *QBF model* for φ is a tree of (partial) truth assignments where (1) each truth assignment assigns values to the variables $x_1, \ldots, x_i$ for some $i \in [n]$, (2) the root is the empty assignment, and for all assignments α in the tree, assigning truth values to the variables $x_1, \ldots, x_i$ for some $i \in [n]$, the following conditions hold: (3) if $i < n$, every child of α agrees with α on the variables $x_1, \ldots, x_i$, and assigns a truth value to x_{i+1} (and to no other variables); (4) if $i = n$, i.e., if α is a total truth assignment on the variables $x_1, \ldots, x_n$, then α satisfies ψ, and α has no children; (5) if $i < n$ i.e., if α does not assign a truth value to x_{i+1}, and $Q_i = \exists$, then α has one child α' that assigns some truth value to x_{i+1}; and (6) if $i < n$ i.e., if α does not assign a truth value to x_{i+1}, and $Q_i = \forall$, then α has two children α_1 and α_2 that assign different truth values to x_{i+1}. It is straightforward to show that a quantified Boolean formula φ is true if and only if there exists a QBF model for φ. Note that this definition of QBF models is a special case of the original definition [178].

For each $i \geq 1$, the decision problem QSAT_i is defined as follows.

QSAT_i

Instance: A quantified Boolean formula $\varphi = \exists X_1 \forall X_2 \exists X_3 \ldots Q_i X_i.\psi$, where Q_i is a universal quantifier if i is even and an existential quantifier if i is odd, and where ψ is quantifier-free.

Question: Is φ true?

For each nonnegative integer $i \leq 0$, the problem QSAT_i is complete for the class Σ_i^{p} under polynomial-time reductions [185, 198]. Moreover, Σ_i^{p}-hardness of QSAT_i holds already when the matrix of the input formula is restricted to 3CNF for odd i, and restricted to 3DNF for even i. For any class $\mathcal{C}$ of propositional formulas, we let $\text{QSAT}_i(\mathcal{C})$ denote the problem QSAT_i restricted to QBFs where the matrix is in $\mathcal{C}$.

2.2.2 Polynomial Space

Next, we briefly consider the complexity class PSPACE, consisting of all decision problems that can be solved by an algorithm that uses a polynomial amount of space (or memory). The amount of *space* that a Turing machine uses for an input $x \in \Sigma^*$ is defined as the number of tape cells to which it writes during the computation. In other words, the class PSPACE consists of all problems that can be solved using space $O(n^c)$, for some constant $c \in \mathbb{N}$. The class PSPACE contains all classes of the PH, that is, for each $i \in \mathbb{N}$ it holds that $\Sigma_i^{\text{p}} \cup \Pi_i^{\text{p}} \subseteq \text{PSPACE}$.

The following variant of the problems QSAT_i—where the maximum number of quantifier alternations is not bounded by a constant—is PSPACE-complete [186].

QSAT

Instance: A quantified Boolean formula $\varphi = \exists X_1 \forall X_2 \exists X_3 \ldots Q_n X_n . \psi$, where Q_i is a universal quantifier if i is even and an existential quantifier if i is odd, and where ψ is quantifier-free.

Question: Is φ true?

2.2.3 Alternating Turing Machines

The class PSPACE and the classes of the PH can also be characterized using *alternating Turing machines*. We use the same notation as Flum and Grohe [85, Appendix A.1].

Let $m \geq 1$ be a positive integer. An *alternating Turing machine (ATM)* with m tapes is a 6-tuple $\mathbb{M} = (S_\exists, S_\forall, \Sigma, \Delta, s_0, F)$, where:

- $S_\exists$ and $S_\forall$ are disjoint sets;
- $S = S_\exists \cup S_\forall$ is the finite, non-empty set of *states*;
- Σ is the finite, non-empty *alphabet*;
- $s_0 \in S$ is the *initial state*;
- $F \subseteq S$ is the set of *accepting states*;
- $\Delta \subseteq S \times (\Sigma \cup \{\$, \square\})^m \times S \times (\Sigma \cup \{\$\})^m \times \{\mathbf{L}, \mathbf{R}, \mathbf{S}\}^m$ is the *transition relation*. The elements of Δ are the *transitions*.
- $\$, \square \notin \Sigma$ are special symbols. "$\$$" marks the left end of any tape. It cannot be overwritten and only allows $\mathbf{R}$-transitions.[1] "$\square$" is the *blank symbol*.

Intuitively, the tapes of our machine are bounded to the left and unbounded to the right. The leftmost cell, the 0-th cell, of each tape carries a "$\$$", and initially, all other tape cells carry the blank symbol. The input is written on the first tape, starting with the first cell, the cell immediately to the right of the "$\$$".

A *configuration* is a tuple $C = (s, x_1, p_1, \ldots, x_m, p_m)$, where $s \in S$, $x_i \in \Sigma^*$, and $p_i \in [0, |x_i| + 1]$ for each $i \in [k]$. Intuitively, $\$x_i \square \square \ldots$ is the sequence of symbols in the cells of tape i, and the head of tape i scans the p_i-th cell. The *initial configuration* for an input $x \in \Sigma^*$ is $C_0(x) = (s_0, x, 1, \epsilon, 1, \ldots, \epsilon, 1)$, where ϵ denotes the empty word.

A *computation step* of $\mathbb{M}$ is a pair (C, C') of configurations such that the transformation from C to C' obeys the transition relation. We omit the formal details. We write $C \to C'$ to denote that (C, C') is a computation step of $\mathbb{M}$. If $C \to C'$, we call C' a *successor configuration* of C. A *halting configuration* is a configuration that has no successor configuration. A halting configuration is *accepting*

---

[1]To formally achieve that "$\$$" marks the left end of the tapes, we require that whenever $(s, (a_1, \ldots, a_m), s', (a'_1, \ldots, a'_m), (d_1, \ldots, d_m)) \in \Delta$, then for all $i \in [m]$ it holds that $a_i = \$$ if and only if $a'_i = \$$ and that $a_i = \$$ implies $d_i = \mathbf{R}$.

if its state is in F. A step $C \to C'$ is *non-deterministic* if there is a configuration $C'' \neq C'$ such that $C \to C''$, and is *existential* if C is an existential configuration. A state $s \in S$ is called *deterministic* if for any $a_1, \ldots, a_m \in \Sigma \cup \{\$, \square\}$, there is at most one $(s, (a_1, \ldots, a_m), s', (a'_1, \ldots, a'_m), (d_1, \ldots, d_m)) \in \Delta$. Similarly, we call a non-halting configuration *deterministic* if its state is deterministic, and *non-deterministic* otherwise.

A configuration is called *existential* if it is not a halting configuration and its state is in $S_\exists$, and *universal* if it is not a halting configuration and its state is in $S_\forall$. Intuitively, in an existential configuration, there must be one possible run that leads to acceptance, whereas in a universal configuration, all runs must lead to acceptance. Formally, a *run* of an ATM $\mathbb{M}$ is a directed tree where each node is labeled with a configuration of $\mathbb{M}$ such that:

- The root is labeled with an initial configuration.
- If a vertex is labeled with an existential configuration C, then the vertex has precisely one child that is labeled with a successor configuration of C.
- If a vertex is labeled with a universal configuration C, then for every successor configuration C' of C the vertex has a child that is labeled with C'.

We often identify nodes of the tree with the configurations with which they are labeled. The run is *finite* if the tree is finite, and *infinite* otherwise. The *length* of the run is the height of the tree. The run is *accepting* if it is finite and every leaf is labeled with an accepting configuration. If the root of a run ρ is labeled with $C_0(x)$, then ρ is a run *with input x*. Any path from the root of a run ρ to a leaf of ρ is called a *computation path*.

The *language (or problem) accepted by* $\mathbb{M}$ is the set $Q_\mathbb{M}$ of all $x \in \Sigma^*$ such that there is an accepting run of $\mathbb{M}$ with initial configuration $C_0(x)$. $\mathbb{M}$ *runs in time $t : \mathbb{N} \to \mathbb{N}$* if for every $x \in \Sigma^*$ the length of every run of $\mathbb{M}$ with input x is at most $t(|x|)$.

A *step* $C \to C'$ is an *alternation* if either C is existential and C' is universal, or vice versa. A run ρ of $\mathbb{M}$ is *ℓ-alternating*, for an $\ell \in \mathbb{N}$, if on every path in the tree associated with ρ, there are less than ℓ alternations between existential and universal configurations. The machine $\mathbb{M}$ is *ℓ-alternating* if every run of $\mathbb{M}$ is ℓ-alternating.

The classes Σ_i^p, Π_i^p and PSPACE can be characterized using ATMs as follows. Let $i \geq 1$. The class Σ_i^p consists of all problems that are decided by an i-alternating polynomial-time ATM $\mathbb{M} = (S_\exists, S_\forall, \Sigma, \Delta, s_0, F)$ such that $s_0 \in S_\exists$. Similarly, the class Π_i^p consists of all problems that are decided by an i-alternating polynomial-time ATM $\mathbb{M} = (S_\exists, S_\forall, \Sigma, \Delta, s_0, F)$ such that $s_0 \in S_\forall$. Finally, the class PSPACE consists of all problems that are decided by a polynomial-time ATM [40].

2.3 Bounded Query Complexity

In the previous section, we considered complexity classes based on non-deterministic Turing machines with an oracle. Next, we consider a number of classes containing problems that can be solved by deterministic Turing machines with access to an oracle. The branch of complexity theory that studies how many queries to an oracle are needed to solve certain problems is often called *bounded query complexity*.

We begin with some classes that can be solved by a constant number of queries to an oracle in NP. The *Boolean Hierarchy* (BH) [35, 41, 126] consists of a hierarchy of complexity classes BH_i for all $i \geq 1$. Each class BH_i can be characterized as the class of problems that can be reduced in polynomial time to the problem BH_i-SAT, which is defined inductively as follows. The problem BH_1-SAT consists of all sequences (φ) of length 1, where φ is a satisfiable propositional formula. For even $i \geq 2$, the problem BH_i-SAT consists of all sequences $(\varphi_1, \ldots, \varphi_i)$ of propositional formulas such that both $(\varphi_1, \ldots, \varphi_{i-1}) \in BH_{(i-1)}$-SAT and φ_i is unsatisfiable. For odd $i \geq 2$, the problem BH_i-SAT consists of all sequences $(\varphi_1, \ldots, \varphi_i)$ of propositional formulas such that $(\varphi_1, \ldots, \varphi_{i-1}) \in BH_{(i-1)}$-SAT or φ_i is satisfiable. The class BH_2 is also denoted by DP, and the problem BH_2-SAT is also denoted by SAT-UNSAT. For each $i \geq 1$, problems in the class BH_i can be solved in deterministic polynomial time using i queries to an oracle in NP.

Next, we briefly discuss two more classes of problems that can be solved in polynomial time by a Turing machine that has access to an NP oracle. The first class that we consider is the class Δ_2^p, consisting of those problems that can be solved by a polynomial-time algorithm with access to an NP oracle that can make a polynomial number of calls to this oracle. However, for many problems less than a polynomial number of calls to the NP oracle are needed. The class Θ_2^p contains problems that can be solved in polynomial time by means of querying the NP oracle only $O(\log n)$ times, where n is the size of the input.

The classes Δ_2^p and Θ_2^p can be characterized as follows using non-deterministic Turing machines with a designated output tape. Let $\mathbb{M}$ be an NTM with multiple tapes, where one of the tapes is used as output tape. We say that the length of the output is bounded by a function $z : \mathbb{N} \to \mathbb{N}$ if for each input $x \in \Sigma^*$, it holds that for each computation path of the machine $\mathbb{M}$ when given input x, the length of the output $y \in \Sigma^*$ is at most $z(n)$, where $n = |x|$ denotes the size of the input. A problem Q is in Δ_2^p if and only if there exists an NTM $\mathbb{M}$ with an output tape such that for any input $x \in \Sigma^*$ it holds that $x \in Q$ if and only if there exists an accepting computation path of $\mathbb{M}$ for input x such that the output for this computation path is lexicographically larger than the output for any other computation path of $\mathbb{M}$ for input x. A similar characterization holds for the class Θ_2^p, with the only difference that the output of $\mathbb{M}$ is bounded by a function that is $O(\log n)$ [184].

The following problem is complete for the class Θ_2^p under polynomial-time reductions [48, 138, 193].

MAX-MODEL

Instance: A satisfiable propositional formula φ, and a variable $w \in Var(\varphi)$.
Question: Is there a model of φ that sets a maximal number of variables in $Var(\varphi)$ to true (among all models of φ) and that sets w to true?

2.4 Non-uniform Complexity

We conclude this chapter by surveying some basic notions and results from an area of complexity theory known as non-uniform complexity, which is related to the investigation of lower bounds for Boolean circuits.

A *(Boolean) circuit* is a directed, acyclic graph, where all nodes are labelled with either a Boolean constant 0 or 1, the name of a variable x, or one of the Boolean operators $\neg$, $\wedge$, or $\vee$. If a node is labelled with a constant or with the name of a variable, it has indegree 0; if it is labelled with $\neg$, it has indegree 1; if it is labelled with $\wedge$ or $\vee$, it has indegree larger than 1. Moreover, there is a single node with outdegree 0, called the *output node*. (We restrict our attention to Boolean circuits with a single output node). The nodes labelled with the name of a variable are called *input nodes* (or simply *variables*). All nodes that are not input nodes are called the *gates* of the circuits. Let $v_1, \ldots, v_n$ be the variables of a circuit C, and let $\alpha : \{v_1, \ldots, v_n\} \rightarrow \mathbb{B}$ be a truth assignment for these variables. The gates of the circuit are then assigned truth values according to the operators with which they are labelled. The truth value that the output gate of the circuit gets under the truth assignment, is denoted by $C[\alpha]$. Moreover, let $\alpha : V \rightarrow \mathbb{B}$ be a truth assignment to a subset $V \subseteq \{v_1, \ldots, v_n\}$ of variables of the circuit C—such an assignment is called a *partial assignment* to the variables in C. We then let $C[\alpha]$ denote the circuit obtained from C by replacing each variable $v \in V$ by a gate labelled with the constant $\alpha(v)$. Possibly, the circuit can be simplified after such an instantiation, e.g., if a gate labelled with $\wedge$ that has an incoming edge from a gate labelled with 0, we can replace it by a gate labelled with 0. Let $\mathcal{C}$ be a class of Boolean circuits. We say that $\mathcal{C}$ is *closed under partial instantiation* if for each circuit $C \in \mathcal{C}$ and for each partial assignment α to the variables in C it holds that $C[\alpha] \in \mathcal{C}$.

Let Q be a problem over the alphabet $\Sigma = \mathbb{B}$. Moreover, let $(C_n)_{n \in \mathbb{N}}$ be a family of Boolean circuits, where for each $n \in \mathbb{N}$, the circuit C_n has exactly n variables $v_1, \ldots, v_n$. We say that the problem Q is decided by $(C_n)_{n \in \mathbb{N}}$ if for each input $x \in \Sigma^*$, consisting of the symbols $x_1 x_2 \ldots x_n$, it holds that $x \in Q$ if and only if $C_n[\alpha] = 1$, where $\alpha : \{v_1, \ldots, v_n\} \rightarrow \mathbb{B}$ is defined by letting $\alpha(v_i) = x_i$, for each $i \in [n]$. Moreover, we say that a family $(C_n)_{n \in \mathbb{N}}$ is of *polynomial size* if there exists a polynomial p such that $|C_n| \leq z(n)$ for each $n \in \mathbb{N}$. The complexity class P/poly consists of all decision problems that are decided by some polynomial-size family of Boolean circuits. The class P/poly is called a *non-uniform class*.

Alternatively, the class P/poly can be characterized using the concept of *polynomial-size advice strings*. That is, a problem Q is in P/poly if and only if there exists some polynomial p and a problem $Q' \in \text{P}$ such that for each $n \in \mathbb{N}$ there exists some *advice string* $\alpha_n \in \Sigma^{p(n)}$ of length $p(n)$ such that for each string $x \in Q$ if and only if $(\alpha_{|x|}, x) \in Q'$.

The class P/poly is a strict superset of P, i.e., $\text{P} \subsetneq \text{P/poly}$. Intuitively, non-uniformity provides additional solving power because the advice strings can be uncomputable. To give an example, consider an uncomputable unary set, such as the following set $S = \{ 1^m : m$ is the index of a Turing machine that halts on the empty input }. Then $S \in$ P/poly, because for each input size n, the advice string α_n simply encodes whether $1^n \in S$ or not. In fact, this argument shows that any unary set $U \subseteq \{1\}^*$ is in P/poly. However, since S is uncomputable, we know that $S \notin$ P.

A central result relating the class P/poly to the classes of the Polynomial Hierarchy is the Karp-Lipton Theorem [127, 128]. This result states that if NP $\subseteq$ P/poly, then $\Sigma_2^p = \Pi_2^p$, and thus the PH collapses at the second level. In other words, the Karp-Lipton Theorem states that SAT cannot be solved by polynomial-size circuits, unless the PH collapses.

Chapter 3
Parameterized Complexity Theory

> *Parameterized complexity is based on a deal with the devil of intractability.*
>
> — Rod Downey and Mike Fellows,
> *Parameterized Complexity* [67]

In Chapt. 2, we went over some of the central notions and results from the theory of classical complexity, that play a role in this thesis. In this chapter, we will give a similar overview of the area of parameterized complexity theory.

We begin by explaining the key idea that underlies the perspective taken in parameterized complexity. Then, we will define the most important parameterized complexity classes that play a role in nearly every parameterized complexity analysis. We then turn to some more parameterized complexity classes that are less commonly used—several of these classes provide the starting point for the theoretical investigations in this thesis. Finally, we consider a number of well-known problems that are complete for various parameterized complexity classes.

This overview of parameterized complexity differs from most other treatments of parameterized complexity in that it places more emphasis on parameterized analogues of complexity classes involving non-determinism and alternation. For more background on parameterized complexity, we refer to other sources [57, 64, 66, 67, 85, 166].

3.1 Fixed-Parameter Tractability

The driving force behind the area of parameterized complexity is the observation that the worst-case perspective classically taken in complexity theory leads to an analysis of the complexity of a problem where the only thing you know about an input is its size in bits, whereas in essentially all cases where you are faced with the task of designing an algorithm to solve a computational problem, you know more about

© Springer-Verlag GmbH Germany, part of Springer Nature 2019
R. de Haan: Parameterized Complexity in the Polynomial Hierarchy, LNCS 11880,
https://doi.org/10.1007/978-3-662-60670-4_3

the inputs than just their size. Parameterized complexity provides tools for a multi-dimensional complexity analysis where you can take such additional information about the input into account.

To explain this idea in more detail, we consider the problem of model checking for the temporal logic LTL. This is a modal logic in which one can express a variety of temporal properties of automated systems. For the model checking problem, the input consists of a model $\mathcal{M}$ of an automated system and an LTL formula φ, and the question is to decide whether the system satisfies the property expressed by the formula. (For a more detailed definition of LTL and its model checking problem, we refer to Chapt. 9.) This problem is PSPACE-complete, and therefore there is no algorithm known that solves the problem in time less than $2^{\Omega(n)}$, where n denotes the input size. However, this classical complexity diagnosis is based on the fact that we know nothing about the relation between the size of the model $\mathcal{M}$ and the size of the formula φ. Now suppose that you are working on an application of this problem, where in all cases the formula φ is much smaller than the model $\mathcal{M}$—for example, the formula φ expresses the property that the automated system never gets stuck in a deadlock state, which can be expressed by an extremely small LTL formula. In this situation the PSPACE-hardness verdict of classical complexity is overshadowed by the fact that the problem can also be solved in time $O(2^k m^c)$, where k denotes the size of the formula φ, m denotes the size of the model $\mathcal{M}$, and c is some constant. In other words, we can use our knowledge that the formula is small for the inputs that we care about to identify algorithms that work well for those inputs.

The concept of identifying information about problem inputs that can be exploited algorithmically is formally captured in parameterized complexity by the notions of parameterized problems and fixed-parameter tractability. A *parameterized problem* Q is a subset of $\Sigma^* \times \mathbb{N}$. For an instance $(x, k) \in \Sigma^* \times \mathbb{N}$, we call x the *main part* of the instance and k the *parameter*. A parameterized problem Q is *fixed-parameter tractable* if there exists a computable function $f : \mathbb{N} \to \mathbb{N}$, a constant $c \in \mathbb{N}$, and an algorithm that decides whether $(x, k) \in Q$ in time $f(k)n^c$, where n denotes the size of x. Such an algorithm is called an *fpt-algorithm*, and this amount of time is called *fixed-parameter tractable time* (or *fpt-time*). The class of all parameterized problems that are fixed-parameter tractable is denoted by FPT. For instance, the model checking problem for LTL is fixed-parameter tractable, when the parameter is the size of the LTL formula in the input.

3.1.1 Polynomial-Time Solvability for Constant Parameter Values

One might think that the concept of fixed-parameter tractability can be characterized by the fact that for any fixed (i.e., constant) value of the parameter, the problem is polynomial-time tractable. While it is true that every fixed-parameter tractable problems admits a polynomial-time algorithm for each fixed value of the parameter, the converse is not true in general. To elaborate on this, we consider the parameterized complexity class XP, that consists of all parameterized problems Q for which there exists a computable function $f : \mathbb{N} \to \mathbb{N}$ and an algorithm that decides whether $(x, k) \in Q$ in time $|x|^{f(k)}$. Each problem $Q \in \text{XP}$ can also be solved in polynomial time for each fixed value of the parameter. However, the order of the polynomial depends on the parameter value k, whereas for fixed-parameter tractable problems the order of the polynomial is independent of the value k. It is therefore not surprising that FPT $\neq$ XP. To see why this difference leads us to consider FPT as a class of tractable problems, and XP not, consider the following two running time bounds: $2^k n^2$ and n^k. Already for the values $n = 1000$ and $k = 10$, for instance, the difference between these two functions is enormous.

3.1.2 Integers as Parameter Values

The use of nonnegative integers as parameter values to capture any kind of structure that could be present in the input might seem restrictive. One can alternatively define parameterized problems as sets $Q \subseteq \Sigma^* \times \Sigma*$, where the parameter value is a string $k \in \Sigma^*$ rather than an integer, and correspondingly adapt the notion of fixed-parameter tractability using functions $f : \Sigma^* \to \mathbb{N}$. This alternative definition does not capture a wider range of problems that are fixed-parameter tractable, because we put no bound on the running time of the functions f. Therefore, one could in principle always encode any parameter value $k \in \Sigma^*$ as an integer.

Similarly, one could combine multiple parameters into a single parameter by taking their sum. Suppose that you have an algorithm that runs in time $f(k_1, k_2)n^c$ for any input $x \in \Sigma^*$ and any two parameter values $k_1, k_2 \in \mathbb{N}$, where n is the size of x, f is some computable function, and c is some constant. Then there exists another computable function f' such that the algorithm runs in time $f'(k)n^c$, where $k = k_1 + k_2$. By a similar argument, one can combine any number of parameters into a single parameter.

The potential to use integers as parameter values to successfully capture structure in problem inputs is illustrated by the following two examples of parameters that have been used fruitfully in many settings. The first example is that of the notion of *treewidth*, which, intuitively, expresses how much a graph is like a tree. Trees have treewidth 1, and the lower the treewidth of a graph, the more it is like a tree. (For a definition of the notion of treewidth, we refer to Sect. 4.2.1). Many NP-hard problems are fixed-parameter tractable when parameterized by the treewidth of a graph representation of the input (see, e.g., [26]). The second example concerns the parameter of *backdoor size* (see, e.g., [88]). Backdoors are typically considered for logic-related problems, and, intuitively, consist of a set of variables that have the property that once they are instantiated, the problem inputs falls into a class of inputs for which the problem can be solved efficiently. For instance, for propositional formulas in 2CNF, the satisfiability problem is polynomial-time solvable, and the problem 3SAT is fixed-parameter tractable when parameterized by the size of the smallest backdoor to 2CNF.

One final thing that we should mention about the general set-up of parameterized complexity theory is that parameters are often formally modelled differently, using (polynomial-time computable) functions $\kappa : \Sigma^* \to \mathbb{N}$. Instead of viewing parameterized problems as sets $Q \subseteq \Sigma^* \times \mathbb{N}$, where the parameter value is given explicitly as part of the input, parameterized problems are then considered as pairs (Q, κ), where $Q \subseteq \Sigma^*$ is a classical decision problem. For each input $x \in \Sigma^*$, the parameter value is then given by $\kappa(x)$. For the purpose of this thesis, the difference between these two different formal ways of capturing parameter values is immaterial.

3.1.3 Alternative Characterizations of FPT

We end this section on fixed-parameter tractability by giving two alternative characterizations of the class FPT. The first of these states that FPT consists of exactly those parameterized problems that can be solved in polynomial-time after performing a *precomputation* that depends on the parameter value only. Formally, FPT consists of those parameterized problems $Q \subseteq \Sigma^* \times \mathbb{N}$, for which there a computable function $f : \mathbb{N} \to \Sigma^*$, and a problem $Q' \subseteq \Sigma^* \times \Sigma^*$ that is in P, such that for all instances $(x, k) \in \Sigma^* \times \mathbb{N}$ it holds that $(x, k) \in Q$ if and only if $(x, f(k)) \in Q'$. In this definition, the function f performs the precomputation. As we will see in the next section in more detail, by replacing the complexity class P in the above definition by another class K, we can construct a parameterized analogue para-K of any classical complexity class K.

The other alternative characterization of FPT uses the concept of *kernelization*. Kernelizations can be seen as preprocessing algorithms with performance guarantees. Formally, a kernelization for a parameterized problem $Q \subseteq \Sigma^* \times \mathbb{N}$ is a polynomial-time algorithm A for which there exist computable functions $f, g : \mathbb{N} \to \mathbb{N}$ such that for any instance $(x, k) \in \Sigma^* \times \mathbb{N}$, the algorithm A, when given (x, k) as input, produces another instance (x', k') such that: (1) $(x, k) \in Q$ if and only if $(x', k') \in Q$; (2) $|x'| \leq f(k)$; and (3) $k' \leq g(k)$. In other words, the kernelization produces an equivalent instance whose size is bounded by a function of the parameter only. Any parameterized problem that is decidable is fixed-parameter tractable if and only if it has a kernelization [166].

3.2 Fixed-Parameter Intractability

Similarly to the traditional theory of computational complexity, the theory of parameterized complexity also offers a completeness theory, along with various intractability classes, that can be used to provide evidence that certain problems are not fixed-parameter tractable. In this section, we will give an overview of the notions of hardness and completeness that are used in parameterized complexity, and we will consider several parameterized intractability classes.

The first prerequisite for a successful completeness theory is an appropriate notion of reductions. In parameterized complexity, this role is filled by *fpt-reductions*. Let $Q \subseteq \Sigma^* \times \mathbb{N}$ and $Q' \subseteq \Sigma^* \times \mathbb{N}$ be two parameterized problems. An fpt-reduction from Q to Q' is a mapping $R : \Sigma^* \times \mathbb{N} \to \Sigma^* \times \mathbb{N}$ from instances of Q to instances of Q' such that there exist a computable function $g : \mathbb{N} \to \mathbb{N}$ such that for all $(x, k) \in \Sigma^* \times \mathbb{N}$ it holds that (1) R is computable in fpt-time, (2) $(x, k) \in Q$ if and only if $R(x, k) \in Q'$, and (3) $k' \leq g(k)$, where $R(x, k) = (x', k')$. The last requirement, intuitively, is there to ensure that the composition of an fpt-reduction and another fpt-algorithm is also an fpt-algorithm. If the value k' is unrestricted, it might be the case that the factor $f(k')$ does not depend only on the original parameter value k. As a result of this requirement, the class FPT is closed under fpt-reductions.

3.2.1 The Classes para-K

As already mentioned in Sect. 3.1.3, for each classical complexity class K, we can construct a parameterized analogue para-K. This works as follows. Let K be a classical complexity class, e.g., NP. The parameterized complexity class para-K is then defined as the class of all parameterized problems $L \subseteq \Sigma^* \times \mathbb{N}$ for which there exist a computable function $f : \mathbb{N} \to \Sigma^*$ and a problem $Q' \subseteq \Sigma^* \times \Sigma^*$ such that $Q \in K$, such that for all instances $(x, k) \in \Sigma^* \times \mathbb{N}$ it holds that $(x, k) \in Q$ if and only if $(x, f(k)) \in Q'$. Intuitively, the class para-K consists of all problems that are in K after a precomputation that only involves the parameter. A common example of such parameterized analogues of classical complexity classes is the parameterized complexity class para-NP. This class can alternatively be defined as the class of parameterized problems that are solvable in fpt-time by a non-deterministic algorithm [84]. Other examples include the classes para-Σ_2^p, para-Π_2^p and para-PSPACE.

Using the classes para-K and the notion of fpt-reductions, one can already provide evidence that certain parameterized problems are not fixed-parameter tractable. For instance, a parameterized problem Q is para-NP-hard if each problem $Q' \in$ para-NP is fpt-reducible to Q. Moreover, a problem is para-NP-complete if it is both para-NP-hard and in para-NP. (These notions of hardness and completeness hold for every parameterized complexity class.) An example of a para-NP-complete problem is a parameterized version SAT(constant) of SAT where the parameter is a fixed constant, that is, SAT(constant) $= \{ (x, 1) : x \in$ SAT $\}$. In fact, for any classical complexity class K and any K-complete problem Q, it holds that the parameterized problem $Q(\text{constant}) = \{ (x, 1) : x \in Q \}$ is para-K-complete. If a para-K-hard parameterized problem is fixed-parameter tractable, then K $=$ P. For example, a para-NP-hard parameterized problem is not fixed-parameter tractable, unless P $=$ NP.

3.2.2 The Weft Hierarchy

For many interesting parameterized problems, it turns out that a more subtle hardness theory is needed to provide evidence that they are not fixed-parameter tractable. These problems are polynomial-time solvable for each constant value of the parameter—that is, they are in XP—and therefore they cannot be para-NP-hard, unless $P = NP$. The most prominent hardness theory that is used to analyze the complexity of such problems consists of the *Weft hierarchy*, containing the classes $W[1] \subseteq W[2] \subseteq \cdots \subseteq W[SAT] \subseteq W[P]$. (All these inclusions are believed to be strict.) The parameterized complexity classes $W[t]$, for $t \in \mathbb{N}^+ \cup \{SAT, P\}$, are based on weighted variants of the satisfiability problem for various classes of Boolean circuits and formulas. We consider Boolean circuits with a single output gate (see Sect. 2.4 for a definition of Boolean circuits). We say that a Boolean circuit is a *formula* if all its gates have outdegree at most 1. Moreover, we distinguish between *small* gates, with indegree at most 2, and *large gates*, with indegree larger than 2. The *depth* of a circuit is the length of a longest path from any variable to the output gate. The *weft* of a circuit is the largest number of large gates on any path from a variable to the output gate. We say that a truth assignment for a Boolean circuit has *weight k* if it sets exactly k of the variables of the circuit to true. We denote the class of Boolean circuits with depth u and weft t by $\text{CIRC}_{t,u}$. We denote the class of all Boolean circuits by CIRC, and the class of all Boolean formulas by FORM. Now, for any class $\mathcal{C}$ of Boolean circuits, we define the following parameterized problem.

WSAT($\mathcal{C}$)
Instance: A Boolean circuit $C \in \mathcal{C}$, and an integer k.
Parameter: k.
Question: Does there exist an assignment of weight k that satisfies C?

The parameterized complexity classes $W[t]$, for $t \in \mathbb{N}^+ \cup \{SAT, P\}$, are defined as follows:

$$W[t] = [\ \{\ \text{WSAT}(\text{CIRC}_{t,u}) : u \geq 1\ \}\ ]^{\text{fpt}}, \text{ for each } t \geq 1;$$
$$W[SAT] = [\ \{\text{WSAT}(\text{FORM})\}\ ]^{\text{fpt}}; \text{ and}$$
$$W[P] = [\ \{\text{WSAT}(\text{CIRC})\}\ ]^{\text{fpt}}.$$

Here the notation $[\ \mathcal{S}\ ]^{\text{fpt}}$ denotes the closure of the set $\mathcal{S}$ of parameterized problems under fpt-reductions—that is, $[\ \mathcal{S}\ ]^{\text{fpt}}$ consists of all parameterized problems that can be fpt-reduced to some problem $S \in \mathcal{S}$. It is widely believed that $\text{FPT} \neq W[1]$, and it turns out that there are many parameterized problems that are complete for $W[1]$ or $W[2]$.

An example of a $W[1]$-complete problem is the problem MULTI-COLORED CLIQUE [80], which is often called MCC for short. Instances for this problem consist of tuples (V, E, k), where k is a positive integer, V is a finite set of vertices partitioned into k subsets $V_1, \ldots, V_k$, and (V, E) is a simple graph. The parameter is k. The question is whether there exists a k-clique in (V, E) that contains a vertex in each V_i—that is, a set $V' = \{v_1, \ldots, v_k\}$ such that $v_i \in V_i$, for each $i \in [k]$, and $\{v_i, v_j\} \in E$, for each $i, j \in [k]$ with $i < j$.

An example of a parameterized problem that is complete for the class W[SAT] is MONOTONE-WSAT(FORM) [2]. In this problem, the input consists of a monotone propositional formula φ—i.e., φ contains no negations—and a positive integer k. The parameter is k, and the question is whether there exists a truth assignment that sets at most k variables in $\mathrm{Var}(\varphi)$ true and that satisfies φ.

3.2.3 The A-Hierarchy and First-Order Logic

Parameterized counterparts for the classes of the Polynomial Hierarchy are provided by the parameterized complexity classes para-Σ_i^{p} and para-Π_i^{p}. However, for many interesting parameterized problems, these classes cannot be used. As explained for the example of para-NP in Sect. 3.2.2, parameterized problems that are in XP cannot be complete for these classes, unless P $=$ NP. Parameterized complexity theory features other intractability classes that are counterparts of the Polynomial Hierarchy that are applicable also for problems inside XP. These classes A[t], for each $t \geq 1$, form the A-hierarchy, and can be best described using first-order logic model checking.

We define the basic concepts of first-order logic. A *(relational) vocabulary* τ is a finite set of relation symbols. Each relation symbol R has an *arity* arity$(R) \geq 1$. A *structure* $\mathcal{A}$ of vocabulary τ, or τ-*structure* (or simply *structure*), consists of a set A called the *domain* (or *universe*) and an interpretation $R^{\mathcal{A}} \subseteq A^{\mathrm{arity}(R)}$ for each relation symbol $R \in \tau$. In first-order logic, formulas are built from a countably infinite set $\{x_1, x_2, \ldots\}$ of variables, relation symbols, existential and universal quantification, and the Boolean operators $\neg$, $\wedge$, and $\vee$. That is, if $R \in \tau$ is a relation symbol of arity a, and $x_1, \ldots, x_a$ are variables, then $R(x_1, \ldots, x_a)$ is a formula. Moreover, if φ_1 and φ_2 are formulas and x is a variable, then $\exists x.\varphi_1$, $\forall x.\varphi_1$, $\neg\varphi_1$, $(\varphi_1 \wedge \varphi_2)$, and $(\varphi_1 \vee \varphi_2)$ are also formulas. We use $(\varphi_1 \to \varphi_2)$ as an abbreviation for $(\neg\varphi_1 \vee \varphi_2)$, and we use $(\varphi_1 \leftrightarrow \varphi_2)$ as an abbreviation for $((\varphi_1 \to \varphi_2) \wedge (\varphi_2 \to \varphi_1))$. For a formula φ, we call the variables occurring in φ that are not bound by any quantifier the *free variables* of φ, and we write Free(φ) to denote the set of free variables in a formula φ. Formally, Free(φ) is defined inductively as follows:

$$\mathrm{Free}(R(x_1, \ldots, x_a)) = \{x_1, \ldots, x_a\},$$
$$\mathrm{Free}(\neg\varphi) = \mathrm{Free}(\varphi),$$
$$\mathrm{Free}(\varphi_1 \wedge \varphi_2) = \mathrm{Free}(\varphi_1) \cup \mathrm{Free}(\varphi_2),$$
$$\mathrm{Free}(\varphi_1 \vee \varphi_2) = \mathrm{Free}(\varphi_1) \cup \mathrm{Free}(\varphi_2),$$
$$\mathrm{Free}(\exists x.\varphi) = \mathrm{Free}(\varphi)\backslash\{x\}, \text{ and}$$
$$\mathrm{Free}(\forall x.\varphi) = \mathrm{Free}(\varphi)\backslash\{x\}.$$

Truth of first-order formulas given a structure and an assignment to the free variables of the formula is defined in the usual way. Let $\mathcal{A}$ be a τ-structure with universe A, let φ be a first-order formula over the vocabulary τ, and let $\alpha : \text{Free}(\varphi) \to A$ be an assignment. We often consider the assignment α as a set of mappings, i.e., $\alpha = \{x \mapsto \alpha(x) : x \in \text{Free}(\varphi)\}$. Then the following conditions define when φ is true in $\mathcal{A}$ given α, written $\mathcal{A}, \alpha \models \varphi$.

$$\mathcal{A}, \alpha \models R(x_1, \ldots, x_a) \quad \text{if and only if} \quad (x_1, \ldots, x_a) \in R^{\mathcal{A}},$$
$$\mathcal{A}, \alpha \models \neg\varphi \quad \text{if and only if} \quad \mathcal{A}, \alpha \not\models \varphi,$$
$$\mathcal{A}, \alpha \models (\varphi_1 \wedge \varphi_2) \quad \text{if and only if} \quad \mathcal{A}, \alpha \models \varphi_1 \text{ and } \mathcal{A}, \alpha \models \varphi_2,$$
$$\mathcal{A}, \alpha \models (\varphi_1 \vee \varphi_2) \quad \text{if and only if} \quad \mathcal{A}, \alpha \models \varphi_1 \text{ or } \mathcal{A}, \alpha \models \varphi_2,$$
$$\mathcal{A}, \alpha \models \exists x.\varphi \quad \text{if and only if} \quad \mathcal{A}, \alpha \cup \{x \mapsto a\} \models \varphi \text{ for some } a \in A,$$
$$\mathcal{A}, \alpha \models \forall x.\varphi \quad \text{if and only if} \quad \mathcal{A}, \alpha \cup \{x \mapsto a\} \models \varphi \text{ for each } a \in A.$$

For more details, we refer to textbooks (see, e.g., [85, Section 4.2]). A *first-order logic sentence* is a first-order logic formula that contains no free variables, i.e., a formula φ such that $\text{Free}(\varphi) = \emptyset$. For sentences φ, we write $\mathcal{A} \models \varphi$ to denote $\mathcal{A}, \emptyset \models \varphi$.

Let $i \geq 1$. We say that a first-order logic sentence φ is a Σ_i sentence if it is of the form $\exists x_1^1, \ldots, x_{\ell_1}^1.\forall x_1^2, \ldots, x_{\ell_2}^2 \ldots Q_i x_1^i, \ldots, x_{\ell_i}^i.\psi$, where $Q_i = \exists$ if i is odd and $Q_i = \forall$ if i is even, and where ψ is quantifier-free. For instance, Σ_2 sentences are of the form $\exists x_1^1, \ldots, x_{\ell_1}^1.\forall x_1^2, \ldots, x_{\ell_2}^2.\psi$. For each $i \geq 1$, the parameterized problem $\text{MC}(\Sigma_i)$ is defined as follows. Inputs consist of a structure $\mathcal{A}$ and a Σ_i sentence φ (over the same signature τ), and the parameter is $|\varphi|$. The question is to decide whether $\mathcal{A} \models \varphi$. The classes A[$t$], for each $t \geq 1$, are defined as follows:

$$A[t] = [\text{MC}(\Sigma_t)]^{\text{fpt}}.$$

So in particular, the problem $\text{MC}(\Sigma_2)$ is A[2]-complete. The problem remains A[2]-hard when (1) $\ell_1 = \ell_2$, (2) τ is a fixed signature containing only binary predicates, and (3) ψ is a disjunction of atoms [85, Lemma 8.10].

3.2.4 The Classes XNP, Xco-NP, $X\Sigma_i^p$, and $X\Pi_i^p$

We conclude our overview of known parameterized classes with several other parameterized analogues of classical complexity classes. Above, we already saw the class XP, which consists of all parameterized problems that are solvable by an algorithm that, for each parameter value k, runs in polynomial-time (where the order of the polynomial is bounded by $f(k)$, for some computable function f). A slight variant of this class is the (non-uniform) class XP_{nu}, that can be defined using the notion of *slices*. Let Q be a parameterized problem. For each positive integer $s \geq 1$, the s-th slice of Q is defined as the classical problem $Q_s = \{x : (x, s) \in Q\}$. Then XP_{nu} consist of those parameterized problems Q such that for each $s \in \mathbb{N}$, the slice Q_s is polynomial-time solvable. This class can be called non-uniform, because the polynomial-time algorithms for the different slices Q_s might not be computable from the values $s \in \mathbb{N}$.

This definition straightforwardly generalizes to arbitrary classical complexity classes. Let K be a classical complexity class. Then, the parameterized complexity class XK_{nu} consists of those parameterized problems Q such that for each $s \in \mathbb{N}$, the slice Q_s is in K [67, 84]. (Originally, the classes XK_{nu} were introduced under the names XK. We use the names XK_{nu} to emphasize the non-uniformity that is present in these classes, and to distinguish these classes from uniform counterparts such as XP.)

As it is often more convenient to work with uniform counterparts of these classes, we will define such counterparts for several cases. We begin with the case where K = NP. The parameterized complexity class XNP consists of all parameterized problems Q for which there exists a computable function $f : \mathbb{N} \to \mathbb{N}$ and an algorithm that decides whether $(x, k) \in Q$ in non-deterministic time $|x|^{f(k)}$, for some computable function f. The second case is that of the dual class, denoted by Xco-NP. That is, Xco-NP consists of all parameterized problems Q such that co-$Q \in$ XNP, where co-$Q = \{(x, k) \in \Sigma^* \times \mathbb{N} : (x, k) \notin Q\}$. More generally, we consider the case where K $= \Sigma_i^p$ or K $= \Pi_i^p$, for some $i \in \mathbb{N}$. We define these classes using alternating Turing machines. The parameterized complexity class $X\Sigma_i^p$ consists of all parameterized problems Q for which there exists an i-alternating ATM $\mathbb{M} = (S_\exists, S_\forall, \Sigma, \Delta, s_0, F)$ with $s_0 \in S_\exists$, that decides whether $(x, k) \in Q$ in time $|x|^{f(k)}$, for some computable function f. Similarly, the parameterized complexity class $X\Pi_i^p$ consists of all parameterized problems Q for which there exists an i-alternating ATM $\mathbb{M} = (S_\exists, S_\forall, \Sigma, \Delta, s_0, F)$ with $s_0 \in S_\forall$, that decides whether $(x, k) \in Q$ in time $|x|^{f(k)}$, for some computable function f.

Beyond Para-NP

Chapter 4
Fpt-Reducibility to SAT

While we still seem to be quite far from resolving the questions
about the computational complexity of SAT, progress on the
engineering side has been nothing short of spectacular.
— Moshe Y. Vardi [191]

In the traditional parameterized complexity literature, the concept of fixed-parameter tractability is commonly used as a desideratum for algorithms that solve a particular problem in its entirety (i.e., fpt-algorithms) [27, 66, 67, 85, 166]. Less attention has been given to reducing one problem in fixed-parameter tractable time to another problem that can be attacked by means of other algorithmic techniques. However, this latter solving strategy has the potential of producing algorithmic methods that work well in practical settings for highly intractable problems, in view of the outstanding performance of modern SAT solvers on many instances occurring in practice [22, 96, 153, 159, 176]—and efficient solvers for other NP-complete problems such as integer linear programming (ILP) and the constraint satisfaction problem (CSP). In this thesis, we carry out a structured theoretical investigation of the possibilities and limits of the technique of solving problems by first reducing them to one or more instances of SAT in fixed-parameter tractable time, and invoking a SAT solver to solve the SAT instances (and thereby solving the instance of the original problem).

Transforming instances of one problem to instances of another problem in fixed-parameter tractable time is a concept that is ubiquitous in parameterized complexity. Algorithms that perform such a transformation are known as fpt-reductions. (Technically, fpt-reductions must satisfy the additional requirement that the parameter value of the resulting instance is bounded by a computable function of the original parameter value.) Fpt-reductions play a vital role in showing fixed-parameter intractability: they are used to show hardness for intractability classes. They are occasionally also used to show fixed-parameter tractability by reducing some problem to another problem that is already known to be fixed-parameter tractable. However,

© Springer-Verlag GmbH Germany, part of Springer Nature 2019
R. de Haan: Parameterized Complexity in the Polynomial Hierarchy, LNCS 11880,
https://doi.org/10.1007/978-3-662-60670-4_4

fpt-reductions have seldom been used in combination with algorithmic techniques developed outside of parameterized complexity.

When taking a worst-case complexity-theoretic point of view and when considering polynomials of arbitrary order, it does not matter to what NP-complete problem one reduces problem instances. For the sake of convenience, we will take SAT as the canonical NP-complete problem that we reduce to, and we will speak informally of "fpt-reductions to SAT."

The technique of using fpt-reductions to SAT makes most sense for problems that are highly intractable, i.e., problems whose complexity lies at the second level of the PH or higher, as problems at the first level of the PH can be encoded into SAT in polynomial time. In our investigation, we therefore focus mainly on problems that lie at higher levels of the PH and problems that are even PSPACE-complete. The solvers that are available for problems at these levels of complexity are generally not as successful as the solvers available for problems at the first level of the PH. Therefore, the investment of computational resources (when using fpt-reductions rather than polynomial-time reductions) could pay off, if more efficient solvers could then be used to solve the problem.

This generic scheme of solving a problem by means of an fpt-reduction to SAT can be embodied in various ways. For instance, there are various formal interpretations of the concept of problem reductions that one could choose. We begin our investigation in this chapter by considering an interpretation of the idea of fpt-reductions to SAT that is based on the most widely used form of problem reductions: *many-to-one* (or *Karp*) *reductions*. In this setting, a fixed-parameter tractable algorithm R takes as input an instance (x, k) of a parameterized problem Q, and outputs an instance x' of SAT such that $(x, k) \in Q$ if and only if $x' \in$ SAT. Equivalently, one can consider this reduction R as an fpt-reduction from Q to the parameterized problem SAT(constant) $= \{ (x, 1) : x \in$ SAT $\}$, where the parameter value $k = 1$ is constant. For the sake of convenience, we will use the following (notational) convention throughout this thesis.

Convention 4.1. *We say that a parameterized problem is* fpt-reducible to SAT *if it is fpt-reducible to* SAT(constant).

It is straightforward to show that SAT(constant) is para-NP-complete—membership follows from the fact that SAT is in NP and hardness follows from the fact that the only slice of SAT(constant) is NP-hard. This leads to the following observation, that shows that the parameterized complexity class para-NP captures exactly those parameterized problems that are many-to-one fpt-reducible to SAT.

Observation 4.2. *A parameterized problem is many-to-one fpt-reducible to SAT if and only if it is in* para-NP.

Most modern SAT solvers are complete solvers, i.e., for every instance x they report one of two possible answers: "$x \in$ SAT" or "$x \notin$ SAT". Therefore, it also makes sense to consider (many-to-one) reductions to UNSAT, the co-problem of SAT. Many-to-one fpt-reductions to UNSAT are defined similarly to many-to-one fpt-reductions to SAT, and a parameterized problem Q is many-to-one fpt-reducible to UNSAT if and only if $Q \in$ para-co-NP.

In this view, para-NP-membership results and para-co-NP-membership results can be considered as positive results for parameterized problems based on problems that are hard for the second level of the PH (or higher). For several problems, such positive results have been developed in the literature. Additionally, there are several results in the literature that can be rephrased as fpt-reductions to SAT. On the other hand, there are of course also problems that do not admit fpt-reductions to SAT.

Outline of This Chapter

In this chapter, we survey to what extent the existing parameterized complexity literature can be used to investigate the possibilities and limits of many-to-one fpt-reductions to SAT.

Firstly, to provide some context on the great performance of SAT solvers in practice, in Sect. 4.1, we give a brief overview of the engineering achievements for SAT solving (and related) algorithms that have been attained over the last few decades.

Then, in Sect. 4.2, we describe several many-to-one fpt-reductions to SAT (or UNSAT), several of which appeared in the literature. We use these results to argue that several productive parameterized complexity techniques—namely, using treewidth and backdoors—can also be used to obtain fpt-reductions to SAT.

In Sect. 4.3, we give a foretaste of the use of more powerful fpt-reductions to SAT (namely Turing reductions). In particular, we consider a problem that can be solved by means of such an fpt-time Turing reduction to SAT.

Finally, in Sect. 4.4, we give several examples of parameterized problems that are unlikely to admit an fpt-reduction to SAT, and illustrate how known parameterized complexity classes (e.g., A[2] and para-Σ_2^p) can be used to support such irreducibility claims using hardness results.

4.1 Modern SAT Solvers

Over the last two decades, enormous progress has been made on the engineering of SAT solving algorithms. As Moshe Vardi put it, this progress has been "nothing short of spectacular" [191]. To provide context for the potential of using fpt-reductions to SAT for obtaining practically useful solving methods, we give a brief (and incomplete) historical overview of the development and success of SAT solving algorithms[1], and we describe some of the most important algorithmic techniques behind their amazing performance. For a more detailed account of the history and working of modern SAT solvers, we refer to the literature [96, 129].

---

[1] We restrict ourselves to *complete* solving methods, i.e., algorithms that give a correct answer for every possible input.

4.1.1   The Success of SAT Solvers

Research in the 1960s and 1970s identified SAT—the satisfiability problem for propositional formulas—as a key problem underlying hundreds of important search problems from a variety of areas in computer science and other domains. Important milestones in this research are the Cook-Levin Theorem [54, 144]—which states that SAT is NP-complete, even when restricted to propositional formulas that are in CNF—and the seminal book by Garey and Johnson [86]—which provides an enormous list of problems that can be reduced to SAT.

These results form the basis for the currently widely held belief that SAT cannot be solved faster than in exponential time, in the worst case. It is easy to see that the problem can be solved in exponential time. The naive truth table algorithm, that goes over all possible truth assignments, runs in time $2^{O(n)}$ for every input, where n denotes the number of variables of the formula. However, already in the 1960s, algorithms were constructed that are much faster in many cases. An important example of such an algorithm is the DPLL algorithm—named after Martin Davis, Hilary Putnam, George Logemann, and Donald W. Loveland—that underlies a large majority of today's efficient SAT solving algorithms. The DPLL algorithm is an improvement on the straightforward backtracking algorithm that traverses the search tree containing all possible truth assignments, and is based on the eager use of *unit propagation* (instantiating unit clauses) and *pure literal elimination* (instantiating variables that occur only positively or only negatively) to the reduced formula corresponding to the current node in the search tree. These techniques efficiently prune the search space, allowing the algorithm to be must faster than the worst-case exponential bound in many cases.

Since the 1990s, numerous important improvements have been made to the basic DPLL algorithm, leading to a much wider range of inputs for which the algorithms work extremely fast. (Below, in Sect. 4.1.2, we discuss several of the principal techniques that feature in modern SAT algorithms.) In fact, the performance of modern SAT solvers is so impressive that it can be seen as a serious challenge to the relevance of the worst-case intractability status of SAT. This progress is stimulated by the yearly International Conference on Theory and Applications of Satisfiability Testing—also named SAT—which hosts a SAT solving competition each year, where SAT solving algorithms compete against each other on a wide range of benchmark instances, many of which find their origin in real-world industrial applications or hard combinatorial problems.

Due to their great performance, SAT solvers can increasingly be used as efficient general-purpose tools to solve hard problems in a plethora of areas in computer science and artificial intelligence. As an illustration, we mention two cases where SAT solvers have been successfully applied to problems that are traditionally not considered to be closely related to propositional satisfiability. The first example is related to software and hardware verification, where the problem is to decide whether an abstract model of an automated system satisfies a desired temporal property (e.g., whether the system will never get stuck in a particular state). For this problem, SAT algorithms are at the basis of one of the most efficient solving methods available [18, 20, 21, 52]. The second example concerns the problem of planning in artificial intelligence, where the task is to find a sequence of actions (given a set of available actions) that achieves a certain goal [130, 131]. (We consider these problems again in Chaps. 9 and 12, respectively.)

4.1.2 Algorithmic Techniques

We continue by discussing some of the most important algorithmic techniques used in modern SAT algorithms. The first valuable technique that we consider is that of *clause learning*. Whenever a conflict is reached, (1) the resolution rule is used on the clauses involved in the conflict to extract information about the cause of the conflict, (2) this information is remembered in the form of a learned clause, i.e., a clause entailed by the formula, and (3) this information is used to prune the search elsewhere in the search space. There are various choices one can make when implementing such *conflict-driven learning*—e.g., what clause to learn, and which clauses to remember—resulting in different learning schemes.

A second influential technique, that is used in combination with clause learning, is the technique of *non- chronological backjumping* [157, 158]. This refers to the practice of jumping back to (and reverting) a decision in the search tree that is not the decision that was made last chronologically, upon reaching a conflict. The choice of decision to revert to is often made using clauses learned from the conflict. Algorithms that employ clause learning in combination with non-chronological backjumping are often referred to as *conflict-driven clause learning (CDCL)* algorithms.

Yet another important method is the use of *randomized restarts* [97], where clause learning algorithms can arbitrarily stop and restart the search, keeping the clauses that were learned so far during the search. Such restarts are a way of exploiting randomization to improve the efficiency of the search algorithms. Most solvers employ aggressive restart strategies, restarting after a very small number of backtracks.

The use of efficient *(lazy) data structures* to implement constraint propagation has also led to rewarding improvements. The most successful example of this is the *two watched literals scheme* [164], where—rather than maintaining a counter for each clause to keep track of the number of literals that are not yet satisfied—two literals are watched in each clause, triggering unit propagation when one of them is falsified. This method allows for lazy updating of the status of clauses, which results in a significant decrease in overhead.

Besides using the advanced methods that we described above, the performance of SAT solving algorithms can be greatly improved by choosing the right *branching heuristics*. Even for the vanilla version of the DPLL algorithm, whenever a decision is to be made, there are many possibilities for choosing which literal to assign which truth value. Branching heuristics can be static (i.e., choosing on the basis of a fixed strategy) or dynamic (i.e., making a choice that depends on the reduced formula corresponding to the current node in the search tree), and often have an enormous effect on the efficiency of the algorithm.

4.1.3   Successful Algorithms for Other NP-complete Problems

We finish Sect. 4.1 by considering a number of other NP-complete problems for which algorithms have been developed that work well in many cases, and briefly describing some of the main algorithmic techniques used to solve these problems. As mentioned above, because we take a worst-case point of view and because we consider polynomials of arbitrary order, the theory that we develop in this thesis applies to any NP-complete problem.

Satisfiability Modulo Theories

The problem of *Satisfiability Modulo Theories (SMT)* [13] consists of finding a satisfying assignment for a Boolean formula where the atomic propositions are statements from some background theory (expressed in first-order logic where certain function and predicate symbols have a predefined meaning). An example of such a background theory is the theory of integer arithmetic—where atomic propositions could be of the form $(x + 5 \leq 2y)$, for instance.

One of the most important solving methods for the problem of SMT involves the combination of DPLL-based search algorithms with dedicated solvers for the background theory at hand. These background theory solvers take as input a collection of statements in the theory, and decide whether this collection of statements is satisfiable. In order for the DPLL-based algorithms to work well in combination with the theory solvers, it is important that the theory solvers have various features. For instance, (1) the theory solvers should be able to provide a model in case a collection of statements is satisfiable, (2) the theory solvers should be able to provide an explanation in case a collection of statements is unsatisfiable (e.g., a proof of unsatisfiability), and (3) the theory solver should be able to perform various types of deductive reasoning (e.g., detecting that a satisfiable collection of statements implies another statement).

Constraint Satisfaction

The *Constraint Satisfaction Problem (CSP)* [175] consists of finding an assignment for a set of variables (each with a finite domain) that satisfies a given set of constraints. The constraints are represented as a list of possible subassignments that are allowed for the variables to which the constraint applies. This way, many types of constraints can be expressed easily, resulting in a convenient method of modelling search problems of many different kinds. The approach of solving search problems by modelling them as instances of the constraint satisfaction problem and subsequently solving them using CSP solvers is also known as *constraint programming*. We consider the constraint satisfaction problem in more detail in Sect. 4.4.2.

One of the most successful solving methods for the constraint satisfaction problem is based on algorithms that traverse a search tree, similarly to DPLL-based algorithms for SAT. Going down a branch in the search tree corresponds to assigning a value to a variable. At each node in the search tree, the problem is simplified by removing (combinations of) variable-value assignments that can not be used in any solution. This simplification process is known as *constraint propagation*, and it is a very general concept that includes many types of reasoning. For example, one could consider unit propagation as a type of constraint propagation. Whenever there are no possible values left for some variable, the algorithm backtracks and goes back up in the search tree. These search algorithms are often improved by means of *look-ahead techniques* that examine part of the search tree before deciding which branch to go down on, in order to anticipate the effects of this branching decision.

Answer Set Programming

Answer Set Programming (ASP) [31, 90, 154] is a form of logic programming that is based on the stable model (or answer set) semantics of logic programs. The problem of finding an answer set for a (propositional) logic program is NP-complete. The stable model semantics of answer set programming allows for a convenient way of modelling search problems that involve the closed-world assumption, for example. We consider a more expressive variant of answer set programming (where the head of rules may include disjunctions) in more detail in Sects. 4.2.3 and 5.1.

Answer set solvers are algorithms for finding and enumerating answer sets for a given logic program. Many of the most successful answer set solvers are based on extensions of DPLL-based SAT solving algorithms, where additional steps are performed to ensure that solutions that are found are answer sets—each answer set of a logic program is also a satisfying assignment of the propositional formula that expresses the logic program, but not vice versa. These answer set solvers also often make use of the techniques of clause learning and non-chronological backjumping.

(Mixed) Integer Linear Programming

Linear programming consists of finding an optimal solution of a linear function, subject to linear inequality constraints—that is, finding a combination of values for variables $x_1, \ldots, x_n$ that both (1) maximizes (or minimizes) the value of $c_1 x_1 + \cdots + c_n x_n$ and that (2) satisfies a given set of constraints of the form $a_1 x_1 + \cdots + a_n x_n \leq b$, where $a_i, b, c_i \in \mathbb{Z}$. If the values for the variables x_i are allowed to be rational numbers, this problem can be solved in polynomial time (for instance using the simplex method [55, Section 29.3]). However, for the case where (some of) the variables are restricted to have integer values, no polynomial-time algorithms are known—this variant is referred to as *(Mixed) Integer Linear Programming (ILP/MILP)* [196]. For this variant, the problem of finding an integer solution that achieves a given value for the linear function is NP-complete.

Many of the most efficient algorithms for (mixed) integer linear programming are based on the following general scheme. This scheme uses relaxations of the integer linear program where the solutions are not required to have integer values—any solution for the program is also a solution for the relaxation, but not necessarily vice versa. The scheme consists of initially taking the (non-integer) relaxation of the integer linear program that consists of the same constraints, and using one of the available polynomial-time algorithms to find an optimal solution for this relaxation. If the solution that is found is an integer solution, then this is an optimal (integer) solution for the original integer linear program. Otherwise, the relaxation is augmented with a linear inequality that is not satisfied by the current (non-integer) solution, but that is satisfied by each integer solution. This process of finding optimal solutions for

the relaxation and refining the relaxation is repeated until an optimal integer solution
is found. There are various choices that one could make for the inequalities that are
introduced to rule out non-integer solutions. For example, one could directly rule
out a particular non-integer value in the current solution—e.g., if the current solution
assigns variable x_1 to 3.2, one could require that $x_1 \leq 3$ or $x_1 \geq 4$. As disjunctions of
inequalities are not expressible in the problem, this yields a search tree. Algorithms
based on adding this type of inequalities are often called *branch-and-bound methods*.
Another type of inequality that is often introduced to rule out non-integer solutions
is based on the facets of the convex hull of integer solutions—adding this type of
inequalities is referred to as the *cutting-plane method*.

4.2   Various Fpt-Reductions to SAT

In this section, we consider fpt-reductions to SAT (or UNSAT) for several parameterized problems. Some of these results have appeared in the literature (albeit not
always interpreted as fpt-reductions to SAT). In particular, we give several examples how the productive parameterized complexity methods of using treewidth and
backdoor size as parameters can be used to obtain fpt-reductions to SAT.

4.2.1   Quantified Boolean Satisfiability

We begin by considering several parameterized variants related to the satisfiability
problem for quantified Boolean formulas (QBFs). Concretely, we consider the problem QSAT parameterized by the number of universal variables, and a parameterized
variant of the problem $QSAT_2(DNF)$—the satisfiability problem of QBFs with an $\exists\forall$
quantifier prefix whose matrix is in DNF.

Quantifier Expansion
The problem of deciding satisfiability of a QBF does not allow a polynomial-time
SAT encoding (unless NP = PSPACE). However, if the number of universal variables
is small, one can use known methods in QBF solving [9, 19] to get an many-to-one
fpt-reduction to SAT. Consider the following parameterized problem.

QSAT(#∀-vars)
Instance: A quantified Boolean formula φ.
Parameter: The number of universally quantified variables of φ.
Question: Is φ true?

Since the problem QSAT is PSPACE-complete in general, the following para-NP-
membership result can be seen as a positive example of a setting where many-to-one
fpt-reduction to SAT can be applied successfully.

Proposition 4.3. QSAT(#∀-vars) *is* para-NP-*complete.*

Proof. The problem of deciding the satisfiability of a propositional formula can be seen as the problem of deciding whether a quantified Boolean formula with only existentially quantified variables is true. Therefore, the problem QSAT(#∀-vars) is NP-hard already for the parameter value $k = 0$. Thus QSAT(#∀-vars) is para-NP-hard.

To show membership in para-NP, we give an fpt-time reduction to the problem of deciding truth of a quantified Boolean formula with only existentially quantified variables. We do so by repeatedly applying universal quantifier expansion [9, 19]. This is a transformation that eliminates the rightmost universally quantified variable from a quantified Boolean formula as follows. Let $\varphi = Q_1 x_1 \ldots Q_{\ell-1} x_{\ell-1} \forall x_\ell \exists x_{\ell+1} \ldots \exists x_m.\psi$ be a quantified Boolean formula, where $Q_i \in \{\exists, \forall\}$ for all $i \in [\ell - 1]$. We eliminate the universally quantified variable x_ℓ by transforming φ into a quantified Boolean formula φ' that is true if and only if φ is true. We introduce a copy x_i' of the variables x_i with $i \in [\ell + 1, m]$, then, we obtain the (quantifier-free) formula ψ' from ψ by replacing each occurrence of a variable $x \in X$ in ψ by the corresponding copy x'. Finally, we define the quantified Boolean formula φ' as follows:

$$\varphi' = Q_1 x_1 \ldots Q_{\ell-1} x_{\ell-1} \exists x_{\ell+1} \ldots \exists x_m \exists x_{\ell+1}' \ldots \exists x_m'.(\psi[x_\ell \mapsto 1] \wedge \psi'[x_\ell \mapsto 0]).$$

It can be readily verified that φ' is true if and only if φ is true.

Each time we perform this transformation, we blow up the quantified Boolean formula with a factor of at most 2. Therefore, eliminating all k universally quantified variables in this manner results in an existentially quantified formula that is at most a factor of 2^k larger than the original formula.

Existential and Universal Treewidth

Next, we investigate the parameterized complexity of a parameterized variant of the problem QSAT$_2$(DNF). This parameterization is related to the incidence treewidth of QDNF formulas. Remember that a QDNF formula is a QBF whose matrix is in DNF.

The graph parameter *treewidth* measures the tree-likeness of a graph, and is defined as follows. A tree decomposition of a graph $G = (V, E)$ is a pair $(\mathcal{T}, (B_t)_{t \in T})$ where $\mathcal{T} = (T, F)$ is a rooted tree and $(B_t)_{t \in T}$ is a family of subsets of V such that:

- for every $v \in V$, the set $B^{-1}(v) = \{ t \in T : v \in B_t \}$ is nonempty and connected in $\mathcal{T}$; and
- for every edge $\{v, w\} \in E$, there is a $t \in T$ such that $v, w \in B_t$.

The *width* of the decomposition $(\mathcal{T}, (B_t)_{t \in T})$ is the number $\max\{ |B_t| : t \in T \} - 1$. The *treewidth* of G is the minimum width over all tree decompositions of G. Let G be a graph and k a nonnegative integer. There is an fpt-algorithm that computes a tree decomposition of G of width k if it exists, and fails otherwise [23]. We call a tree decomposition $(\mathcal{T}, (B_t)_{t \in T})$ *nice* if every node $t \in T$ is of one of the following four types:

- *leaf node*: t has no children and $|B_t| = 1$;
- *introduce node*: t has one child t' and $B_t = B_{t'} \cup \{v\}$ for some vertex $v \notin B_{t'}$;
- *forget node*: t has one child t' and $B_t = B_{t'} \backslash \{v\}$ for some vertex $v \in B_{t'}$; or
- *join node*: t has two children t_1, t_2 and $B_t = B_{t_1} = B_{t_2}$.

Given any graph G and a tree decomposition of G of width k, a nice tree decomposition of G of width k can be computed in polynomial time [132].

Many hard problems are fixed-parameter tractable when parameterized by the treewidth of a graph associated with the input [26, 100]. By associating the following graph with a QDNF formulas one can use treewidth also as a parameter for problems where the input is a QDNF formula.

The *incidence graph* of a QDNF formula φ is the bipartite graph where one side of the partition consists of the variables and the other side consists of the terms. A variable and a term are adjacent if the variable appears (positively or negatively) in the term. The *incidence treewidth* of φ, in symbols itw(φ), is the treewidth of the incidence graph of φ. It is well known that checking the truth of a QDNF formula φ whose number of quantifier alternations is bounded by a constant is fixed-parameter tractable when parameterized by itw(φ) (this can easily be shown using Courcelle's Theorem [100]).

Bounding the treewidth of the entire incidence graph is very restrictive. Instead, we investigate whether bounding the treewidth of certain subgraphs of the incidence graph is sufficient to reduce the complexity. To this aim, we define the *existential incidence treewidth* of a QDNF formula φ, in symbols $\exists$-itw(φ), as the treewidth of the incidence graph of φ after deletion of all universal variables. The *universal incidence treewidth*, in symbols $\forall$-itw(φ), is the treewidth of the incidence graph of φ after deletion of all existential variables.

The existential and universal treewidth can be small for formulas whose incidence treewidth is arbitrarily large. Take for instance a QDNF formula φ whose incidence graph is an $n \times n$ grid, as in Fig. 4.1. In this example, $\exists$-itw(φ) = $\forall$-itw(φ) = 2 (since after the deletion of the universal or the existential variables the incidence graph becomes a collection of trivial path-like graphs), but itw(φ) = n [24]. Hence, a tractability result in terms of existential or universal incidence treewidth would apply to a significantly larger class of instances than a tractability result in terms of incidence treewidth.

In this section, we characterize the complexity of checking the satisfiability of QDNF formulas with an $\exists\forall$ prefix, parameterized by $\forall$-itw. That is, we consider the following parameterized variant of QSAT$_2$(DNF).

QSAT$_2$($\forall$-itw)
Instance: A quantified Boolean formula $\varphi = \exists X.\forall Y.\psi$, where ψ is a quantifier-free formula in DNF, with universal incidence treewidth k.
Parameter: k.
Question: Is φ true?

For this problem, we assume that a tree decomposition of width k is given as part of the input. We may assume this without loss of generality, since a tree decomposition can be computed in fpt-time [23].

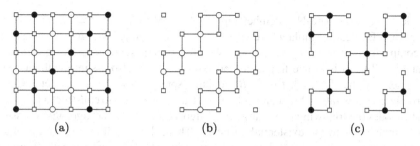

Fig. 4.1 Incidence graph of a QDNF formula (a). Universal variables are drawn with black round shapes, existential variables with white round shapes, and terms are drawn with square shapes. Both deleting the universal variables (b) and the existential variables (c) significantly decreases the treewidth of the incidence graph.

(We return to the problem $\mathrm{QSAT}_2(\mathrm{DNF})$ in Sect. 4.4, where we analyze the complexity of a natural counterpart of $\mathrm{QSAT}_2(\forall\text{-itw})$—the problem $\mathrm{QSAT}_2(\mathrm{DNF})$ parameterized by the existential incidence treewidth $\exists\text{-itw}$.)

We show that the problem $\mathrm{QSAT}_2(\forall\text{-itw})$ is para-NP-complete.

Theorem 4.4. $\mathrm{QSAT}_2(\forall\text{-itw})$ *is para-NP-complete.*

Proof. We show para-NP-hardness by showing that the problem is already NP-hard when restricted to instances where the parameter value is 1 [84]. We reduce from SAT, and the idea behind this reduction is to reduce the instance of SAT to an instance of QSAT_2 whose matrix is in DNF by using the standard Tseitin transformation, resulting in tree-like interactions between the universally quantified variables.

Let φ be a propositional formula whose satisfiability we want to decide, with $\mathrm{Var}(\varphi) = X$. Assume without loss of generality that φ contains only the connectives $\neg$ and $\wedge$. Equivalently, we want to determine whether the QBF $\psi = \exists X.\forall\emptyset.\varphi$ is true. We can transform ψ into an equivalent QBF $\psi' = \exists X.\forall Y.\chi$ whose matrix is in DNF by using the standard Tseitin transformation [188] as follows.

Let $\mathrm{Sub}(\varphi) = \{r_1, \ldots, r_s\}$ be the set of subformulas of φ. We let $Y = \{y_1, \ldots, y_s\}$ contain one propositional variable y_i for each $r_i \in \mathrm{Sub}(\varphi)$. Then, we define $\psi' = \exists X.\forall Y.\chi$, where $\chi = r_\varphi \vee \bigvee_{i \in [u]} \chi_i$, and:

$$\chi_i = \begin{cases} (r_i \wedge r_j) \vee (\neg r_i \wedge \neg r_j) & \text{if } r_i = \neg r_j; \\ (r_i \wedge \neg x) \vee (\neg r_i \wedge x) & \text{if } r_i = x \in X; \\ (r_i \wedge \neg r_j) \vee (r_i \wedge \neg r_k) \vee (\neg r_i \wedge r_j \wedge r_k) & \text{if } r_i = r_j \wedge r_k. \end{cases}$$

It is straightforward to verify that ψ and ψ' are equivalent. Moreover, $\mathrm{IG}(Y, \chi)$ is a tree, and thus has treewidth 1. Therefore, it is easy to construct a tree decomposition $(\mathcal{T}, (B_t)_{t \in T})$ of $\mathrm{IG}(Y, \chi)$ of width 1 in polynomial time. Then $(\mathcal{T}, (B_t)_{t \in T})$ and ψ' together are an instance of $(\mathrm{QSAT}_2(\forall\text{-itw}))_1$ such that ψ' is true if and only if φ is satisfiable.

Next, we show para-NP-membership of $\text{QSAT}_2(\forall\text{-itw})$. Let $\varphi = \exists X.\forall Y.\psi$ be a quantified Boolean formula where $\psi = \delta_1 \vee \cdots \vee \delta_u$, and let $(\mathcal{T}, (B_t)_{t \in T})$ be a tree decomposition of $\text{IG}(Y, \psi)$ of width k. We may assume without loss of generality that $(\mathcal{T}, (B_t)_{t \in T})$ is a nice tree decomposition. We may also assume without loss of generality that for each $t \in T$, B_t contains some $y \in Y$. We construct a CNF formula φ' that is satisfiable if and only if φ is true. The idea is to construct a formula that encodes the following guess-and-check algorithm. Firstly, the algorithm guesses an assignment γ to the existential variables. Then, the algorithm uses a dynamic programming approach using the tree decomposition to decide whether the formula instantiated with γ is valid. This dynamic programming approach is widely used to solve problems for instances where some graph representing the structure of the instance has small treewidth (see, e.g., [25]).

Next, we show how to encode this guess-and-check algorithm into a formula φ' that is satisfiable if and only if the algorithm accepts. We let $\text{Var}(\varphi') = X \cup Z$ where $Z = \{ z_{t,\alpha,i} : t \in T, \alpha : \text{Var}(t) \to \mathbb{B}, i \in [u] \}$. Intuitively, the variables $z_{t,\alpha,i}$ represent whether at least one assignment extending α (to the variables occurring in nodes t' below t) violates the term δ_i of ψ. We then construct φ' as follows by using the structure of the tree decomposition. For all $t \in T$, all $\alpha : \text{Var}(t) \to \mathbb{B}$, all $i \in [u]$, and each literal $l \in \delta_i$ such that $\text{Var}(l) \in X$, we introduce the clause:

$$(\bar{l} \to z_{t,\alpha,i}). \tag{4.1}$$

Then, for all $t \in T$, all $\alpha : \text{Var}(t) \to \mathbb{B}$, and all $i \in [u]$ such that for some $l \in \delta_i$ it holds that $\text{Var}(l) \in Y$ and $\alpha(l) = 0$, we introduce the clause:

$$(z_{t,\alpha,i}). \tag{4.2}$$

Next, let $t \in T$ be any introduction node with child t', and let $\alpha : \text{Var}(t') \to \mathbb{B}$ be an arbitrary assignment. For any assignment $\alpha' : \text{Var}(t) \to \mathbb{B}$ that extends α, and for each $i \in [u]$, we introduce the clause:

$$(z_{t',\alpha,i} \to z_{t,\alpha',i}). \tag{4.3}$$

Then, let $t \in T$ be any forget node with child t', and let $\alpha : \text{Var}(t) \to \mathbb{B}$ be an arbitrary assignment. For any assignment $\alpha' : \text{Var}(t') \to \mathbb{B}$ that extends α, and for each $i \in [u]$, we introduce the clause:

$$(z_{t',\alpha',i} \to z_{t,\alpha,i}). \tag{4.4}$$

Next, let $t \in T$ be any join node with children t_1, t_2, and let $\alpha : \text{Var}(t) \to \mathbb{B}$ be an arbitrary assignment. For each $i \in [u]$, we introduce the clauses:

$$(z_{t_1,\alpha,i} \to z_{t,\alpha,i}) \text{ and } (z_{t_2,\alpha,i} \to z_{t,\alpha,i}). \tag{4.5}$$

Finally, for the root node $t_{root} \in T$ and for each $\alpha : \text{Var}(t_{root}) \to \mathbb{B}$ we introduce the clause:

$$\bigvee_{i \in [u]} \neg z_{t_{root}, \alpha, i}. \tag{4.6}$$

It is straightforward to verify that φ' contains $O(2^k |T|)$ clauses.

We now show that φ is true if and only if φ' is satisfiable.

($\Rightarrow$) Assume that there exists an assignment $\beta : X \to \mathbb{B}$ such that $\forall Y. \psi[\beta]$ is true. We construct an assignment $\gamma : Z \to \mathbb{B}$ such that $\beta \cup \gamma$ satisfies φ'. Let C be the set of clauses $(z_{t,\alpha,i})$ such that φ' contains a clause $(\bar{l} \to z_{t,\alpha,i})$ for which $\beta(\bar{l}) = 1$. Since C together with the clauses (4.2–4.5) forms a definite Horn formula χ, we can compute its unique subset-minimal model (by unit-propagation). Let γ be this subset-minimal model of χ. We show that γ also satisfies the clauses (4.6). Let $\alpha : \text{Var}(t_{root}) \to \mathbb{B}$ be an arbitrary assignment. Since $\forall Y. \psi[\beta]$ is true, we know that there exists an assignment $\alpha' : Y \to \mathbb{B}$ extending α such that $\psi[\alpha' \cup \beta]$ is true, i.e., for some $i \in [u]$ the term δ_i is satisfied by $\alpha' \cup \beta$. This assignment α' induces a family $(\alpha_t)_{t \in T}$ of assignments as follows: for each $t \in T$, the assignment α_t is the restriction of α' to the variables $\text{Var}(t)$. It is straightforward to verify that $\gamma(z_{t,\alpha_t,i}) = 0$ for all $t \in T$. Therefore, since $\alpha = \alpha_{t_{root}}$, the clause $\bigvee_{i \in [u]} (\neg z_{t_{root}, \alpha, i})$ is satisfied. Since α was arbitrary, we know that γ satisfies all clauses in (4.6). Thus, $\beta \cup \gamma$ satisfies φ'.

($\Leftarrow$) Assume that there is an assignment $\beta : X \to \mathbb{B}$ and an assignment $\gamma : Z \to \mathbb{B}$ such that $\beta \cup \gamma$ satisfies φ'. We show that $\forall Y. \psi[\beta]$ is true. Let $\alpha' : Y \to \mathbb{B}$ be an arbitrary assignment. We show that for some $i \in [u]$, $\alpha' \cup \beta$ satisfies the term δ_i, and thus that $\psi[\alpha' \cup \beta]$ is true. The assignment α' induces a family $(\alpha_t)_{t \in T}$ of assignments as follows: for each $t \in T$, the assignment α_t is the restriction of α' to the variables $\text{Var}(t)$. Since γ satisfies the clauses (4.6), we know that there exists some $i \in [u]$ such that $\gamma(z_{t_{root}, \alpha_{t_{root}}, i}) = 0$. It is straightforward to verify that $\gamma(z_{t,\alpha_t,i}) = 0$ for all $t \in T$ then; otherwise, the clauses (4.3–4.5) would force $\gamma(z_{t_{root}, \alpha_{t_{root}}, i})$ to be 1. Then, by the clauses (4.1–4.2) we know that α' must satisfy δ_i. Since α' was arbitrary, we know that $\forall Y. \psi[\beta]$ is true, and thus that φ is true.

Instead of incidence graphs, one can also use *primal graphs* to model the structure of QDNF formulas (see, e.g., [8, 169]). In the primal graph of a QDNF φ, the vertices are the variables of φ, and two variables are connected by an edge if and only if they appear together in some term in the matrix of φ. The parameters of primal treewidth, universal primal treewidth, and existential primal treewidth are defined analogously to the corresponding parameters based on the incidence graph. The parameter incidence treewidth is more general than primal treewidth in the sense that small primal treewidth implies small incidence treewidth [135, 177], but the converse does not hold in general. Therefore, Theorem 4.4 also holds when universal primal treewidth is used as parameter.

4.2.2   Backdoors for Abductive Reasoning

The following example that we consider is related to two notions of backdoors for the problem of (propositional) abductive reasoning. An *abduction instance* $\mathcal{P}$ consists of a tuple (V, H, M, T), where V is a finite set of *variables*, $H \subseteq V$ is a finite set of *hypotheses*, $M \subseteq V$ is a set of *manifestations*, and T is the *theory*, which is a CNF formula. It is required that $M \cap H = \emptyset$. A set $S \subseteq H$ is a *solution* (or *explanation*) of $\mathcal{P}$ if (i) $T \cup S$ is consistent and (ii) $T \cup S \models M$. The problem ABDUCTION then consists of deciding whether a given abduction instance has a solution (that is not larger than a given maximum size).

ABDUCTION
Input: An abduction instance $\mathcal{P} = (V, H, M, T)$, and a positive integer m.
Question: Does there exist a solution S of $\mathcal{P}$ of size at most m?

This problem is Σ_2^p-complete in general [72]. However, when restricted to Horn formulas, the problem ABDUCTION is NP-complete [181]. Also, when restricted to Krom (2CNF) formulas, the problem ABDUCTION is NP-complete ([168], Lemma 61). Pfandler, Rümmele and Szeider have showed that the problem admits a many-to-one fpt-reduction to SAT, when parameterized by the distance to Horn or Krom [173]. This notion of distance is formalized as follows, using the concept of backdoors, which was introduced by Williams et al. [197]. Backdoors have been used before in the context of parameterized complexity. For an example, see the survey article of Gaspers and Szeider about the use of backdoors for satisfiability [88].

Let $\mathcal{C}$ be a class of CNF formulas, called the *base class*, e.g., the class of Horn formulas or the class of Krom formulas. A *strong $\mathcal{C}$-backdoor* of a CNF formula φ is a set $B \subseteq \text{Var}(\varphi)$ of variables such that $\varphi[\tau] \in \mathcal{C}$ for every truth assignment $\tau : B \to \mathbb{B}$. Deciding if a CNF formula φ has a strong Horn-backdoor or a strong Krom-backdoor of size at most k can be decided in fixed-parameter tractable time [88, 167]. Moreover, in case φ has such a strong backdoor of size at most k, such a backdoor B can be computed in fixed-parameter tractable time.

To state the many-to-one fpt-reductions to SAT developed by Pfandler et al. [173], we consider the parameterizations that compute the size of the smallest strong Horn-backdoor or Krom-backdoor of a given CNF formula φ. These values can be computed in fixed-parameter tractable time (with respect to their own values). The problem ABDUCTION parameterized by the size of the smallest Horn-backdoor of T and by the size of the smallest Krom-backdoor of T, we denote by ABDUCTION(Horn-bd-size) and ABDUCTION(Krom-bd-size), respectively.

Proposition 4.5 ([173, Theorem 4]).   ABDUCTION(Horn-bd-size) *is* para-NP-*complete*.

Proposition 4.6 ([173, Theorem 9]).   ABDUCTION(Krom-bd-size) *is* para-NP-*complete*.

We will consider more parameterizations of the problem of propositional abduction in Sect. 8.2.

4.2.3 Backdoors for Disjunctive Answer Set Programming

The next example that we consider also concerns the use of backdoors to obtain an fpt-reduction to SAT. This example involves the logic programming setting of answer set programming (ASP) [31, 90, 154]. In particular, we will consider the consistency problem for disjunctive answer set programs. (In Chap. 5, we will return to the consistency problem for disjunctive answer set programs, and we will use it as a running example throughout Chap. 6).

Let A be a countably infinite set of atoms. A *disjunctive logic program* (or simply: a *program*) P is a finite set of rules of the form $r = (a_1 \vee \cdots \vee a_k \leftarrow b_1, \ldots, b_m, not\ c_1, \ldots, not\ c_n)$, for $k, m, n \geq 0$, where all a_i, b_j and c_l are atoms from A. A rule is called *disjunctive* if $k > 1$, and it is called *normal* if $k \leq 1$ (note that we only call rules with strictly more than one disjunct in the head disjunctive).

We let At(P) denote the set of all atoms occurring in P. By *literals* we mean atoms a or their negations $not\ a$. The *(Gelfond-Lifschitz) reduct* of a program P with respect to a set M of atoms, denoted P^M, is the program obtained from P by: (i) removing rules with $not\ a$ in the body, for each $a \in M$, and (ii) removing literals $not\ a$ from all other rules [91]. An *answer set* A of a program P is a subset-minimal model of the reduct P^A.

The consistency problem for disjunctive answer set programming is defined as follows, and involves deciding whether a given program has an answer set.

ASP-CONSISTENCY
Instance: A disjunctive logic program P.
Question: Does P have an answer set?

The problem is Σ_2^p-complete in general [71].

We consider a parameterized variant of this problem that has been studied by Fichte and Szeider [82]. For this variant, the parameter is based on the notion of backdoors to normality for disjunctive logic programs. A set B of atoms is a *normality-backdoor* for a program P if deleting the atoms $b \in B$ (and their negations) from the rules of P results in a normal program. Deciding if a program P has a normality-backdoor of size at most k can be decided in fixed-parameter tractable time [82]. Moreover, in case P has a normality-backdoor of size at most k, such a backdoor B can be computed in fixed-parameter tractable time.

We consider the following parameterized problem.

ASP-CONSISTENCY(norm.bd-size)
Instance: A disjunctive logic program P.
Parameter: The size of the smallest normality-backdoor for P.
Question: Does P have an answer set?

This problem admits an fpt-reduction to SAT.

Proposition 4.7 ([82]). ASP-CONSISTENCY(norm.bd-size) *is para-NP-complete.*

4.2.4 Bounded Model Checking

The final example of a result that can be seen as a manifestation of the general scheme of many-to-one fpt-reductions to SAT, is an fpt-reduction to UNSAT, and concerns the model checking problem of linear-time temporal logic (LTL). This is a temporal logic that is widely used in the area of software and hardware verification, for instance, to specify desired system behavior. The model checking problem for LTL is the problem of deciding whether a given system satisfies the property specified by a given LTL formula. This problem is central for the task of verification, and is well-known to be PSPACE-complete in general (see, e.g., [11]).

When parameterized by the size of the logic formula, the problem is fixed-parameter tractable when the description of the system is spelled-out explicitly (see, e.g., [11, 85]). However, as the size of the systems grows extremely rapidly when more features are added to it (this is known as the *state explosion problem*), systems are often described succinctly using (generic or domain-specific) description formalisms—in this case, the system is said to be described *symbolically*. When the system is described symbolically, the LTL model checking problem parameterized by the size of the logic formula is para-PSPACE-complete (see Proposition 9.2 in Chap. 9).

For this symbolic model checking problem of LTL, parameterized by the size of the logic formula, a result has been developed in the literature that can be seen as a many-to-one fpt-reduction to UNSAT. Kroening et al. [139] identified a restricted sublanguage of LTL and a restricted class of structures for which the problem becomes para-co-NP-complete (see Proposition 9.3).

We consider the (symbolic) model checking problem for LTL, parameterized by the size of the logic formula, in more detail in Chap. 9.

4.3 Sneak Preview: Fpt-Time Turing Reductions to SAT

In this section, we briefly consider the idea of using fpt-time Turing reductions to SAT to solve problems at higher levels of the Polynomial Hierarchy. We will return to this idea in later chapters (most prominently in Chap. 7).

The parameterized problem that we examine here is related to the topic of judgment aggregation in the area of computational social choice. Judgment aggregation studies the question of how a group of individuals can make consistent collective judgments on a logically related set of issues. Such a set of issues is often modelled using agendas. An *agenda* is a finite, nonempty set Φ of propositional formulas that does not contain any doubly-negated formulas and that is closed under complementation, i.e., $\Phi = \{\varphi_1, \ldots, \varphi_n, \neg\varphi_1, \ldots, \neg\varphi_n\}$, where for each $i \in [n]$, the formula φ_i does not have negation as its outermost logical connective.

There are several procedures that can be used to construct a collective outcome given a set of individual judgments for a particular agenda. Arguably the simplest such procedure is the *majority rule*, that chooses between φ_i and $\neg\varphi_i$, for each i, on the basis of which of these is judged to be true by more individuals. However, not for every agenda Φ this procedure is guaranteed to result in a logically consistent outcome, even if the individual judgments are required to be logically consistent.

It turns out that those agendas for which the majority rule is guaranteed to produce a consistent outcome are exactly those agendas that satisfy the property that every inconsistent subset of the agenda itself contains an inconsistent subset of size at most 2 [75, 165]. We consider the problem of deciding whether a given agenda satisfies this property—denoted by MAJORITY-SAFETY. (We consider the problem MAJORITY-SAFETY in more detail in Chap. 11, where we analyze the complexity of several parameterized variants of the problem).

In general the problem is Π_2^p-complete [75]. However, whenever parameterized by the number of formulas in the agenda (each of which can be of unbounded size), we show that the problem is solvable by an fpt-time Turing reduction to SAT. That is, it is solvable by an fpt-algorithm that has access to a SAT oracle. Moreover, for any input with parameter value k, this fpt-algorithm queries the SAT oracle only $2^{O(k)}$ times. (For more details on fpt-time Turing reductions to SAT, we refer to Chap. 7.)

Concretely, we consider the following parameterized problem.

MAJORITY-SAFETY(agenda-size)
Instance: An agenda Φ containing $2k$ formulas.
Parameter: k.
Question: Does every inconsistent subset of Φ itself contain an inconsistent subset of size at most 2?

We describe how MAJORITY-SAFETY(agenda-size) can be solved in fpt-time by an algorithm that queries a SAT oracle at most $2^{O(k)}$ times. (This result features as Proposition 11.6 in Chap. 11.) The fpt-algorithm A works as follows. It iterates over each subset $\Phi' \subseteq \Phi$ of size at least 3, that does not contain as subset $\{\varphi_i, \neg\varphi_i\}$ for any $i \in [n]$. There are at most 2^k such subsets Φ'. Then, for each such subset Φ', the algorithm A verifies that Φ' is not minimally inconsistent. If at any point, some minimally inconsistent Φ' of size at least 3 is found, the algorithm rejects the input. Otherwise, after having iterated over all subsets without having found a minimally inconsistent subset of size at least 3, the algorithm accepts the input.

For each Φ', the algorithm A decides as follows whether Φ' is minimally inconsistent, i.e., whether it contains no inconsistent proper subset $\Phi'' \subsetneq \Phi'$. Firstly, by a single query to the SAT oracle, it checks whether Φ' is inconsistent. Then, by using another oracle query, it checks whether there is a proper subset $\Phi'' \subseteq \Phi'$ that is inconsistent. This is the case if and only if the following propositional formula χ is unsatisfiable:

$$\chi = \bigwedge_{\psi \in \Phi} \left[\bigwedge_{\psi' \in \Phi \setminus \{\psi\}} \psi' \right]^{\psi},$$

where the notation $[\varphi]^\psi$ denotes the propositional formula obtained from φ by replacing each occurrence of a variable $x \in \text{Var}(\varphi)$ in φ by a copy x^ψ. Then, the set Φ' is minimally inconsistent if and only if the first oracle query returns "unsatisfiable" and the second oracle query returns "satisfiable", i.e., if Φ' is inconsistent and contains no inconsistent proper subset $\Phi'' \subseteq \Phi'$.

For each Φ', the algorithm uses 2 queries to the SAT oracle, so overall, the algorithm A uses $2^{O(k)}$ oracle queries. Moreover, the algorithm A runs in time $2^{O(k)}n^c$, where n denotes the input size and c is some constant.

4.4 Irreducibility

Clearly, not every parameterized variant of a problem that is hard for the second level of the PH admits an fpt-reduction to SAT. In this section, we consider two examples of parameterized problems that can be shown (under various complexity-theoretic assumptions) not to be fpt-reducible to SAT, by means of hardness results for known parameterized complexity classes.

4.4.1 Hardness for para-Σ_2^p

As a first extreme example, consider a Σ_2^p-complete problem (such as QSAT_2) with a (trivial) constant parameterization, i.e., for each instance the parameter value k is a fixed constant $c \in \mathbb{N}$. This parameterized problem is para-Σ_2^p-complete, and does not admit an fpt-reduction to SAT unless the PH collapses. Since the parameter value k is constant, such an fpt-reduction would also be a polynomial-time reduction from QSAT_2 to SAT.

Another, less trivial, example of a parameterized variant of QSAT_2 that does not admit an fpt-reduction to SAT unless the PH collapses, for similar reasons, is the following natural counterpart of $\text{QSAT}_2(\forall\text{-itw})$. Here the parameter is the existential incidence treewidth of any QDNF formula.

$\text{QSAT}_2(\exists\text{-itw})$
Instance: A quantified Boolean formula $\varphi = \exists X.\forall Y.\psi$, where ψ is a quantifier-free formula in DNF, with existential incidence treewidth k.
Parameter: k.
Question: Is φ true?

We show para-Σ_2^p-completeness for $\text{QSAT}_2(\exists\text{-itw})$.

Proposition 4.8. $\text{QSAT}_2(\exists\text{-itw})$ *is para-Σ_2^p-complete.*

Proof. Membership in para-Σ_2^p is obvious. To show para-Σ_2^p-hardness, it suffices to show that the problem is already Σ_2^p-hard when the parameter value is restricted to 1 [84]. We show this by means of a reduction from QSAT_2. The idea of this reduction is to introduce for each existentially quantified variable x a corresponding universally quantified variable z_x that is used to represent the truth value assigned to x. Each of the existentially quantified variables then only directly interacts with universally quantified variables.

Take an arbitrary instance of QSAT$_2$, specified by $\varphi = \exists X.\forall Y.\psi(X, Y)$, where $\psi(X, Y)$ is in DNF. We introduce a new set $Z = \{ z_x : x \in X \}$ of variables. It is straightforward to verify that $\varphi = \exists X.\forall Y.\psi(X, Y)$ is equivalent to the formula $\exists Z.\forall X.\forall Y.\chi$, where:

$$\chi = \bigvee_{x \in X} \left((x \wedge \neg z_x) \vee (\neg x \wedge z_x) \right) \vee \psi(Z, Y).$$

Also, clearly, IG(Z, χ') consists only of isolated paths of length 2, and thus has treewidth 1. Thus, the unparameterized problem consisting of all yes-instances of QSAT$_2(\exists$-itw$)$, where the input formula is in DNF and the parameter value is 1, is Σ_2^p-hard. This proves that QSAT$_2(\exists$-itw$)$ is para-Σ_2^p-hard.

The proof of Proposition 4.8 also shows that QSAT$_2$(DNF) is para-Σ_2^p-hard when parameterized by the existential primal treewidth of the QBF.

4.4.2 Hardness for A[2]

There are also parameterized problems that do not admit an fpt-reduction to SAT (under reasonable complexity-theoretic assumptions), but whose hardness is not as blatant as in the previous severe examples. We show how existing tools from parameterized complexity theory can be used to rule out fpt-reductions to SAT in such more subtle cases, where hardness for para-Σ_2^p or para-Π_2^p cannot be used.

Concretely, we consider a problem that is Σ_2^p-complete, together with a natural parameterization of the problem. We characterize the complexity of this parameterized problem using an existing parameterized variant of the PH. Namely, we show that the problem is A[2]-complete. Moreover, we preview a result from Chap. 14 that hardness for A[2] can be used to give evidence that a problem does not admit an fpt-reduction to SAT.

The problem that we consider involves deciding whether a given set of constraints contains a small unsatisfiable subset. In order to define the problem, we introduce several relevant notions from the area of *constraint satisfaction*.

Let D be a finite set of values (called the *domain*). An *n-ary relation* on D is a set of n-tuples of elements from D. Let V be a countably infinite set of variables. A *constraint of arity n* is a pair (S, R) where $S = (v_1, \ldots, v_n)$ is a sequence of variables from V and R is an n-ary relation (called the *constraint relation*). An *assignment* $\alpha : V' \to D$ is a mapping defined on a set $V' \subseteq V$ of variables. An assignment $\alpha : V' \to D$ *satisfies* a constraint $C = ((v_1, \ldots, v_n), R)$ if Var$(C) \subseteq V'$ and $(\alpha(v_1), \ldots, \alpha(v_n)) \in R$. If the domain D is not explicitly given, we can derive it from any set $\mathcal{I}$ of constraints by taking the set of all values occurring in the constraint relation of any constraint in $\mathcal{I}$.

An assignment $\alpha : \text{Var}(\mathcal{I}) \rightarrow D$ is a *solution* for a finite set $\mathcal{I}$ of constraints if it simultaneously satisfies all the constraints in $\mathcal{I}$. A finite set $\mathcal{I}$ of constraints is *satisfiable* if there exists a solution for it; it is *unsatisfiable* otherwise.

Now, we consider the following parameterized problem of deciding whether a given CSP instance has an unsatisfiable subset of size at most k.

SMALL-CSP-UNSAT-SUBSET
Instance: A CSP instance $\mathcal{I}$, and a positive integer k.
Parameter: k.
Question: Is there an unsatisfiable subset $\mathcal{I}' \subseteq \mathcal{I}$ of size k?

We firstly show that the unparameterized variant of the problem is Σ_2^p-complete—this problem is denoted by CSP-UNSAT-SUBSET. In order to do so, we consider the problem CNF-UNSAT-SUBSET. In this latter problem, instances consist of CNF formulas φ and a positive integer $m \geq 1$, and the question is to decide if there is an unsatisfiable subset $\varphi' \subseteq \varphi$ of size m. The problem CNF-UNSAT-SUBSET is Σ_2^p-complete [145]. We start by showing that Σ_2^p-hardness holds already for a restricted fragment of inputs.

Proposition 4.9. *CNF-UNSAT-SUBSET is Σ_2^p-hard, even when restricted to CNF formulas containing only clauses of size at most* 3.

Proof. Let (φ, m) be an instance of CNF-UNSAT-SUBSET. We show how to transform (φ, m) into an equivalent instance (φ', m') in polynomial time, where φ' contains only clauses of size at most 3.

Let u be the maximum size of any clause appearing in φ. We firstly transform φ into a formula φ_1 that contains an unsatisfiable subset of size $m_1 = m + 1$ if and only if φ contains an unsatisfiable subset of size m. Moreover, each unsatisfiable subset of φ_1 must contain one particular (unit) clause. Let $z \notin \text{Var}(\varphi)$ be a fresh variable. We let $\varphi_1 = \{\{z\}\} \cup \{c \cup \{\overline{z}\} : c \in \varphi\}$. It is straightforward to show that each unsatisfiable subset of φ_1 must contain the clause $\{z\}$, and that there exists a natural correspondence between unsatisfiable subset of φ (of size m) and unsatisfiable subset of φ_1 (of size m_1).

Next, we transform φ_1 into a formula φ_2 that has only clauses that are either unit clauses, or clauses of size $u + 1$. We introduce u fresh variables $z_1, \ldots, z_u \notin \text{Var}(\varphi_1)$. Then, for each clause $c \in \varphi_1$ of size $\ell \in [u + 1]$, we add the clause $c' = c \cup \{\overline{z_\ell}, \overline{z_{\ell+1}}, \ldots, \overline{z_u}\}$ to φ_2. Moreover, we add the unit clauses $c_1, \ldots, c_u$, where $c_i = \{z_i\}$ for each $i \in [u]$. to φ_2.

Firstly, we show that each unsatisfiable subset ψ of φ_2 must contain the unit clauses $c_1, \ldots, c_u$. To derive a contradiction, suppose that this is not the case, that is, there is some $i \in [u]$ such that $c_i \notin \psi$. We know that ψ must contain the clause $\{z, \overline{z_1}, \ldots, \overline{z_u}\}$. Moreover, we know that all other clauses in ψ contain the literal $\overline{z}$. Then any truth assignment that sets z to 0 and that sets z_i to 0 satisfies ψ, which is a contradiction with our assumption that ψ is unsatisfiable. Therefore, we can conclude that each unsatisfiable subset of φ_2 contains all of the clauses $c_1, \ldots, c_u$.

Moreover, there exists a natural correspondence between unsatisfiable subsets of φ_1 of size m_1 and unsatisfiable subsets of φ_2 of size $m_2 = m_1 + u$. Therefore, (φ_1, m_1) and (φ_2, m_2) are equivalent instances of CNF-UNSAT-SUBSET. Moreover, we know that each unsatisfiable subset $\psi \subseteq \varphi_2$ of size m_2 contains exactly u unit clauses and m_1 clauses of size exactly $u + 1$.

We can now transform (φ_2, m_2) into an equivalent instance (φ', m') as follows. Each clause $c = \{l_1, \ldots, l_{u+1}\} \in \varphi_2$ of size $u + 1$, we replace by the clauses $\{l_1, y_1^c\}$, $\{\overline{y_1^c}, l_2, y_2^c\}$, ..., $\{\overline{y_{u-1}^c}, l_u, y_u^c\}$, and $\{\overline{y_u^c}, l_{u+1}\}$, where the variables $y_1^c, \ldots, y_u^c$ are fresh variables. Clearly, φ' contains only clauses of size at most 3. Moreover, it is straightforward to verify that φ_2 contains an unsatisfiable subset of size m_2 if and only if φ' contains an unsatisfiable subset of size $m' = (u + 1)m_1 + u = (u + 1)(m + 1) + u$. Therefore, φ contains an unsatisfiable subset of size m if and only if φ' contains an unsatisfiable subset of size m'.

This result also provides a polynomial-time reduction from the Σ_2^p-complete problem of CNF-UNSAT-SUBSET to the problem CSP-UNSAT-SUBSET, by expressing each clause of size 3 as a constraint over 3 variables and with a constraint relation containing 7 tuples. Given that a proof of Σ_2^p-membership is routine, this gives us the following result.

Corollary 4.10. CSP-UNSAT-SUBSET *is* Σ_2^p-*complete*.

Having shown that the unparameterized problem CSP-UNSAT-SUBSET is Σ_2^p-complete, we can now turn our attention to the parameterized problem SMALL-CSP-UNSAT-SUBSET and show that it is A[2]-complete. This result follows directly from Lemmas 4.12 and 4.13, that we will prove below.

Proposition 4.11. SMALL-CSP-UNSAT-SUBSET *is* A[2]-*complete*.

This is one of the first A[2]-completeness results in the literature, and arguably the first A[2]-completeness result for a problem that did not feature in the theoretical development of the A-hierarchy (see [85], Chapter 8).

In Chap. 14, we will give evidence that parameterized problems that are A[2]-hard do not admit an fpt-reduction to SAT. In particular, if any A[2]-hard parameterized problem is in para-NP (i.e, if A[2] $\subseteq$ para-NP), then there is a subexponential-time reduction from the Σ_2^p-complete problem QSAT$_2$(3DNF) to SAT (Theorem 14.1 and Corollary 14.2). A subexponential-time reduction is a reduction that runs in time $2^{o(n)}$, where n denotes the number of variables, i.e., a reduction that runs in time $2^{n/s(n)}$, for some unbounded, nondecreasing, computable function s. In other words, under the assumption that such subexponential-time reductions from QSAT$_2$(3DNF) to SAT do not exist, we can use A[2]-hardness to rule out many-to-one fpt-reductions to SAT. In Chap. 14, we also show how A[2]-hardness rules out many-to-one fpt-reductions to UNSAT under a similar assumption.

We conclude this section by giving a detailed proof that SMALL-CSP-UNSAT-SUBSET is A[2]-complete. In Lemma 4.12, we show membership, and in Lemma 4.13, we show hardness.

Lemma 4.12. SMALL-CSP-UNSAT-SUBSET *is in* A[2].

Proof. We show membership in A[2] by fpt-reducing the problem to $MC(\Sigma_2)$. Let $(\mathcal{I}, k)$ be an instance of SMALL-CSP-UNSAT-SUBSET, with $\mathcal{I} = \{C_1, \ldots, C_m\}$, and $C_i = (S_i, R_i)$ for each $i \in [m]$. Moreover, let u be the maximum number of tuples in any of the constraint relations R_i.

We construct an instance $(\mathcal{A}, \varphi)$ of $MC(\Sigma_2)$ (over a fixed signature τ) as follows. We define the universe A of $\mathcal{A}$ as follows:

$$A = \{c_1, \ldots, c_m\} \cup \{t_1, \ldots, t_u\}.$$

Intuitively, the elements c_i represent the constraints C_i, and the elements t_j represent tuples in constraint relations R_i.

We introduce unary relations C and T to τ. Intuitively, C encodes whether an element represents a constraint, and T encodes whether an element represents a tuple. We let:

$$C^{\mathcal{A}} = \{c_1, \ldots, c_m\} \quad \text{and} \quad T^{\mathcal{A}} = \{t_1, \ldots, t_u\}.$$

We introduce a binary relation I to τ that intuitively encodes whether a constraint C_i has at least j tuples. We let:

$$I^{\mathcal{A}} = \{(c_i, t_j) : |R_i| \geq j\}.$$

Finally, we introduce a 4-ary relation N to τ, that intuitively represents whether a tuple t_{j_1} in the constraint relation R_{i_1} and a tuple t_{j_2} in the constraint relation R_{i_2} are non-conflicting. (Moreover, it ensures that there is a j_1-th tuple in R_{i_1} and a j_2-th tuple in R_{i_2}.) We let:

$$N^{\mathcal{A}} = \{(c_{i_1}, t_{j_1}, c_{i_2}, t_{j_2}) : i_1 \in [m], j_1 \in [|R_{i_1}|], i_2 \in [m], j_2 \in [|R_{i_2}|], \text{ the} \\ j_1\text{-th tuple in } R_{i_1} \text{ and the } j_2\text{-th tuple in } R_{i_2} \text{ are not conflicting}\}.$$

Then, we define the first-order logic sentence φ as follows:

$$\varphi = \exists x_1, \ldots, x_k. \forall y_1, \ldots, y_k.$$

$$\bigwedge_{i \in [k]} C(x_i) \wedge \left(\bigwedge_{i \in [k]} (T(y_i) \wedge I(x_i, y_i)) \right) \rightarrow \left(\bigvee_{i,j \in [k], i < j} \neg N(x_i, y_i, x_j, y_j) \right).$$

Clearly $|\varphi| \leq f(k)$ for some function f.

Any assignment $\alpha : \{x_1, \ldots, x_k\} \rightarrow \{c_1, \ldots, c_m\}$ then naturally corresponds to a subset $\mathcal{I}' \subseteq \mathcal{I}$ of size k. Also, any subsequent assignment $\beta : \{y_1, \ldots, y_k\} \rightarrow \{t_1, \ldots, t_u\}$ that for each $i \in [k]$ assigns y_i to a sufficiently small value t_j (that is, $j \in [|R_\ell|]$ where $\alpha(x_i) = c_\ell$) naturally corresponds to a choice of a tuple t_j in the constraint relation R_i of each constraint $C_i \in \mathcal{I}'$. Using these correspondences, it is straightforward to verify that $(\mathcal{I}, k) \in$ SMALL-CSP-UNSAT-SUBSET if and only if $\mathcal{A} \models \varphi$.

Finally, we show A[2]-hardness for the parameterized problem SMALL-CSP-UNSAT-SUBSET.

Lemma 4.13. SMALL-CSP-UNSAT-SUBSET *is A[2]-hard.*

Proof. We show A[2]-hardness by means of an fpt-reduction from MC(Σ_2). Take an arbitrary instance of MC(Σ_2), consisting of a structure $\mathcal{A}$ with universe A (over the fixed binary signature τ), and a first-order formula $\varphi = \exists x_1, \ldots, x_k.\forall y_1, \ldots, y_k.\psi$, where ψ is a disjunction of atoms. We construct a CSP instance $\mathcal{I}$ and an integer k', such that $(\mathcal{I}, k') \in$ SMALL-CSP-UNSAT-SUBSET if and only if $\mathcal{A} \models \varphi$.

We define $k' = \binom{k}{2} + k + 1$. We let the domain of $\mathcal{I}$ be $D = A \cup \{0, 1, \ldots, k\}$. We may assume without loss of generality that $A \cap \{0, 1, \ldots, k\} = \emptyset$. We introduce the following variables:

$$\begin{aligned}
\mathrm{Var}(\mathcal{I}) &= V \cup W \cup X \cup Z, \text{ where} \\
V &= \{v\}, \\
W &= \{w_i : i \in [k]\}, \\
X &= \{x_{i,a} : i \in [k], a \in A\}, \\
Y &= \{y_i : i \in [k]\}, \text{ and} \\
Z &= \{z_{i,j} : i, j \in [k], i < j\}.
\end{aligned}$$

The constraints in $\mathcal{I}$ are defined as follows. For each $i, j \in [k]$ with $i < j$ and each $a_1, a_2 \in A$, we introduce a constraint C_{i,j,a_1,a_2} for which $\mathrm{Var}(C_{i,j,a_1,a_2}) = \{v, x_{i,a_1}, x_{j,a_2}, y_i, y_j, z_{i,j}\}$. The constraint relation has the following $k^2|A|^2 + 1$ tuples. For each $i' \in [0, k]$ such that $i \neq i'$, for each $j' \in [0, k]$ such that $j \neq j'$, and for each $a_3, a_4 \in A$, there is a tuple $\overline{r}_{i',j',a_3,a_4}$. Each such tuple $\overline{r}_{i',j',a_3,a_4}$ sets x_{i,a_1} to j', sets x_{j,a_2} to i', sets y_i to a_3, sets y_j to a_4, and sets v to 1. Moreover, it sets $z_{i,j}$ to 1 if at least one atom in ψ containing only variables among x_i, x_j, y_i, y_j is satisfied by the partial assignment $\alpha = \{x_i \mapsto a_1, x_j \mapsto a_2, y_i \mapsto a_3, y_j \mapsto a_4\}$; otherwise it sets $z_{i,j}$ to 0. In addition, there is a tuple that sets all variables in $\mathrm{Var}(C_{i,j,a_1,a_2})$ to 0. In particular, this constraint rules out that x_{i,a_1} is set to j and that x_{j,a_2} is set to i.

Then, for each $i \in [k]$ and each $a \in A$, we introduce a constraint $C_{i,a}$ for which $\mathrm{Var}(C_{i,a}) = \{x_{i,a}, w_i\}$. The constraint relation has $2k + 1$ tuples. For each $j \in [k]$ and each $b \in \mathbb{B}$, there is a tuple that sets $x_{i,a}$ to j and w_i to b. In addition, there is a tuple that sets $x_{i,a}$ to 0 and that sets w_i to 1. In other words, this constraint enforces that w_i is set to 1 if $x_{i,a}$ cannot be set to any value $j > 0$.

Finally, we introduce a constraint C_0 for which $\mathrm{Var}(C_0) = V \cup W \cup Z$. The constraint relation has $(2^k - 1) \cdot 2^{k''} + 1$ tuples, where $k'' = \binom{k}{2}$. For each assignment $\rho : W \cup Z \to \mathbb{B}$ such that at for at least one $i \in [k]$ it holds that $\rho(w_i) = 1$, there is a tuple that sets all variables according to ρ, and that sets v to 0. Moreover, there is an additional tuple that sets all variables $w_i \in W$ to 1, that sets all variables $z_{i,j} \in Z$ to 0, and that sets v to 1. In other words, C_0 is satisfied if and only if either (1) at least one w_i is set to 0, or (2) all w_i are set to 1 and all $z_{i,j}$ are set to 1.

Before we show the correctness of this reduction, we argue that each unsatisfiable subset $\mathcal{I}' \subseteq \mathcal{I}$ must include C_0. Suppose that this is not the case. Then the assignment that sets all variables $w_i \in W$ to 1 and that sets all other variables to 0 satisfies $\mathcal{I}'$.

Also, we argue that each unsatisfiable subset $\mathcal{I}' \subseteq \mathcal{I}$ must include some constraint C_{i,a_i} for each $i \in [k]$, and some $a_i \in A$. Suppose that this is not the case. Moreover, let $i_1, \ldots, i_m \in [k]$ with $i_1 < \cdots < i_m$ be the indices for which $\mathcal{I}'$ does contain some constraint $C_{i_\ell, a_{i_\ell}}$. We know that there is some $i_0 \in [k]$ such that $\mathcal{I}'$ contains no constraint $C_{i_0,a}$ (for $a \in A$). Then consider the assignment $\alpha : \mathrm{Var}(\mathcal{I}) \to D$ that sets w_{i_ℓ} to 1 for each $\ell \in [m]$, and that sets all other variables to 0. Then α satisfies all constraints C_{i,j,a_1,a_2} and $C_{i,a}$ in $\mathcal{I}'$. Since α sets at least one variable w_i to 1, we also know that α satisfies C_0. Therefore, α must satisfy all constraints in $\mathcal{I}'$, which is a contradiction with our assumption that $\mathcal{I}'$ is unsatisfiable.

Finally, we argue that for each unsatisfiable subset $\mathcal{I}' \subseteq \mathcal{I}$ of size at most k' there must be some assignment $\alpha : \{1, \ldots, k\} \to A$ such that $\mathcal{I}'$ contains the constraints $C_{i,\alpha(i)}$ for each $i \in [k]$, and the constraints $C_{i,j,\alpha(i),\alpha(j)}$ for each $i, j \in [k]$ with $i < j$. Suppose that this is not the case. By a straightforward counting argument, we then know that it must be the case that for some $i_0 \in [k]$, and some $j_0 \in [k]$ such that $i_0 \neq j_0$, there is some $C_{i_0,a_{i_0}} \in \mathcal{I}'$ but for no a_{j_0} the constraint $C_{i_0,j_0,a_{i_0},a_{j_0}}$ is in $\mathcal{I}'$. (We implicitly assume that $j_0 > i_0$; the case where $j_0 < i_0$ is entirely analogous.) Then consider the assignment $\alpha : \mathrm{Var}(\mathcal{I}) \to D$ that sets $x_{i_0,a_{i_0}}$ to j_0, that sets w_{i_0} to 0, and sets all other variables in $\mathrm{Var}(\mathcal{I})$ to 0. The only constraints that are conflicting with $\alpha(x_{i_0,a_{i_0}}) = j_0$ are constraints $C_{i_0,j_0,a_{i_0},a_{j_0}}$, which by assumption are not contained in $\mathcal{I}'$. Therefore, it is readily verified that α satisfies all constraints in $\mathcal{I}'$. This is a contradiction with our assumption that $\mathcal{I}'$ is unsatisfiable.

We are now ready to prove that $\mathcal{I}$ has an unsatisfiable subset $\mathcal{I}'$ of size at most k' if and only if $\mathcal{A} \models \varphi$, i.e., that $(\mathcal{I}, k') \in$ SMALL-CSP-UNSAT-SUBSET if and only if $(\mathcal{A}, \varphi) \in \mathrm{MC}(\Sigma_2)$.

($\Rightarrow$) Suppose that there is an unsatisfiable subset $\mathcal{I}' \subseteq \mathcal{I}$ of at most k' constraints. As we showed above, then $C_0 \in \mathcal{I}'$. Moreover there exists some assignment $\alpha : \{1, \ldots, k\} \to A$ such that $\mathcal{I}'$ contains the constraints $C_{i,\alpha(i)}$ for each $i \in [k]$, and the constraints $C_{i,j,\alpha(i),\alpha(j)}$ for each $i, j \in [k]$ with $i < j$. Now consider the assignment $\alpha' : X \to A$ defined by letting $\alpha'(x_i) = \alpha(i)$, for each $i \in [k]$. We claim that $\mathcal{A}, \alpha' \models \forall y_1, \ldots, y_k.\psi$. Take an arbitrary assignment $\beta' : Y \to A$. We show that $\mathcal{A}, \alpha' \cup \beta' \models \psi$. Consider the assignment $\beta : \mathrm{Var}(\mathcal{I}) \to D$ defined by letting $\beta(y_i) = \beta'(y_i)$ for all $y_i \in Y$, letting $\beta(x_{i,a}) = 0$ for all $x_{i,a} \in X$, letting $\beta(v) = 1$, letting $\beta(z_{i,j}) = 0$ for all $z_{i,j} \in Z$, and letting $\beta(w_i) = 1$ for all $w_i \in W$. Since $\mathcal{I}'$ is unsatisfiable, we know that β cannot satisfy all constraints in $\mathcal{I}'$. The only possible contradiction is that some constraint $C_{i_0,j_0,\alpha(i_0),\alpha(j_0)} \in \mathcal{I}'$ does not allow z_{i_0,j_0} to be set to 0. By construction of $C_{i_0,j_0,\alpha(i_0),\alpha(j_0)}$, this is only the case if $\alpha' \cup \beta'$ satisfies some atom in ψ. Therefore, we can conclude that $\mathcal{A}, \alpha' \cup \beta' \models \psi$. Since β' was arbitrary, we know that $\mathcal{A}, \alpha' \models \forall y_1, \ldots, y_k.\psi$. Therefore, $\mathcal{A} \models \varphi$, and thus $(\mathcal{A}, \varphi) \in \mathrm{MC}(\Sigma_2)$.

($\Leftarrow$) Conversely, suppose that $\mathcal{A} \models \varphi$, that is, that there exists some assignment $\alpha : \{x_1, \ldots, x_k\} \to A$ such that $\mathcal{A}, \alpha \models \forall y_1, \ldots, y_k.\psi$. Consider the following subset $\mathcal{I}' \subseteq \mathcal{I}$ of constraints:

$$\mathcal{I}' = \{C_0\} \cup \{\, C_{i,\alpha(x_i)} : i \in [k]\,\} \cup \{\, C_{i,j,\alpha(x_i),\alpha(x_j)} : i, j \in [k], i < j\,\}.$$

Clearly, $|\mathcal{I}'| = k'$. We claim that $\mathcal{I}'$ is unsatisfiable. To derive a contradiction, suppose that $\mathcal{I}'$ is satisfiable, i.e., that there is an assignment $\beta' : \mathrm{Var}(\mathcal{I}) \to D$ that satisfies all constraints in $\mathcal{I}'$. Take an arbitrary $i \in [k]$. Since $\mathcal{I}'$ contains the constraints $C_{i,j,\alpha(x_i),\alpha(x_j)}$ for all $j > i$, and the constraints $C_{j,i,\alpha(x_j),\alpha(x_i)}$ for all $j < i$, we know that β' must set $x_{i,\alpha(x_i)}$ to 0, and thus since $\mathcal{I}'$ contains the constraint $C_{i,\alpha(x_i)}$, we know that β' must set w_i to 1. Then, since $\mathcal{I}'$ contains C_0, we know that β' sets all $z_{i,j} \in Z$ to 0, and that β' sets v to 1. Moreover, since $\mathcal{I}'$ contains the constraints $C_{i,j,\alpha(x_i),\alpha(x_j)}$ for all $i, j \in [k]$ with $i < j$, we know that $\beta'(y_i) \in A$, for all $y_i \in Y$. Then, consider the restriction $\beta : \{y_1, \dots, y_k\} \to A$ of β' to the variables in Y, i.e., $\beta(y_i) = \beta'(y_i)$ for all $y_i \in Y$. Since $\mathcal{A}, \alpha \models \forall y_1, \dots, y_k.\psi$, we know that $\mathcal{A}, \alpha \cup \beta \models \psi$, that is, there must be some atom R in ψ that is satisfied by $\alpha \cup \beta$. Let $i_0, j_0 \in [k]$ with $i_0 < j_0$ be indices such that the variables in R are among $\{x_{i_0}, x_{j_0}, y_{i_0}, y_{j_0}\}$ (we know such i_0, j_0 exist, because R is unary or binary). Then by construction of $C_{i_0,j_0,\alpha(x_{i_0}),\alpha(x_{j_0})}$, we know that β' is forced to set z_{i_0,j_0} to 1. This is a contradiction with our previous conclusion that $\beta'(z_{i,j}) = 0$ for all $z_{i,j} \in Z$. Therefore, we can conclude that $\mathcal{I}'$ is unsatisfiable, and that $(\mathcal{I}, k') \in$ SMALL-CSP-UNSAT-SUBSET.

Summary

In this chapter, we introduced the concept of fpt-reducibility to SAT (or to other problems in NP) as a new, more permissive notion of tractability for the parameterized complexity analysis of problems at higher levels of the PH. We discussed the formal interpretation of this tractability notion in the framework of parameterized complexity theory. Moreover, we surveyed to what extent results and tools from the literature can be used to investigate the possibilities and limits of fpt-reductions to SAT. We consider several cases where known techniques can be used to obtain fixed-parameter tractable reductions to SAT for different problems. Additionally, we showed how hardness for known parameterized complexity classes can be employed in some cases to show that fpt-reductions to SAT are not possible.

Notes

Propositions 4.3 and 4.8 and Theorem 4.4 appeared in a paper in the proceedings of SAT 2014 [112]. Propositions 4.9 and 4.11 and Corollary 4.10 were shown in an article appearing in the ACM Transactions on Computational Logic [108].

Chapter 5
The Need for a New Completeness Theory

> *If I had an hour to solve a problem, I'd spend 55 minutes thinking about the problem and 5 minutes thinking about solutions.*
>
> — Albert Einstein

In Chap. 4, we showed that known parameterized complexity classes such as para-NP and para-co-NP can be used to characterize many-to-one fpt-reductions to SAT and UNSAT. Moreover, we illustrated how hardness for known parameterized complexity classes can be used to give evidence that many-to-one fpt-reductions to SAT or UNSAT do not exist. We did so by showing for several parameterized problems that they are complete for one of the known parameterized complexity classes A[2] and para-Σ_2^p.

In this chapter, we argue that the known parameterized complexity classes— the ones mentioned above, as well as familiar classes such as W[1] and XP, and other classes such as para-Π_2^p, XNP, and Xco-NP—do not suffice to characterize the complexity of many natural parameterized variants of problems that lie at higher levels of the PH. We do so by considering various parameterized variants of the Σ_2^p-complete consistency problem for disjunctive answer set programming as an example (we already briefly considered this problem in Sect. 4.2.3). We show that several of these parameterized variants are complete for known parameterized complexity classes, but that other parameterized variants are not complete for any known class (under various complexity-theoretic assumptions).

Because these natural parameterized problems are not complete for any of the known classes, additional parameterized complexity classes are called for to adequately characterize the complexity of such problem. In other words, the results in this chapter provide the motivation for developing a new completeness theory, which we will do in Chap. 6. We will use the example of the consistency problem for disjunctive answer set programming as a running example throughout Chap. 6.

© Springer-Verlag GmbH Germany, part of Springer Nature 2019
R. de Haan: Parameterized Complexity in the Polynomial Hierarchy, LNCS 11880,
https://doi.org/10.1007/978-3-662-60670-4_5

Outline of This Chapter:

We begin in Sect. 5.1 by considering several natural parameterized variants of the consistency problem for disjunctive answer set programming. Moreover, for several of these parameterized variants, we show that their complexity can be characterized using the known parameterized complexity classes para-NP, para-co-NP and para-Σ_2^p.

Then, in Sect. 5.2, we turn to the remaining parameterizations for the consistency problem for disjunctive answer set programming, and we argue that these problems are not complete for any of the known parameterized complexity classes. In particular, we show that these parameterized problems cannot be complete for any of the classes depicted above the top dashed gray line in Fig. 5.1 (or for any superset of these classes), and that these problems cannot be complete for any of the classes depicted below the bottom dashed gray line in Fig. 5.1 (or for any subset of these classes). In other words, any parameterized complexity class for which these parameterized problems are complete must lie between the dashed gray lines, in the dashed area.

5.1   Running Example: Disjunctive Answer Set Programming

We begin by considering the consistency problem for disjunctive answer set programming [31, 90, 154]—that we already briefly considered in Sect. 4.2.3—in more detail. For the reader's convenience, we repeat the basic definitions of disjunctive answer set programming here, before defining several parameterized variants of the consistency problem that we did not yet consider in Sect. 4.2.3.

Let A be a countably infinite set of atoms. A *disjunctive logic program* (or simply: a *program*) P is a finite set of rules of the form $r = (a_1 \vee \cdots \vee a_k \leftarrow b_1, \ldots, b_m,$ *not* $c_1, \ldots,$ *not* $c_n)$, for $k, m, n \geq 0$, where all a_i, b_j and c_l are atoms from A. A rule is called *disjunctive* if $k > 1$, and it is called *normal* if $k \leq 1$ (note that we only call rules with strictly more than one disjunct in the head disjunctive). A rule is called *negation-free* if $n = 0$. A program is called normal if all its rules are normal, and called negation-free if all its rules are negation-free. A rule is called a *constraint* if $k = 0$. A rule is called *dual-normal* if it is either a constraint, or $m \leq 1$.

We let $\mathrm{At}(P)$ denote the set of all atoms occurring in P. By *literals* we mean atoms a or their negations *not* a. With $\mathrm{NF}(r)$ we denote the rule $(a_1 \vee \cdots \vee a_k \leftarrow b_1, \ldots, b_m)$. The *(Gelfond-Lifschitz) reduct* of a program P with respect to a set M of atoms, denoted P^M, is the program obtained from P by: (i) removing rules with *not* a in the body, for each $a \in M$, and (ii) removing literals *not* a from all other rules [91]. An *answer set* A of a program P is a subset-minimal model of the reduct P^A. The answer sets of a program P can alternatively be characterized as follows (see, e.g., [81, Proposition 3]). A set $M \subseteq \mathrm{At}(P)$ is an answer set of P if and only if it is a model of P_c and an answer set of P_r, where P_c is the program consisting of all

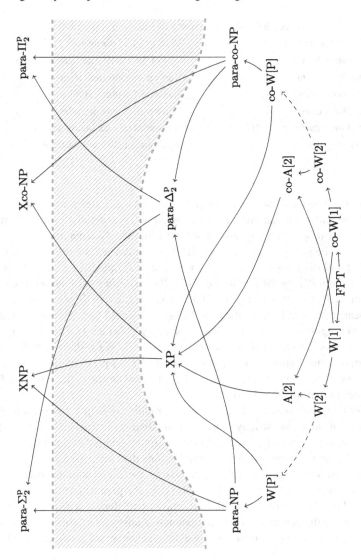

Fig. 5.1 Known parameterized complexity classes.

constraints in P, and $P_r = P \setminus P_c$. We give an example of a program and one of its answer sets below, in Example 5.1.

The consistency problem for disjunctive answer set programming is defined as follows, and involves deciding whether a given program has an answer set.

ASP-CONSISTENCY
Instance: A disjunctive logic program P.
Question: Does P have an answer set?

The problem is Σ_2^p-complete in general [71].

Table 5.1 Parameterizations for ASP-CONSISTENCY (abbreviated ASP-CONS).

Name of parameterized problem	Description of parameter
ASP-CONS(norm.bd-size)	Size of the smallest normality-backdoor
ASP-CONS(#cont.atoms)	Number of contingent atoms
ASP-CONS(#cont.rules)	Number of contingent rules
ASP-CONS(#disj.rules)	Number of disjunctive rules
ASP-CONS(#non-dual-normal.rules)	Number of non-dual-normal rules
ASP-CONS(max.atom.occ.)	Maximum number of times that any atom occurs

Two of the parameterizations for this problem that we consider are related to atoms that must be part of any answer set of a program P. We identify a subset Comp(P) of *compulsory atoms*, that any answer set must include. Given a program P, we let Comp(P) be the smallest set such that: (i) if ($w \leftarrow not\ w$) is a rule of P, then $w \in$ Comp(P); and (ii) if ($b \leftarrow a_1, \ldots, a_n$) is a rule of P, and $a_1, \ldots, a_n \in$ Comp(P), then $b \in$ Comp(P). We then let the set Cont(P) of *contingent atoms* be those atoms that occur in P but are not in Comp(P). We call a rule *contingent* if all the atoms that appear in the head are contingent. In fact, we could use any polynomial time computable algorithm A that computes for every program P a set Comp$_A$(P) of atoms that must be included in any answer set of P. We restrict ourselves to the algorithm described above that computes Comp(P). As parameterizations of ASP-CONSISTENCY, we consider the number of contingent atoms in the program P, and the number of contingent rules of P.

Additionally, we consider as parameters the number of disjunctive rules of P (i.e., the number of rules with strictly more than one disjunct in the head), the number of rules of P that are not dual-normal, and the maximum number of times that any atom occurs in P. An overview of all parameterizations that we consider for the problem ASP-CONSISTENCY, together with the names of the corresponding parameterized problems, can be found in Table 5.1. In this table, the parameter that we considered in Sect. 4.2.3—the size of the smallest normality-backdoor—is also included.

To illustrate the central concepts in disjunctive answer set programming and the parameters for ASP-CONSISTENCY that we defined above, we consider the following example.

Example 5.1. Let P be the program consisting of the following rules over the atoms $a, \ldots, f$:

$$a \leftarrow \qquad\qquad (r_1)$$
$$b \vee c \leftarrow a \qquad\qquad (r_2)$$
$$d \leftarrow b, not\ c \qquad\qquad (r_3)$$
$$e \leftarrow a, not\ b \qquad\qquad (r_4)$$
$$f \leftarrow a \qquad\qquad (r_5)$$

To see that $M = \{a, b, d, f\}$ is an answer set of P, consider the reduct P^M:

$$
\begin{aligned}
a &\leftarrow \\
b \vee c &\leftarrow a \\
d &\leftarrow b \\
f &\leftarrow a
\end{aligned}
$$

The set M is a model of P^M, and no strict subset $M' \subsetneq M$ is a model of P^M.

The size of the smallest normality-backdoor for P is 1 (the set $B = \{b\}$ is a normality-backdoor, for example). The program P contains 4 contingent atoms (the atoms b, c, d, e) and 3 contingent rules (the rules r_2–r_4). The program P contains 1 disjunctive rule (the rule r_2), and 1 non-dual-normal rule (the rule r_5). Finally, the maximum number of times that any atom occurs is 4 (the atom a occurs 4 times). ⊣

5.1.1 Parameterized Variants Where Known Theory Suffices

Next, we show for several of the parameterized variants of ASP-CONSISTENCY that they are complete for known parameterized complexity classes. As we have already seen in Sect. 4.2.3, the problem is para-NP-complete when parameterized by the size of the smallest normality-backdoor of the program.

Proposition 5.2 ([82]). ASP-CONSISTENCY(norm.bd-size) *is* para-NP-*complete*.

Next, we show that the problem is complete for para-co-NP when parameterized by the number of contingent atoms of the program.

Proposition 5.3. ASP-CONSISTENCY(#cont.atoms) *is* para-co-NP-*complete*.

Proof. Hardness for para-co-NP follows from a reduction by Eiter and Gottlob [71, Theorem 3]. They give a reduction from QSAT$_2$ to ASP-CONSISTENCY. We can view this reduction as a polynomial-time reduction from UNSAT to the slice of ASP-CONSISTENCY(#cont.atoms) where the parameter value is 0. Namely, considering an instance of UNSAT as an instance of QSAT$_2$ with no existentially quantified variables, the reduction results in an equivalent instance of ASP-CONSISTENCY that has no contingent atoms. Therefore, we can conclude that ASP-CONSISTENCY(#cont.atoms) is para-co-NP-hard.

We show membership in para-co-NP. Let P be a program that contains k contingent atoms. We describe an fpt-reduction to SAT for the dual problem whether P has no answer set. This is then also an fpt-reduction from ASP-CONSISTENCY(#cont.atoms) to UNSAT. Since each answer set of P must contain all atoms in $\mathrm{Comp}(P)$, there are only 2^k candidate answer sets that we need to consider, namely $N \cup \mathrm{Comp}(P)$ for each $N \subseteq \mathrm{Cont}(P)$. For each such set $M_N = N \cup \mathrm{Comp}(P)$ it can be checked in deterministic polynomial time whether M_N is a model of P^{M_N}, and it can be checked by an NP-algorithm whether M_N is not a minimal model of P^{M_N} (namely, a counterexample consisting of a model $M' \subsetneq M_N$ of P^{M_N} can be found in non-deterministic polynomial time). Therefore, by the NP-completeness of SAT, for each $N \subseteq \mathrm{Cont}(P)$, there exists a propositional formula φ_N that is satisfiable if and only if M_N is not a minimal model of P^{M_N}. Moreover, we can construct such a formula φ_N in polynomial time, for each $N \subseteq \mathrm{Cont}(P)$. All together, it holds that for no $N \subseteq \mathrm{Cont}(P)$ the set $N \cup \mathrm{Comp}(P)$ is an answer set if and only if the disjunction $\bigvee_{N \subseteq \mathrm{Cont}(P)} \varphi_N$ is satisfiable.

Finally, we show that the problem is para-Σ_2^p-complete when parameterized by the maximum number of times that any atom occurs in the program.

Proposition 5.4. ASP-CONSISTENCY(max.atom.occ.) *is* para-Σ_2^p-*complete.*

Proof. We consider the following polynomial-time transformation on disjunctive logic programs. Let P be an arbitrary program, and let x be an atom of P that occurs $\ell \geq 3$ times. We introduce new atoms x_j for all $j \in [\ell]$. We replace each occurrence of x in P by a unique atom x_j. Then, to P, we add the rules $(x_{j+1} \leftarrow x_j)$, for all $j \in [\ell - 1]$, and the rule $(x_1 \leftarrow x_\ell)$. We call the resulting program P'. It is straightforward to verify that P has an answer set if and only if P' has an answer set.

The transformation described above can be repeatedly applied to ensure that each atom occurs at most 3 times. This gives us a polynomial-time reduction from an arbitrary instance of the Σ_2^p-hard problem ASP-CONSISTENCY to an instance where each atom occurs at most 3 times. Therefore, ASP-CONSISTENCY(max.atom.occ.) is para-Σ_2^p-hard. Membership in para-Σ_2^p follows from Σ_2^p-membership of the problem ASP-CONSISTENCY. □

5.2 Motivating New Theory

In this section, we argue that the remaining parameterized variants of ASP-CONSISTENCY—the parameterized problems ASP-CONSISTENCY(#cont.rules), ASP-CONSISTENCY(#disj.rules) and ASP-CONSISTENCY(#non-dual-normal.rules)—are not complete for any known parameterized complexity classes. We do so by showing that these problems are not complete for any of the parameterized complexity classes that are depicted in Fig. 5.1 on page 73 (under various complexity-theoretic assumptions). (For some technical results that we use in this argument and that need meticulous preparations to establish, we defer to Chap. 14.)

We describe the general lines of the arguments that we use to rule out completeness for known parameterized complexity classes, before we work out the required proofs in detail.

Firstly, we rule out membership in para-NP and para-co-NP, and any parameterized complexity class contained in either of them. Membership in para-NP or para-co-NP would mean that there exists a many-to-one fpt-reduction to SAT or UNSAT, respectively. To get an intuition why membership in these classes is highly unlikely, one can attempt to construct an fpt-reduction to SAT or UN-SAT for ASP-CONSISTENCY(#cont.rules), ASP-CONSISTENCY(#disj.rules) or ASP-CONSISTENCY(#non-dual-normal.rules), and the profound obstacles that would have to be overcome to succeed in this become immediately apparent. Nevertheless, we provide more tangible evidence in Chaps. 6 and 14 that these problems are not contained in para-NP or para-co-NP. For the case of ASP-CONSISTENCY(#cont.rules), for instance, see Theorem 6.3 and Proposition 6.32 in Chap. 6 and Corollary 14.12 in Chap. 14. These results show that ASP-CONSISTENCY(#cont.rules) is not in para-NP, unless NP = co-NP, and that ASP-CONSISTENCY(#cont.rules) is not in para-co-NP, unless there exists a subexponential-time reduction from $QSAT_2$ to UNSAT.

Then, to rule out membership in XP—or any parameterized complexity classes contained in XP, such as the classes of the Weft hierarchy or A[2]—we argue in Sect. 5.2.1 that the three parameterized variants of ASP-CONSISTENCY that we consider are each hard for either para-NP or para-co-NP. If any problem is both hard for either para-NP or para-co-NP and in XP, then P = NP. Therefore, these parameterized problems are not in XP, unless P = NP.

We can rule out membership in para-Δ_2^p, by using results that we provide in Chaps. 6 and 14. Membership in para-Δ_2^p amounts to the existence of an fpt-time Turing reduction to SAT. The results in Chaps. 6 and 14 imply that such an fpt-time Turing reduction to SAT does not exist for the parameterized problems that we consider, assuming that various subexponential-time Turing reductions from the second level to the first level of the PH do not exist. To take the case of ASP-CONSISTENCY(#cont.rules) as example again, Theorem 6.3 in Chap. 6 implies that ASP-CONSISTENCY(#cont.rules) is A[2]-hard, and therefore by Theorem 14.6 in Chap. 14 the problem is not in para-Δ_2^p, unless there exists a subexponential-time Turing reduction from $QSAT_2$(3DNF) to SAT.

The arguments in the previous paragraphs show that the parameterized problems ASP-CONSISTENCY(#cont.rules), ASP-CONSISTENCY(#disj.rules) and ASP-CONSISTENCY(#non-dual-normal.rules) cannot be complete for any parameterized complexity class that is depicted below the bottom dashed gray line in Fig. 5.1. Next, we argue that they can also not be complete for any of the classes depicted above the top dashed gray line in Fig. 5.1.

To rule out hardness for para-Σ_2^p or para-Π_2^p, we show in Sect. 5.2.2 that each of the parameterized problems is in XNP or in Xco-NP. If there is any problem that is both hard for either para-Σ_2^p or para-Π_2^p and in XNP or Xco-NP, then NP = co-NP [84, Proposition 8]. Therefore, the parameterized problems are not hard for para-Σ_2^p or para-Π_2^p, unless NP = co-NP.

Finally, we can rule out hardness for XNP or Xco-NP under the assumption that there exists no subexponential-time reduction from $QSAT_t$(3CNF $\cup$ 3DNF) to $QSAT_2$, for any $t \geq 3$. This follows by results in Chaps. 6 and 14. For the case of ASP-CONSISTENCY(#cont.rules), for example, this follows from Theorem 6.3 in Chap. 6 and Corollary 14.8 in Chap. 14.

By eliminating the possibility that these remaining parameterized variants of ASP-CONSISTENCY are complete for any known parameterized complexity class, we motivate the development of new parameterized complexity tools to characterize the complexity of these problems and similar parameterized variants of problems at higher levels of the PH. The completeness theory that we define and work out in Chap. 6 will be this new toolbox.

5.2.1 Hardness for para-NP and para-co-NP

In this section, we point out known proofs from the literature that can be used to show that each of the parameterized problems ASP-CONSISTENCY(#cont.rules), ASP-CONSISTENCY(#disj.rules) and ASP-CONSISTENCY(#non-dual-normal.rules) is hard for either para-NP or para-co-NP.

The parameterized problem ASP-CONSISTENCY(#cont.rules) is para-co-NP-hard. This hardness result can be shown straightforwardly by using the reduction in a proof of Eiter and Gottlob [71, Theorem 3] as a reduction from UNSAT to the problem ASP-CONSISTENCY restricted to instances without contingent rules. Moreover, the parameterized problem ASP-CONSISTENCY(#disj.rules) is para-NP-hard. This follows from the fact that deciding whether a program without disjunctive rules has an answer set is NP-hard [155]. Finally, the parameterized problem ASP-CONSISTENCY(#non-dual-normal.rules) is para-NP-hard, because the consistency problem restricted to disjunctive programs without non-dual-normal rules is NP-hard [187].

5.2.2 Membership in XNP and Xco-NP

Finally, we show that the parameterized problems ASP-CONSISTENCY(#cont.rules), ASP-CONSISTENCY(#disj.rules) and ASP-CONSISTENCY(#non-dual-normal.rules) are all contained in either XNP or Xco-NP.

Firstly, we show that when the number of disjunctive rules is bounded by a fixed constant the problem ASP-CONSISTENCY is in NP. In order to prove this, we use the following lemma.

Lemma 5.5. *Let P be a negation-free disjunctive logic program, and let M be a minimal model of P, i.e., there is no model $M' \subsetneq M$ of P. Then there exists a subset $R \subseteq P$ of disjunctive rules and a mapping $\mu : R \to \mathrm{At}(P)$ such that:*

- *for each $r \in R$, the value $\mu(r)$ is an atom in the head of r; and*
- *$M = M_\mu$, where M_μ is the smallest set such that:*
  - *$\mathrm{Rng}(\mu) \subseteq M_\mu$, and*
  - *if $b_1, \ldots, b_m \in M_\mu$, and $(a \leftarrow b_1, \ldots, b_m) \in P$, then $a \in M_\mu$.*

Proof. We give an indirect proof. Assume that M is a minimal model of P, but there exist no suitable R and μ. We will derive a contradiction. Since M is a model of P, we know that for each disjunctive rule r_i either holds (i) that M does not satisfy the body, or (ii) that M satisfies an atom a_i in the head. We construct the set R and the mapping μ as follows. For each r_i, we let $r_i \in R$ and $\mu(r_i) = a_i$ if and only if M satisfies an atom a_i in the head of the disjunctive rule r_i.

Clearly, $\mathrm{Rng}(\mu) \subseteq M$. Define $A_0 = \mathrm{Rng}(\mu)$. For each $i \in \mathbb{N}$, we define A_{i+1} as follows:

$$A_{i+1} = A_i \cup \{ a : (a \leftarrow b_1, \ldots, b_m) \in P, b_1, \ldots, b_m \in A_i \}.$$

We show by induction on i that $A_i \subseteq M$ for all $i \in \mathbb{N}$. Clearly, $A_0 \subseteq M$. Assume that $A_i \subseteq M$, and let $a \in A_{i+1} \backslash A_i$ be an arbitrary atom. This can only be the case if $b_1, \ldots, b_m \in A_i$ and $(a \leftarrow b_1, \ldots, b_m) \in P$. However, since M is a model of P, also $a \in M$. Therefore, $A_{i+1} \subseteq M$.

Now, define $M' = \bigcup_{i \in \mathbb{M}} A_i$. We have that $M' \subseteq M$. We show that M' is a model of P. Clearly, M' satisfies all normal rules of P. Let $r = (a_1 \vee \cdots \vee a_n \leftarrow b_1, \ldots, b_n)$ be a disjunctive rule of P, and assume that M' does not satisfy r. Then it must be the case that $b_1, \ldots, b_m \in M'$ and $a_1, \ldots, a_n \notin M'$. However, since $M' \subseteq M$, we know that M satisfies the body of r. Therefore, by construction of R and μ, we know that $a_i \in M$, for some $i \in [n]$, and thus $r \in R$ and $\mu(r) = a_i$, and so $a_i \in A_0 \subseteq M'$. This is a contradiction with our previous conclusion that $a_1, \ldots, a_n \notin M'$. Therefore, we can conclude that M' is a model of P.

If $M' \subsetneq M$, we have a contradiction with the fact that M is a minimal model of P. Otherwise, if $M' = M$, then we have a contradiction with the fact that M cannot be represented in the way described above by suitable R and μ. This concludes our proof.

With this technical result in place, we are now ready to show that ASP-CONSISTENCY is in NP when the number of disjunctive rules is bounded by a constant.

Proposition 5.6. *Let d be a fixed positive integer. The restriction of* ASP-CONSISTENCY *to programs containing at most d disjunctive rules is in* NP.

Proof. We sketch a guess-and-check algorithm A that solves the problem. Let P be a disjunctive logic program with at most d disjunctive rules. The algorithm A guesses a subset $M \subseteq \mathrm{At}(P)$. All that remains is to verify that M is a minimal model of P^M. We employ Lemma 5.5 to do so. Concretely, the algorithm iterates over all combinations of a suitable set R and a suitable mapping μ, as defined in Lemma 5.5. There are at most $O((|\mathrm{At}(P)| + 1)^d)$ of these, which is polynomial in the input size. For each such mapping μ, we compute the set M_μ. Moreover, we check if M_μ is a model of P^M. If it is a model of P^M, we check if $M_\mu \subsetneq M$. If this is the case, we know that M is not a minimal model of P^M, and the algorithm A rejects the input. Otherwise, the algorithm continues.

Then, after having gone over all possible mappings μ (and the corresponding sets M_μ) without having found some $M_\mu \subsetneq M$ that is a model of P^M, the algorithm accepts. In this case, we can safely conclude that M is a minimal model of P^M. If this were not the case, by definition, we know that there is a model $M' \subsetneq M$ of P^M. But then, by Lemma 5.5, we know that $M' = M_\mu$ for some μ, which is a contradiction with the fact that we found no $M_\mu \subsetneq M$ that is a model of P^M. Therefore, the algorithm A accepts if and only if M is a minimal model of P^M, and thus it accepts if and only if P has an answer set.

Next, we show that when the number of contingent rules is bounded by a fixed constant the problem ASP-CONSISTENCY is in co-NP. Again, to show this, we use a technical lemma.

Lemma 5.7. *Let P be a disjunctive logic program, and let M be an answer set of P. Then there exists a subset $R \subseteq P$ of contingent rules and a mapping μ : $R \to \mathrm{Cont}(P)$ such that for each $r \in R$ it holds that $\mu(r)$ occurs in the head of r, and $M = \mathrm{Comp}(P) \cup \mathrm{Rng}(\mu)$.*

Proof. We show that $M = \mathrm{Rng}(\mu) \cup \mathrm{Comp}(P)$ for some subset $R \subseteq P$ of contingent rules and some mapping $\mu : R \to \mathrm{Cont}(P)$ such that for each $r \in R$ it holds that $\mu(r)$ occurs in the head of r. Since M is an answer set of P, we know that for each contingent rule r of P it holds that either (i) $\mathrm{NF}(r) \notin P^M$, or (ii) $\mathrm{NF}(r) \in P^M$ and M satisfies the body of $\mathrm{NF}(r)$, or (iii) $\mathrm{NF}(r) \in P^M$ and M does not satisfy the body of $\mathrm{NF}(r)$. Recall that $\mathrm{NF}(r)$ is the rule r where all negative literals are removed. We construct the subset $R \subseteq P$ of contingent rules and the mapping $\mu : R \to \mathrm{Cont}(P)$ as follows. For each r, if case (i) or (iii) holds, we let $r \notin R$. If case (ii) holds for rule r, then since M is a model of P^M, we know that there exists some $d \in M \cap \mathrm{Cont}(P)$ such that d occurs in the head of $\mathrm{NF}(r)$. We then let $r \in R$ and we let $\mu(r) = d$. Clearly, $\mathrm{Rng}(\mu) \subseteq M$, so $\mathrm{Rng}(\mu) \cup \mathrm{Comp}(P) \subseteq M$. Now, to show that $M \subseteq \mathrm{Rng}(\mu) \cup \mathrm{Comp}(P)$, assume the contrary, i.e., assume that there exists some $d \in (M \cap \mathrm{Cont}(P))$ such that $d \notin \mathrm{Rng}(\mu)$. Then $M \setminus \{d\}$ is a model of P^M, and therefore M is not a subset-minimal model of P^M. This is a contradiction with our assumption that M is an answer set of P. Therefore, $M = \mathrm{Rng}(\mu) \cup \mathrm{Comp}(P)$. $\qquad\square$

We are now ready to show that ASP-CONSISTENCY is in co-NP when the number of contingent rules is bounded by a constant.

Proposition 5.8. *Let d be a fixed positive integer. The restriction of ASP-CONSISTENCY to programs containing at most d contingent rules is in co-NP.*

Proof. We sketch a guess-and-check algorithm A that decides whether programs P containing at most d contingent rules have no answer set. Let P be an arbitrary program. Then any answer set can be represented by means of a suitable subset $R \subseteq P$ and a suitable mapping μ, as described in Lemma 5.7. Let $M_{R,\mu}$ denote the possible answer set corresponding to R and μ. There are at most $O((|P| + 1)^d)$ candidate sets $M_{R,\mu}$. The algorithm A guesses a subset $M'_{R,\mu} \subsetneq M_{R,\mu}$ for each such R and μ. Then, it verifies whether for all $M'_{R,\mu}$ it holds that $M'_{R,\mu}$ is a model of $P^{M_{R,\mu}}$, and it accepts if and only if this is the case. Therefore, it accepts if and only if P has no answer set.

We show that when the number of non-dual-normal is bounded by a fixed constant the problem ASP-CONSISTENCY is in NP. Once again, we use a technical lemma.

Lemma 5.9. *Let P be a negation-free disjunctive logic program, and let M be a maximal model of P, i.e., there is no model $M' \supsetneq M$ of P. Then there exists a subset $R \subseteq P$ of non-dual-normal rules and a mapping $\mu : R \to \text{At}(P)$ such that:*

- *for each $r \in R$, the value $\mu(r)$ is an atom in the body of r; and*
- *$M = \text{At}(P) \backslash E_\mu$, where E_μ is the smallest set such that:*

  - *$\text{Rng}(\mu) \subseteq E_\mu$, and*
  - *if $b_1, \ldots, b_m \in E_\mu$, and $(b_1 \vee \cdots \vee b_m \leftarrow a) \in P$, then $a \in E_\mu$.*

Proof. We give an indirect proof. Assume that M is a maximal model of P, but there exist no suitable R and μ. We will derive a contradiction. Since M is a model of P, we know that for each non-dual-normal rule r_i either holds (i) that M does not satisfy an atom b_i in the body, or (ii) that M satisfies the head. We construct the set R and the mapping μ as follows. For each r_i, we let $r_i \in R$ and $\mu(r_i) = b_i$ if and only if M does not satisfy some atom b_i in the body of the disjunctive rule r_i.

Let $E = \text{At}(P) \backslash M$. Clearly, $\text{Rng}(\mu) \subseteq E$. Define $E_0 = \text{Rng}(\mu)$. For each $i \in \mathbb{N}$, we define E_{i+1} as follows:

$$E_{i+1} = E_i \cup \{ a : (b_1 \vee \cdots \vee b_m \leftarrow a) \in P, b_1, \ldots, b_m \in E_i \}.$$

We show by induction on i that $E_i \subseteq E$ for all $i \in \mathbb{N}$. Clearly, $E_0 \subseteq E$. Assume that $E_i \subseteq E$, and let $a \in E_{i+1} \backslash E_i$ be an arbitrary atom. This can only be the case if $b_1, \ldots, b_m \in E_i$ and $(b_1 \vee \cdots \vee b_m \leftarrow a) \in P$. However, since $M = \text{At}(P) \backslash E$ is a model of P, we get that $a \notin M$ and thus that $a \in E$. Therefore, $E_{i+1} \subseteq E$.

Now, define $E' = \bigcup_{i \in \mathbb{M}} E_i$. We have that $E' \subseteq E$. We show that $M' = \text{At}(P) \backslash E'$ is a model of P. Clearly, M' satisfies all dual-normal rules of P. Let $r = (a_1 \vee \cdots \vee a_n \leftarrow b_1, \ldots, b_n)$ be a non-dual-normal rule of P, and assume that M' does not satisfy r. Then it must be the case that $b_1, \ldots, b_m \in M'$ and $a_1, \ldots, a_n \notin M'$. In other words, $b_1, \ldots, b_m \notin E'$ and $a_1, \ldots, a_n \in E'$. However, since $E' \subseteq E$, we know that $a_1, \ldots, a_n \in E$, and thus that M does not satisfy the head of r. Therefore, by construction of R and μ, we know that $b_i \in E$, for some $i \in [n]$, and thus $r \in R$ and $\mu(r) = b_i$, and so $b_i \in E_0 \subseteq E'$. This is a contradiction with our previous conclusion that $b_1, \ldots, b_m \notin E'$. Therefore, we can conclude that M' is a model of P.

Since $E' \subseteq E$, we know that $M' \supseteq M$. If $M' \supsetneq M$, we have a contradiction with the fact that M is a maximal model of P. Otherwise, if $M' = M$, then we have a contradiction with the fact that M cannot be represented in the way described above by suitable R and μ. This concludes our proof.

Now that we have established this technical lemma, we show that ASP-CONSI STENCY is in NP when the number of non-dual-normal rules is bounded by a constant.

Proposition 5.10. *Let d be a fixed positive integer. The restriction of ASP-CONSIS TENCY to programs containing at most d non-dual-normal rules is in NP.*

Proof. We sketch a guess-and-check algorithm A that solves the problem. Let P be a disjunctive logic program with at most d non-dual-normal rules. The algorithm A guesses a subset $M \subseteq \mathrm{At}(P)$, and verifies that M satisfies all constraints in P, i.e., all rules of the form $(\leftarrow b_1, \ldots, b_m, not\ c_1, \ldots, not\ c_n)$. All that remains is to verify that M is a minimal model of $(P')^M$, where P' is obtained from P by removing all constraints. In order to do so, we use the fact that M is a minimal model of $(P')^M$ if and only if for all $m \in M$ the following program $P_{m,M}$ has no models:

$$P_{m,M} = (P')^M \cup \{ \leftarrow a : a \in \mathrm{At}(P) \backslash M \} \cup \{ \leftarrow m \}.$$

(This construction using the programs $P_{m,M}$ has also been used by Fichte et al. [81].)

The algorithm A proceeds as follows. It iterates over all $m \in M$, and for each m it computes the program $P_{m,M}$. Clearly, each such $P_{m,M}$ has at most d non-dual-normal rules (and only constraints with one atom in the body). We now employ Lemma 5.9 to decide if $P_{m,M}$ has a model. The algorithm iterates over all combinations of a suitable set R and a suitable mapping μ, as defined in Lemma 5.9. There are at most $O((|\mathrm{At}(P)| + 1)^d)$ of these, which is polynomial in the input size. For each such mapping μ, the algorithm computes the set E_μ. Moreover, the algorithm checks whether $M_\mu = \mathrm{At}(P) \backslash E_\mu$ is a model of $P_{m,M}$. If this is the case, then we know that $P_{m,M}$ has some model, and therefore that M is not a minimal model of P^M, and the algorithm rejects the input. If M_μ is not a model of $P_{m,M}$, the algorithm continues.

Then, after having gone over all possible $m \in M$ and all possible mappings μ without having found some M_μ that is a model of $P_{m,M}$, the algorithm accepts. In this case, we can safely conclude that M is a minimal model of P^M. If this were not the case, there would be some $m \in M$ such that $P_{m,M}$ has a model (and in particular, $P_{m,M}$ has a maximal model M'). But then, by Lemma 5.9, we know that $M' = M_\mu$ for some suitable μ, which is a contradiction with the fact that we found no M_μ that is a model of $P_{m,M}$. Therefore, the algorithm A accepts if and only if M is a minimal model of P^M, and thus it accepts if and only if P has an answer set.

The algorithms given in the proofs of Propositions 5.6, 5.10 and 5.8 are the same for each positive value d of the parameter (only the running times differ for different parameter values). Therefore, we directly get the following membership results in XNP and Xco-NP.

Corollary 5.11. *The parameterized problems* ASP-CONSISTENCY(#disj.rules) *and* ASP-CONSISTENCY(#non-dual-normal.rules) *are in* XNP.

Corollary 5.12. *The parameterized problem* ASP-CONSISTENCY(#cont.rules) *is in* Xco-NP.

Summary

In this chapter, we argued for the need to introduce new parameterized complexity classes between the first and second level of the PH. To do so, we considered the consistency problem for disjunctive answer set programs, which is an important problem that comes up in the declarative programming paradigm of answer set programming. We showed that several natural parameterized variants of this problem are not complete for any of the parameterized complexity classes that are known from the literature (under various complexity-theoretic assumptions). From this, we concluded that new classes are needed to adequately characterize the parameterized complexity of these problems.

Notes

The results in this chapter appeared in a paper in the proceedings of KR 2014 [116, 117].

Chapter 6
A New Completeness Theory

> *Alright, it's time to define some complexity classes.*
> *(Then again, when isn't it time?)*
> — Scott Aaronson, *Quantum Computing Since*
> *Democritus* [1]

In Chap. 5, we identified several natural parameterized variants of our running example (the consistency problem for disjunctive answer set programming) whose complexity cannot be adequately characterized using the known parameterized complexity classes. That is, we argued that these parameterized complexity problems are not complete for any of the known parameterized complexity classes. In this chapter, we define and develop novel parameterized complexity classes that can be used to characterize the complexity of these (and other) problems. The new classes are based on weighted variants of the satisfiability problem for quantified Boolean formulas, analogously to the way the classes of the Weft hierarchy are based on weighted variants of propositional satisfiability.

Outline of This Chapter

In Sect. 6.1, we introduce the new parameterized complexity classes. In particular, we define two hierarchies $\Sigma_2^p[k*, t]$ and $\Sigma_2^p[*k, t]$ of parameterized complexity classes (for $t \in \mathbb{N}^+ \cup \{\text{SAT}, \text{P}\}$), dubbed "$k$-$*$" and "$*$-$k$" for the combinations of weighted and unrestricted quantifier blocks in their definitions.

Then, in Sect. 6.2, we have a closer look at the classes of the k-$*$ hierarchy. We show that this hierarchy in fact collapses to a single parameterized complexity class—this class we call $\Sigma_2^p[k*]$. We show that one of the parameterizations of our running example is complete for this class. We also provide alternative characterizations of $\Sigma_2^p[k*]$, using (1) first-order logic model checking and (2) alternating Turing machines.

In Sect. 6.3, we have a closer look at the classes of the $*$-k hierarchy. We provide a normalization result for the first level of this hierarchy, showing that the canonical

© Springer-Verlag GmbH Germany, part of Springer Nature 2019
R. de Haan: Parameterized Complexity in the Polynomial Hierarchy, LNCS 11880,
https://doi.org/10.1007/978-3-662-60670-4_6

problem for $\Sigma_2^p[*k, 1]$ is already complete for this class when restricted to quantified Boolean formulas whose matrix is in 3DNF. We also provide a normalization result for the top level $\Sigma_2^p[*k, P]$ of the hierarchy, and we give an alternative characterization of $\Sigma_2^p[*k, P]$ using alternating Turing machines. We then show that two parameterizations of our running example are complete for the class $\Sigma_2^p[*k, P]$.

Finally, in Sect. 6.4, we investigate how the new complexity classes relate to several relevant (parameterized) complexity classes known from the literature. In particular, we present results that the classes $\Sigma_2^p[k*]$ and $\Sigma_2^p[*k, t]$ are different from the classes para-NP, para-co-NP, para-Σ_2^p, and para-Π_2^p (under various complexity-theoretic assumptions).

6.1   New Parameterized Complexity Classes

In this section, we define several hierarchies of parameterized complexity classes. These classes are based on weighted variants of $QSAT_2$, the satisfiability problem for quantified Boolean formulas with an $\exists\forall$ quantifier prefix. An instance of the problem $QSAT_2$ has both an existential quantifier and a universal quantifier block. Therefore, there are several ways of restricting the weight of assignments. Restricting the weight of assignments to the existential quantifier block results in the k-$*$ hierarchy, and restricting the weight of assignments to the universal quantifier block results in the $*$-k hierarchy. Moreover, restricting the weight of assignments to both quantifier blocks simultaneously results in a hierarchy of classes (dubbed "k-k") that are closely related to the classes of the A-hierarchy.

6.1.1   The Hierarchies $\Sigma_2^p[k*, t]$ and $\Sigma_2^p[*k, t]$

The hierarchies of classes $\Sigma_2^p[k*, t]$ and $\Sigma_2^p[*k, t]$ are based on the following two parameterized decision problems. Let $\mathcal{C}$ be a class of Boolean circuits. The problem $\Sigma_2^p[k*]$-WSAT($\mathcal{C}$) provides the foundation for the k-$*$ hierarchy.

$\Sigma_2^p[k*]$-WSAT($\mathcal{C}$)
Instance: A Boolean circuit $C \in \mathcal{C}$ over two disjoint sets X and Y of variables, and a positive integer k.
Parameter: k.
Question: Does there exist a truth assignment α to X of weight k such that for all truth assignments β to Y the assignment $\alpha \cup \beta$ satisfies C?

Similarly, the problem $\Sigma_2^p[*k]$-WSAT($\mathcal{C}$) provides the foundation for the $*$-k hierarchy.

$\Sigma_2^P[*k]$-WSAT($\mathcal{C}$)

Instance: A Boolean circuit $C \in \mathcal{C}$ over two disjoint sets X and Y of variables, and a positive integer k.

Parameter: k.

Question: Does there exist a truth assignment α to X such that for all truth assignments β to Y of weight k the assignment $\alpha \cup \beta$ satisfies C?

For the sake of convenience, instances to these two problems consisting of a circuit C over sets X and Y of variables and a positive integer k, we will denote by $(\exists X.\forall Y.C, k)$.

We now define the following parameterized complexity classes, that together form the k-$*$ hierarchy:

$$\Sigma_2^P[k*, t] = [\, \{\, \Sigma_2^P[k*]\text{-WSAT}(\text{CIRC}_{t,u}) : u \geq 1 \,\} \,]^{\text{fpt}},$$

$$\Sigma_2^P[k*, \text{SAT}] = [\, \Sigma_2^P[k*]\text{-WSAT}(\text{FORM}) \,]^{\text{fpt}}, \text{ and}$$

$$\Sigma_2^P[k*, \text{P}] = [\, \Sigma_2^P[k*]\text{-WSAT}(\text{CIRC}) \,]^{\text{fpt}}.$$

Similarly, we define the classes of the $*$-k hierarchy as follows:

$$\Sigma_2^P[*k, t] = [\, \{\, \Sigma_2^P[*k]\text{-WSAT}(\text{CIRC}_{t,u}) : u \geq 1 \,\} \,]^{\text{fpt}},$$

$$\Sigma_2^P[*k, \text{SAT}] = [\, \Sigma_2^P[*k]\text{-WSAT}(\text{FORM}) \,]^{\text{fpt}}, \text{ and}$$

$$\Sigma_2^P[*k, \text{P}] = [\, \Sigma_2^P[*k]\text{-WSAT}(\text{CIRC}) \,]^{\text{fpt}}.$$

Remember that the notation $[\, \cdot \,]^{\text{fpt}}$ denotes the class of all parameterized problems that are fpt-reducible to the referenced (set of) problem(s). Remember also that $\text{CIRC}_{t,u}$ denotes the class of all Boolean circuits of weft t and depth u, that FORM denotes the class of all Booelan circuits that represent a propositional formula, and that CIRC denotes the class of all Boolean circuits. These definitions are entirely analogous to those of the parameterized complexity classes $W[t]$ of the W-hierarchy [66, 67].

By definition of the classes $\Sigma_2^P[k*, t]$ and $\Sigma_2^P[*k, t]$, we directly get the following inclusions:

$$\Sigma_2^P[k*, 1] \subseteq \Sigma_2^P[k*, 2] \subseteq \cdots \subseteq \Sigma_2^P[k*, \text{SAT}] \subseteq \Sigma_2^P[k*, \text{P}], \text{ and}$$

$$\Sigma_2^P[*k, 1] \subseteq \Sigma_2^P[*k, 2] \subseteq \cdots \subseteq \Sigma_2^P[*k, \text{SAT}] \subseteq \Sigma_2^P[*k, \text{P}].$$

Dual to the classical complexity class Σ_2^P is its co-class Π_2^P, whose canonical complete problem is complementary to the problem QSAT$_2$. Similarly, we can define dual classes for each of the parameterized complexity classes in the k-$*$ and $*$-k hierarchies. These co-classes are based on problems complementary to the problems $\Sigma_2^P[k*]$-WSAT and $\Sigma_2^P[*k]$-WSAT, i.e., these problems have as yes-instances exactly the no-instances of $\Sigma_2^P[k*]$-WSAT and $\Sigma_2^P[*k]$-WSAT, respectively. Equivalently, these complementary problems can be considered as variants of $\Sigma_2^P[k*]$-WSAT and

$\Sigma_2^P[*k]$-WSAT where the existential and universal quantifiers are swapped. These dual classes are denoted by $\Pi_2^P[k*]$-WSAT and $\Pi_2^P[*k]$-WSAT. We use a similar notation for the dual complexity classes, e.g., we denote co-$\Sigma_2^P[*k, t]$ by $\Pi_2^P[*k, t]$.

6.1.2   Another Hierarchy

Similarly to the definition of the complexity classes of the k-$*$ and $*$-k hierarchies, one can define weighted variants of the problem QSAT$_2$ with weight restrictions on both quantifier blocks. This results in the parameterized complexity classes $\Sigma_2^P[kk, t]$, whose definition is based on the following parameterized complexity problem. Let $\mathcal{C}$ be a class of Boolean circuits. The problem $\Sigma_2^P[kk]$-WSAT$(\mathcal{C})$ provides the foundation for the k-k hierarchy.

$\Sigma_2^P[kk]$-WSAT$(\mathcal{C})$
Instance: A Boolean circuit $C \in \mathcal{C}$ over two disjoint sets X and Y of variables, and a positive integer k.
Parameter: k.
Question: Does there exist a truth assignment α to X of weight k such that for all truth assignments β to Y of weight k the assignment $\alpha \cup \beta$ satisfies C?

The classes $\Sigma_2^P[kk, t]$, for $t \in \mathbb{N}^+ \cup \{\text{SAT}, \text{P}\}$ are then defined as follows:

$$\Sigma_2^P[kk, t] = [\ \{\ \Sigma_2^P[kk]\text{-WSAT}(\text{CIRC}_{t,u}) : u \geq 1\ \}\ ]^{\text{fpt}},$$

$$\Sigma_2^P[kk, \text{SAT}] = [\ \Sigma_2^P[kk]\text{-WSAT}(\text{FORM})\ ]^{\text{fpt}}, \text{ and}$$

$$\Sigma_2^P[kk, \text{P}] = [\ \Sigma_2^P[kk]\text{-WSAT}(\text{CIRC})\ ]^{\text{fpt}}.$$

The complexity class $\Sigma_2^P[kk, \text{SAT}]$ has been defined and considered by Gottlob, Scarcello and Sideri [101] under the name $\Sigma_2\text{W}[\text{SAT}]$. Also, for each $t \in \mathbb{N}^+$, variants of the problems $\Sigma_2^P[kk, t]$ have been studied in the literature (see, e.g., [85, Chapter 8]). Based on these problems, the parameterized complexity classes A[2,t] (for $t \geq 1$) have been defined. These classes generalize A[2], because A[2] = A[2,1]. Due to fact that the classes A[2,t] and the classes $\Sigma_2^P[kk, t]$ are defined in a very similar way—in fact, the canonical problems for the classes A[2,t] are a special case of the problems $\Sigma_2^P[kk]$-WSAT$(\text{CIRC}_{t,u})$—it is straightforward to verify that for all $t \geq 1$ it holds that A[2,t] $\subseteq \Sigma_2^P[kk, t]$.

Moreover, it can also routinely be proved that for each $t \in \mathbb{N}^+ \cup \{\text{SAT}, \text{P}\}$ it holds that $\Sigma_2^P[kk, t] \subseteq \Sigma_2^P[k*, t]$ and that $\Sigma_2^P[kk, t] \subseteq \Sigma_2^P[*k, t]$. Therefore, we directly get the following result (that we state without proof), that relates A[2] and the classes of the k-$*$ and $*$-k hierarchies.

Proposition 6.1. *Let* $t \in \mathbb{N}^+ \cup \{\text{SAT}, \text{P}\}$. *Then* A[2] $\subseteq \Sigma_2^P[k*, t]$ *and* A[2] $\subseteq \Sigma_2^P[*k, t]$.

6.2 The Parameterized Complexity Class $\Sigma_2^p[k*]$

In this section, we consider the classes $\Sigma_2^p[k*, t]$ of the k-$*$ hierarchy in more detail. It turns out that this hierarchy collapses entirely into a single parameterized complexity class, that we denote by $\Sigma_2^p[k*]$. We start by showing this collapse. Then, we show that $\Sigma_2^p[k*]$ can be used to characterize the complexity of one of the parameterized problems in our running example. In particular, we show that ASP-CONSISTENCY (#cont.rules) is $\Sigma_2^p[k*]$-complete. Finally, we characterize the class $\Sigma_2^p[k*]$ using first-order logic model checking and alternating Turing machines.

6.2.1 Collapse

We begin with showing that the classes of the k-$*$ hierarchy all coincide. We do so by showing that $\Sigma_2^p[k*, 1] = \Sigma_2^p[k*, P]$.

Theorem 6.2 (Collapse of the k-$*$ hierarchy). $\Sigma_2^p[k*, 1] = \Sigma_2^p[k*, P]$. *Moreover,* $\Sigma_2^p[k*]$-WSAT(3DNF) *is complete for this class.*

Proof. Since by definition $\Sigma_2^p[k*, 1] \subseteq \Sigma_2^p[k*, 2] \subseteq \ldots \subseteq \Sigma_2^p[k*, P]$, it suffices to show that $\Sigma_2^p[k*, P] \subseteq \Sigma_2^p[k*, 1]$. We show this by giving an fpt-reduction from $\Sigma_2^p[k*]$-WSAT(CIRC) to $\Sigma_2^p[k*]$-WSAT(3DNF). Since 3DNF $\subseteq$ CIRC$_{1,3}$, this suffices. We remark that this reduction is based on the standard Tseitin transformation that transforms arbitrary Boolean formulas into 3CNF by means of additional variables.

Let (φ, k) be an instance of $\Sigma_2^p[k*]$-WSAT(CIRC) with $\varphi = \exists X.\forall Y.C$. Assume without loss of generality that C contains only binary conjunctions and negations. Let o denote the output gate of C. We construct an instance (φ', k) of $\Sigma_2^p[k*]$-WSAT(3DNF) as follows. The formula φ' will be over the set of variables $X \cup Y \cup Z$, where $Z = \{ z_r : r \in \text{Nodes}(C) \}$. For each $r \in \text{Nodes}(C)$, we define a subformula χ_r. We distinguish three cases. If $r = r_1 \wedge r_2$, then we let $\chi_r = (z_r \wedge \neg z_{r_1}) \vee (z_r \wedge \neg z_{r_2}) \vee (z_{r_1} \wedge z_{r_2} \wedge \neg z_r)$. If $r = \neg r_1$, then we let $\chi_r = (z_r \wedge z_{r_1}) \vee (\neg z_r \wedge \neg z_{r_1})$. If $r = w$, for some $w \in X \cup Y$, then we let $\chi_r = (z_r \wedge \neg w) \vee (\neg z_r \wedge w)$. Now we define $\varphi' = \exists X.\forall Y \cup Z.\psi$, where $\psi = \bigvee_{r \in \text{Nodes}(C)} \chi_r \vee z_o$. We prove the correctness of this reduction.

($\Rightarrow$) Assume that $(\varphi, k) \in \Sigma_2^p[k*]$-WSAT(CIRC). This means that there exists an assignment $\alpha : X \rightarrow \mathbb{B}$ of weight k such that $\forall Y.C[\alpha]$ evaluates to true. We show that $(\varphi', k) \in \Sigma_2^p[k*]$-WSAT(3DNF), by showing that $\forall Y \cup Z.\psi[\alpha]$ evaluates to true. Let $\beta : Y \cup Z \rightarrow \mathbb{B}$ be an arbitrary assignment to the variables $Y \cup Z$, and let β' be the restriction of β to the variables Y. We distinguish two cases: either (i) for each $r \in \text{Nodes}(C)$ it holds that $\beta(z_r)$ coincides with the value that gate r gets in the circuit C given assignment $\alpha \cup \beta'$, or (ii) this is not the case. In case (i), by the fact that $\alpha \cup \beta'$ satisfies C, we know that $\beta(z_o) = 1$, and therefore $\alpha \cup \beta$ satisfies ψ. In case (ii), we know that for some gate $r \in \text{Nodes}(C)$, the value of $\beta(z_r)$ does not coincide with the value assigned to r in C given the assignment $\alpha \cup \beta'$. We may assume without loss of generality that for all parent nodes r' of r it holds that $\beta(z_{r'})$ coincides with the value assigned to r' by $\alpha \cup \beta'$. In this case, there is some term of χ_r that is satisfied by $\alpha \cup \beta$. From this we can conclude that $(\varphi', k) \in \Sigma_2^p[k*]$-WSAT(3DNF).

($\Leftarrow$) Assume that $(\varphi', k) \in \Sigma_2^p[k*]$-WSAT(3DNF). This means that there exists some assignment $\alpha : X \to \mathbb{B}$ of weight k such that $\forall Y \cup Z.\psi[\alpha]$ evaluates to true. We now show that $\forall Y.C[\alpha]$ evaluates to true as well. Let $\beta' : Y \to \mathbb{B}$ be an arbitrary assignment to the variables Y. Define $\beta'' : Z \to \mathbb{B}$ as follows. For any $r \in \mathrm{Nodes}(C)$, we let $\beta''(z_r)$ be the value assigned to the node r in the circuit C by the assignment $\alpha \cup \beta'$. We then let $\beta = \beta' \cup \beta''$. Now, since $\forall Y \cup Z.\psi[\alpha]$ evaluates to true, we know that $\alpha \cup \beta$ satisfies ψ. By construction of β, we know that $\alpha \cup \beta$ does not satisfy the term χ_r for any $r \in \mathrm{Nodes}(C)$. Therefore, we know that $\beta(z_o) = 1$. By construction of β, this implies that $\alpha \cup \beta'$ satisfies C. Since β' was arbitrary, we can conclude that $\forall Y.C[\alpha]$ evaluates to true, and therefore that $(\varphi, k) \in \Sigma_2^p[k*]$-WSAT(CIRC).

As mentioned above, in order to simplify notation, we will use $\Sigma_2^p[k*]$ to denote the class $\Sigma_2^p[k*, 1] = \ldots = \Sigma_2^p[k*, \mathrm{P}]$. Also, for the sake of convenience, by a slight abuse of notation, we will often denote the problems $\Sigma_2^p[k*]$-WSAT(CIRC) and $\Sigma_2^p[k*]$-WSAT(FORM) by $\Sigma_2^p[k*]$-WSAT.

6.2.2 Answer Set Programming and Completeness for $\Sigma_2^p[k*]$

Next, we turn to one of the parameterized problems in our running example: ASP-CONSISTENCY(#cont.rules). We show that this parameterized problem is complete for the class $\Sigma_2^p[k*]$.

Theorem 6.3. ASP-CONSISTENCY(#cont.rules) *is* $\Sigma_2^p[k*]$*-complete.*

This result follows directly from Lemmas 6.4 and 6.5, that we show below.

Lemma 6.4. ASP-CONSISTENCY(#cont.rules) *is* $\Sigma_2^p[k*]$*-hard.*

Proof. We give an fpt-reduction from $\Sigma_2^p[k*]$-WSAT(3DNF). This reduction is a parameterized version of a reduction of Eiter and Gottlob [71, Theorem 3]. Let (φ, k) be an instance of $\Sigma_2^p[k*]$-WSAT(3DNF), where $\varphi = \exists X.\forall Y.\psi$, $X = \{x_1, \ldots, x_n\}$, $Y = \{y_1, \ldots, y_m\}$, $\psi = \delta_1 \vee \cdots \vee \delta_u$, and $\delta_\ell = l_1^\ell \wedge l_2^\ell \wedge l_3^\ell$ for each $\ell \in [u]$. We construct a disjunctive program P. We consider the sets X and Y of variables as atoms. In addition, we introduce fresh atoms $v_1, \ldots, v_n, z_1, \ldots, z_m, w$, and x_i^j for each $j \in [k]$ and $i \in [n]$. We let P consist of the rules described as follows:

$$x_1^j \vee \cdots \vee x_n^j \leftarrow \qquad\qquad\qquad\qquad\qquad \text{for } j \in [k]; \qquad (6.1)$$

$$\leftarrow x_i^j, x_i^{j'} \qquad\qquad \text{for } i \in [n], \text{ and } j, j' \in [k] \text{ with } j < j'; \qquad (6.2)$$

$$y_i \vee z_i \leftarrow \qquad\qquad\qquad\qquad\qquad \text{for } i \in [m]; \qquad (6.3)$$

$$y_i \leftarrow w \qquad\qquad\qquad\qquad\qquad \text{for } i \in [m]; \qquad (6.4)$$

$$z_i \leftarrow w \qquad\qquad\qquad\qquad\qquad \text{for } i \in [m]; \qquad (6.5)$$

$$w \leftarrow z_i, z_i \qquad\qquad\qquad\qquad\qquad \text{for } i \in [m]; \qquad (6.6)$$

$$x_i \leftarrow w \qquad\qquad\qquad\qquad\qquad \text{for } i \in [n]; \qquad (6.7)$$

$$x_i \leftarrow x_i^j \qquad\qquad\qquad\qquad \text{for } i \in [n] \text{ and } j \in [k]; \qquad (6.8)$$

$$v_i \leftarrow w \qquad\qquad\qquad\qquad\qquad \text{for } i \in [n]; \qquad (6.9)$$

$$v_i \leftarrow not\ x_i^1, \ldots, not\ x_i^k \qquad\qquad\qquad \text{for } i \in [n]; \qquad (6.10)$$

$$w \leftarrow \sigma(l_1^\ell), \sigma(l_2^\ell), \sigma(l_3^\ell) \qquad\qquad\qquad \text{for } \ell \in [u]; \qquad (6.11)$$

$$w \leftarrow not\ w. \qquad\qquad\qquad\qquad\qquad\qquad\qquad\qquad (6.12)$$

Here we let $\sigma(x_i) = x_i$ and $\sigma(\neg x_i) = v_i$ for each $i \in [n]$; and we let $\sigma(y_j) = y_j$ and $\sigma(\neg y_j) = z_j$ for each $j \in [m]$. Intuitively, v_i corresponds to $\neg x_i$, and z_j corresponds to $\neg y_j$. The main difference with the reduction of Eiter and Gottlob [71] is that we use the rules in (6.1), (6.2), (6.8) and (6.10) to let the variables x_i and v_i represent an assignment of weight k to the variables in X. The rules in (6.7) and (6.9) ensure that the atoms v_i and x_i are compulsory. It is straightforward to verify that $\mathrm{Comp}(P) = \{w\} \cup \{x_i, v_i : i \in [n]\} \cup \{y_i, z_i : i \in [m]\}$. Notice that P has exactly k contingent rules, namely the rules in (6.1). We show that $(\varphi, k) \in \Sigma_2^P[k*]$-WSAT(3DNF) if and only if P has an answer set.

($\Rightarrow$) Assume that there exists an assignment $\alpha : X \to \mathbb{B}$ of weight k such that $\forall Y.\psi[\alpha]$ is true. Let $\{x_{i_1}, \ldots, x_{i_k}\}$ denote the set $\{x_i : i \in [n], \alpha(x_i) = 1\}$. We construct an answer set M of P. We let $M = \{x_{i_\ell}^\ell : \ell \in [k]\} \cup \mathrm{Comp}(P)$. The reduct P^M consists of Rules (6.1)–(6.9) and (6.11), together with rules $(v_i \leftarrow)$ for all $i \in [n]$ such that $\alpha(x_i) = 0$. We show that M is a minimal model of P^M. We proceed indirectly and assume to the contrary that there exists a model $M' \subsetneq M$ of P^M. By Rule (6.8) and the rules $(v_i \leftarrow)$, we know that for all $i \in [n]$ it holds that $x_i \in M'$ if $\alpha(x_i) = 1$, and $v_i \in M'$ if $\alpha(x_i) = 0$. If $x_{i_\ell}^\ell \notin M'$ for any $\ell \in [k]$, then M' is not a model of P^M. Therefore, $\{x_{i_1}^1, \ldots, x_{i_k}^k\} \subseteq M'$. Also, it holds that $w \notin M'$. To show this, assume the contrary, i.e., assume that $w \in M'$. Then, by Rules (6.4), (6.5), (6.7) and (6.9), it follows that $M' = M$, which contradicts our assumption that $M' \subsetneq M$. By Rules (6.3) and (6.6), we know that $|\{y_i, z_i\} \cap M'| = 1$ for each $i \in [m]$. We define the assignment $\beta : Y \to \mathbb{B}$ by letting $\beta(y_i) = 1$ if and only if $y_i \in M'$, for all $i \in [m]$. Since $\forall Y.\psi[\alpha]$ is true, we know that $\alpha \cup \beta$ satisfies some term δ_ℓ of ψ. It is straightforward to verify that $\sigma(l_1^\ell), \sigma(l_2^\ell), \sigma(l_3^\ell) \in M'$. Therefore, by Rule (6.11), $w \in M'$, which is a contradiction with our assumption that $w \notin M'$. From this we can conclude that M is a minimal model of P^M, and thus that M is an answer set of P.

($\Leftarrow$) Assume that M is an answer set of P. Clearly, $\mathrm{Comp}(P) \subseteq M$. Also, since Rules (6.1) and (6.2) are rules of P^M, M must contain atoms $x_{i_1}^1, \ldots, x_{i_k}^k$

for some $i_1, \ldots, i_k$. We know that P^M contains Rules (6.1)–(6.9) and (6.11), as well as the rules $(v_i \leftarrow)$ for all $i \in [n]$ such that for no $j \in [k]$ it holds that $x_i^j \in M'$. We define the assignment $\alpha : X \to \mathbb{B}$ by letting $\alpha(x_i) = 1$ if and only if $i \in \{i_1, \ldots, i_k\}$. The assignment α has weight k. We show that $\forall Y.\psi[\alpha]$ is true. Let $\beta : Y \to \mathbb{B}$ be an arbitrary assignment. Construct the set $M' \subsetneq M$ by letting $M' = (M \cap \{x_i^j : i \in [n], j \in [k]\}) \cup \{x_i : i \in [n], \alpha(x_i) = 1\} \cup \{v_i : i \in [n], \alpha(v_i) = 0\} \cup \{y_i : i \in [m], \beta(y_i) = 1\} \cup \{z_i : i \in [m], \beta(y_i) = 0\}$. Since M is a minimal model of P^M and $M' \subsetneq M$, we know that M' cannot be a model of P^M. Clearly, M' satisfies Rules (6.1)–(6.9), and all rules of P^M of the form $(v_i \leftarrow)$. Thus there must be some instantiation of Rule (6.11) that M' does not satisfy. This means that there exists some $\ell \in [u]$ such that $\sigma(l_1^\ell), \sigma(l_2^\ell), \sigma(l_3^\ell) \in M'$. By construction of M', this means that $\alpha \cup \beta$ satisfies δ_ℓ, and thus satisfies ψ. Since β was chosen arbitrarily, we can conclude that $\forall Y.\psi[\alpha]$ is true, and therefore $(\varphi, k) \in \Sigma_2^p[k*]$-WSAT(3DNF).

Next, we show $\Sigma_2^p[k*]$-membership.

Lemma 6.5. ASP-CONSISTENCY(#cont.rules) *is in* $\Sigma_2^p[k*]$.

Proof. We show membership in $\Sigma_2^p[k*]$ by reducing ASP-CONSISTENCY(#cont.rules) to $\Sigma_2^p[k*]$-WSAT. Let P be a program, where $r_1, \ldots, r_k$ are the contingent rules of P and where $At(P) = \{d_1, \ldots, d_n\}$. We construct a quantified Boolean formula $\varphi = \exists X.\forall Y.\forall Z.\forall W.\psi$ such that $(\varphi, k) \in \Sigma_2^p[k*]$-WSAT if and only if P has an answer set.

In order to do so, we firstly construct a Boolean formula $\psi_P(z_1, \ldots, z_n, z_1', \ldots, z_n')$ (or, for short: ψ_P) over variables $z_1, \ldots, z_n, z_1', \ldots, z_n'$ such that for any $M \subseteq At(P)$ and any $M' \subseteq At(P)$ holds that M is a model of $P^{M'}$ if and only if $\psi_P[\alpha_M \cup \alpha_{M'}]$ evaluates to true, where $\alpha_M : \{z_1, \ldots, z_n\} \to \mathbb{B}$ is defined by letting $\alpha_M(z_i) = 1$ if and only if $d_i \in M$, and $\alpha_{M'} : \{z_1', \ldots, z_n'\} \to \mathbb{B}$ is defined by letting $\alpha_{M'}(z_i') = 1$ if and only if $d_i \in M'$, for all $i \in [n]$. We define:

$$\psi_P = \bigwedge_{r \in P} \left(\psi_r^1 \vee \psi_r^2 \right) ;$$
$$\psi_r^1 = \left(z_{i_1^3}' \vee \cdots \vee z_{i_c^3}' \right) ; \text{ and}$$
$$\psi_r^2 = \left(\left(z_{i_1^1} \vee \cdots \vee z_{i_a^1} \right) \leftarrow \left(z_{i_1^2} \wedge \cdots \wedge z_{i_b^2} \right) \right) ,$$
where $r = (d_{i_1^1} \vee \cdots \vee d_{i_a^1} \leftarrow d_{i_1^2}, \ldots, d_{i_b^2}, not\ d_{i_1^3}, \ldots, not\ d_{i_c^3})$.

It is easy to verify that ψ_P satisfies the required property.

We now introduce the set X of existentially quantified variables of φ. For each contingent rule r_i of P we let $a_1^i, \ldots, a_{\ell_i}^i$ denote the atoms that occur in the head of r_i. For each r_i, we introduce variables $x_0^i, x_1^i, \ldots, x_{\ell_i}^i$, i.e., $X = \{x_j^i : i \in [k], j \in [0, \ell_i]\}$. Furthermore, for each atom d_i, we add universally quantified variables y_i, z_i and w_i, i.e., $Y = \{y_i : i \in [n]\}$, $Z = \{z_i : i \in [n]\}$, and $W = \{w_i : i \in [n]\}$.

We then construct ψ as follows:

$$\psi = \psi_X \wedge \left(\psi_Y^1 \vee \psi_W \vee \psi_{\min}\right) \wedge \left(\psi_Y^1 \vee \psi_Y^2\right);$$

$$\psi_X = \bigwedge_{i \in [k]} \left(\bigvee_{j \in [\ell_i]} x_j^i \wedge \bigwedge_{j,j' \in [0,\ell_i], j < j'} (\neg x_j^i \vee \neg x_{j'}^i) \right);$$

$$\psi_Y^1 = \bigvee_{i \in [k]} \bigvee_{j \in [\ell_i]} \psi_y^{i,j} \vee \bigvee_{d_i \in \mathrm{Cont}(P)} \psi_y^{d_i} \vee \bigvee_{d_i \in \mathrm{Comp}(P)} \neg y_i;$$

$$\psi_y^{d_m} = \begin{cases} (y_m \wedge \neg x_{j_1}^{i_1} \wedge \cdots \wedge \neg x_{j_u}^{i_u}) & \begin{array}{l} \text{if } \{\, x_j^i : i \in [k], j \in [\ell_i], a_j^i = d_m \,\} \\ \quad = \{a_{j_1}^{i_1}, \ldots, a_{j_u}^{i_u}\}, \end{array} \\[2ex] \bot & \begin{array}{l} \text{if } \{\, x_j^i : i \in [k], j \in [\ell_i], a_j^i = d_m \,\} \\ \quad = \emptyset; \end{array} \end{cases}$$

$$\psi_y^{i,j} = (x_j^i \wedge \neg y_m) \text{ where } a_j^i = d_m;$$

$$\psi_Y^2 = \psi_P(y_1, \ldots, y_n, y_1, \ldots, y_n);$$

$$\psi_W = \bigvee_{i \in [n]} (w_i \leftrightarrow (y_i \leftrightarrow z_i));$$

$$\psi_{\min} = \psi_{\min}^1 \vee \psi_{\min}^2 \vee \psi_{\min}^3;$$

$$\psi_{\min}^1 = \bigvee_{i \in [n]} (z_i \wedge \neg y_i);$$

$$\psi_{\min}^2 = (\neg w_1 \wedge \cdots \wedge \neg w_m); \text{ and}$$

$$\psi_{\min}^3 = \neg\psi_P(z_1, \ldots, z_n, y_1, \ldots, y_n).$$

The idea behind this construction is the following. The variables in X represent guessing at most one atom in the head of each contingent rule to be true. By Lemma 5.7, such a guess represents a possible answer set $M \subseteq \mathrm{At}(P)$. The formula ψ_X ensures that for each $i \in [k]$, exactly one x_j^i is set to true. The formula ψ_Y^1 filters out every assignment in which the variables Y are not set corresponding to M. The formula ψ_Y^2 filters out every assignment corresponding to a candidate $M \subseteq \mathrm{At}(P)$ such that $M \not\models P$. The formula ψ_W filters out every assignment such that w_i is not set to the value $(y_i \oplus z_i)$. The formula $\psi_{\min}^1$ filters out every assignment where the variables Z correspond to a set M' such that $M' \not\subseteq M$. The formula $\psi_{\min}^2$ filters out every assignment where the variables Z correspond to the set M, by referring to the variables w_i. The formula $\psi_{\min}^3$, finally, ensures that in every remaining assignment, the variables Z do not correspond to a set $M' \subseteq M$ such that $M' \models P$. We now formally prove that P has an answer set if and only if $(\varphi, k) \in \Sigma_2^P[k*]$-WSAT.

($\Rightarrow$) Assume that P has an answer set M. By Lemma 5.7, we know that there exist some subset $R \subseteq \{r_1, \ldots, r_k\}$ and a mapping $\mu : R \to \mathrm{Cont}(P)$ such that for each $r \in R$, $\mu(r)$ occurs in the head of r, and such that $M = \mathrm{Rng}(\mu) \cup \mathrm{Comp}(P)$. We construct the mapping $\alpha_\mu : X \to \mathbb{B}$ by letting $\alpha_\mu(x_j^i) = 1$ if and only if $r_i \in R$ and $\mu(r_i) = a_j^i$, and letting $\alpha_\mu(x_0^i) = 1$ if and only if $r_i \notin R$. Clearly, α_μ has weight k

and satisfies ψ_X. We show that $\forall Y.\forall Z.\forall W.\psi$ evaluates to true. Let $\beta : Y \cup Z \cup W \to \mathbb{B}$ be an arbitrary assignment. We let $M_Y = \{ d_i : i \in [m], \beta(y_i) = 1 \}$, and $M_Z = \{ d_i : i \in [m], \beta(z_i) = 1 \}$. We distinguish a number of cases:

(i)   either $M_Y \neq M$,
(ii) or the previous case does not apply and for some $i \in [m]$, $\beta(w_i) \neq (\beta(y_i) \oplus \beta(z_i))$,
(iii) or all previous cases do not apply and $M_Z \nsubseteq M_Y$,
(iv) or all previous cases do not apply and $M_Z = M_Y$,
(v)   or all previous cases do not apply and $M_Z \not\models P^M$,
(vi) or all previous cases do not apply and $M_Z \models P^M$.

The following is now straightforward to verify. In case (i), $\alpha \cup \beta$ satisfies ψ_Y^1. Thus, $\alpha \cup \beta$ satisfies ψ. In all further cases, we know that $\alpha \cup \beta$ satisfies ψ_Y^2, since $M_Y = M$, and $M \models P^M$. In case (ii), $\alpha \cup \beta$ satisfies ψ_W. In case (iii), $\alpha \cup \beta$ satisfies $\psi_{\min}^1$. In case (iv), $\alpha \cup \beta$ satisfies $\psi_{\min}^2$. In case (v), $\alpha \cup \beta$ satisfies $\psi_{\min}^3$. In case (vi), we get a direct contradiction from the facts that $M_Z \subsetneq M$, that $M_Z \models P^M$, and that M is a subset-minimal model of P^M. We can thus conclude, that in any case $\alpha \cup \beta$ satisfies ψ. Therefore, $\forall Y \cup Z \cup W.\psi$ evaluates to true, and thus $(\varphi, k) \in \Sigma_2^p[k*]$-WSAT.

($\Leftarrow$) Assume that $(\varphi, k) \in \Sigma_2^p[k*]$-WSAT. This means that there exists an assignment $\beta : X \to \mathbb{B}$ of weight k such that $\forall Y \cup Z \cup W.\psi[\alpha]$ evaluates to true. We construct $M = \mathrm{Comp}(P) \cup \{ d_m : i \in [k], j \in [\ell_i], \alpha(x_j^i) = 1, a_j^i = d_m \}$. We show that M is an answer set of P. Construct an assignment $\beta_1 : Y \cup Z \cup W \to \mathbb{B}$ as follows. We let $\beta_1(y_i) = 1$ if and only if $d_i \in M$. For all $y \in Z \cup W$, the assignment $\beta_1(y)$ is arbitrary. We know that $\alpha \cup \beta_1$ satisfies ψ, and thus that $\alpha \cup \beta_1$ satisfies $(\psi_Y^1 \vee \psi_Y^2)$. It is straightforward to verify that $\alpha \cup \beta_1$ does not satisfy ψ_Y^1. Therefore $\alpha \cup \beta_1$ satisfies ψ_Y^2. From this, we can conclude that $M \models P^M$.

Now we show that M is a subset-minimal model of P^M. Let $M' \subseteq \mathrm{At}(P)$ be an arbitrary set such that $M' \subsetneq M$. We show that $M' \not\models P^M$. We construct an assignment $\beta_2 : Y \cup Z \cup W \to \mathbb{B}$ as follows. For all $y_i \in Y$, we let $\beta_2(y_i) = 1$ if and only if $d_i \in M$. For all $z_i \in Z$, let $\beta_2(z_i) = 1$ if and only if $d_i \in M'$. For all $w_i \in W$, let $\beta_2(w_i) = \beta_2(y_i) \oplus \beta_2(z_i)$. It is straightforward to verify that $\alpha \cup \beta_2$ satisfies $\neg\psi_Y^1, \neg\psi_W, \neg\psi_{\min}^1$, and $\neg\psi_{\min}^2$. Therefore, since $\alpha \cup \beta_2$ satisfies ψ, we know that $\alpha \cup \beta_2$ satisfies $\psi_{\min}^3$. Therefore, we know that $M' \not\models P^M$. This concludes our proof that M is a subset-minimal model of P^M, and thus we can conclude that M is an answer set of P.

6.2.3 Additional Characterizations of $\Sigma_2^p[k*]$

Finally, we provide a number of different equivalent characterizations of $\Sigma_2^p[k*]$. In particular, we characterize $\Sigma_2^p[k*]$ using a parameterized model checking problem for first-order logic formulas and using alternating Turing machines.

First-order Model Checking Characterization

We begin with giving an equivalent characterization of the class $\Sigma_2^p[k*]$ in terms of model checking of first-order logic formulas. The perspective of first-order logic model checking is also used in parameterized complexity theory to define the classes A[t] [85]. Consider the following parameterized model checking problem for first-order logic formulas (with an $\exists\forall$ quantifier prefix) with k existential variables.

$\Sigma_2^p[k*]$-MC
Instance: A first-order logic sentence $\varphi = \exists x_1, \ldots, x_k.\forall y_1, \ldots, y_n.\psi$ over a vocabulary τ, where ψ is quantifier-free, and a finite τ-structure $\mathcal{A}$.
Parameter: k.
Question: Is it the case that $\mathcal{A} \models \varphi$?

We show that this problem is complete for the class $\Sigma_2^p[k*]$.

Theorem 6.6. $\Sigma_2^p[k*]$-MC *is* $\Sigma_2^p[k*]$-*complete.*

This completeness result follows Lemmas 6.7 and 6.8, that we prove below.

Lemma 6.7. $\Sigma_2^p[k*]$-MC *is in* $\Sigma_2^p[k*]$.

Proof. We show $\Sigma_2^p[k*]$-membership of $\Sigma_2^p[k*]$-MC by giving an fpt-reduction to $\Sigma_2^p[k*]$-WSAT. Let $(\varphi, \mathcal{A})$ be an instance of $\Sigma_2^p[k*]$-MC, where $\varphi = \exists x_1, \ldots, x_k.\forall y_1, \ldots, y_n.\psi$ is a first-order logic sentence over vocabulary τ, and $\mathcal{A}$ is a τ-structure with domain A. We assume without loss of generality that ψ contains only connectives $\wedge$ and $\neg$.

We construct an instance (φ', k) of $\Sigma_2^p[k*]$-WSAT, where φ is of the form $\exists X'.\forall Y'.\psi'$. We define:

$$X' = \{ x'_{i,a} : i \in [k], a \in A \}, \text{ and}$$
$$Y' = \{ y'_{j,a} : j \in [n], a \in A \}.$$

In order to define ψ', we will use the following auxiliary function μ on subformulas of ψ:

$$\mu(\chi) = \begin{cases} \mu(\chi_1) \wedge \mu(\chi_2) & \text{if } \chi = \chi_1 \wedge \chi_2, \\ \neg\mu(\chi_1) & \text{if } \chi = \neg\chi_1, \\ \displaystyle\bigvee_{i \in [u]} \left(\psi_{z_1, a_1^i} \wedge \cdots \wedge \psi_{z_m, a_m^i} \right) & \text{if } \chi = R(z_1, \ldots, z_m) \text{ and} \\ & R^{\mathcal{A}} = \{(a_1^1, \ldots, a_m^1), \ldots, (a_1^u, \ldots, a_m^u)\}, \end{cases}$$

where for each $z \in X \cup Y$ and each $a \in A$ we define:

$$\psi_{z,a} = \begin{cases} x'_{i,a} & \text{if } z = x_i, \\ y'_{j,a} & \text{if } z = y_j. \end{cases}$$

Now, we define ψ' as follows:

$$\psi' = \psi'_{\text{unique-}X'} \wedge \left(\psi'_{\text{unique-}Y'} \rightarrow \mu(\psi) \right), \text{ where}$$

$$\psi'_{\text{unique-}X'} = \bigwedge_{i \in [k]} \left(\bigvee_{a \in A} x'_{i,a} \wedge \bigwedge_{\substack{a,a' \in A \\ a \neq a'}} (\neg x'_{i,a} \vee \neg x'_{i,a'}) \right), \text{ and}$$

$$\psi'_{\text{unique-}Y'} = \bigwedge_{j \in [n]} \left(\bigvee_{a \in A} y'_{j,a} \wedge \bigwedge_{\substack{a,a' \in A \\ a \neq a'}} (\neg y'_{j,a} \vee \neg y'_{j,a'}) \right).$$

We show that $(\mathcal{A}, \varphi) \in \Sigma^{\text{p}}_2[k*]$-MC if and only if $(\varphi', k) \in \Sigma^{\text{p}}_2[k*]$-WSAT.

($\Rightarrow$) Assume that there exists an assignment $\alpha : \{x_1, \ldots, x_k\} \rightarrow A$ such that $\mathcal{A}$, $\alpha \models \forall y_1, \ldots, y_n.\psi$. We define the assignment $\alpha' : X' \rightarrow \mathbb{B}$ where $\alpha'(x'_{i,a}) = 1$ if and only if $\alpha(x_i) = a$. Clearly, α' has weight k. Also, note that α' satisfies $\psi'_{\text{unique-}X'}$. Now, let $\beta' : Y' \rightarrow \mathbb{B}$ be an arbitrary assignment. We show that $\alpha' \cup \beta'$ satisfies ψ'. We distinguish two cases: either (i) for each $j \in [n]$, there is a unique $a_j \in A$ such that $\beta'(y'_{j,a_j}) = 1$, or (ii) this is not the case. In case (i), $\alpha' \cup \beta'$ satisfies $\psi'_{\text{unique-}Y'}$, so we have to show that $\alpha' \cup \beta'$ satisfies $\mu(\psi)$. Define the assignment $\beta : \{y_1, \ldots, y_n\} \rightarrow A$ by letting $\beta(y_j) = a_j$. We know that $\mathcal{A}, \alpha \cup \beta \models \psi$. It is now straightforward to show by induction on the structure of ψ that for each subformula χ of ψ holds that $\alpha' \cup \beta'$ satisfies $\mu(\chi)$ if and only if $\mathcal{A}, \alpha \cup \beta \models \chi$. We then know in particular that $\alpha' \cup \beta'$ satisfies $\mu(\psi)$. In case (ii), we know that $\alpha' \cup \beta'$ does not satisfy $\psi'_{\text{unique-}Y'}$, and therefore $\alpha' \cup \beta'$ satisfies ψ'. This concludes our proof that $(\varphi', k) \in \Sigma^{\text{p}}_2[k*]$-WSAT.

($\Leftarrow$) Assume that there exists an assignment $\alpha : X' \rightarrow \mathbb{B}$ of weight k such that $\forall Y'.\psi'[\alpha]$ is true. Since $\psi'_{\text{unique-}X'}$ contains only variables in X', we know that α satisfies $\psi'_{\text{unique-}X'}$. From this, we can conclude that for each $i \in [k]$, there is some unique $a_i \in A$ such that $\alpha(x'_{i,a_i}) = 1$. Now, define the assignment $\alpha' : \{x_1, \ldots, x_k\} \rightarrow A$ by letting $\alpha'(x_i) = a_i$.

We show that $\mathcal{A}, \alpha' \models \forall y_1, \ldots, y_n.\psi$. Let $\beta' : \{y_1, \ldots, y_n\} \rightarrow A$ be an arbitrary assignment. We define $\beta : Y' \rightarrow \mathbb{B}$ by letting $\beta(y'_{i,a}) = 1$ if and only if $\beta(y_i) = a$. It is straightforward to verify that β satisfies $\psi'_{\text{unique-}Y'}$. We know that $\alpha \cup \beta$ satisfies ψ', so therefore $\alpha \cup \beta$ satisfies $\mu(\psi)$. It is now straightforward to show by induction on the structure of ψ that for each subformula χ of ψ holds that $\alpha \cup \beta$ satisfies $\mu(\chi)$ if and only if $\mathcal{A}, \alpha' \cup \beta' \models \chi$. We then know in particular that $\mathcal{A}, \alpha' \cup \beta' \models \psi$. This concludes our proof that $(\mathcal{A}, \varphi) \in \Sigma^{\text{p}}_2[k*]$-MC.

Next, we show $\Sigma^{\text{p}}_2[k*]$-hardness.

Lemma 6.8. $\Sigma^{\text{p}}_2[k*]$-MC is $\Sigma^{\text{p}}_2[k*]$-hard.

Proof. We show $\Sigma^{\text{p}}_2[k*]$-hardness by giving an fpt-reduction from $\Sigma^{\text{p}}_2[k*]$-WSAT (DNF). Let (φ, k) specify an instance of $\Sigma^{\text{p}}_2[k*]$-WSAT, where $\varphi = \exists X.\forall Y.\psi$, $X = \{x_1, \ldots, x_n\}$, $Y = \{y_1, \ldots, y_m\}$, $\psi = \delta_1 \vee \cdots \vee \delta_u$, and for each $\ell \in [u]$, $\delta_\ell = l_1^\ell \vee l_2^\ell \vee l_3^\ell$. We construct an instance $(\mathcal{A}, \varphi')$ of $\Sigma^{\text{p}}_2[k*]$-MC. In order to do so, we first fix the following vocabulary τ (which does not depend on the instance (φ, k)): it contains unary relation symbols D, X and Y, and binary relation symbols C_1, C_2, C_3 and O. We construct the domain A of $\mathcal{A}$ as follows:

$$A = X \cup Y \cup \{ \delta_\ell : \ell \in [u] \} \cup \{\star\}.$$

Then, we define:

$$
\begin{aligned}
D^{\mathcal{A}} &= \{ \delta_\ell : \ell \in [u] \}; \\
X^{\mathcal{A}} &= X; \\
Y^{\mathcal{A}} &= Y \cup \{\star\}; \\
C_d^{\mathcal{A}} &= \{ (\delta_\ell, z) : \ell \in [u], z \in X \cup Y, l_d^\ell \in \{x, \neg x\} \} \qquad \text{for } d \in [3]; \text{ and} \\
O^{\mathcal{A}} &= \{ (\delta_\ell, \delta_{\ell'}) : \ell, \ell' \in [u], \ell < \ell' \}.
\end{aligned}
$$

Intuitively, the relations D, X and Y serve to distinguish the various subsets of the domain A. The relations C_d, for $d \in [3]$, encode (part of) the structure of the matrix ψ of the formula φ. The relation O encodes a linear ordering on the terms δ_ℓ.

We now define the formula φ' as follows:

$$\varphi' = \exists u_1, \ldots, u_k. \forall v_1, \ldots, v_m. \forall w_1, \ldots, w_u. \chi,$$

where we define χ to be of the following form:

$$\chi = \chi_{\text{proper}}^U \wedge ((\chi_{\text{proper}}^V \wedge \chi_{\text{exact}}^W) \to \chi_{\text{sat}}).$$

We will define the subformulas of χ below. Intuitively, the assignment of the variables u_i will correspond to an assignment $\alpha : X \to \mathbb{B}$ of weight k that sets a variable $x \in X$ to true if and only if some u_i is assigned to x. Similarly, any assignment of the variables v_i will correspond to an assignment $\beta : Y \to \{0, 1\}$ that sets $y \in Y$ to true if and only if some v_i is assigned to y. The variables w_ℓ will function to refer to the elements $\delta_\ell \in A$.

The formula χ_{proper}^U ensures that the variables $u_1, \ldots, u_k$ select exactly k different elements from X. We define:

$$\chi_{\text{proper}}^U = \bigwedge_{i \in [k]} X(u_i) \wedge \bigwedge_{i,i' \in [k], i < i'} (u_i \neq u_{i'}).$$

For the sake of clarity, we use the formula χ_{proper}^V to check whether each variable v_i is assigned to a value in $Y \cup \{\star\}$. We define:

$$\chi_{\text{proper}}^V = \bigwedge_{i \in [m]} Y(v_i).$$

Next, the formula χ_{exact}^W encodes whether the variables $w_1, \ldots, w_u$ get assigned exactly to the elements $\delta_1, \ldots, \delta_u$ (and also in that order, i.e., w_ℓ gets assigned δ_ℓ for each $\ell \in [u]$). We let:

$$\chi_{\text{exact}}^{W} = \bigwedge_{\ell \in [u]} D(w_\ell) \wedge \bigwedge_{\ell, \ell' \in [u], \ell < \ell'} O(w_\ell, w_{\ell'}).$$

Finally, we can turn to the formula χ_{sat}, which represents whether the assignments α and β represented by the assignment to the variables u_i and v_j satisfies ψ. We define:

$$\chi_{\text{sat}} = \bigvee_{\ell \in [u]} \chi_{\text{sat}}^{\ell},$$

where we let:

$$\chi_{\text{sat}}^{\ell} = \chi_{\text{sat}}^{\ell, 1} \wedge \chi_{\text{sat}}^{\ell, 2} \wedge \chi_{\text{sat}}^{\ell, 3},$$

and for each $d \in [3]$ we let:

$$\chi_{\text{sat}}^{\ell, d} = \begin{cases} \bigvee_{j \in [k]} C_d(w_\ell, u_j) & \text{if } l_d^{\ell} = x \in X, \\ \bigwedge_{j \in [k]} \neg C_d(w_\ell, u_j) & \text{if } l_d^{\ell} = \neg x \text{ for some } x \in X, \\ \bigvee_{j \in [m]} C_d(w_\ell, v_j) & \text{if } l_d^{\ell} = y \in Y, \\ \bigwedge_{j \in [m]} \neg C_d(w_\ell, v_j) & \text{if } l_d^{\ell} = \neg y \text{ for some } y \in X. \end{cases}$$

Intuitively, for each $\ell \in [u]$ and each $d \in [3]$, the formula $\chi_{\text{sat}}^{\ell, d}$ will be satsified by the assignments to the variables u_i and v_i if and only if the corresponding assignments α to X and β to Y satisfy l_d^{ℓ}.

It is straightforward to verify that the instance $(\mathcal{A}, \varphi')$ can be constructed in polynomial time. We show that $(\varphi, k) \in \Sigma_2^{\text{p}}[k*]\text{-WSAT(3DNF)}$ if and only if $(\mathcal{A}, \varphi') \in \Sigma_2^{\text{p}}[k*]\text{-MC}$.

$(\Rightarrow)$ Assume that there exists an assignment $\alpha : X \to \mathbb{B}$ of weight k such that $\forall Y. \psi[\alpha]$ is true. We show that $\mathcal{A} \models \varphi'$. Let $\{x \in X : \alpha(x) = 1\} = \{x_{i_1}, \ldots, x_{i_k}\}$. We define the assignment $\mu : \{u_1, \ldots, u_k\} \to A$ by letting $\mu(x_j) = x_{i_j}$ for all $j \in [k]$. It is straightforward to verify that $\mathcal{A}, \mu \models \psi_{\text{proper}}^{U}$. Now, let $\nu : \{y_1, \ldots, y_m, w_1, \ldots, w_u\} \to A$ be an arbitrary assignment. We need to show that $\mathcal{A}, \mu \cup \nu \models \chi$, and we thus need to show that $\mathcal{A}, \mu \cup \nu \models (\chi_{\text{proper}}^{V} \wedge \chi_{\text{exact}}^{W}) \to \chi_{\text{sat}}$. We distinguish several cases: either (i) $\nu(y_i) \notin Y \cup \{\star\}$ for some $i \in [m]$, or (ii) the above is not the case and $\nu(w_\ell) \neq \delta_\ell$ for some $\ell \in [u]$, or (iii) neither of the above is the case. In case (i), it is straightforward to verify that $\mathcal{A}, \mu \cup \nu \models \neg \chi_{\text{proper}}^{V}$. In case (ii), it is straightforward to verify that $\mathcal{A}, \mu \cup \nu \models \neg \chi_{\text{exact}}^{W}$. Consider case (iii). We construct the assignment $\beta : Y \to \mathbb{B}$ by letting $\beta(y) = 1$ if and only if $\nu(v_i) = y$ for some $i \in [m]$. We know that $\alpha \cup \beta$ satisfies ψ, and thus in particular that $\alpha \cup \beta$ satisfies some term δ_ℓ. It is now straightforward to verify that $\mathcal{A}, \mu \cup \nu \models \chi_{\text{sat}}^{\ell}$, and thus that $\mathcal{A}, \mu \cup \nu \models \chi_{\text{sat}}$. This concludes our proof that $\mathcal{A} \models \varphi'$.

($\Leftarrow$) Assume that $\mathcal{A} \models \varphi'$. We show that $(\varphi, k) \in \Sigma_2^p[k*]$-WSAT(3DNF). We know that there exists an assignment $\mu : \{u_1, \ldots, u_k\} \to A$ such that $\mathcal{A}, \mu \models \forall u_1, \ldots, u_m. \forall w_1, \ldots, w_u. \chi$. Since $\mathcal{A}, \mu \models \chi_{\text{proper}}^U$, we know that μ assigns the variables u_i to k different values $x \in X$. Define $\alpha : X \to \mathbb{B}$ by letting $\alpha(x) = 1$ if and only if $\mu(u_i) = x$ for some $i \in [k]$. Clearly, α has weight k. Now, let $\beta : Y \to \mathbb{B}$ be an arbitrary assignment. Construct the assignment $\nu : \{v_1, \ldots, v_m, w_1, \ldots, w_u\}$ as follows. For each $i \in [m]$, we let $\nu(v_i) = y_i$ if $\beta(y_i) = 1$, and we let $\nu(v_i) = \star$ otherwise. Also, for each $\ell \in [u]$, we let $\nu(w_\ell) = \delta_\ell$. It is straightforward to verify that $\mathcal{A}, \mu \cup \nu \models \chi_{\text{proper}}^V \wedge \chi_{\text{exact}}^W$. Therefore, we know that $\mathcal{A}, \mu \cup \nu \models \chi_{\text{sat}}$, and thus that for some $\ell \in [u]$ it holds that $\mathcal{A}, \mu \cup \nu \models \chi_{\text{sat}}^{\ell}$. It is now straightforward to verify that $\alpha \cup \beta$ satisfies δ_ℓ. Since β was arbitrary, this concludes our proof that $(\varphi, k) \in \Sigma_2^p[k*]$-WSAT. $\square$

The problem $\Sigma_2^p[k*]$-MC takes the relational vocabulary τ over which the structure $\mathcal{A}$ and the first-order logic sentence φ are defined as part of the input. However, the proof of Lemma 6.8 shows that the problem $\Sigma_2^p[k*]$-MC is $\Sigma_2^p[k*]$-hard already when the vocabulary τ is fixed and contains only unary and binary relation symbols.

Another Weighted Satisfiability Characterization

Next, we show that for the canonical problem $\Sigma_2^p[k*]$-WSAT, it does not matter whether we require the weight of truth assignments to the existential variables to be exactly k or at most k. Formally, we consider the problem $\Sigma_2^p[k*]$-WSAT$^{\leq k}$, that is defined as follows.

$\Sigma_2^p[k*]$-WSAT$^{\leq k}$

Instance: A quantified Boolean formula $\phi = \exists X. \forall Y. \psi$, where ψ is quantifier-free, and an integer k.
Parameter: k.
Question: Does there exist an assignment α to X with weight at most k, such that for all truth assignments β to Y the assignment $\alpha \cup \beta$ satisfies ψ?

We show that $\Sigma_2^p[k*]$-WSAT$^{\leq k}$ is $\Sigma_2^p[k*]$-complete. This characterization will serve as a technical lemma for the results in Sect. 6.2.3.

Proposition 6.9. $\Sigma_2^p[k*]$-WSAT$^{\leq k}$ *is* $\Sigma_2^p[k*]$-complete.

Proof. Firstly, to show membership in $\Sigma_2^p[k*]$, we give an fpt-reduction from $\Sigma_2^p[k*]$-WSAT$^{\leq k}$ to $\Sigma_2^p[k*]$-WSAT. Let (φ, k) be an instance of $\Sigma_2^p[k*]$-WSAT, with $\varphi = \exists X. \forall Y. \psi$. We construct an instance (φ', k) of $\Sigma_2^p[k*]$-WSAT. Let $X' = \{x_1', \ldots, x_k'\}$. Now define $\varphi' = \exists X \cup X'. \forall Y. \psi$. We show that $(\varphi, k) \in \Sigma_2^p[k*]$-WSAT$^{\leq k}$ if and only if $(\varphi', k) \in \Sigma_2^p[k*]$-WSAT.

($\Rightarrow$) Assume that $(\varphi, k) \in \Sigma_2^p[k*]$-WSAT$^{\leq k}$. This means that there exists an assignment $\alpha : X \to \mathbb{B}$ of weight $\ell \leq k$ such that $\forall Y. \psi[\alpha]$ evaluates to true. Define the assignment $\alpha' : X' \to \mathbb{B}$ as follows. We let $\alpha'(x_i') = 1$ if and only if $i \in [k] - \ell$.

Then the assignment $\alpha \cup \alpha'$ has weight k, and $\forall Y.\psi[\alpha \cup \alpha']$ evaluates to true. Therefore, $(\varphi', k) \in \Sigma_2^p[k*]$-WSAT.

($\Leftarrow$) Assume that $(\varphi', k) \in \Sigma_2^p[k*]$-WSAT. This means that there exists an assignment $\alpha : X \cup X' \to \mathbb{B}$ of weight k such that $\forall Y.\psi[\alpha]$ evaluates to true. Now let α' be the restriction of α to the set X of variables. Clearly, α' has weight at most k. Also, since ψ contains no variables in X', we know that $\forall Y.\psi[\alpha']$ evaluates to true. Therefore, $(\varphi, k) \in \Sigma_2^p[k*]$-WSAT$^{\leq k}$.

Then, to show $\Sigma_2^p[k*]$-hardness, we give an fpt-reduction from $\Sigma_2^p[k*]$-WSAT to $\Sigma_2^p[k*]$-WSAT$^{\leq k}$. Let (φ, k) be an instance of $\Sigma_2^p[k*]$-WSAT, where $\varphi = \exists X.\forall Y.\psi$, and $X = \{x_1, \dots, x_n\}$. We construct an instance (φ', k') of $\Sigma_2^p[k*]$-WSAT$^{\leq k}$. Let $C = \{c_j^i : i \in [k], j \in [n]\}$ be a set of fresh propositional variables. Intuitively, we can think of the variables c_j^i as being placed in a matrix with k columns and n rows: variable c_j^i is positioned in the i-th column and in the j-th row. We ensure that in each column, exactly one variable is set to true (see ψ_{col} below), and that in each row, at most one variable is set to true (see ψ_{row} below). This way, any satisfying assignment must set exactly k variables in the matrix to true, in different rows. Next, we ensure that if any variable in the j-th row is set to true, that x_j is set to true (see ψ_{corr} below). This way, we know that exactly k variables x_j must be set to true in any satisfying assignment.

Formally, we define:

$$\varphi' = \exists X \cup C.\forall Y.\psi';$$

$$k' = 2k;$$

$$\psi' = \psi_{\mathrm{col}} \wedge \psi_{\mathrm{row}} \wedge \psi_{\mathrm{corr}} \wedge \psi;$$

$$\psi_{\mathrm{col}} = \bigwedge_{j \in [n]} \left(\bigvee_{i \in [k]} c_j^i \wedge \bigwedge_{i,i' \in [k], i < i'} (\neg c_j^i \vee \neg c_j^{i'}) \right);$$

$$\psi_{\mathrm{row}} = \bigwedge_{i \in [k]} \bigwedge_{j,j' \in [n], j < j'} (\neg c_j^i \vee \neg c_{j'}^i); \text{ and}$$

$$\psi_{\mathrm{corr}} = \bigwedge_{i \in [k]} \bigwedge_{j \in [n]} c_j^i \to x_j.$$

Any assignment $\alpha : X \cup C \to \mathbb{B}$ that satisfies $\psi_{\mathrm{col}} \wedge \psi_{\mathrm{row}}$ must set the variables $c_{j_1}^1, \dots, c_{j_k}^k$ to true, for some $j_1, \dots, j_k \in [n]$, $j_1 < \cdots < j_k$. Furthermore, if α satisfies ψ_{corr}, it must also set $x_{j_1}, \dots, x_{j_k}$ to true.

It is now easy to show that $(\varphi, k) \in \Sigma_2^p[k*]$-WSAT if and only if $(\varphi', k') \in \Sigma_2^p[k*]$-WSAT$^{\leq k}$. Let $\alpha : X \to \mathbb{B}$ be an assignment of weight k such that $\forall Y.\psi[\alpha]$ is true, where $\{x_i : i \in [n], \alpha(x_i) = 1\} = \{x_{j_1}, \dots, x_{j_k}\}$. Then consider the assignment $\gamma : C \to \mathbb{B}$ where $\gamma(c_j^i) = 1$ if and only if $j = j_i$. Then the assignment $\alpha \cup \gamma$ has weight k', and has the property that $\forall Y.\psi'[\alpha \cup \gamma]$ is true.

Conversely, let $\gamma : X \cup C \to \mathbb{B}$ be an assignment of weigth k' such that $\forall Y.\psi'[\gamma]$ is true. Then the restriction α of γ to the variables X has weight k, and has the property that $\forall Y.\psi[\alpha]$ is true.

Then, before we continue with characterizing the parameterized complexity class $\Sigma_2^p[k*]$ using alternating Turing machines, we make a brief digression. We consider another variant of the weighted satisfiability problem $\Sigma_2^p[k*]$-WSAT, where the truth assignments to the existentially quantified variables are not restricted to have weight at most k, but to have weight at least k. In contrast to the former restriction, the latter restriction results in a problem that is para-Σ_2^p-complete.

Formally, we consider the problem $\Sigma_2^p[k*]$-WSAT$^{\geq k}$, that is defined as follows.

$\Sigma_2^p[k*]$-WSAT$^{\geq k}$

Instance: A quantified Boolean formula $\phi = \exists X.\forall Y.\psi$, and an integer k.
Parameter: k.
Question: Does there exist an assignment α to X with weight at least k, such that for all truth assignments β to Y the assignment $\alpha \cup \beta$ satisfies ψ?

We show that the problem $\Sigma_2^p[k*]$-WSAT$^{\geq k}$ is para-Σ_2^p-complete.

Proposition 6.10. $\Sigma_2^p[k*]$-WSAT$^{\geq k}$ *is para-Σ_2^p-complete.*

Proof. To show para-Σ_2^p-membership, we reduce the problem to QSAT$_2$. Let (φ, k) be an instance of $\Sigma_2^p[k*]$-WSAT$^{\geq k}$, where $\varphi = \exists X.\forall Y.\psi$. We construct an instance φ' of QSAT$_2$ as follows.

We let Z be a set of fresh variables. Then, let χ be a propositional formula on the variables $Y \cup Z$ that is unsatisfiable if and only if less than k variables in Y are set to true. This formula χ is straightforward to construct, and we omit the details of the construction here. Then, we let $\varphi' = \exists X.\forall Y \cup Z.\psi'$, where $\psi' = \chi \to \psi$. We claim that $(\varphi, k) \in \Sigma_2^p[k*]$-WSAT$^{\geq k}$ if and only if $\varphi' \in$ QSAT$_2$.

($\Rightarrow$) Assume that $(\varphi, k) \in \Sigma_2^p[k*]$-WSAT$^{\geq k}$, i.e., that there exists a truth assignment $\alpha : X \to \mathbb{B}$ such that for all truth assignments $\beta : Y \to \mathbb{B}$ of weight at least k it holds that $\psi[\alpha \cup \beta]$ is true. We show that $\varphi' \in$ QSAT$_2$. We show that $\forall Y \cup Z.\psi'[\alpha]$ is true. Let $\gamma : Y \cup Z \to \mathbb{B}$ be an arbitrary truth assignment. We distinguish two cases: either (i) γ satisfies χ or (ii) this is not the case. In case (i), we know that χ is satisfied, and thus that γ sets at least k variables in Y to true. Therefore, we know that $\psi[\alpha \cup \gamma]$ is true, and thus that $\alpha \cup \gamma$ satisfies ψ'. In case (ii), $\alpha \cup \gamma$ satisfies ψ' because it does not satisfy χ. Then, since γ was arbitrary, we know that $\varphi' \in$ QSAT$_2$.

($\Leftarrow$) Conversely, assume that $\varphi' \in$ QSAT$_2$, i.e., that there exists an truth assignment $\alpha : X \to \mathbb{B}$ such that for all truth assignments $\gamma : Y \cup Z \to \mathbb{B}$ it holds that $\psi'[\alpha \cup \gamma]$ is true. We show that $(\varphi, k) \in \Sigma_2^p[k*]$-WSAT$^{\geq k}$. In particular, we show that for all truth assignments $\beta : Y \to \mathbb{B}$ of weight at least k it holds that $\psi[\alpha \cup \beta]$ is true. Let $\beta : Y \to \mathbb{B}$ be a truth assignment of weight at least k. Since, by construction, χ is satisfiable if and only if at least k variables in Y are set to true, we know that we can extend the assignment β to a truth assignment $\gamma : Y \cup Z \to \mathbb{B}$ that satisfies χ. Then, since $\psi'[\alpha \cup \gamma]$ is true, and since γ satisfies χ, we know that $\alpha \cup \gamma$ satisfies ψ.

Moreover, since ψ contains only variables in $X \cup Y$, and since γ coincides with β on the variables in Y, we know that $\psi[\alpha \cup \beta]$ is true. Since β was arbitrary, we can conclude that $(\varphi, k) \in \Sigma_2^p[k*]\text{-WSAT}^{\geq k}$.

Then, to show para-Σ_2^p-hardness, it suffices to show that the problem is already Σ_2^p-hard when the parameter value is restricted to 1 [84]. We give a polynomial-time reduction from QSAT_2 to the slice of $\Sigma_2^p[k*]\text{-WSAT}^{\geq k}$ where $k = 1$. Let $\varphi = \exists X.\forall Y.\psi$ be an instance of QSAT_2. We let $\varphi' = \exists X.\forall Y'.\psi$, where $Y' = Y \cup \{y_0\}$ for a fresh variable y_0. Moreover, we let $k = 1$. We claim that $\varphi \in \text{QSAT}_2$ if and only if $(\varphi', k) \in \Sigma_2^p[k*]\text{-WSAT}^{\geq k}$.

($\Rightarrow$) Assume that $\varphi \in \text{QSAT}_2$, i.e., that there exists a truth assignment $\alpha : X \to \mathbb{B}$ such that $\forall Y.\psi[\alpha]$ is true. We show that for all truth assignments $\beta : Y' \to \mathbb{B}$ of weight at least 1 it holds that $\psi[\alpha \cup \beta]$ is true. Let $\beta : Y' \to \mathbb{B}$ be an arbitrary such truth assignment. Clearly, since ψ contains only variables in Y and since $\forall Y.\psi[\alpha]$ is true, we know that $\psi[\alpha \cup \beta]$ is true. Thus, we can conclude that $(\varphi, k) \in \Sigma_2^p[k*]\text{-}$ $\text{WSAT}^{\geq k}$.

($\Leftarrow$) Conversely, assume that $(\varphi, k) \in \Sigma_2^p[k*]\text{-WSAT}^{\geq k}$, i.e., that there exists a truth assignment $\alpha : X \to \mathbb{B}$ such that for all truth assignments $\beta : Y' \to \mathbb{B}$ of weight at least 1 it holds that $\psi[\alpha \cup \beta]$ is true. We show that $\forall Y.\psi[\alpha]$ is true. Let $\beta : Y \to \mathbb{B}$ be an arbitrary truth assignment. Construct the truth assignment $\beta' : Y' \to \mathbb{B}$ as follows. On the variables in Y, β and β' coincide, and $\beta'(y') = 1$. Clearly, β' has weight at least 1, and thus $\psi[\alpha \cup \beta']$ is true. Since ψ does not contain the variable y_0, we then know that $\psi[\alpha \cup \beta]$ is true as well. Then, since β was arbitrary, we can conclude that $\varphi \in \text{QSAT}_2$.

Alternating Turing Machine Characterization

Finally, we give a characterization of the class $\Sigma_2^p[k*]$ by means of alternating Turing machines.

We consider two particular types of ATMs. An $\exists\forall$-*Turing machine* (or simply $\exists\forall$-machine) is a 2-alternating ATM $(S_\exists, S_\forall, \Sigma, \Delta, s_0, F)$, where $s_0 \in S_\exists$. Let $\ell, t \geq 1$ be positive integers. We say that an $\exists\forall$-machine $\mathbb{M}$ *halts (on the empty string) with existential cost ℓ and universal cost t* if:

- there is an accepting run of $\mathbb{M}$ with input ϵ,
- each computation path of $\mathbb{M}$ contains at most ℓ existential configurations and at most t universal configurations.

Let P be a parameterized problem. An $\Sigma_2^p[k*]$-*machine for P* is a $\exists\forall$-machine $\mathbb{M}$ such that there exists a computable function f and a polynomial p such that

- $\mathbb{M}$ decides P in time $f(k) \cdot p(|x|)$; and
- and for all instances (x, k) of P and each computation path R of $\mathbb{M}$ with input (x, k), at most $f(k) \cdot \log|x|$ of the existential configurations of R are nondeterministic.

We say that a parameterized problem P is *decided by some $\Sigma_2^p[k*]$-machine* if there exists a $\Sigma_2^p[k*]$-machine for P.

Let $m \in \mathbb{N}$ be a positive integer. We consider the following parameterized problem.

$\Sigma_2^p[k*]$-TM-HALT^m
Instance: An $\exists\forall$-machine $\mathbb{M}$ with m tapes, and positive integers $k, t \geq 1$.
Parameter: k.
Question: Does $\mathbb{M}$ halt on the empty string with existential cost k and universal cost t?

Moreover, we consider the following parameterized problem.

$\Sigma_2^p[k*]$-TM-HALT*
Instance: An $\exists\forall$-machine $\mathbb{M}$ with m tapes, and positive integers $k, t \geq 1$.
Parameter: k.
Question: Does $\mathbb{M}$ halt on the empty string with existential cost k and universal cost t?

Note that for $\Sigma_2^p[k*]$-TM-HALT^m, the number m of tapes of the $\exists\forall$-machines in the input is a fixed constant, whereas for $\Sigma_2^p[k*]$-TM-HALT*, the number of tapes is given as part of the input.

The parameterized complexity class $\Sigma_2^p[k*]$ can then be characterized by alternating Turing machines as follows. These results can be seen as an analogue to the Cook-Levin Theorem for the complexity class $\Sigma_2^p[k*]$.

Theorem 6.11. *The problem $\Sigma_2^p[k*]$-TM-HALT* is $\Sigma_2^p[k*]$-complete, and so is the problem $\Sigma_2^p[k*]$-TM-HALT^m for each $m \in \mathbb{N}^+$.*

Theorem 6.12. *$\Sigma_2^p[k*]$ is exactly the class of parameterized decision problems P that are decided by some $\Sigma_2^p[k*]$-machine.*

Proof (Proof of Theorems 6.11 and 6.12). In order to show these results, we will use the following statements. We show how the results follow from these statements. We then present the statements (with a detailed proof) as Propositions 6.15 and 6.17–6.19.

(i) Let A and B be parameterized problems. If B is decidable by some $\Sigma_2^p[k*]$-machine with m tapes, and if $A \leq_{fpt} B$, then A is decidable by some $\Sigma_2^p[k*]$-machine with m tapes (Proposition 6.15).

(ii) $\Sigma_2^p[k*]$-TM-HALT$^* \leq_{fpt} \Sigma_2^p[k*]$-MC (Proposition 6.17).

(iii) For any parameterized problem P that is decidable by some $\Sigma_2^p[k*]$-machine with m tapes, it holds that $P \leq_{fpt} \Sigma_2^p[k*]$-TM-HALT^{m+1} (Proposition 6.18).

(iv) There is an $\Sigma_2^p[k*]$-machine with a single tape that decides $\Sigma_2^p[k*]$-WSAT$^{\leq k}$ (Proposition 6.19).

In addition to these statements, we will need one result known from the literature (Corollary 6.14, which follows from Proposition 6.13).

(v) $\Sigma_2^p[k*]$-TM-HALT$^2 \leq_{fpt} \Sigma_2^p[k*]$-TM-HALT^1 (Corollary 6.14).

To see that these statements imply the desired results, observe the following.

Together, (iii) and (iv) imply that $\Sigma_2^p[k*]$-WSAT$^{\leq k} \leq_{\text{fpt}} \Sigma_2^p[k*]$-TM-HALT^2. Clearly, for all $m \geq 2$, $\Sigma_2^p[k*]$-TM-HALT$^2 \leq_{\text{fpt}} \Sigma_2^p[k*]$-TM-HALT^m. This gives us $\Sigma_2^p[k*]$-hardness of $\Sigma_2^p[k*]$-TM-HALT^m, for all $m \geq 2$. $\Sigma_2^p[k*]$-hardness of $\Sigma_2^p[k*]$-TM-HALT^1 follows from Corollary 6.14, which implies that there is an fpt-reduction from $\Sigma_2^p[k*]$-TM-HALT^2 to $\Sigma_2^p[k*]$-TM-HALT^1. This also implies that $\Sigma_2^p[k*]$-TM-HALT* is $\Sigma_2^p[k*]$-hard. Then, by (ii), and since $\Sigma_2^p[k*]$-MC is in $\Sigma_2^p[k*]$ by Theorem 6.6, we obtain $\Sigma_2^p[k*]$-completeness of $\Sigma_2^p[k*]$-TM-HALT* and $\Sigma_2^p[k*]$-TM-HALT^m, for each $m \geq 1$. This proves Theorem 6.11. It remains to prove Theorem 6.12.

By (ii) and (iii), and by transitivity of fpt-reductions, we have that any parameterized problem P that is decided by an $\Sigma_2^p[k*]$-machine is fpt-reducible to $\Sigma_2^p[k*]$-WSAT, and thus is in $\Sigma_2^p[k*]$. Conversely, let P be any parameterized problem in $\Sigma_2^p[k*]$. Then, by $\Sigma_2^p[k*]$-hardness of $\Sigma_2^p[k*]$-WSAT$^{\leq k}$, we know that $P \leq_{\text{fpt}} \Sigma_2^p[k*]$-WSAT$^{\leq k}$. By (i) and (iv), we know that P is decidable by some $\Sigma_2^p[k*]$-machine with a single tape. From this we conclude that $\Sigma_2^p[k*]$ is exactly the class of parameterized problems P decided by some $\Sigma_2^p[k*]$-machine.

Firstly, we state the result known from the literature that we used in the proof of Theorems 6.11 and 6.12.

Proposition 6.13 ([123, Theorems 8.9 and 8.10]). *Let $m \geq 1$ be a (fixed) positive integer. For each ATM $\mathbb{M}$ with m tapes, there exists an ATM $\mathbb{M}'$ with 1 tape such that:*

- *$\mathbb{M}$ and $\mathbb{M}'$ are equivalent, i.e., they accept the same language;*
- *$\mathbb{M}'$ simulates n steps of $\mathbb{M}$ using $O(n^2)$ steps; and*
- *$\mathbb{M}'$ simulates existential steps of $\mathbb{M}$ using existential steps, and simulates universal steps of $\mathbb{M}$ using universal steps.*

Corollary 6.14. $\Sigma_2^p[k*]$-TM-HALT$^2 \leq_{\text{fpt}} \Sigma_2^p[k*]$-TM-HALT^1.

Next, we give detailed proofs of the statements (i)–(iv) that were used in the proof of Theorems 6.11 and 6.12 (Propositions 6.15 and 6.17–6.19). We begin with proving the first statement.

Proposition 6.15. *Let A and B be parameterized problems, and let $m \in \mathbb{N}$ be a positive integer. If B is decided by some $\Sigma_2^p[k*]$-machine with m tapes and if $A \leq_{\text{fpt}} B$, then A is decided by some $\Sigma_2^p[k*]$-machine with m tapes.*

Proof. Let R be the fpt-reduction from A to B, and let M be an algorithm that decides B and that can be implemented by an $\Sigma_2^p[k*]$-machine with m tapes. Clearly, the composition of R and M is an algorithm that decides A—by Proposition 6.13, we may assume that the fpt-reduction R is implemented by a deterministic Turing machine with 1 tape. It is straightforward to verify that the composition of R and M can be implemented by an $\Sigma_2^p[k*]$-machine with m tapes.

In order to prove the second statement that we used in the proof of Theorems 6.11 and 6.12, we prove the following technical lemma.

Lemma 6.16. *Let* $\mathbb{M}$ *be an* $\exists\forall$*-machine with* m *tapes and let* $k, t \in \mathbb{N}$. *We can construct an* $\exists\forall$*-machine* $\mathbb{M}'$ *with* m *tapes (in time polynomial in* $|(\mathbb{M}, k, t)|$*) such that the following are equivalent:*

- *there is an accepting run* ρ *of* $\mathbb{M}'$ *with input* ϵ *and each computation path in* ρ *contains* exactly k *existential configurations and* exactly t *universal configurations*
- $\mathbb{M}$ *halts on* ϵ *with existential cost* k *and universal cost* t.

Proof. Let $\mathbb{M} = (S_\exists, S_\forall, \Sigma, \Delta, s_0, F)$ be an $\exists\forall$-machine with m tapes. Now construct $\mathbb{M} = (S'_\exists, S'_\forall, \Sigma, \Delta', s_0, F')$ as follows:

$$S'_\exists = \{\, s_i : s \in S_\exists, 1 \le i \le k + t \,\},$$
$$S'_\forall = \{\, s_i : s \in S_\forall, 1 \le i \le k + t \,\},$$
$$\Delta' = \{\, (s_i, \overline{a}, s'_{i+1}, \overline{a}', \overline{d}) : (s, \overline{a}, s', \overline{a}', \overline{d}) \in \Delta, 1 \le i \le k + t - 1 \,\} \cup$$
$$\{\, (s_i, \overline{a}, s_{i+1}, \overline{a}, \mathbf{S}^m) : s \in S_\exists \cup S_\forall, \overline{a} \in \Sigma^m \,\}, \text{ and}$$
$$F' = \{\, f_{k+t} : f \in F \,\}.$$

To see that $\mathbb{M}'$ satisfies the required properties, it suffices to see that for each (accepting) computation path $C_1 \to \ldots \to C_{k'+t'}$ of $\mathbb{M}$ with input ϵ that contains existential configurations $C_1, \ldots, C_{k'}$ and universal configurations $C_{k'+1}, \ldots, C_{k'+t'}$ for $1 \le k' \le k$ and $1 \le t' \le t$, it holds that

$$C_1^1 \to \ldots \to C_{k'}^{k'} \to C_{k'}^{k'+1} \to \ldots \to C_{k'}^k \to C_{k'+1}^{k+1} \to \ldots \to C_{k'+t'}^{k+t'}$$
$$\to C_{k'+t'}^{k+t'+1} \to \ldots \to C_{k'+t'}^{k+t}$$

is an (accepting) computation path of $\mathbb{M}'$ with input ϵ, where for each $1 \le i \le k + t$ and where for each $1 \le j \le k' + t'$ we let C_j^i be the configuration $(s_i, x_1, p_1, \ldots, x_m, p_m)$, where $C_j = (s, x_1, p_1, \ldots, x_m, p_m)$.

With this technical lemma in place, we can now prove the second statement that we used in the proof of Theorems 6.11 and 6.12.

Proposition 6.17. $\Sigma_2^p[k*]$-TM-HALT$^* \le_{fpt} \Sigma_2^p[k*]$-MC

Proof. Let $(\mathbb{M}, k, t)$ be an instance of $\Sigma_2^p[k*]$-TM-HALT*, where k and t are positive integers, and $\mathbb{M} = (S_\exists, S_\forall, \Sigma, \Delta, s_0, F)$ is an $\exists\forall$-machine with m tapes. We construct in fpt-time an instance $(\mathcal{A}, \varphi)$ of $\Sigma_2^p[k*]$-MC, such that $(\mathbb{M}, k, t) \in \Sigma_2^p[k*]$-TM-HALT* if and only if $(\mathcal{A}, \varphi) \in \Sigma_2^p[k*]$-MC. By Lemma 6.16, it suffices to construct $(\mathcal{A}, \varphi)$ in such a way that $(\mathcal{A}, \varphi) \in \Sigma_2^p[k*]$-MC if and only if there exists an accepting run ρ of $\mathbb{M}$ with input ϵ such that each computation path of ρ contains exactly k existential configurations and exactly t universal configurations.

We construct $\mathcal{A}$ to be a τ-structure with a domain A. We will define the vocabulary τ below. The domain A of $\mathcal{A}$ is defined as follows:

$$A = S_\exists \cup S_\forall \cup \Sigma \cup \{\$, \square\} \cup \{\mathbf{L}, \mathbf{R}, \mathbf{S}\} \cup \{0, \ldots, \max\{m, k + t - 1\}\} \cup T,$$

where T is the set of tuples $(a_1, \ldots, a_m) \in (\Sigma \cup \{\$, \square\})^m$ and of tuples $(d_1, \ldots, d_m)$ $\in \{\mathbf{L}, \mathbf{R}, \mathbf{S}\}^m$ occurring in transitions of Δ. Observe that $|A| = O(k + t + |\mathbb{M}|)$.

We now describe the relation symbols in τ and their interpretation in $\mathcal{A}$. The vocabulary τ contains the 5-ary relation symbol D (intended as "transition relation"), and the ternary relation symbol P (intended as "projection relation"), with the following interpretations:

$$D^{\mathcal{A}} = \Delta, \text{ and}$$
$$P^{\mathcal{A}} = \{ (j, \overline{b}, b_j) : 1 \leq j \leq m, \overline{b} \in T, \overline{b} = (b_1, \ldots, b_m) \}.$$

Moreover, τ contains the unary relation symbols R_{tape}, R_{cell}, R_{blank}, R_{end}, R_{symbol}, R_{init}, R_{acc}, R_{left}, R_{right}, R_{stay}, $R_{\exists}$, $R_{\forall}$, R_i for each $1 \leq i \leq k + t - 1$, and R_a for each $a \in \Sigma$, which are interpreted in $\mathcal{A}$ as follows:

$$R_{\text{tape}}^{\mathcal{A}} = \{1, \ldots, m\}, R_{\text{cell}}^{\mathcal{A}} = \{1, \ldots, k + t\}, R_{\text{blank}}^{\mathcal{A}} = \{\square\}, R_{\text{end}}^{\mathcal{A}} = \{\$\},$$

$$R_{\text{symbol}}^{\mathcal{A}} = \Sigma, R_{\text{init}}^{\mathcal{A}} = \{s_0\}, R_{\text{acc}}^{\mathcal{A}} = F, R_{\text{left}}^{\mathcal{A}} = \{\mathbf{L}\}, R_{\text{right}}^{\mathcal{A}} = \{\mathbf{R}\}, R_{\text{stay}}^{\mathcal{A}} = \{\mathbf{S}\},$$

$$R_{\exists}^{\mathcal{A}} = S_{\exists}, R_{\forall}^{\mathcal{A}} = S_{\forall}, R_i^{\mathcal{A}} = \{i\} \text{ for each } 1 \leq i \leq k + t - 1,$$

$$\text{and } R_a^{\mathcal{A}} = \{a\} \text{ for each } a \in \Sigma.$$

The formula φ that we will construct contains variables $z_{\square}$, $z_{\$}$, z_{init}, z_{left}, z_{right}, z_{stay}, $z_1, \ldots, z_{k+t}, z_{a_1}, \ldots, z_{a_{|\Sigma|}}$, where $\Sigma = \{a_1, \ldots, a_{|\Sigma|}\}$, that we will use to refer to elements of the singleton relations of $\mathcal{A}$. We define a formula $\psi_{\text{constants}}$ that is intended to provide a fixed interpretation of some variables that we can use to refer to the elements of the singleton relations of $\mathcal{A}$:

$$\psi_{\text{constants}} = R_{\text{end}}(z_{\$}) \wedge R_{\text{blank}}(z_{\square}) \wedge R_{\text{left}}(z_{\text{left}}) \wedge R_{\text{right}}(z_{\text{right}}) \wedge$$

$$R_{\text{stay}}(z_{\text{stay}}) \wedge \bigwedge_{0 \leq i \leq k + t - 1} R_i(z_i) \wedge \bigwedge_{a \in \Sigma} R_a(z_a).$$

The formula φ that we will construct aims to express that there exist k transitions (from existential states), such that for any sequence of $t - 1$ transitions (from universal states), the entire sequence of transitions results in an accepting state. It will contain variables $s_i, t_i, s_i', t_i', d_i$, for $1 \leq i \leq k + t - 1$.

The formula φ will also contain variables $p_{i,j}$ and $q_{i,j,\ell}$, for each $k + 1 \leq i \leq k + t$, each $1 \leq j \leq m$ and each $1 \leq \ell \leq k + t$. The variables $p_{i,j}$ will encode the position of the tape head for tape j at the i-th configuration in the computation path, and the variables $q_{i,j,\ell}$ will encode the symbol that is at cell ℓ of tape j at the i-th configuration in the computation path.

The position of the tape heads and the contents of the tapes for configurations 1 to k in the computation path, will not be encoded by means of variables, but by means of the formulas $\psi_{\text{symbol},i}$ and $\psi_{\text{position},i}$, which we define below. Intuitively, the reason for this is that the number of existentially quantified variables in the formula φ

has to be bounded by a function of k, and the total amount of information that we need to encode is not bounded by any function of k. The size of subformulas of φ does not need to be bounded by a function of k, so we can encode this information using formulas $\psi_{\text{symbol},i}$ and $\psi_{\text{position},i}$. However, the size of the formulas $\psi_{\text{symbol},i}$ and $\psi_{\text{position},i}$ grows exponentially in i. Therefore, we can only use this encoding for configurations up to k. This is the reason why we use variables $p_{i,j}$ and $q_{i,j,\ell}$ to encode the configurations $k + 1 \le i \le k + t$.

We define φ as follows:

$$
\begin{aligned}
\varphi = {} &\exists s_1, t_1, s_1', t_1', d_1, \ldots, s_k, t_k, s_k', t_k', d_k. \\
&\forall z_0, z_\square, z_\$, z_{\text{init}}, z_{\text{left}}, z_{\text{right}}, z_{\text{stay}}, z_1, \ldots, z_{k+t}, z_{a_1}, \ldots, z_{a_{|\Sigma|}}. \\
&\forall s_{k+1}, t_{k+1}, s_{k+1}', t_{k+1}', d_{k+1}, \ldots, s_{k+t-1}, t_{k+t-1}, s_{k+t-1}', t_{k+t-1}', d_{k+t-1}. \\
&\forall p_{k+1,1}, \ldots, p_{k+t,m}, q_{k+1,1,1}, \ldots, q_{k+t,m,k+t}.\psi,
\end{aligned}
$$

$$
\psi = \psi_{\text{constants}} \rightarrow \left(\psi_{\exists\text{-states}} \wedge \psi_{\exists\text{-tapes}} \wedge \left((\psi_{\forall\text{-states}} \wedge \psi_{\forall\text{-tapes}}) \rightarrow \psi_{\text{accept}} \right) \right),
$$

$$
\psi_{\exists\text{-states}} = (s_1 = z_{\text{init}}) \wedge \bigwedge_{1 \le i \le k} D(s_i, t_i, s_i', t_i', d_i) \wedge \bigwedge_{1 \le i \le k-1} \left((s_{i+1} = s_i') \wedge R_\exists(s_{i+1}) \right),
$$

$$
\psi_{\forall\text{-states}} = \bigwedge_{k+1 \le i \le k+t-1} D(s_i, t_i, s_i', t_i', d_i) \wedge \bigwedge_{k \le i \le k+t-2} \left((s_{i+1} = s_i') \wedge R_\forall(s_{i+1}) \right), \text{ and}
$$

$$
\psi_{\text{accept}} = R_{\text{acc}}(s_{k+t-1}'),
$$

where we define the formulas $\psi_{\exists\text{-tapes}}$ and $\psi_{\forall\text{-tapes}}$ below. In order to do so, for each $1 \le i \le k + 1$ we define the quantifier-free formulas

$$
\psi_{\text{symbol},i}(w, p, a, \overline{v}_i) \quad \text{and} \quad \psi_{\text{position},i}(w, p, \overline{v}_i),
$$

with $\overline{v}_i = s_1, t_1, s_1', t_1', d_1, \ldots, s_{i-1}, t_{i-1}, s_{i-1}', t_{i-1}', d_{i-1}$. Intuitively:

- $\psi_{\text{symbol},i}(w, p, a, \overline{v}_i)$ represents whether the p-th cell of the w-th tape contains the symbol a, whenever the sequence of transitions in $\overline{v}_i$ has been carried out starting with empty tapes; and
- $\psi_{\text{position},i}(w, p, \overline{v}_i)$ represents whether the head of the w-th tape is at position p, whenever the sequence of transitions in $\overline{v}_i$ has been carried out starting with empty tapes.

In particular, for $i = 1$, we define formulas $\psi_{\text{symbol},1}(w, p, a)$ and $\psi_{\text{position},1}(w, p)$, because $\overline{v}_1$ is the empty sequence.

We define $\psi_{\text{symbol},i}(w, p, a, \overline{v}_i)$ and $\psi_{\text{position},i}(w, p, \overline{v}_i)$ simultaneously by induction on i as follows:

$$\psi_{\text{symbol},1}(w, p, a) = R_{\text{tape}}(w) \wedge R_{\text{cell}}(p) \wedge$$
$$(p = z_0 \rightarrow a = z_\$) \wedge (p \neq z_0 \rightarrow a = z_\square),$$

$$\psi_{\text{position},1}(w, p) = R_{\text{tape}}(w) \wedge (p = z_1),$$

$$\psi_{\text{symbol},i+1}(w, p, a, \overline{v}_{i+1}) = R_{\text{tape}}(w) \wedge R_{\text{cell}}(p) \wedge$$
$$((\psi_{\text{position},i}(w, p, \overline{v}_i) \wedge P(w, t_i', a)) \vee$$
$$(\neg\psi_{\text{position},i}(w, p, \overline{v}_i) \wedge \psi_{\text{symbol},i}(w, p, a, \overline{v}_i))),$$

$$\psi_{\text{position},i+1}(w, p, \overline{v}_{i+1}) = R_{\text{tape}}(w) \wedge (\psi_{\text{left},i+1}(w, p, \overline{v}_{i+1}) \vee \psi_{\text{right},i+1}(w, p, \overline{v}_{i+1}) \vee$$
$$\psi_{\text{stay},i+1}(w, p, \overline{v}_{i+1})),$$

$$\psi_{\text{left},i+1}(w, p, \overline{v}_{i+1}) = P(w, d_i, z_{\text{left}}) \wedge \bigvee_{1 \leq j \leq i+1} (\psi_{\text{position},i}(w, z_j, \overline{v}_i) \wedge (p = z_{j-1})),$$

$$\psi_{\text{right},i+1}(w, p, \overline{v}_{i+1}) = P(w, d_i, z_{\text{right}}) \wedge \bigvee_{1 \leq j \leq i+1} (\psi_{\text{position},i}(w, z_j, \overline{v}_i) \wedge (p = z_{j+1})),$$

and

$$\psi_{\text{stay},i+1}(w, p, \overline{v}_{i+1}) = P(w, d_i, z_{\text{stay}}) \wedge \bigvee_{1 \leq j \leq i+1} (\psi_{\text{position},i}(w, z_j, \overline{v}_i) \wedge (p = z_j)).$$

Note that for each $1 \leq i \leq k$, the size of the formulas $\psi_{\text{symbol},i}(w, p, a, \overline{v}_i)$ and $\psi_{\text{position},i}(w, p, \overline{v}_i)$ only depends on k. We can now define $\psi_{\exists\text{-tapes}}$:

$$\psi_{\exists\text{-tapes}} = \forall w. \forall p. \forall a. \bigwedge_{1 \leq i \leq k} ((\psi_{\text{position},i}(w, p, \overline{v}_i) \wedge \psi_{\text{symbol},i}(w, p, a, \overline{v}_i)) \rightarrow P(w, t_i, a)).$$

Intuitively, the formulas $\psi_{\exists\text{-states}}$ and $\psi_{\exists\text{-tapes}}$ together represent whether the transitions specified by $s_i, t_i, s_i', t_i', d_i$, for $1 \leq i \leq k$, together constitute a valid (partial) computation path.

Next, we define the formula $\psi_{\forall\text{-tapes}}$:

$$\psi_{\forall\text{-tapes}} = \psi_{\forall\text{-tapes-1}} \wedge \psi_{\forall\text{-tapes-2}} \wedge \psi_{\forall\text{-tapes-3}} \wedge \psi_{\forall\text{-tapes-4}} \wedge \psi_{\forall\text{-tapes-5}},$$

$$\psi_{\forall\text{-tapes-1}} = \bigwedge_{\substack{k < i \le k+t \\ 1 \le j \le m}} \left(R_{\text{cell}}(p_{i,j}) \wedge \bigwedge_{1 \le \ell \le k+t} R_{\text{symbol}}(q_{i,j,\ell}) \right),$$

$$\psi_{\forall\text{-tapes-2}} = \bigwedge_{1 \le j \le m} \left(\begin{array}{l} \bigwedge_{\substack{1 \le \ell \le k+1 \\ a \in \Sigma}} ((q_{k+1,j,\ell} = z_a) \leftrightarrow \psi_{\text{symbol},k+1}(z_j, z_{k+1}, z_a, \overline{v}_{k+1})) \wedge \\ \bigwedge_{k+2 \le \ell \le k+t} (q_{k+1,j,\ell} = z_\square) \end{array} \right),$$

$$\psi_{\forall\text{-tapes-3}} = \bigwedge_{\substack{1 \le j \le m \\ 1 \le i \le k+1}} ((p_{k+1,j} = z_i) \leftrightarrow \psi_{\text{position},k+1}(z_j, z_i, \overline{v}_{k+1})),$$

$$\psi_{\forall\text{-tapes-4}} = \bigwedge_{\substack{1 \le j \le m \\ k < i < k+t \\ 1 \le \ell \le k+t}} \left(\begin{array}{l} (P(z_j, d_i, z_{\text{left}}) \wedge (p_{i,j} = z_\ell)) \to (p_{i+1,j} = z_{\ell-1}) \wedge \\ (P(z_j, d_i, z_{\text{right}}) \wedge (p_{i,j} = z_\ell)) \to (p_{i+1,j} = z_{\ell+1}) \wedge \\ (P(z_j, d_i, z_{\text{stay}}) \wedge (p_{i,j} = z_\ell)) \to (p_{i+1,j} = z_\ell) \wedge \end{array} \right), \quad \text{and}$$

$$\psi_{\forall\text{-tapes-5}} = \bigwedge_{\substack{1 \le j \le m \\ k < i < k+t \\ 1 \le \ell \le k+t \\ a \in \Sigma}} \left(\begin{array}{l} ((p_{i,j} \ne z_\ell) \to (q_{i+1,j,\ell} = q_{i,j,\ell})) \wedge \\ ((p_{i,j} = z_\ell) \wedge P(z_j, t_i, z_a) \to (q_{i,j,\ell} = z_a)) \wedge \\ ((p_{i,j} = z_\ell) \wedge P(z_j, t_i', z_a) \to (q_{i+1,j,\ell} = z_a)) \end{array} \right).$$

Intuitively, the formulas $\psi_{\forall\text{-states}}$ and $\psi_{\forall\text{-tapes}}$ together represent whether the transitions specified by $s_i, t_i, s_i', t_i', d_i$, for $k + 1 \le i \le k + t - 1$, together constitute a valid (partial) computation path, extending the computation path represented by the transitions $s_i, t_i, s_i', t_i', d_i$, for $1 \le i \le k$.

It is straightforward to verify that φ is (logically equivalent to a formula) of the right form, containing $k' = 5k$ existentially quantified variables. (Not all quantifiers in φ occur as outermost operators in the formula, but one can easily move them outwards.) Also, it is now straightforward to verify that $(\mathbb{M}, k, t) \in \Sigma_2^P[k*]$-TM-HALT* if and only if $(\mathcal{A}, \varphi) \in \Sigma_2^P[k*]$-MC.

We now turn our attention to the third statement that we used in the proof of Theorems 6.11 and 6.12.

Proposition 6.18. *For any parameterized problem P that is decided by some $\Sigma_2^P[k*]$-machine with m tapes, it holds that $P \le_{\text{fpt}} \Sigma_2^P[k*]$-TM-HALT^{m+1}.*

Proof. Let P be a parameterized problem, and let $\mathbb{M} = (S_\exists, S_\forall, \Sigma, \Delta, s_0, F)$ be an $\Sigma_2^P[k*]$-machine with m tapes that decides it, i.e., there exists some computable function f and some polynomial p such that for any instance (x, k) of P we have that any computation path of $\mathbb{M}$ with input (x, k) has length at most $f(k) \cdot p(|x|)$ and contains at most $f(k) \cdot \log |x|$ nondeterministic existential configurations. Moreover, let $S = S_\exists \cup S_\forall$. We show how to construct in fpt-time for each instance (x, k) of P an $\exists\forall$-machine $\mathbb{M}^{(x,k)}$ with $m + 1$ tapes, and positive integers $k', t \in \mathbb{N}$ such that $\mathbb{M}^{(x,k)}$ accepts the empty string with existential cost k' and universal cost t if and only if $\mathbb{M}$ accepts (x, k).

The idea of this construction is the following. We add to Σ a fresh symbol $\sigma_{(C_1,\ldots,C_u)}$ for each $u \leq \lceil \log |x| \rceil$ and each sequence of possible "transitions" $T_1, \ldots, T_u$ of $\mathbb{M}$. The machine $\mathbb{M}^{(x,k)}$ starts with nondeterministically writing down $f(k)$ symbols $\sigma_{(T_1,\ldots,T_{\lceil \log |x| \rceil})}$ to tape $m + 1$ (stage 1). We will choose k' in such a way (see below) so that this can be done using k' nondeterministic existential steps. Then, using universal steps, it writes down the input (x, k) to its first tape (stage 2). It continues with simulating the existential steps in the execution of $\mathbb{M}$ with input (x, k) (stage 3): each deterministic existential step can simply be performed by a deterministic universal step, and each nondeterministic existential step can be simulated by "reading off" the next configuration from the symbols on tape $m + 1$, and transitioning into this configuration (if this step is allowed by Δ). Finally, the machine $\mathbb{M}^{(x,k)}$ simply performs the universal steps in the execution of $\mathbb{M}$ with input (x, k) (stage 4).

Let (x, k) be an arbitrary instance of P. We construct the machine $\mathbb{M}^{(x,k)} = (S'_\exists, S'_\forall, \Sigma', \Delta', s'_0, F')$. We split the construction of $\mathbb{M}^{(x,k)}$ into several steps that correspond to the various stages in the execution of $\mathbb{M}^{(x,k)}$ described above. We begin with defining Σ':

$$\Sigma' = \Sigma \cup \{ \sigma_{(T_1,\ldots,T_u)} : 0 \leq u \leq \lceil \log |x| \rceil, \ 1 \leq i \leq u, \\ T_i \in S \times \Sigma^m \times \{\mathbf{L}, \mathbf{R}, \mathbf{S}\}^m \}.$$

Observe that for each $s \in S$ and each $\overline{a} \in \Sigma^m$, each $T_u = (s', \overline{a}', \overline{d})$ corresponds to a tuple $(s, \overline{a}, s', \overline{a}', \overline{d})$ that may or may not be contained in Δ, i.e., a "possible transition." Note that also $\sigma_{()} \in \Sigma'$, where $()$ denotes the empty sequence. Moreover, it is straightforward to verify that $|\Sigma'| = |\Sigma| + O(|x| \cdot |S|)$, since m is a constant.

We now construct the formal machinery that executes the first stage of the execution of $\mathbb{M}^{(x,k)}$. We let:

$$S_{1,\exists} = \{s_{1,\mathrm{guess}}, s_{1,\mathrm{done}}\}, \text{ and}$$
$$\Delta'_1 = \{ (s_{1,\mathrm{guess}}, \overline{a}, s_{1,\mathrm{guess}}, \overline{a}', \overline{d}) : \overline{a} = \square^{m+1}, \overline{a}' = \square^m \sigma_{(T_1,\ldots,T_{\lceil \log |x| \rceil})},$$
$$1 \leq i \leq \lceil \log |x| \rceil, T_i \in S \times \Sigma^m \times \{\mathbf{L}, \mathbf{R}, \mathbf{S}\}^m, \overline{d} = \mathbf{S}^m \mathbf{R} \} \cup$$
$$\{ (s_{1,\mathrm{guess}}, \overline{a}, s_{1,\mathrm{done}}, \overline{a}, \overline{d}) : \overline{a} = \square^{m+1}, \overline{d} = \mathbf{S}^m \mathbf{L} \} \cup$$
$$\{ (s_{1,\mathrm{done}}, \overline{a}, s_{1,\mathrm{done}}, \overline{a}, \overline{d}) : \overline{a} \in \{\square\}^m \times \Sigma', \overline{d} = \mathbf{S}^m \mathbf{L} \} \cup$$
$$\{ (s_{1,\mathrm{done}}, \overline{a}, s_{2,0}, \overline{a}, \overline{d}) : \overline{a} = \square^m \$, \overline{d} = \mathbf{S}^m \mathbf{R} \},$$

where we will define $s_{2,0} \in S'_\forall$ below ($s_{2,0}$ will be the first state of the second stage of $\mathbb{M}^{(x,k)}$). Furthermore, we let:

$$s'_0 = s_{1,\mathrm{guess}}.$$

The intuition behind the above construction is that state $s_{1,\text{guess}}$ can be used as many times as necessary to write a symbol $\sigma_{(T_1,\ldots,T_{\lceil\log|x|\rceil})}$ to the $(m+1)$-th tape, for some sequence $T_1,\ldots,T_{\lceil\log|x|\rceil}$ of "guessed transitions." Then, the state $s_{1,\text{done}}$ moves the tape head of tape $m+1$ back to the first position, in order to continue with the second stage of the execution of $\mathbb{M}^{(x,k)}$.

We continue with the definition of those parts of $\mathbb{M}^{(x,k)}$ that perform the second stage of the execution of $\mathbb{M}^{(x,k)}$, i.e., writing down the input (x,k) to the first tape. Let the sequence $(\sigma_1,\ldots,\sigma_n) \in \Sigma^n$ denote the representation of (x,k) using the alphabet Σ. We define:

$$S_{2,\forall} = \{s_{2,i} : 1 \le i \le n\} \cup \{s_{2,n+1} = s_{2,\text{done}}\}, \text{ and}$$
$$\Delta_2' = \{(s_{2,i}, \overline{a}, s_{2,i+1}, \overline{a}', \overline{d}) : 1 \le i \le n, \overline{a} \in \square^m \sigma, \sigma \in \Sigma', \overline{a} = \sigma_i \square^{m-1} \sigma, \overline{d} = \mathbf{RS}^m \}$$
$$\cup \{(s_{2,\text{done}}, \overline{a}, s_{2,\text{done}}, \overline{a}, \overline{d}) : \overline{a} \in \Sigma \times \{\square\}^{m-1} \times \Sigma', \overline{d} = \mathbf{LS}^m \}$$
$$\cup \{(s_{2,\text{done}}, \overline{a}, s_{3,0}, \overline{a}, \overline{d}) : \overline{a} = \{\$\} \times \{\square\}^{m-1} \times \Sigma', \overline{d} = \mathbf{RS}^m \},$$

where we will define $s_{3,0} \in S_\forall'$ below ($s_{3,0}$ will be the first state of the third stage of $\mathbb{M}^{(x,k)}$). Intuitively, each state $s_{2,i}$ writes the i-th symbol of the representation of (x,k) (that is, symbol σ_i) to the first tape, and state $s_{2,n+1} = s_{2,\text{done}}$ moves the tape head of the first tape back to the first position. Note that the states in $S_{2,\forall}$ are deterministic.

Next, we continue with the definition of those parts of $\mathbb{M}^{(x,k)}$ that perform the third stage of the execution of $\mathbb{M}^{(x,k)}$, i.e., simulating the existential steps in the execution of $\mathbb{M}$ with input (x,k). We define:

$$S_{3,\forall} = S_\exists, \text{ and}$$
$$\Delta_3' = \bigcup \{\Delta_{3,s}' : s \in S_\exists\},$$

where for each $s \in S_\exists$ we define the set $\Delta_{3,s}'$ as follows:

$$\Delta_{3,s}' = \bigcup \{\Delta_{3,s,\overline{a}}' : \overline{a} \in \Sigma^m\},$$

and where for each $s \in S_\exists$ and each $\overline{a} \in \Sigma^m$ we define:

$$\Delta_{(s,\overline{a})} = \{(s', \overline{a}', \overline{d}) : (s, \overline{a}, s', \overline{a}', \overline{d}) \in \Delta\},$$

$$\Delta_{3,s,\overline{a}}' = \begin{cases} \{(s, \overline{a}\sigma', s', \overline{a}'\sigma', \overline{d}\mathbf{S}) : \sigma' \in \Sigma'\} & \text{if } \Delta_{(s,\overline{a})} = \{(s', \overline{a}', \overline{d})\}, \\[6pt] \{(s, \overline{a}\sigma_{(T_1,\ldots,T_u)}, s', \overline{a}'\sigma_{(T_2,\ldots,T_u)}, \overline{d}\mathbf{S}) : \\ \quad 1 \le u \le \lceil\log|x|\rceil, \\ \quad T_1 = (s', \overline{a}', \overline{d}), (s, \overline{a}, s', \overline{a}', \overline{d}) \in \Delta\} \cup & \text{if } |\Delta_{(s,\overline{a})}| > 1, \\ \{(s, \overline{a}\sigma_{()}, s, \overline{a}\square, \mathbf{S}^m\mathbf{R})\} \\[6pt] \emptyset & \text{otherwise.} \end{cases}$$

Observe that there exist transitions from states in $S_{3,\forall}$ to states in $S_\forall$; this will be unproblematic, since we will have that $S_\forall \subseteq S_\forall'$ (see below). Intuitively, each state in $S_\exists$ that is deterministic in $\mathbb{M}$ simply performs its behavior from $\mathbb{M}$ on the first m tapes, and ignores tape $m+1$. Each state in $S_\exists$ that would lead to nondeterministic behavior in $\mathbb{M}$, performs the transition T_1 that is written as first "possible transition" in the currently read symbol $\sigma_{(T_1,\ldots,T_u)}$ on tape $m+1$ (if this transition is allowed by Δ), and removes T_1 from tape $m+1$ (by replacing $\sigma_{(T_1,\ldots,T_u)}$ by $\sigma_{(T_2,\ldots,T_u)}$). Note that the states in $S_{3,\forall}$ are deterministic.

We continue with formally defining the part of $\mathbb{M}^{(x,k)}$ that performs stage 4, i.e., performing the (possibly nondeterministic) universal steps in the execution of $\mathbb{M}$ with input (x, k). We define:

$$S_{4,\forall} = S_\forall, \text{ and}$$
$$\Delta_4' = \{ (s, \overline{a}\delta', s', \overline{a}'\delta', \overline{d}\mathbf{S}) : s \in S_\forall, \overline{a} \in \Sigma^m, (s, \overline{a}, s', \overline{a}', \overline{d}) \in \Delta \}.$$

Intuitively, each state in $S_\forall$ simply performs its behavior from $\mathbb{M}$ on the first m tapes, and ignores tape $m+1$. Note that the states in $S_{4,\forall}$ may be nondeterministic.

We conclude our definition of $\mathbb{M}^{(x,k)} = (S_\exists', S_\forall', \Sigma', \Delta', s_0', F')$:

$$S_\exists' = S_{1,\exists},$$
$$S_\forall' = S_{2,\forall} \cup S_{3,\forall} \cup S_{4,\forall},$$
$$\Delta' = \Delta_1' \cup \Delta_2' \cup \Delta_3' \cup \Delta_4',$$
$$s_0' = s_{1,\text{guess}} \text{ (as mentioned above), and}$$
$$F' = F.$$

Finally, we define k' and t:

$$k' = 2f(k) + 2 \qquad \text{and} \qquad t = 2|(x, k)| + f(k) \cdot (p(|x|) + 1) + 2.$$

Intuitively, $\mathbb{M}'$ needs $k' = 2f(k) + 2$ existential steps to write down $f(k)$ symbols $\sigma_{(T_1,\ldots,T_{\lceil \log|x| \rceil})}$ and return the tape head of tape $m+1$ to the first position. It needs $2|(x, k)| + 2$ steps to write the input (x, k) to the first tape and return the tape head of tape 1 to the first position. It needs at most $f(k) \cdot p(|x|) + f(k)$ steps to simulate the existential steps in the execution of $\mathbb{M}$ with input (x, k), and to perform the universal steps in the execution of $\mathbb{M}$ with input (x, k).

This concludes our construction of the instance $(\mathbb{M}^{(x,k)}, k', t)$ of $\Sigma_2^p[k*]$-TM-HALT^{m+1}. It is straightforward to verify that $(x, k) \in P$ if and only if $(\mathbb{M}^{(x,k)}, k', t) \in \Sigma_2^p[k*]$-TM-HALT^{m+1}, by showing that $\mathbb{M}$ accepts (x, k) if and only if $(\mathbb{M}^{(x,k)}, k', t) \in \Sigma_2^p[k*]$-TM-HALT^{m+1}.

Finally, we prove the fourth statement that we used in the proof of Theorems 6.11 and 6.12.

Proposition 6.19. *There is a* $\Sigma_2^p[k*]$-*machine with a single tape that decides* $\Sigma_2^p[k*]$-WSAT$^{\leq k}$.

Proof. We describe a $\Sigma_2^p[k*]$-machine $\mathbb{M}$ with 1 tape for $\Sigma_2^p[k*]$-WSAT$^{\leq k}$. We will not spell out the machine $\mathbb{M} = (S_\exists, S_\forall, \Sigma, \Delta, s_0, F)$ in full detail, but describe $\mathbb{M}$ in such detail that the working of $\mathbb{M}$ is clear and writing down the complete formal description of $\mathbb{M}$ can be done straightforwardly.

We assume that instances (φ, k) are encoded as strings $\sigma_1 \sigma_2 \ldots \sigma_n$ over an alphabet $\Sigma' \subseteq \Sigma$. We denote the representation of an instance (φ, k) using the alphabet Σ' by $\mathrm{Repr}(\varphi, k)$. Also, for any Boolean formula $\psi(Z)$ over variables Z and any (partial) assignment $\gamma : Z \to \{0, 1\}$, we let $\mathrm{Repr}(\psi, \gamma)$ denote the representation (using alphabet Σ) of the formula ψ, where each variable z in the domain of γ is replaced by the constant value $\gamma(z)$.

Let (φ, k) be an instance of $\Sigma_2^P[k*]\text{-WSAT}^{\leq k}$, where $\varphi = \exists X . \forall Y . \psi$, $X = \{x_1, \ldots, x_n\}$, and $Y = \{y_1, \ldots, y_m\}$. In the initial configuration of $\mathbb{M}$, the tape contains the word $\mathrm{Repr}(\varphi, k)$. We construct $\mathbb{M}$ in such a way that it proceeds in seven stages. Intuitively, in stage 1, $\mathbb{M}$ adds $\square\mathrm{Repr}(\psi, \emptyset)$ to the right of the tape contents—here $\emptyset$ denotes the empty assignment and ψ is the quantifier-free part of φ. We will refer to this word $\mathrm{Repr}(\psi, \emptyset)$ as the representation of ψ. In stage 2, it appends the word $(\square 1 \ldots 1)$, containing $\lceil \log n \rceil = u$ times the symbol 1, k times to the right of the tape contents. Next, in stage 3, $\mathbb{M}$ (nondeterministically) overwrites each such word $(\square 1 \ldots 1)$ by $(\square b_1, \ldots, b_u)$, for some bits $b_1, \ldots, b_u \in \{0, 1\}$. Then, in stage 4, it repeatedly reads some word $(\square b_1, \ldots, b_u)$ written at the rightmost part of the tape, and in the representation of ψ, written as "second word" on the tape, instantiates variable x_i to the value 1, where $b_1 \ldots b_u$ is the binary representation of i. After stage 4, at most k variables x_i are instantiated to 1. Then, in stage 5, $\mathbb{M}$ instantiates the remaining variables x_i in the representation of ψ to the value 0. These first five stages are all implemented using states in $S_\exists$. The remaining two stages are implemented using states in $S_\forall$. In stage 6, $\mathbb{M}$ nondeterministically instantiates each variable y_j in the representation of ψ to some truth value 0 or 1. Finally, in stage 7, the machine verifies whether the fully instantiated formula ψ evaluates to true or not, and accepts if and only if the formula ψ evaluates to true.

We now give a more detailed description of the seven stages of $\mathbb{M}$, by describing what each stage does to the tape contents, and by giving bounds on the number of steps that each stage needs. In the initial configuration, the tape contents w_0 are as follows (we omit trailing blank symbols):

$$w_0 = \$\mathrm{Repr}(\varphi, k).$$

In stage 1, $\mathbb{M}$ transforms the tape contents w_0 to the following contents w_1:

$$w_1 = \$\mathrm{Repr}(\varphi, k)\square\mathrm{Repr}(\psi, \emptyset),$$

where $\emptyset$ denotes the empty assignment to the variables $X \cup Y$. This addition to the tape contents can be done by means of $O(|\mathrm{Repr}(\varphi, k)|)$ deterministic existential steps.

Next, in stage 2, $\mathbb{M}$ adds to the tape contents k words of the form $(\square 1 \ldots 1)$, each containing $\lceil \log n \rceil$ times the symbol 1, resulting in the tape contents w_2 after stage 2:

$$w_2 = \$\mathrm{Repr}(\varphi, k)\square\mathrm{Repr}(\psi, \emptyset)\underbrace{\square\overbrace{1 \ldots 1}^{\lceil \log n \rceil}}_{\text{word 1}}\underbrace{\square\overbrace{1 \ldots 1}^{\lceil \log n \rceil}}_{\text{word 2}}\square \ldots \underbrace{\square\overbrace{1 \ldots 1}^{\lceil \log n \rceil}}_{\text{word } k}.$$

This addition to the tape contents can be done by means of $O(k \cdot |\text{Repr}(\varphi, k)|^2)$ deterministic existential steps.

Then, in stage 3, $\mathbb{M}$ proceeds nondeterministically. It replaces each word of the form $(\square 1 \dots 1)$ that were written to the tape in stage 2 by a word of the form $(\square b_1, \dots, b_u)$, for some bits $b_1, \dots, b_u \in \{0, 1\} \subseteq \Sigma$. Here we let $u = \lceil \log n \rceil$. Resultingly, the tape contents w_3 after stage 3 are:

$$w_3 = \$\text{Repr}(\varphi, k)\square\text{Repr}(\psi, \emptyset)\square b_1^1 \dots b_u^1 \square b_1^2 \dots b_u^2 \square \dots \square b_1^k \dots b_u^k,$$

where for each $1 \leq i \leq k$ and each $1 \leq j \leq u$, $b_j^i \in \{0, 1\}$. This transformation of the tape contents can be done by means of $O(k \cdot \lceil \log n \rceil)$ nondeterministic existential steps.

In stage 4, $\mathbb{M}$ repeatedly performs the following transformation of the tape contents, until all words $\square b_1^i \dots b_u^i$ are removed. The tape contents w_3' before each such transformation are as follows:

$$w_3' = \$\text{Repr}(\varphi, k)\square\text{Repr}(\psi, \alpha)\square b_1^1 \dots b_u^1 \square b_1^2 \dots b_u^2 \square \dots \square b_1^\ell \dots b_u^\ell,$$

for some partial assignment $\alpha : X \to \{0, 1\}$, and some $1 \leq \ell \leq k$. Each such transformation functions in such a way that the tape contents w_3'' afterwards are:

$$w_3'' = \$\text{Repr}(\varphi, k)\square\text{Repr}(\psi, \alpha')\square b_1^1 \dots b_u^1 \square b_1^2 \dots b_u^2 \square \dots \square b_1^{\ell-1} \dots b_u^{\ell-1},$$

where the bit string $b_1^\ell \dots b_u^\ell$ is the binary representation of the integer $i \leq 2^u$, and where the assignment α' is defined for all $1 \leq j \leq n$, by:

$$\alpha'(x_j) = \begin{cases} \alpha(x_j) & \text{if } \alpha(x_j) \text{ is defined,} \\ 1 & \text{if } \alpha(x_j) \text{ is undefined and } j = i, \\ \text{undefined} & \text{otherwise.} \end{cases}$$

Each such transformation can be implemented by means of $O(|\text{Repr}(\psi, \alpha)|^2 \cdot k\lceil \log n \rceil)$ deterministic existential steps. After all the k transformation of stage 4 are performed, the tape contents w_4 are thus as follows:

$$w_4 = \$\text{Repr}(\varphi, k)\square\text{Repr}(\psi, \alpha_{\text{pos}}),$$

where $\alpha_{\text{pos}} : X \to \{0, 1\}$ is the partial assignment such that $\text{Dom}(\alpha_{\text{pos}}) = \{i_1, \dots, i_k\}$, and $\alpha_{\text{pos}}(x_{i_j}) = 1$ for each $1 \leq j \leq k$, where for each $1 \leq j \leq k$, the integer i_j is such that $b_1^j \dots b_u^j$ is the binary representation of i_j. The operations in stage 4 can be implemented by means of $O(|\text{Repr}(\psi, \emptyset)|^2 \cdot k^2 \lceil \log n \rceil)$ deterministic existential steps.

Next, in stage 5, the machine $\mathbb{M}$ transforms the tape contents by modifying the word $\text{Repr}(\psi, \alpha_{\text{pos}})$, resulting in w_5:

$$w_5 = \$\mathrm{Repr}(\varphi, k)\square\mathrm{Repr}(\psi, \alpha),$$

where the complete assignment $\alpha' : X \to \{0, 1\}$ is defined as follows:

$$\alpha(x) = \begin{cases} \alpha_{\mathrm{pos}}(x) & \text{if } x \in \mathrm{Dom}(\alpha_{\mathrm{pos}}), \\ 0 & \text{otherwise.} \end{cases}$$

This can be done using $O(|\mathrm{Repr}(\psi, \alpha_{\mathrm{pos}})|)$ nondeterministic existential steps. Note that the assignment α has weight at most k.

Now, in stage 6, the machine $\mathbb{M}$ alternates to universal steps. It nondeterministically transforms the tape contents using $O(|\mathrm{Repr}(\psi, \alpha)|)$ nondeterministic universal steps, resulting in the tape contents w_6:

$$w_6 = \$\mathrm{Repr}(\varphi, k)\square\mathrm{Repr}(\psi, \alpha \cup \beta),$$

for some complete assignment $\beta : Y \to \{0, 1\}$.

Finally, in stage 7, $\mathbb{M}$ checks whether the assignment $\alpha \cup \beta$ satisfies the formula ψ. This check can be done by means of $O(|\mathrm{Repr}(\psi, \alpha \cup \beta)|)$ deterministic universal steps. The machine $\mathbb{M}$ accepts if and only if $\alpha \cup \beta$ satisfies ψ.

It is straightforward to verify that there exists a computable function f and a polynomial p such that each computation path of $\mathbb{M}$ with input (φ, k) has length at most $f(k) \cdot p(|\varphi|)$ and contains at most $f(k) \cdot \log |\varphi|$ nondeterministic existential configurations. Also, it is straightforward to verify that $\mathbb{M}$ accepts an input (φ, k) if and only if $(\varphi, k) \in \Sigma_2^p[k*]\text{-WSAT}^{\leq k}$. This concludes our proof that the $\Sigma_2^p[k*]$-machine $\mathbb{M}$ decides $\Sigma_2^p[k*]\text{-WSAT}^{\leq k}$.

This concludes our detailed treatment of the proof of Theorems 6.11 and 6.12.

6.3 The $\Sigma_2^p[*k, t]$ Hierarchy

We now turn our attention to the $*$-k hierarchy. Unlike in the k-$*$ hierarchy, in the canonical quantified Boolean satisfiability problems of the $*$-k hierarchy, we cannot add auxiliary variables to the second quantifier block whose truth assignment is not restricted. Therefore, the proof technique used to show Theorem 6.2 cannot be used to show a collapse of the $*$-k hierarchy. Due to the similarity to the W-hierarchy, it is plausible that the classes of the $*$-k hierarchy are distinct.

In this section, we give normalization results for the classes $\Sigma_2^p[*k, 1]$ and $\Sigma_2^p[*k, P]$. We also give a characterization of the class $\Sigma_2^p[*k, P]$ in terms of alternating Turing machines. In addition, we show completeness results for the class $\Sigma_2^p[*k, P]$ for two parameterized variants of our running example: ASP-CONSISTENCY(#disj.rules) and ASP-CONSISTENCY(#non-dual-normal.rules).

6.3.1   A Normalization Result for $\Sigma_2^p[*k, 1]$

We begin with showing normalization results for (the canonical problem of) the parameterized complexity class $\Sigma_2^p[*k, 1]$. In particular, we show that the problem $\Sigma_2^p[k*]$-WSAT is already $\Sigma_2^p[*k, 1]$-hard when the input circuits are restricted to formulas in c-DNF, for any constant $c \geq 2$. For the sake of convenience, we switch our perspective to the co-problem $\Pi_2^p[*k]$-WSAT when stating and proving the following results. Because we can make heavy use of the original normalization proof for the class W[1] by Downey and Fellows [65–67] to prove this normalization result, we provide only proof sketches.

Lemma 6.20. *For any $u \geq 1$, it holds that* $\Pi_2^p[*k]$-WSAT($\mathrm{CIRC}_{1,u}$) $\leq_{\mathrm{fpt}} \Pi_2^p[*k]$-WSAT($s$-CNF)*, where* $s = 2^u + 1$.

Proof (sketch). The reduction is completely analogous to the reduction used in the proof of Downey and Fellows [65, Lemma 2.1], where the presence of universally quantified variables is handled in four steps. In Steps 1 and 2, in which only the form of the circuit is modified, no changes are needed. In Step 3, universally quantified variables can be handled exactly as existentially quantified variables. Step 4 can be performed with only a slight modification, the only difference being that universally quantified variables appearing in the input circuit will also appear in the resulting clauses that verify whether a given product-of-sums or sum-of-products is satisfied. It is straightforward to verify that this reduction with the mentioned modifications works for our purposes.

Theorem 6.21. $\Pi_2^p[*k]$-WSAT(2CNF) *is* $\Pi_2^p[*k, 1]$-*complete.*

Proof (sketch). Clearly $\Pi_2^p[*k]$-WSAT(2CNF) is in $\Pi_2^p[*k, 1]$, since a 2CNF formula can be considered as a constant-depth circuit of weft 1. To show that $\Pi_2^p[*k]$-WSAT(2CNF) is $\Pi_2^p[*k, 1]$-hard, we give an fpt-reduction from $\Pi_2^p[*k]$-WSAT ($\mathrm{CIRC}_{1,u}$) to $\Pi_2^p[*k]$-WSAT(2CNF), for arbitrary $u \geq 1$. By Lemma 6.20, we know that we can reduce $\Pi_2^p[*k]$-WSAT($\mathrm{CIRC}_{1,u}$) to $\Pi_2^p[*k]$-WSAT(s-CNF), for $s = 2^u + 1$. We continue the reduction in multiple steps. In each step, we let C denote the circuit resulting from the previous step, and we let Y denote the universally quantified and X the existentially quantified variables of C, and we let k denote the parameter value. We describe the last two steps only briefly, since these are completely analogous to constructions in the work of Downey and Fellows [67].

Step 1: contracting the universally quantified variables.
This step transforms C into a CNF formula C' such that each clause contains at most one variable in Y and such that (C, k) is a yes-instance if and only if (C', k) is a yes-instance. We introduce new universally quantified variables Y' containing a variable y'_A for each set A of literals over Y of size at least 1 and at most s. Now, it is straightforward to construct a set D of polynomially many ternary clauses over Y and Y' such that the following property holds. An assignment α to $Y \cup Y'$ satisfies D if and only if for each subset $A = \{l_1, \dots, l_b\}$ of literals over Y (of size at least 1 and

at most s) it holds that $\alpha(y_A') = 1$ if and only if $\alpha(l_j) = 1$ for some $j \in \{1, \ldots, b\}$. Note that we do not directly add the set D of clauses to the formula C'.

We introduce $k - 1$ new existentially quantified variables $x_1^*, \ldots, x_{k-1}^*$. We add binary clauses to C' that enforce that the variables $x_1^*, \ldots, x_{k-1}^*$ all get the same truth assignment. Also, we add binary clauses to C' that enforce that each $x \in X$ is set to false if x_1^* is set to true.

We introduce $|D|$ existentially quantified variables, including a variable x_d' for each clause $d \in D$. Let X' denote the new existentially quantified variables that we introduced, i.e., $X' = \{x_1^*, \ldots, x_{k-1}^*\} \cup \{x_d' : d \in D\}$. Then, for each $d \in D$, we add the following clauses to C'. Let $d = (l_1, l_2, l_3)$, where each l_i is a literal over $Y \cup Y'$. We add the clauses $(\neg x_d' \vee \neg l_1)$, $(\neg x_d' \vee \neg l_2)$ and $(\neg x_d' \vee \neg l_3)$, enforcing that the clause d cannot be satisfied if x_d' is set to true.

We then modify the clauses of C as follows. Let $c = (l_1^x \vee \cdots \vee l_{s_1}^x \vee l_1^y \vee \cdots \vee l_{s_2}^y)$ be a clause of C, where $l_1^x, \ldots, l_{s_1}^x$ are literals over X, and $l_1^y, \ldots, l_{s_2}^y$ are literals over Y. We replace c by the clause $(l_1^x \vee \cdots \vee l_{s_1}^x \vee x_1^* \vee y_B')$, where $B = \{l_1^y, \ldots, l_{s_2}^y\}$. Clauses c of C that contain no literals over the variables Y remain unchanged.

The idea of this reduction is the following. If x_1^* is set to true, then exactly one of the variables x_d' must be set to true, which can only result in an satisfying assignment if the clause $d \in D$ is not satisfied. Therefore, if an assignment α to the variables $Y \cup Y'$ does not satisfy D, there is a satisfying assignment of weight k that sets both x_1^* and x_d' to true, for some $d \in D$ that is not satisfied by α. Otherwise, we know that the value α assigns to variables y_A' corresponds to the value α assigns to $\bigvee_{a \in A} a$, for each $A \subseteq \text{Lit}(Y)$ with $1 \leq |A| \leq s$. Then any satisfying assignment of weight k for C is also a satisfying assignment of weight k for C'.

We formally prove that (C, k) is a yes-instance if and only if (C', k) is a yes-instance.

($\Rightarrow$) Suppose that for each $\alpha : Y \to \{0, 1\}$ there is a truth assignment $\beta : X \to \{0, 1\}$ of weight k such that $C[\alpha \cup \beta]$ is true. We show that (C', k) is a yes-instance. Take an arbitrary truth assignment $\alpha' : Y \cup Y' \to \{0, 1\}$. We distinguish two cases: either (i) α' does not satisfy D or (ii) α' satisfies D. In case (i), we know that there is some clause $d \in D$ that is not satisfied by α'. Consider the truth assignment $\beta_d' : X \to \{0, 1\}$ that sets the variables $x_1^*, \ldots, x_{k-1}^*$ and x_d' to 1, and all remaining variables to 0. It is straightforward to verify that β_d' has weight k and that $C[\alpha' \cup \beta_d']$ is true. For case (ii), we consider the restriction α of α' to the variables Y. We know that there is some $\beta : X \to \{0, 1\}$ of weight k such that $C[\alpha \cup \beta]$ is true. We construct the truth assignment $\beta' : X \cup X' \to \{0, 1\}$ by letting $\beta(x) = \beta'(x)$ for all $x \in X$, and letting $\beta(x') = 0$ for all $x' \in X'$. Clearly, β has weight k. Moreover, by construction of C', since $C[\alpha \cup \beta]$ is true and because α satisfies D, we know that $C'[\alpha' \cup \beta']$ is true.

($\Leftarrow$) Conversely, suppose that for each $\alpha' : Y \cup Y' \to \{0, 1\}$ there is a truth assignment $\beta' : X \cup X' \to \{0, 1\}$ of weight k such that $C'[\alpha' \cup \beta']$ is true. We show that (C, k) is a yes-instance. Take an arbitrary truth assignment $\alpha : Y \to \{0, 1\}$. Consider a truth assignment $\alpha' : Y \cup Y' \to \{0, 1\}$ that extends α and that satisfies D. We know that there exists some truth assignment $\beta' : X \cup X' \to \{0, 1\}$ of weight k such that $C'[\alpha' \cup \beta']$ is true. Moreover, by construction of C', since α' satisfies D, we know that $\beta'(x'_d) = 0$ for all $d \in D$. Then, we also know that $\beta'(x_1^\star) = 0$. If this were not the case, β' would be of weight $k - 1$, which contradicts our assumption that β' has weight k. We then consider the restriction β of β' to the variables in X. We know that β has weight k. Moreover, by construction of C', since $C'[\alpha' \cup \beta']$ is true, and because α' satisfies D, we know that $C[\alpha \cup \beta]$ is true.

Step 2: making C antimonotone in X.

This step transforms C into a circuit C' that has only negative occurrences of existentially quantified variables, and transforms k into k' depending only on k, such that (C, k) is a yes-instance if and only if (C', k') is a yes-instance. The reduction is completely analogous to the reduction in the proof of Downey and Fellows [67, Theorem 10.6].

Step 3: contracting the existentially quantified variables.

This step transforms C into a circuit C' in CNF that contains only clauses with two variables in X and no variables in Y and clauses with one variable in X and one variable in Y, and transforms k into k' depending only on k, such that (C, k) is a yes-instance if and only if (C', k') is a yes-instance. The reduction is completely analogous to the reduction in the proof of Downey and Fellows [67, Theorem 10.7].

Corollary 6.22. *For any fixed integer $r \geq 2$, the problem $\Sigma_2^p[*k]$-WSAT(r-DNF) is $\Sigma_2^p[*k, 1]$-complete.*

6.3.2   A Normalization Result for $\Sigma_2^p[*k, P]$

Next, we provide a normalization result for $\Sigma_2^p[*k, P]$. In order to do so, we will need some definitions. Let C be a quantified Boolean circuit over two disjoint sets X and Y of variables that is in negation normal form. We say that C is *monotone in the variables in Y* if the only negation nodes that occur in the circuit C' have variables in X as inputs, i.e., the variables in Y can appear only positively in the circuit. Then, the following restriction of $\Sigma_2^p[*k]$-WSAT is already $\Sigma_2^p[*k, P]$-hard.

Proposition 6.23. *The problem $\Sigma_2^p[k*]$-WSAT is $\Sigma_2^p[*k, P]$-hard, even when restricted to quantified circuits that are in negation normal form and that are monotone in the universal variables.*

Proof. We give an fpt-reduction from the problem $\Sigma_2^p[k*]$-WSAT to the problem $\Sigma_2^p[k*]$-WSAT restricted to circuits that are monotone in the universal variables. Let (C, k) be an instance of $\Sigma_2^p[k*]$-WSAT, where C is a quantified Boolean circuit over the set X of existential variables and the set Y of universal variables, where $X = \{x_1, \ldots, x_n\}$ and where $Y = \{y_1, \ldots, y_m\}$. We construct an equivalent instance (C', k) of $\Sigma_2^p[k*]$-WSAT where C' is a quantified Boolean circuit over the set X of existential variables and the set Y' of universal variables, and where the circuit C' is monotone in Y'. We may assume without loss of generality that C is in negation normal form. If this is not the case, we can simply transform C into an equivalent circuit that has this property using the De Morgan rule. The form of the circuit C is depicted in Fig. 6.1.

This construction bears some resemblance to the construction used in a proof by Flum and Grohe [85, Theorem 3.14]. The plan is to replace the variables in Y by k copies of them, grouped in sets $Y^1, \ldots, Y^k$ of new variables. Each assignment of weight k to the new variables that sets a copy of a different variable to true in each set Y^i corresponds exactly to an assignment of weight k to the original variables in Y. Moreover, we will ensure that each assignment of weight k to the new variables that does not set a copy of a different variable to true in each set Y^i satisfies the newly constructed circuit. Using these new variables we can then construct internal nodes y_j and y'_j that, for each assignment to the new input nodes Y', evaluate to the truth value assigned to y_j and $\neg y_j$, respectively, by the corresponding truth assignment to the original input nodes Y.

We will describe this construction in more detail. The construction is also depicted in Fig. 6.2. We let $Y' = \{ y_j^i : i \in [k], j \in [m] \}$. We introduce a number of new internal nodes. For each $j \in [m]$, we introduce an internal node y_j, that is the disjunction of the input nodes y_j^i, for $i \in [k]$. That is, the internal node y_j is true if and only if y_j^i is true for some $i \in [k]$. Intuitively, this node y_j corresponds to the input node y_j in the original circuit C. Moreover, we introduce an internal node $y'_{j,i}$ for each $j \in [m]$ and each $i \in [k]$, that is the disjunction of $y_{j'}^i$, for each $j' \in [m]$ such that $j \neq j'$. That is, the node $y'_{j,i}$ is true if and only if $y_{j'}^i$ is true for some j' that is different from j. Then, we introduce the node y'_j, for each $j \in [m]$, that is the conjunction of the nodes $y'_{j,i}$ for $i \in [k]$. That is, the node y'_j is true if and only if for each $i \in [k]$ there is some $j' \neq j$ for which the input node y_j^i is true. Intuitively, this node y'_j corresponds to the negated input node $\neg y_j$ in the original circuit C.

Also, for each $i \in [k]$ and each $j, j' \in [m]$ with $j < j'$, we add an internal node $z_i^{j,j'}$ that is the conjunction of the input nodes y_j^i and $y_{j'}^i$. Then, for each $i \in [k]$ we add the internal node z_i that is the conjunction of all nodes $z_i^{j,j'}$, for $j, j' \in [m]$, $j < j'$. Intuitively, z_i is true if and only if at least two input nodes in the set Y_i are set to true. In addition, we add a subcircuit B that acts on the nodes $y'_1, \ldots, y'_m$, and that is satisfied if and only if at least $m - k + 1$ of the nodes y'_j are set to true. It is straightforward to construct such a circuit B in polynomial time. Then, we add the subcircuit C with input nodes $x_1, \ldots, x_n$, negated input nodes $\neg x_1, \ldots, \neg x_n$, where the input nodes $y_1, \ldots, y_m$ are identified with the internal nodes $y_1, \ldots, y_m$ in the newly constructed circuit C', and where the negated input nodes $\neg y_1, \ldots, \neg y_m$ are identified with the internal nodes $y'_1, \ldots, y'_m$ in the newly constructed circuit C'. Finally, we let the output node be the disjunction of the nodes $z_1, \ldots, z_k$ and the output nodes of the subcircuits C and B. Since C is a circuit in negation normal form, the

Fig. 6.1 The original quantified Boolean circuit C in the proof of Proposition 6.23.

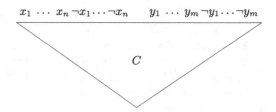

$$x_1 \cdots x_n \; \neg x_1 \cdots \neg x_n \qquad y_1 \cdots y_m \; \neg y_1 \cdots \neg y_m$$

circuit C' is monotone in Y'. We claim that for each assignment $\alpha : X \to \mathbb{B}$ it holds that the circuit $C[\alpha]$ is satisfied by all assignments of weight k if and only if $C'[\alpha]$ is satisfied by all assignments of weight k.

($\Rightarrow$) Let $\alpha : X \to \mathbb{B}$ be an arbitrary truth assignment. Assume that $C[\alpha]$ is satisfied by all truth assignments $\beta : Y \to \mathbb{B}$ of weight k. We show that $C'[\alpha]$ is satisfied by all truth assignments $\beta' : Y' \to \mathbb{B}$ of weight k. Let $\beta' : Y' \to \mathbb{B}$ be an arbitrary truth assignment of weight k. We distinguish several cases: either (i) for some $i \in [k]$ there are some $j, j' \in [m]$ with $j < j'$ such that $\beta'(y_j^i) = \beta'(y_{j'}^i) = 1$, or (ii) for each $i \in [k]$ there is exactly one ℓ_i such that $\beta'(y_{\ell_i}^i) = 1$ and for some $i, i' \in [k]$ with $i < i'$ it holds that $\ell_i = \ell_{i'}$, or (iii) for each $i \in [k]$ there is exactly one ℓ_i such that $\beta'(y_{\ell_i}^i) = 1$ and for each $i, i' \in [k]$ with $i < i'$ it holds that $\ell_i \neq \ell_{i'}$. In case (i), we know that the assignment β' sets the node $z_i^{j,j'}$ to true. Therefore, β' sets the node z_i to true, and thus satisfies the circuit $C'[\alpha]$. In case (ii), we know that β' sets y_j' to true for at least $m - k + 1$ different values of j. Therefore, β' satisfies the subcircuit B, and thus satisfies $C'[\alpha]$. Finally, in case (iii), we know that β' sets exactly k different internal nodes y_j to true, and for each $j \in [m]$ sets the internal node y_j' to true if and only if it sets y_j to false. Then, since $C[\alpha]$ is satisfied by all truth assignments of weight k, we know that β' satisfies the subcircuit C, and thus satisfies $C'[\alpha]$. Since β' was arbitrary, we can conclude that $C'[\alpha]$ is satisfied by all truth assignments $\beta' : Y' \to \mathbb{B}$ of weight k.

($\Leftarrow$) Let $\alpha : X \to \mathbb{B}$ be an arbitrary truth assignment. Assume that $C'[\alpha]$ is satisfied by all truth assignments $\beta' : Y' \to \mathbb{B}$ of weight k. We show that $C[\alpha]$ is satisfied by all truth assignments $\beta : Y \to \mathbb{B}$ of weight k. Let $\beta : Y \to \mathbb{B}$ be an arbitrary truth assignment of weight k. We now define the truth assignment $\beta' : Y' \to \mathbb{B}$ as follows. Let $\{y_{\ell_1}, \ldots, y_{\ell_k}\} = \{ y_j : j \in [m], \beta(y_j) = 1 \}$. For each $i \in [k]$ and each $j \in [m]$ we let $\beta'(y_j^i) = 1$ if and only if $j = \ell_i$. Clearly, β' has weight k. Moreover, the assignment β' sets the nodes $z_1, \ldots, z_k$ to false. Furthermore, it is the case that β' sets the internal node y_j in C' to true for exactly those $j \in [m]$ for which $\beta(y_j) = 1$, and it sets the internal node y_j' in C' to true for exactly those $j \in [m]$ for which $\beta(y_j) = 0$. Thus, β' sets (the output node of) the subcircuit B to false. Therefore, since β' satisfies the circuit $C'[\alpha]$, we can conclude that β' satisfies the subcircuit C, and thus that β satisfies $C[\alpha]$. Since β was arbitrary, we can conclude that $C[\alpha]$ is satisfied by all truth assignments $\beta : Y \to \mathbb{B}$ of weight k.

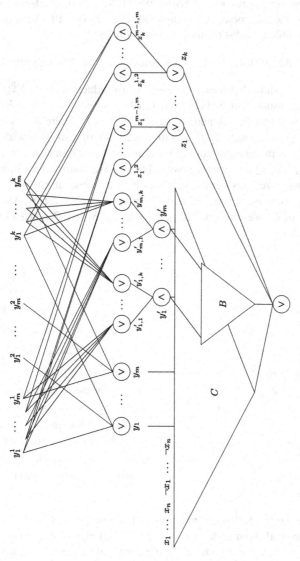

Fig. 6.2 The constructed quantified Boolean circuit C' in the proof of Proposition 6.23.

6.3.3 Answer Set Programming and $\Sigma_2^p[*k, P]$-Completeness

In this section, we show that two parameterized variants of our running example—
ASP-CONSISTENCY(#disj.rules) and ASP-CONSISTENCY(#non-dual-normal.rules)—
are complete for the parameterized complexity class $\Sigma_2^p[*k, P]$. We begin with the
parameterized problem ASP-CONSISTENCY(#disj.rules).

Theorem 6.24. ASP-CONSISTENCY(#disj.rules) *is* $\Sigma_2^p[*k, P]$-*complete.*

Proof. In order to show hardness, we give an fpt-reduction from $\Sigma_2^p[*k]$-WSAT.
Let (C, k) be an instance of $\Sigma_2^p[*k]$-WSAT, where C is a quantified Boolean circuit over existential variables X and universal variables Y, where $X = \{x_1, \dots, x_n\}$,
and where $Y = \{y_1, \dots, y_m\}$. By Proposition 6.23, we may assume without loss of
generality that C is in negation normal form and that it is monotone in Y, i.e., that
the variables $y_1, \dots, y_m$ occur only positively in C. We construct a disjunctive program P as follows. We consider the variables in X and Y as atoms. In addition, we
introduce fresh atoms $v_1, \dots, v_n, w$, and y_i^j for all $j \in [k]$, $i \in [m]$. Also, for each
internal node g of C, we introduce a fresh atom z_g. We let P consist of the rules
described as follows:

$$
\begin{array}{lrr}
x_i \leftarrow not\ v_i & \text{for } i \in [n]; & (6.13) \\
v_i \leftarrow not\ x_i & \text{for } i \in [n]; & (6.14) \\
y_1^j \vee \cdots \vee y_m^j \leftarrow & \text{for } j \in [k]; & (6.15) \\
y_i \leftarrow y_i^j & \text{for } i \in [m]; & (6.16) \\
y_i^j \leftarrow w & \text{for } i \in [m] \text{ and } j \in [k]; & (6.17) \\
w \leftarrow y_i^j, y_i^{j'} & \text{for } i \in [m],\ j, j' \in [k] \text{ and } j < j'; & (6.18) \\
z_g \leftarrow w & \text{for each internal node } g \text{ of } C; & (6.19) \\
z_g \leftarrow \sigma(g_1), \dots, \sigma(g_u) & \text{for each conjunction node } g \text{ of } C; & \\
& \text{with inputs } g_1, \dots, g_u & (6.20) \\
z_g \leftarrow \sigma(g_i) & \text{for each disjunction node } g \text{ of } C & \\
& \text{with inputs } g_1, \dots, g_u, \text{ and each } i \in [u]; & (6.21) \\
w \leftarrow z_o & \text{where } o \text{ is the output node of } C; & (6.22) \\
w \leftarrow not\ w. & & (6.23)
\end{array}
$$

Here, we define the following mapping σ from nodes of C to variables in V.
For each non-negated input node $x_i \in X$, we let $\sigma(x_i) = x_i$. For each negated input
node $\neg x_i$, for $x_i \in X$, we let $\sigma(\neg x_i) = v_i$. For each input node $y_j \in Y$, we let $\sigma(y) =
y_j$. For each internal node g, we let $\sigma(g) = z_g$. Intuitively, v_i corresponds to $\neg x_i$. One
of the main differences with the reduction of Eiter and Gottlob [71] is that we use the
rules in (6.15)–(6.18) to let the variables y_i represent an assignment of weight k to
the variables in Y. Another main difference is that we encode an arbitrary Boolean
circuit, rather than just a DNF formula, in the rules (6.19)–(6.22). Note that P has k

disjunctive rules, namely the rules (6.15). We show that $(C, k) \in \Sigma_2^p[*k]$-WSAT if and only if P has an answer set.

($\Rightarrow$) Assume there exists an assignment $\alpha : X \to \mathbb{B}$ such that for each assignment $\beta : Y \to \mathbb{B}$ of weight k it holds that $\alpha \cup \beta$ satisfies C. We show that $M = \{ x_i : \alpha(x_i) = 1, i \in [n] \} \cup \{ y_i^j, y_i : i \in [m], j \in [k] \} \cup \{ z_g : g$ an internal node of $C \} \cup \{w\}$ is an answer set of P. We have that P^M consists of Rules (6.15)–(6.22), the rules $(x_i \leftarrow)$ for all $i \in [n]$ such that $\alpha(x_i) = 1$, and the rules $(v_i \leftarrow)$ for all $i \in [n]$ such that $\alpha(x_i) = 0$. Clearly, M is a model of P^M. We show that M is a minimal model of P^M. Assume $M' \subsetneq M$ is a minimal model of P^M. If M' does not coincide with M on the atoms x_i and v_i, then M' is not a model of P^M. If $w \in M'$, then by Rules (6.16), (6.17) and (6.19), $M' = M$. Therefore, $w \notin M'$. By Rule (6.15), we have that $y_{i_1}^1, y_{i_2}^2, \ldots, y_{i_k}^k \in M'$, for some $i_1, \ldots, i_k \in [m]$ with $i_1, \ldots, i_k$. By Rule (6.18), we know that $i_1, \ldots, i_k$ are all different, since otherwise it would have to hold that $w \in M'$. By Rule (6.16), it holds that $y_{i_1}, \ldots, y_{i_k} \in M'$. By minimality of M', we know that $\{ i \in [m] : i \notin \{i_1, \ldots, i_k\} \} \cap M' = \emptyset$. Define the assignment $\gamma : X \cup Y \to \mathbb{B}$ by letting $\gamma(x_i) = 1$ if and only if $x_i \in M'$ and $\gamma(y_i) = 1$ if and only if $y_i \in M'$. Clearly, γ coincides with α on X, and γ assigns exactly k variables y_j to true. Now, since $(C, k) \in \Sigma_2^p[*k]$-WSAT, we know that γ satisfies C. Using the Rules (6.20) and (6.21), we can show by an inductive argument that for each internal node g of C that is set to true by γ it must hold that $z_g \in M'$. Since γ satisfies C, it sets the output node o of C to true, and thus by Rule (6.22), we know that $w \in M'$. This is a contradiction. From this we can conclude that no model $M' \subsetneq M$ of P^M exists, and thus M is an answer set of P.

($\Leftarrow$) Assume P has an answer set M. We know that $w \in M$, since otherwise $(w \leftarrow)$ would be a rule of P^M, and then M would not be a model of P^M. Then, by Rules (6.16) and (6.17), also $y_i, y_i^j \in M$ for all $i \in [m]$ and $j \in [k]$. We show that for each $i \in [n]$ it holds that $|M \cap \{x_i, v_i\}| = 1$. Assume that for some $i \in [n]$, $M \cap \{x_i, v_i\} = \emptyset$. Then $(x_i \leftarrow)$ and $(v_i \leftarrow)$ would be rules of P^M, and then M would not be a model of P^M, which is a contradiction. Assume instead that $\{x_i, v_i\} \subseteq M$. Then P^M would contain no rules with x_i and v_i in the head, and hence M would not be a minimal model of P^M, which is a contradiction.

We now construct an assignment $\alpha : X \to \mathbb{B}$ such that for all assignments $\beta : Y \to \mathbb{B}$ of weight k it holds that $\psi[\alpha \cup \beta]$ evaluates to true. Define α by letting $\alpha(x_i) = 1$ if and only if $x_i \in M$. Now let β be an arbitrary truth assignment to Y of weight k. We show that $\alpha \cup \beta$ satisfies C. We proceed indirectly, and assume to the contrary that $\alpha \cup \beta$ does not satisfy C. We construct a model $M' \subsetneq M$ of P^M. We let $y_{i_1}, \ldots, y_{i_k}$ denote the k variables y_i such that $\beta(y_i) = 1$. We let M' consist of $(M \cap \{ x_i, v_i : i \in [n] \})$, of $\{ y_{i_\ell}^\ell, y_{i_\ell} : \ell \in [k] \}$, and of $\{ z_g : g$ an internal node of C set to true by $\alpha \cup \beta \}$. For all rules of P^M other than Rules (6.20)–(6.22), it is clear that M' satisfies them. It holds that M' also satisfies Rules (6.20)–(6.21), since M' contains z_g for all internal nodes g of C that are satisfied by $\alpha \cup \beta$. Then, since $\alpha \cup \beta$ does not satisfy the output node o of C, we know that $z_o \notin M'$, and thus M' satisfies Rule (6.22). It then holds that $M' \subsetneq M$ is a model of P^M, which is a contradiction with the fact that M is an answer set of P. From this we can conclude that $\alpha \cup \beta$

satisfies C. Since β was arbitrary, we know this holds for all truth assignments β to Y of weight k. Therefore, $(C, k) \in \Sigma_2^p[*k]$-WSAT.

To show membership in $\Sigma_2^p[*k, P]$, we give an fpt-reduction to $\Sigma_2^p[k*]$-WSAT. Let P be an instance of ASP-CONSISTENCY(#disj.rules), where P is a disjunctive logic program that contains k disjunctive rules and that contains atoms $\{a_1, \ldots, a_n\}$. Let $r_1, \ldots, r_k$ be the disjunctive rules of P, where the head of rule r_i is $a_i^1 \vee \cdots \vee a_i^{\ell_i}$, for each $i \in [k]$. Moreover, let $\ell = \sum_{i \in [k]} \ell_i + k$. We sketch an algorithm that takes as input two bitstrings: one string $\overline{x} = x_1 \ldots x_n$, of length n, and one string $\overline{z} = z_1 \ldots z_\ell$ of length ℓ containing exactly k 1's. Moreover, we consider the string $\overline{z}$ as the concatenation of the strings $\overline{z}_1, \ldots, \overline{z}_k$, where $\overline{z}_i = z_{i,0} \ldots z_{i,\ell_i}$. Firstly, the algorithm checks (0) whether each z_i contains exactly one 1. If this checks fails, the algorithm accepts the input. Otherwise, the algorithm constructs the set $M = \{a_i : x_i = 1\}$, and it checks (1) whether M is a model of the reduct P^M. Moreover, the algorithm constructs the subset R of disjunctive rules that is defined as follows. For each $i \in [k]$, it holds that $r_i \in R$ if and only if $z_{i,0} = 0$. In addition, it constructs the mapping $\mu : R \rightarrow \text{At}(P)$, by letting $\mu(r_i) = a_i^j$ for the unique $j \in [\ell_i]$ for which $z_{i,j} = 1$. The algorithm constructs the set M_μ, as defined in Lemma 5.5. Then, the algorithm checks (2) whether M_μ is not a strict subset of M. The algorithm accepts the input if and only if both checks (1) and (2) pass. We can choose this algorithm so that it runs in polynomial time.

Now, by Lemma 5.5, it is straightforward to verify that $P \in$ ASP-CONSISTENCY-(#disj.rules) if and only if there exists some strings $\overline{x}$ such that for all strings $\overline{z}$ containing exactly k 1's the algorithm accepts. Since the algorithm runs in polynomial time, we can construct in polynomial time a quantified Boolean circuit C over the set X of existential variables and the set Z of universal variables, with the property that the algorithm accepts for some strings $\overline{x}$ and for all suitable strings $\overline{z}$ if and only if there is a truth assignment $\alpha : X \rightarrow \mathbb{B}$ such that for all truth assignments $\beta : Z \rightarrow \mathbb{B}$ of weight k the assignment $\alpha \cup \beta$ satisfies C. In other words, $(C, k) \in \Sigma_2^p[*k]$-WSAT if and only if $P \in$ ASP-CONSISTENCY(#disj.rules).

In the above proof, to show $\Sigma_2^p[*k, P]$-membership, we (implicitly) appealed to a version of the Cook-Levin Theorem to transform a polynomial-time computation (that uses a non-deterministically guessed input) into a circuit of polynomial size. In Sect. 6.3.4, we will prove an analogue of the Cook-Levin Theorem for the class $\Sigma_2^p[*k, P]$ in terms of alternating Turing machines, that might be more convenient for showing membership in the class $\Sigma_2^p[*k, P]$.

For the problem ASP-CONSISTENCY(#disj.rules), $\Sigma_2^p[*k, P]$-hardness holds even in the case where each atom occurs only a constant number of times in the input program.

Corollary 6.25. *Let $\ell \geq 3$. ASP-CONSISTENCY(#disj.rules) is $\Sigma_2^p[*k, P]$-hard, even when each atom occurs at most ℓ times.*

Proof. The reduction in the proof of Proposition 5.4 can be seen as an fpt-reduction that maps any instance of ASP-CONSISTENCY(#disj.rules) to another instance of ASP-CONSISTENCY(#disj.rules) in which each atom occurs at most ℓ times.

Next, we show $\Sigma_2^P[*k, \text{P}]$-completeness for the parameterized problem ASP-CONSISTENCY(#non-dual-normal.rules).

Theorem 6.26. *The problem* ASP-CONSISTENCY(#non-dual-normal.rules) *is* $\Sigma_2^P[*k, \text{P}]$*-complete.*

Proof. To show hardness, we give an fpt-reduction from $\Sigma_2^P[*k]$-WSAT. Let (C, k) be an instance of $\Sigma_2^P[*k]$-WSAT, where $C = \exists X.\forall Y.C$, $X = \{x_1, \ldots, x_n\}$, and $Y = \{y_1, \ldots, y_m\}$. By Proposition 6.23, we may assume without loss of generality that C is monotone in Y, i.e., that the variables $y_1, \ldots, y_m$ occur only positively in C. We construct a disjunctive program P as follows. We consider the variables X and Y as atoms. In addition, we introduce fresh atoms $v_1, \ldots, v_n$, w, and y_i^j for all $j \in [k], i \in [m]$. Also, for each internal node g of C, we introduce a fresh atom z_g. We let P consist of the following rules:

$$x_i \leftarrow not \; v_i \qquad \text{for } i \in [n]; \tag{6.24}$$

$$v_i \leftarrow not \; x_i \qquad \text{for } i \in [n]; \tag{6.25}$$

$$w \leftarrow y_1^j, \ldots, y_m^j \qquad \text{for } j \in [k]; \tag{6.26}$$

$$y_i^j \leftarrow y_i \qquad \text{for } i \in [m], j \in [k]; \tag{6.27}$$

$$y_i^j \leftarrow w \qquad \text{for } i \in [m], j \in [k]; \tag{6.28}$$

$$z_g \leftarrow w \qquad \text{for each internal node } g \text{ of } C; \tag{6.29}$$

$$z \leftarrow w \tag{6.30}$$

$$y_i \leftarrow w \qquad \text{for } i \in [m] \tag{6.31}$$

$$y_i^j \vee y_i^{j'} \leftarrow z \qquad \text{for } i \in [m], j, j' \in [k], j < j'; \tag{6.32}$$

$$\sigma(g_1) \vee \cdots \vee \sigma(g_u) \leftarrow z_g \qquad \text{for each conjunction node } g \text{ of } C \text{ with inputs } g_1, \ldots, g_u; \tag{6.33}$$

$$\sigma(g_i) \leftarrow z_g \qquad \text{for each disjunction node } g \text{ of } C \text{ with inputs } g_1, \ldots, g_u, \text{ and each } i \in [u]; \tag{6.34}$$

$$z_o \leftarrow z \qquad \text{where } o \text{ is the output node of } C; \tag{6.35}$$

$$w \vee z \leftarrow \tag{6.36}$$

$$w \leftarrow not \; w. \tag{6.37}$$

Here, we define the following mapping σ from nodes of C to variables in V. For each non-negated input node $x_i \in X$, we let $\sigma(x_i) = v_i$. For each negated input node $\neg x_i$, for $x_i \in X$, we let $\sigma(\neg x_i) = x_i$. For each input node $y_j \in Y$, we let $\sigma(y) = y_j$. For each internal node g, we let $\sigma(g) = z_g$. Intuitively, v_i corresponds to $\neg x_i$. Note, however, that σ negates literals over variables in X. Note that P has k rules that are not dual-Horn, namely the rules (6.26). We show that $(\varphi, k) \in \Sigma_2^p[*k]$-WSAT if and only if P has an answer set.

($\Rightarrow$) Assume there exists an assignment $\alpha : X \to \mathbb{B}$ such that for each assignment $\beta : Y \to \mathbb{B}$ of weight k it holds that $\alpha \cup \beta$ satisfies $\mathcal{C}$. We show that $M = \{x_i : \alpha(x_i) = 1, i \in [n]\} \cup \{y_i^j, y_i : i \in [m], j \in [k]\} \cup \{z_g : g \text{ an internal node of } \mathcal{C}\} \cup \{w, z\}$ is an answer set of P. We have that P^M consists of Rules (6.26)–(6.36), the rules $(x_i \leftarrow)$ for all $i \in [n]$ such that $\alpha(x_i) = 1$, and the rules $(v_i \leftarrow)$ for all $i \in [n]$ such that $\alpha(x_i) = 0$. Clearly, M is a model of P^M. We show that M is a minimal model of P^M. Assume $M' \subsetneq M$ is a minimal model of P^M. If M' does not coincide with M on the atoms x_i and v_i, then M' is not a model of P^M. If $w \in M'$, then by Rules (6.28)–(6.31), $M' = M$. Therefore, $w \notin M'$. By Rule (6.26), we have that $y_{i_1}^1, y_{i_2}^2, \ldots, y_{i_k}^k \notin M'$, for some $i_1, \ldots, i_k \in [m]$. By Rule (6.32) and (6.36), we know that $i_1, \ldots, i_k$ are all different, since otherwise it would have to hold that $w \in M'$. By Rule (6.27), it holds that $y_{i_1}, \ldots, y_{i_k} \notin M'$. Define the assignment $\gamma : X \cup Y \to \mathbb{B}$ by letting $\gamma(x_i) = 1$ if and only if $x_i \in M'$ and $\gamma(y_i) = 1$ if and only if $y_i \in M'$. Clearly, γ coincides with α on X, and γ assigns exactly k variables y_j to true. Now, since $(C, k) \in \Sigma_2^p[*k]$-WSAT, we know that γ satisfies $\mathcal{C}$. Using the Rules (6.33) and (6.34), we can show by an inductive argument that for each internal node g of $\mathcal{C}$ that is forced to true by γ it must hold that $z_g \notin M'$. Since γ satisfies $\mathcal{C}$, it forces the output node o of $\mathcal{C}$ to be true, and thus by Rule (6.35), we know that $z \notin M'$. Then, by Rule (6.36), we know that $w \in M'$, which is a contradiction. From this we can conclude that no model $M' \subsetneq M$ of P^M exists, and thus M is an answer set of P.

($\Leftarrow$) Assume P has an answer set M. We know that $w \in M$, since otherwise $(w \leftarrow)$ would be a rule of P^M, and then M would not be a model of P^M. Then, by Rules (6.28)–(6.31), also $y_i, y_i^j \in M$ for all $i \in [m]$ and $j \in [k]$, $z_g \in M$ for all internal nodes g of $\mathcal{C}$, and $z \in M$. We show that for each $i \in [n]$ it holds that $|M \cap \{x_i, v_i\}| = 1$. Assume that for some $i \in [n]$, $M \cap \{x_i, v_i\} = \emptyset$. Then $(x_i \leftarrow)$ and $(v_i \leftarrow)$ would be rules of P^M, and then M would not be a model of P^M, which is a contradiction. Assume instead that $\{x_i, v_i\} \subseteq M$. Then P^M would contain no rules with x_i and v_i in the head, and hence M would not be a minimal model of P^M, which is a contradiction.

We now construct an assigment $\alpha : X \to \mathbb{B}$ such that for all assignments $\beta : Y \to \mathbb{B}$ of weight k it holds that $\psi[\alpha \cup \beta]$ evaluates to true. Define α by letting $\alpha(x_i) = 1$ if and only if $x_i \in M$. Now let β be an arbitrary truth assignment to Y of weight k. We show that $\alpha \cup \beta$ satisfies $\mathcal{C}$. We proceed indirectly, and assume to the contrary that $\alpha \cup \beta$ does not satisfy $\mathcal{C}$. We construct a model $M' \subsetneq M$ of P^M. We let $y_{i_1}, \ldots, y_{i_k}$ denote the k variables y_i such that $\beta(y_i) = 1$. We let M' consist of $(M \cap \{x_i, v_i : i \in [n]\})$, of $\{y_i^j, y_i : i \in [m], j \in [k]\} \setminus \{y_{i_\ell}^\ell, y_{i_\ell} : \ell \in [k]\}$, of $\{z_g : g \text{ an internal node of } \mathcal{C} \text{ not satisfied by } \alpha \cup \beta\}$, and of $\{z\}$. For all rules of P^M other than Rules (6.33)–(6.35), it is clear that M' satisfies them. It holds that M' also satisfies Rules (6.33)–(6.34), since M' does not contain z_g for all internal nodes g of $\mathcal{C}$ that are satisfied by $\alpha \cup \beta$. Then, since $\alpha \cup \beta$ does not satisfy the output node o of $\mathcal{C}$, we know that $z_o \in M'$, and thus M' satisfies Rule (6.35). It then holds that $M' \subsetneq M$ is a model of P^M, which is a contradiction with the fact that M is an answer set of P. From this we can conclude that $\alpha \cup \beta$ satisfies $\mathcal{C}$. Since β was arbitrary, we know this holds for all truth assignments β to Y of weight k. Therefore, $(C, k) \in \Sigma_2^p[*k]$-WSAT.

We defer the proof of membership in $\Sigma_2^p[*k, P]$ until after we provided an additional characterization of $\Sigma_2^p[*k, P]$ in terms of alternating Turing machines.

Also for ASP-CONSISTENCY(#non-dual-normal.rules), $\Sigma_2^p[*k, P]$-hardness holds even in the case where each atom occurs only a constant number of times in the input program.

Corollary 6.27. *Let* $\ell \geq 3$.   ASP-CONSISTENCY(#non-dual-normal.rules)   *is* $\Sigma_2^p[*k, P]$-*hard, even when each atom occurs at most ℓ times.*

Proof. The reduction in the proof of Proposition 5.4 can be seen as an fpt-reduction that maps instances of ASP-CONSISTENCY(#non-dual-normal.rules) to another instance of ASP-CONSISTENCY(#non-dual-normal.rules) in which each atom occurs at most ℓ times.

6.3.4 Alternating Turing Machine Characterization

In this section, we give an alternative characterization of $\Sigma_2^p[*k, P]$ in terms of ATMs. In particular, we show that $\Sigma_2^p[*k, P]$ contains those parameterized decision problems that can be decided by a certain class of alternating Turing machines. This characterization can be used conveniently to show membership in $\Sigma_2^p[*k, P]$, which we illustrate by showing $\Sigma_2^p[*k, P]$-membership for ASP-CONSISTENCY-(#non-dual-normal.rules) —thereby completing the proof of Theorem 6.26.

We consider the following restrictions on ATMs.

Definition 6.28. *Let Q be a parameterized problem. An $\Sigma_2^p[*k, P]$-machine for Q is a $\exists\forall$-machine $\mathbb{M}$ such that there exists a computable function f and a polynomial p such that:*

1. *$\mathbb{M}$ decides P in time $f(k)p(|x|)$; and*
2. *for all instances (x, k) of Q and each computation path R of $\mathbb{M}$ with input (x, k), at most $f(k) \log |x|$ of the universal configurations of R are non-deterministic.*

We say that a parameterized problem Q is decided by some $\Sigma_2^p[*k, P]$-machine if there exists a $\Sigma_2^p[*k, P]$-machine for Q.

Proposition 6.29. *Let Q be a parameterized problem. Then $Q \in \Sigma_2^p[*k, P]$ if and only if Q is decided by some $\Sigma_2^p[*k, P]$-machine.*

Proof. The result follows directly from Lemmas 6.30 and 6.31.

We begin by showing that all problems that are decided by some $\Sigma_2^p[*k, \mathrm{P}]$-machine are in $\Sigma_2^p[*k, \mathrm{P}]$.

Lemma 6.30. *If a parameterized problem Q is decided by some $\Sigma_2^p[*k, \mathrm{P}]$-machine, then $Q \in \Sigma_2^p[*k, \mathrm{P}]$.*

Proof (sketch). We describe a way of constructing, for each instance (x, k) of Q, an instance (φ, k') of $\Sigma_2^p[*k]$-WSAT(CIRC) that is a yes-instance if and only if $(x, k) \in Q$. Our construction is based on the well-known proof of the Cook-Levin Theorem [54, 144].

We begin with some observations. Let $\mathbb{M}$ be an $\Sigma_2^p[*k, \mathrm{P}]$-machine for Q. We may assume without loss of generality that for any non-deterministic transition of $\mathbb{M}$, there are exactly two possible ways of proceeding. Any run of $\mathbb{M}$ on input (x, k) can be specified entirely by indicating what non-deterministic choices $\mathbb{M}$ makes. Given (a representation of) these non-deterministic choices, determining whether this run of $\mathbb{M}$ is an accepting run can be done in fpt-time in (x, k)—simply by simulating $\mathbb{M}$ using the given choices. In other words, to decide whether $\mathbb{M}$ accepts an input (x, k), we need to decide whether there exists some sequence s_1 of non-deterministic choices for the existential phase of the computation, such that for all sequences s_2 of non-deterministic choices for the universal phase of the computation it holds that the run of $\mathbb{M}$ on (x, k) that is specified by (s_1, s_2) is an accepting run.

By definition of $\Sigma_2^p[*k, \mathrm{P}]$-machines, we know that there is some computable function f and some constant c such that for each input (x, k) with $|x| = n$,

1. $\mathbb{M}$ runs in time $f(k)n^c$ and
2. $\mathbb{M}$ makes at most $f(k) \log n$ non-deterministic choices in the universal phase of the computation.

Therefore, in particular, in the existential phase of the computation $\mathbb{M}$ makes at most $f(k)n^c$ non-deterministic choices. We can encode all possibilities for the $2^{f(k)n^c}$ different possible combinations of choices that $\mathbb{M}$ makes in the existential phase of the computation using $f(k)n^c$ existential variables of φ. Moreover, since there are at most $2^{f(k) \log n} = n^{f(k)}$ different possible combinations of choices that $\mathbb{M}$ makes in the universal phase of the computation, we can encode these possibilities using n universal variables, whose assignments are restricted to set only $k' = f(k)$ variables to true.

The circuit C then simulates the behavior of $\mathbb{M}$ on input (x, k) with the non-deterministic behavior given by (s_1, s_2) as specified by the assignment to the variables. Since such a simulation can be done in fpt-time, we know we can encode this simulation in a circuit C that can be constructed in fpt-time. Then, (C, k') is a yes-instance of $\Sigma_2^p[*k]$-WSAT(CIRC) if and only if $\mathbb{M}$ accepts (x, k), which is the case if and only if $(x, k) \in Q$.

Next, we show the converse statement. That is, we show that each problem that is in $\Sigma_2^p[*k, \mathrm{P}]$ is decided by some $\Sigma_2^p[*k, \mathrm{P}]$-machine.

Lemma 6.31. *Let Q be a parameterized problem. If $Q \in \Sigma_2^p[*k, \mathrm{P}]$, then Q is decided by some $\Sigma_2^p[*k, \mathrm{P}]$-machine.*

Proof (sketch). We describe the $\Sigma_2^p[*k, \mathrm{P}]$-machine $\mathbb{M}$ that decides Q. Let (x, k) be an instance of Q. The computation of $\mathbb{M}$ proceeds in two stages. The first stage only involves deterministic computation. Because $Q \in \Sigma_2^p[*k, \mathrm{P}]$, we know that there exists an fpt-reduction R from Q to $\Sigma_2^p[*k]$-WSAT(CIRC). In the first stage, $\mathbb{M}$ computes $R(x, k) = (C, k')$. This can be done in (deterministic) time $f(k)n^c$, for some computable function f and some constant c, where $n = |x|$. Moreover, we know that $k' \leq g(k)$ for some computable function g, and that $|C| \leq f(k)n^c$.

In the second stage, $\mathbb{M}$ decides whether $(C, k') \in \Sigma_2^p[*k]$-WSAT(CIRC). Let X be the set of existential variables of C, and Y the set of universal variables of C. In the existential phase of the computation, $\mathbb{M}$ guesses a truth assignment to the variables in X. Since $|X| \leq f(k)n^c$, this can be done using at most $f(k)n^c$ (existential) non-deterministic steps. Then, in the universal phase of the computation, $\mathbb{M}$ guesses a truth assignment to the variables in Y of weight at most k'. Since $|Y| \leq f(k)n^c$ and $k' \leq g(k)$, this can be done using at most $g(k)\log(f(k)n^c) = g(k)f(k) + g(k)c\log n \leq cg(k)^2 f(k)\log n$ (universal) non-deterministic steps. Finally, $\mathbb{M}$ verifies whether the guessed truth assignments to the variables in X and Y satisfy the circuit.

It is readily verified that $\mathbb{M}$ correctly decides whether $(C, k') \in \Sigma_2^p[*k]$-WSAT (CIRC), and thus whether $(x, k) \in Q$. Moreover, $\mathbb{M}$ satisfies the bounds on the number of (existential and universal) non-deterministic steps, so $\mathbb{M}$ is an $\Sigma_2^p[*k, \mathrm{P}]$-machine for Q.

Using this alternative characterization of $\Sigma_2^p[*k, \mathrm{P}]$, we can now conveniently finish the proof of Theorem 6.26.

Proof (Proof of Theorem 6.26 (continued)). We show membership in $\Sigma_2^p[*k, \mathrm{P}]$ for ASP-CONSISTENCY(#non-dual-normal.rules), by describing an $\Sigma_2^p[*k, \mathrm{P}]$-machine $\mathbb{M}$ for the problem. The algorithm computed by the machine $\mathbb{M}$ is entirely similar to the algorithm in the proof of Proposition 5.10 in Chap. 4. Let P be a disjunctive logic program with k non-dual-Horn rules. Firstly, in the existential phase of the computation, the machine $\mathbb{M}$ guesses a subset $M \subseteq \mathrm{At}(P)$. Then, in order to verify whether M is a minimal model of P^M, it verifies whether for all $m \in M$ the program $P_{m,M}$ (as defined in the proof of Proposition 5.10) has no models. It does so as follows. In the universal phase of the computation, the machine $\mathbb{M}$ non-deterministically guesses some $m \in M$. This can be done using $\log|\mathrm{At}(P)|$ non-deterministic choices. Then, similarly to the proof of Proposition 5.10, it verifies that there is no suitable set R and suitable mapping μ, as defined in Lemma 5.9, such that the set M_μ is a model of $P_{m,M}$. It does so by guessing a suitable set R and a suitable mapping μ, still in the universal phase of the computation. Since there are at most $O((|\mathrm{At}(P)| + 1)^k)$ possible combinations of R and μ, this can be done using $k\log(|\mathrm{At}(P)| + 1)$ non-deterministic choices. Then, it computes M_μ and checks if M_μ is a model of $P_{m,M}$. If this is the case, the machine $\mathbb{M}$ rejects the input. Otherwise, it accepts.

Similarly to the proof of Proposition 5.10, we then get that $\mathbb{M}$ accepts the input if and only if P has an answer set. Moreover, it is readily verified that $\mathbb{M}$ satisfies the bounds on the number of (existential and universal) non-deterministic steps, so $\mathbb{M}$ is an $\Sigma_2^p[*k, P]$-machine for ASP-CONSISTENCY(#non-dual-normal.rules).

It is an interesting topic for future research to investigate machine characterizations for the classes $\Sigma_2^p[*k, 1]$ and $\Sigma_2^p[*k, 2]$. Such characterizations could be similar to the characterization of the class $\Sigma_2^p[k*]$ using the problems $\Sigma_2^p[k*]$-TM-HALT^m (Theorem 6.11), which is in turn similar to machine characterizations for the classes W[1] and W[2] [36, 39, 85].

6.4 Relation to Known Parameterized Complexity Classes

Finally, we relate the classes of the k-$*$ and $*$-k hierarchies to known (parameterized) complexity classes. In particular, we give evidence that these classes differ from the parameterized complexity classes para-NP, para-co-NP, para-Σ_2^p and para-Π_2^p.

6.4.1 Relation of $\Sigma_2^p[k*]$ to Other Classes

We begin with investigating the relation of $\Sigma_2^p[k*]$ to known (parameterized) complexity classes.

It is straightforward to see that $\Sigma_2^p[k*] \subseteq$ para-Σ_2^p. In polynomial time, any formula $\exists X.\forall Y.\psi$ can be transformed into a quantified Boolean formula with a $\exists\forall$ quantifier prefix that is true if and only if for some assignment α of weight k to the variables X the formula $\forall Y.\psi[\alpha]$ is true. Also, trivially, para-co-NP $\subseteq \Sigma_2^p[k*]$, because the para-co-NP-complete parameterized problem UNSAT(constant) can be seen as a special case of $\Sigma_2^p[k*]$-WSAT. To summarize, we observe the following inclusions:

$$\text{para-co-NP} \subseteq \Sigma_2^p[k*] \subseteq \text{para-}\Sigma_2^p \quad \text{and} \quad \text{para-NP} \subseteq \Pi_2^p[k*] \subseteq \text{para-}\Pi_2^p.$$

This immediately leads to the following result.

Proposition 6.32. *If* $\Sigma_2^p[k*] \subseteq$ *para-NP, then* NP = co-NP.

It is also not difficult to see that $\Sigma_2^p[k*] \subseteq$ Xco-NP. This is witnessed by the straightforward brute-force algorithm to solve $\Sigma_2^p[k*]$-WSAT that tries out all $\binom{n}{k} = O(n^k)$ assignments of weight k to the existentially quantified variables (and that uses non-determinism to handle the assignment to the universally quantified variables).

A natural question to ask is whether para-NP $\subseteq \Sigma_2^p[k*]$. Since para-NP $\subseteq$ Xco-NP implies NP = co-NP [84, Proposition 8], this is unlikely. We give a direct proof of this statement.

Proposition 6.33. *If* para-NP $\subseteq \Sigma_2^p[k*]$, *then* NP = co-NP.

Proof. Assume that para-NP $\subseteq \Sigma_2^P[k*]$. The parameterized problem $Q = \{ (\varphi, 1) : \varphi \in \text{SAT} \}$ is in para-NP. Then also $Q \in \Sigma_2^P[k*]$. This means that there is an fpt-reduction R from Q to $\Sigma_2^P[k*]$-WSAT. We construct a polynomial-time reduction S from SAT to UNSAT. Let φ be an instance of SAT. The reduction R maps $(\varphi, 1)$ to an instance $(\varphi', f(1))$ of $\Sigma_2^P[k*]$-WSAT, where f is some computable function and $\varphi' = \exists X.\forall Y.\psi$, such that $(\varphi', f(1)) \in \Sigma_2^P[k*]$-WSAT if and only if $\varphi \in \text{SAT}$. Note that $f(1)$ is a constant, since f is a fixed function. To emphasize this, we let $c = f(1)$, and we will use c to denote $f(1)$. By definition, we know that $(\varphi', c) \in \Sigma_2^P[k*]$-WSAT if and only if for some truth assignment α to the variables X of weight c, the formula $\forall Y.\psi[\alpha]$ is true. Let $\text{ta}(X, c)$ denote the set of all truth assignments to X of weight c. We then get that $\varphi \in \text{SAT}$ if and only if the formula $\forall Y.\chi$ is true, where χ is defined as follows:

$$\chi = \bigvee_{\alpha \in \text{ta}(X,c)} \psi[\alpha]$$

It is straightforward to verify that the mapping $\varphi \mapsto \neg\chi$ is a polynomial-time reduction from SAT to UNSAT.

This implies that $\Sigma_2^P[k*]$ is likely to be a strict subset of para-Σ_2^P.

Corollary 6.34. *If* $\Sigma_2^P[k*] = \text{para-}\Sigma_2^P$, *then* $\text{NP} = \text{co-NP}$.

The following result shows another way in which the class $\Sigma_2^P[k*]$ relates to the complexity class co-NP. Let P be a parameterized decision problem, and let $c \geq 1$ be an integer. Recall that the *c-th slice of* P, denoted by P_c, is the (unparameterized) decision problem $\{ x : (x, c) \in P \}$.

Proposition 6.35. *Let* P *be a parameterized problem complete for* $\Sigma_2^P[k*]$, *and let* $c \geq 1$. *Then* P_c *is in* co-NP. *Moreover, there exists some integer* $d \geq 1$ *such that* $P_1 \cup \cdots \cup P_d$ *is* co-NP-*complete*.

Proof. We show co-NP-membership of P_c, by constructing a polynomial-time reduction S from P_c to UNSAT. Since $P \in \Sigma_2^P[k*]$, we know that there exists an fpt-reduction R from P to $\Sigma_2^P[k*]$-WSAT(Φ). Therefore, there exist computable functions f and g and a polynomial p such that for all instances (x, k) of P, $R(x, k) = (x', k')$ is computable in time $f(k) \cdot p(|x|)$ and $k' \leq g(k)$. We describe the reduction S. Let x be an arbitrary instance of P_c. We know R maps (x, c) to (φ, k'), for some $k' \leq g(c)$, where $\varphi = \exists X.\forall Y.\psi$. Note that k' is bounded by a constant $g(c) = d$. Let $\text{ta}(X, k')$ denote the set of all truth assignment to X of weight k'. Then, for each $\alpha \in \text{ta}(X, k')$, we let Y^α be the set containing of a copy y^α of each variable $y \in Y$, and we let $Y' = \bigcup_{\alpha \in \text{ta}(X,k')} Y^\alpha$. We then get that φ is equivalent to the formula $\forall Y'.\chi$, where $\chi = \bigvee_{\alpha \in \text{ta}(X,k')} (\psi[\alpha])^\alpha$. Here, $(\psi[\alpha])^\alpha$ denotes the formula $\psi[\alpha]$ where each variable $y \in Y$ is replaced by its copy y^α. Also, the size of χ is polynomial in the size of ψ. We then let $S(x) = \neg\chi$. It is straightforward to verify that S is a correct polynomial-time reduction from P_c to UNSAT.

We show that there exists a function f such that for any positive integer $s \geq 1$, there is a polynomial-time reduction from UNSAT to $P_1 \cup \cdots \cup P_{f(s)}$. Then, in particular, $P_1 \cup \cdots \cup P_{f(1)}$ is co-NP-complete. Let $s \geq 1$ be an arbitrary integer. We construct the reduction S. There is a trivial polynomial-time reduction S from UNSAT to $(\Sigma_2^p[k*]\text{-WSAT})_s$, that maps a Boolean formula φ over a set of variables Y to the instance $(\exists\{x_1, \ldots, x_s\}.\forall Y.\neg\chi, s)$ of $\Sigma_2^p[k*]\text{-WSAT}$. Since P is $\Sigma_2^p[k*]$-complete, there exists an fpt-reduction R from $\Sigma_2^p[k*]\text{-WSAT}$ to P. From this, we know that there exists a nondecreasing and unbounded function f such that for each instance (x, k) of $\Sigma_2^p[k*]\text{-WSAT}$ it holds that $k' \leq f(k)$, where $R(x, k) = (x', k')$. This reduction R is a polynomial-time reduction from $(\Sigma_2^p[k*]\text{-WSAT})_s$ to $P_1 \cup \cdots \cup P_{f(s)}$. Composing the polynomial-time reductions S and R, we obtain a reduction from UNSAT to $P_1 \cup \cdots \cup P_{f(s)}$.

Finally, we relate $\Sigma_2^p[k*]$ to the classes para-co-NP and para-Π_2^p. We already observed that para-co-NP $\subseteq \Sigma_2^p[k*]$. In Chap. 14, we give evidence that this inclusion is strict. Concretely, we show that if para-co-NP $= \Sigma_2^p[k*]$, then there is a subexponential-time reduction from QSAT$_2$ to UNSAT (Corollary 14.12).

The class $\Sigma_2^p[k*]$ is likely to be incomparable to para-Π_2^p (w.r.t. set inclusion). It is not difficult to see that if para-$\Pi_2^p \subseteq \Sigma_2^p[k*]$, then $\Sigma_2^p = \Pi_2^p$. For the other direction, we show in Chap. 14 that if $\Sigma_2^p[k*] \subseteq$ para-Π_2^p, then there is a subexponential-time reduction from QSAT$_2$(3DNF) to co-QSAT$_2$ (Proposition 14.4).

6.4.2   Relation of $\Sigma_2^p[*k, t]$ to Other Classes

Next, we continue with relating the classes of the $*$-k hierarchy to known (parameterized) complexity classes. Similarly to the case of k-$*$, we can observe the following inclusions:

$$\text{para-NP} \subseteq \Sigma_2^p[*k, 1] \subseteq \cdots \subseteq \Sigma_2^p[*k, P] \subseteq \text{para-}\Sigma_2^p$$

and

$$\text{para-co-NP} \subseteq \Pi_2^p[*k, 1] \subseteq \cdots \subseteq \Pi_2^p[*k, P] \subseteq \text{para-}\Pi_2^p.$$

This immediately leads to the following result.

Proposition 6.36. *If* $\Sigma_2^p[*k, 1] \subseteq$ para-co-NP, *then* NP $=$ co-NP.

It is also not so difficult to see that $\Sigma_2^p[*k, P] \subseteq$ XNP. This is witnessed by the straightforward brute-force algorithm to solve $\Sigma_2^p[*k]\text{-WSAT}$ that tries out all $\binom{n}{k} = O(n^k)$ assignments of weight k to the universally quantified variables (and that uses non-determinism to handle the assignment to the existentially quantified variables).

A natural question to ask is whether para-co-NP is contained in any of the classes $\Sigma_2^p[*k, t]$. Since para-co-NP $\subseteq$ XNP implies NP $=$ co-NP [84, Proposition 8], this is unlikely. We show how to prove this result directly.

Proposition 6.37. *If* para-co-NP $\subseteq \Sigma_2^p[k*, P]$, *then* NP = co-NP.

Proof (sketch). With an argument similar to the one in the proof of Proposition 6.33, a polynomial-time reduction from UNSAT to SAT can be constructed. An additional technical observation needed for this case is that SAT is in NP also when the input is a Boolean circuit (rather than a propositional formula).

This immediately gives us the following separation.

Corollary 6.38. *If* $\Sigma_2^p[k*, P] =$ para-Σ_2^p, *then* NP = co-NP.

Similarly to the class $\Sigma_2^p[k*]$, the classes $\Pi_2^p[*k, t]$ relate to the complexity class co-NP in the following way.

Proposition 6.39. *Let P be a parameterized problem that is contained in* $\Pi_2^p[*k, P]$ *and that is hard for* $\Pi_2^p[*k, 1]$*, and let* $c \geq 1$*. Then* P_c *is in* co-NP. *Moreover, there exists some integer* $d \geq 1$ *such that* $P_1 \cup \cdots \cup P_d$ *is* co-NP-complete.

Proof. The proof of this proposition is similar to the proof of Proposition 6.35. We show co-NP-membership of P_c, by constructing a polynomial-time reduction S from P_c to UNSAT. Since $P \in \Pi_2^p[*k, P]$, we know that there exists an fpt-reduction R from P to $\Pi_2^p[*k]$-WSAT(Γ). Therefore, there exist computable functions f and g and a polynomial p such that for all instances (x, k) of P, $R(x, k) = (x', k')$ is computable in time $f(k) \cdot p(|x|)$ and $k' \leq g(k)$. We describe the reduction S. Let x be an arbitrary instance of P_c. We know R maps (x, c) to (φ, k'), for some $k' \leq g(c)$, where $\varphi = \exists X. \forall Y. C$. Note that k' is bounded by a constant $g(c) = d$. Let ta(X, k') denote the set of all truth assignment to X of weight k'. We then get that φ is equivalent to the quantified circuit $\forall Y. C'$, where $C' \equiv \bigvee_{\alpha \in \text{ta}(X, k')} C[\alpha]$. Also, the size of C' is polynomial in the size of C. It is straightforward to construct a propositional formula ψ that is valid if and only if C' is valid. We then let $S(x) = \neg \psi$. Then, S is a correct polynomial-time reduction from P_c to UNSAT.

We show that there exists a function f such that for any positive integer $s \geq 1$, there is a polynomial-time reduction from UNSAT to $P_1 \cup \cdots \cup P_{f(s)}$. Then, in particular, $P_1 \cup \cdots \cup P_{f(1)}$ is co-NP-complete. Let $s \geq 1$ be an arbitrary integer. We construct the reduction S. There is a trivial polynomial-time reduction S from UNSAT to $(\Pi_2^p[*k]$-WSAT(3DNF))$_s$, that maps a Boolean formula χ in 3CNF over a set of variables Y to the instance $(\forall Y. \exists \{x_1, \ldots, x_s\}. \neg \chi, s)$ of $\Pi_2^p[*k]$-WSAT(3DNF). Since P is $\Pi_2^p[*k, 1]$-hard, there exists an fpt-reduction R from $\Pi_2^p[*k]$-WSAT(3DNF) to P. From this, we know that there exists a nondecreasing and unbounded function f such that for each instance (x, k) of $\Pi_2^p[*k]$-WSAT(3DNF) it holds that $k' \leq f(k)$, where $R(x, k) = (x', k')$. This reduction R is a polynomial-time reduction from $(\Pi_2^p[*k]$-WSAT(3DNF))$_s$ to $P_1 \cup \cdots \cup P_{f(s)}$. Composing the polynomial-time reductions S and R, we obtain a reduction from UNSAT to $P_1 \cup \cdots \cup P_{f(s)}$.

Finally, we relate the classes $\Sigma_2^p[*k, t]$ to the classes para-NP and para-Π_2^p. We already observed that para-NP $\subseteq \Sigma_2^p[*k, 1]$. In Chap. 14, we give evidence that this inclusion is strict. Concretely, we show that if para-NP $= \Sigma_2^p[*k, 1]$, then there is a subexponential-time reduction from $\text{QSAT}_2(\text{3DNF})$ to SAT (Corollary 14.2). Also, if para-NP $= \Sigma_2^p[*k, 2]$, then there is a subexponential-time reduction from $\text{QSAT}_2(\text{DNF})$ to SAT (Corollary 14.10).

The classes $\Sigma_2^p[*k, t]$ are likely to be incomparable to para-Π_2^p (w.r.t. set inclusion). It is not difficult to see that if para-$\Pi_2^p \subseteq \Sigma_2^p[*k, \text{P}]$, then $\Sigma_2^p = \Pi_2^p$. For the other direction, we show in Chap. 14 that if $\Sigma_2^p[*k, 1] \subseteq$ para-Π_2^p, then there is a subexponential-time reduction from $\text{QSAT}_2(\text{3DNF})$ to co-QSAT_2 (Proposition 14.4).

6.4.3   Relation Between $\Sigma_2^p[k*]$ and $\Sigma_2^p[*k, t]$

Above, we related the classes $\Sigma_2^p[k*]$ and $\Sigma_2^p[*k, t]$ to previously known complexity classes. A natural question that arises when comparing the parameterized complexity classes $\Sigma_2^p[k*]$ and $\Sigma_2^p[*k, t]$ to other classes, is what the relation is between the classes $\Sigma_2^p[k*]$ and $\Sigma_2^p[*k, t]$ themselves. We will address this question in Sect. 14.3, because in order to satisfactorily answer this question, we need to consider some technical machinery in more detail—which we will do in Chap. 14. In particular, we will show that these classes are different from each other—under various complexity-theoretic assumptions.

Summary

In this chapter, we developed the new parameterized complexity classes that we motivated in Chap. 5. In particular, we defined two hierarchies of parameterized complexity classes $\Sigma_2^p[k*, t]$ and $\Sigma_2^p[*k, t]$, that are based on weighted variants of the quantified Boolean satisfiability problem QSAT_2. We showed that the hierarchy containing the classes $\Sigma_2^p[k*, t]$ collapses to a single class $\Sigma_2^p[k*]$. We provided a foundation for future completeness results for the new classes by showing that the inputs for the canonical weighted satisfiability problems that underlie these classes can be transformed into several normal forms. Also, we gave alternative characterizations of the newly developed classes—among others in terms of alternating Turing machines— and we related these classes to several relevant parameterized complexity classes that are known from the literature. Moreover, we showed that the parameterized variants of the consistency problem for disjunctive answer set programming—whose complexity we showed cannot be characterized adequately using classes known from the literature—are complete for several of the introduced parameterized complexity classes.

Notes

The results in Sects. 6.2.3 and 6.2.3 appeared in a paper in the proceedings of SOF-SEM 2015 [113]. The results in Sect. 6.3.4 appeared in a paper in the proceedings of IJCAI 2015 [109]. The remaining results in this chapter appeared in a paper in the proceedings of KR 2014 [116, 117]. Many of the results in this chapter also appeared in an article appearing in the Journal of Computer and System Sciences [114].

In previous work [77, 78, 105, 109, 111–113, 116, 117] the class $\Sigma_2^p[k*]$ appeared under the names $\exists^k\forall$ and $\exists^k\forall^*$. Similarly, the classes $\Sigma_2^p[*k, t]$ appeared under the names $\exists\forall^k\text{-W}[t]$ and $\exists^*\forall^k\text{-W}[t]$.

We would like to thank Hubie Chen for suggesting to use first-order model checking to obtain an alternative characterization of the class $\Sigma_2^p[k*]$.

Chapter 7
Fpt-Algorithms with Access to a SAT Oracle

"The Answer... Is... Forty-two," said Deep Thought, with infinite majesty and calm.

— Douglas Adams,
The Hitchhiker's Guide to the Galaxy [4]

In Chap. 4, we introduced the idea of fpt-reductions to SAT. Moreover, in Chaps. 4–6, we discussed one interpretation of this generic scheme, and we started a theoretical investigation about the limits and possibilities of this type of fpt-reductions to SAT. These fpt-reductions to SAT are based on many-to-one (or Karp) reductions. That is, these reductions transform an instance (x, k) of a parameterized problem Q to a single equivalent instance φ of SAT—in other words, $(x, k) \in Q$ if and only if $\varphi \in$ SAT.

The practical motivation for considering fpt-reductions to SAT is that they could serve as a theoretical starting point for efficient solving methods, due to the excellent performance of modern SAT solvers. For instances with a small parameter value, an fpt-reduction to SAT can be used to efficiently encode the instance into an instance φ of SAT, and subsequently solving the problem by invoking the SAT solver on φ. However, for this solving strategy, there is no evident reason for such a strict bound on the number of calls to the SAT solver (allowing only a single call). On the contrary, research in classical complexity indicates that increasing the number of calls to a SAT solver strictly increases the solving power [42, 119, 138, 193]. Moreover, the approach of calling a SAT solver multiple times to solve various problems, has been shown to be quite effective in practice (see, e.g., [15, 70, 156]).

In this chapter, we consider other formalizations of the scheme of fpt-reductions to SAT, where the restriction on the number of calls to the SAT solver is relaxed. Namely, we consider fpt-reductions to SAT that are based on *Turing reductions*. In this setting, an fpt-reduction to SAT (for a parameterized problem Q) is a fixed-parameter tractable algorithm R that decides Q, and that can make multiple queries to a *SAT oracle*. A SAT oracle is a black box machine that returns the answer to a *query* "$\varphi \in$ SAT?" in a single time step.

© Springer-Verlag GmbH Germany, part of Springer Nature 2019
R. de Haan: Parameterized Complexity in the Polynomial Hierarchy, LNCS 11880,
https://doi.org/10.1007/978-3-662-60670-4_7

The theoretical investigation in this chapter can serve as a theoretical starting point for solving strategies that are more powerful than strategies based on many-to-one fpt-reductions to SAT, but can still be efficient in practice.

Despite the efficiency of modern SAT solvers, SAT solvers do not behave exactly like SAT oracles. Most notably, unlike the idealized SAT oracles, a SAT solver in practice needs more than a single time step to decide for an instance φ whether $\varphi \in$ SAT. For this reason, we investigate the concept of fpt-time Turing reductions to SAT using various bounds on the number of oracle queries.

Outline of this Chapter

In Sect. 7.1, we begin with reviewing several parameterized complexity classes, known from the literature, consisting of those problems that can be solved by a fixed-parameter tractable algorithm that can query a SAT oracle (for various bounds on the number of oracle queries that can be made).

Absent from this inventory of complexity classes for fpt-time Turing reductions to SAT is the class consisting of all problems solvable in fpt-time using $f(k)$ calls to a SAT oracle, where f is some computable function and k is the parameter value. In Sect. 7.2, we introduce this absent class—and name it $\mathrm{FPT}^{\mathrm{NP}}[\mathrm{few}]$. Moreover, we show that it can also be seen as a parameterized variant of the Boolean Hierarchy. We consider several examples of parameterized problems that can be solved in fpt-time using $f(k)$ SAT queries, and we show that these problems are complete for $\mathrm{FPT}^{\mathrm{NP}}[\mathrm{few}]$ (under fpt-reductions).

In Sect. 7.3, we show how hardness for $\mathrm{FPT}^{\mathrm{NP}}[\mathrm{few}]$—and hardness for the classes $\mathrm{A}[2]$, $\Sigma_2^{\mathrm{p}}[k*]$, and $\Sigma_2^{\mathrm{p}}[*k, t]$—can be used to obtain lower bounds on the number of SAT queries made by any fpt-algorithm that solves a problem.

Then, in Sect. 7.4, we consider another alternative characterization of the class $\mathrm{FPT}^{\mathrm{NP}}[\mathrm{few}]$, based on bounded optimization problems. Moreover, we illustrate this characterization using another parameterized problems that is complete for $\mathrm{FPT}^{\mathrm{NP}}[\mathrm{few}]$.

Finally, in Sect. 7.5, we consider an extension of $\mathrm{FPT}^{\mathrm{NP}}[\mathrm{few}]$ based on a more powerful oracle model where the SAT oracles return a satisfying assignment for yes-answers. We show that for the case of decision problems this extension does not yield more power, but for the case of search problems it does.

7.1 Known Parameterized Complexity Classes

We begin with surveying parameterized complexity classes known from the literature that consist of those parameterized problems solvable by means of fixed-parameter tractable algorithms that can query a SAT oracle, for various bounds on the number of oracle queries. These known parameterized complexity classes are all of the form para-K, where K is a classical complexity class [84] (see Chap. 3). Let K be a classical complexity class. Then, intuitively, para-K consists of those problems Q

that are in K after a *precomputation* on the parameter value k. That is, a parameterized problem Q is in para-K if there is a computable function $f : \mathbb{N} \to \Sigma^*$ and a problem $Q' \in K$ such that for all instances (x, k) of Q it holds that $(x, k) \in Q$ if and only if $(x, f(k)) \in Q'$.

For the sake of convenience, we repeat the definition of Turing machines with oracle access (again, see Chap. 3). Let O be a decision problem, e.g., $O = \mathrm{SAT}$. A Turing machine $\mathbb{M}$ with an O *oracle* is a Turing machine with a dedicated *oracle tape* and dedicated states q_{query}, q_{yes} and q_{no}. Whenever $\mathbb{M}$ is in the state q_{query}, it does not proceed according to the transition relation, but instead it transitions into the state q_{yes} if the oracle tape contains a string x that is a yes-instance for the problem O (i.e., if $x \in O$), and it transitions into the state q_{no} otherwise (i.e., if $x \notin O$). We will also often speak of algorithms that query an oracle. In this case, we mean an algorithm implemented by a Turing machine with an oracle.

The first parameterized complexity class that we consider is based on the Boolean Hierarchy (BH) [35, 41, 126]. The classical complexity class BH consists of all decision problems that are decided by some polynomial-time algorithm that queries a SAT oracle a constant number of times. (Equivalently, BH can be characterized as the class of problems that can be expressed as a Boolean combination of sets in NP.) Correspondingly, the parameterized complexity class para-BH consists of those parameterized problems that are decided by some fpt-algorithm that queries a SAT oracle a constant number of times.

Next, we consider the class para-Δ_2^p. The classical complexity class Δ_2^p consists of all decision problems that are decided by a polynomial-time algorithms that can query a SAT oracle. Correspondingly, the parameterized complexity class para-Δ_2^p consists of those parameterized problems that are decided by an fpt-algorithm with a SAT oracle. Both for Δ_2^p and para-Δ_2^p, the number of queries to the SAT oracle is only bounded by the running time of the algorithm.

Then, one can consider various restrictions of the class Δ_2^p. Let $z : \mathbb{N} \to \mathbb{N}$ be a function. Then the classical complexity class $\Delta_2^p[z(n)]$ consists of all problems that are decided by some polynomial-time algorithm that, for any instance x of size n, queries a SAT oracle at most $z(n)$ times. A commonly considered restriction of Δ_2^p is the class $\Theta_2^p = \bigcup_{c \in \mathbb{N}} \Delta_2^p[c \log n]$ [171]—that is, the class of problems that are decided by a polynomial-time algorithm that queries a SAT oracle $O(\log n)$ times. The class para-Θ_2^p then consists of those parameterized problems that are decided by an fpt-algorithm that queries a SAT oracle $f(k) \log n$ times, for some computable function f.

In order to compare the amount of SAT queries allowed for fpt-algorithms witnessing membership in the classes para-BH, para-Θ_2^p and para-Δ_2^p in a uniform setting, we consider the parameterized complexity classes $\mathrm{FPT}^{\mathrm{NP}}[g(n, k)]$ for functions $g : \mathbb{N}^2 \to \mathbb{N}$.

Definition 7.1. *Let $g : \mathbb{N}^2 \to \mathbb{N}$ be a function. Then $FPT^{NP}[g(n, k)]$ is defined as the class of all parameterized problems Q for which there exists a deterministic Turing machine $\mathbb{M}$, with an oracle $O \in NP$, such that $\mathbb{M}$ decides Q in fpt-time, and for* *each instance (x, k) of Q, the machine $\mathbb{M}$ makes at most $g(n, k)$ oracle queries when executed on input (x, k), where $n = |x|$.*

Using the classes $\text{FPT}^{\text{NP}}[g(n, k)]$, we can then express the classes para-BH, para-Θ_2^p, and para-Δ_2^p as follows.

Observation 7.2. *The following equations hold.*

$$\text{para-BH} = \bigcup_{c \in \mathbb{N}} \text{FPT}^{\text{NP}}[c]$$

$$\text{para-}\Theta_2^p = \bigcup_{f \text{ computable}} \text{FPT}^{\text{NP}}[f(k) \log n]$$

$$\text{para-}\Delta_2^p = \bigcup_{\substack{f \text{ computable} \\ c \in \mathbb{N}}} \text{FPT}^{\text{NP}}[f(k)n^c]$$

The various classes discussed above only differ with respect to the number of SAT queries that are allowed. From a practical point of view, the number of calls to a SAT solver may seem to be relatively insignificant, assuming that the queries are easy for the solver, and the solver can reuse information from previous calls [17, 122, 195]. The technique where a SAT solver is called multiple times, and where the solver reuses information computed in previous calls, is called *incremental SAT solving*. For a theoretical worst-case model, however, one must assume that all queries involve hard SAT instances, and that no information from previous queries can be reused. Therefore, in a theoretical analysis, it makes sense to study the number of SAT queries made by fpt-time algorithms.

Finally, we briefly consider some special cases of the class para-BH. One can consider the class $\text{FPT}^{\text{NP}}[c]$, for any constant $c \in \mathbb{N}$. Similar restrictions, for fixed constants c, have been considered for the class BH (we return to these in Sect. 7.2). We show that the classes $\text{FPT}^{\text{NP}}[c]$ are strictly larger than para-NP and para-co-NP (unless NP = co-NP), even for the case where $c = 1$. This indicates that fpt-time Turing reductions to SAT are more powerful than many-to-one fpt-reductions to SAT.

Proposition 7.3. *It holds that para-NP $\cup$ para-co-NP $\subseteq$ FPT^{NP}[1]. Moreover, this inclusion is strict, unless NP = co-NP.*

Proof. The inclusion para-NP $\cup$ para-co-NP $\subseteq$ FPT^{NP}[1] can be shown routinely. To see that the inclusion is strict (unless NP = co-NP), suppose that para-NP $\cup$ para-co-NP = FPT^{NP}[1]. Consider the following problem Q:

$$Q = \{ (x, 0) : x \in \text{UNSAT} \} \cup \{ (x, 1) : x \in \text{SAT} \}.$$

Clearly, $Q \in \text{FPT}^{\text{NP}}[1]$. Then, by assumption, $Q \in$ para-NP or $Q \in$ para-co-NP. Suppose that $Q \in$ para-NP. The case for $Q \in$ para-co-NP is entirely analogous. Then the para-co-NP-complete problem $Q' = \{ (x, 0) : x \in \text{UNSAT} \}$ is also in para-NP, since $Q' \subseteq Q$. From this, it immediately follows that NP = co-NP.

7.2 The Parameterized Complexity Class FPT^{NP}[few]

In the inventory of parameterized complexity classes corresponding to various notions of fpt-time Turing reductions to SAT, that we considered in Sect. 7.1, one natural parameterized complexity class is missing. This is the class of parameterized problems that are decided by an fpt-algorithm that queries a SAT oracle $f(k)$ times, for some computable function f. In this section, we consider this class, which we name FPT^{NP}[few]. Moreover, we show that it can be seen as a parameterized variant of the Boolean Hierarchy.

Definition 7.4. *The parameterized complexity class* FPT^{NP}[few] *is defined as follows:*

$$\text{FPT}^{\text{NP}}[\text{few}] = \bigcup_{f \text{ computable}} \text{FPT}^{\text{NP}}[f(k)]$$

Put differently, FPT^{NP}[few] *consists of all parameterized problems Q that can be decided by an fpt-algorithm that has access to an oracle $O \in$ NP such that for each instance (x, k) of Q, the algorithm makes at most $f(k)$ oracle queries, for some computable function f.*

It is straightforward to verify that FPT^{NP}[few] is closed under fpt-reductions. Moreover, clearly, para-BH $\subseteq$ FPT^{NP}[few] $\subseteq$ para-Θ_2^p $\subseteq$ para-Δ_2^p.

For the unparameterized case, it matters whether queries to the NP oracle are adaptive or parallel—that is, whether the algorithm has to write down all its oracle queries before the answer to any of the queries is given (parallel), or whether an answer to one oracle query can be used to construct the next oracle query (adaptive). (We will define the notion of parallel oracle queries formally in Sect. 7.5.) This is essentially the difference between the classes Θ_2^p and Δ_2^p [33, 121, 133]. Note that this distinction is not relevant for problems in FPT^{NP}[few], since $f(k)$ adaptive oracle queries can straightforwardly be simulated by means of $2^{f(k)+1}$ parallel queries. Without loss of generality, assume that the oracle $O = \text{SAT}$. For each possible partition of the $f(k)$ adaptive queries into a set Y of queries with a yes-answer and a set N of queries with a no-answer, one can construct two propositional formulas $\varphi_{\text{yes}} = \bigwedge_{y \in Y} \varphi_y$ and $\varphi_{\text{no}} = \bigvee_{n \in N} \varphi_n$ such that $\varphi_{\text{yes}} \in \text{SAT}$ and $\varphi_{\text{no}} \notin \text{SAT}$ if and only if the partition into the sets Y and N corresponds to the answers that the oracle gives to the queries. Here the formulas φ_y and φ_n are all pairwise disjoint, and can be constructed by simulating the algorithm (that makes adaptive query queries) using the answers specified by Y and N. This way, one can determine the answers to all the $f(k)$ adaptive oracle queries using $2^{f(k)+1}$ parallel oracle queries.

7.2.1   A Parameterized Variant of the Boolean Hierarchy

Next, we introduce a parameterized variant BH(level) of the Boolean hierarachy, and show that it coincides with $\text{FPT}^{\text{NP}}[\text{few}]$. To define the class BH(level), we consider the following parameterized decision problem, that is based on the canonical problems $\text{BH}_i\text{-SAT}$ of the classes BH_i in the Boolean Hierarchy. Remember that the Boolean Hierarchy is defined using a hierarchy of complexity classes BH_i, for all $i \geq 1$: $\text{BH} = \bigcup_{i \in \mathbb{N}} \text{BH}_i$. Each of these classes BH_i can be characterized as the class of problems that can be reduced to the problem $\text{BH}_i\text{-SAT}$, which is defined inductively as follows. The problem $\text{BH}_1\text{-SAT}$ consists of all sequences (φ), where $\varphi \in \text{SAT}$. For even $i \geq 2$, the problem $\text{BH}_i\text{-SAT}$ consists of all sequences $(\varphi_1, \ldots, \varphi_i)$ of propositional formulas such that both $(\varphi_1, \ldots, \varphi_{i-1}) \in \text{BH}_{(i-1)}\text{-SAT}$ and $\varphi_i \in \text{UNSAT}$. For odd $i \geq 2$, the problem $\text{BH}_i\text{-SAT}$ consists of all sequences $(\varphi_1, \ldots, \varphi_i)$ of propositional formulas such that $(\varphi_1, \ldots, \varphi_{i-1}) \in \text{BH}_{(i-1)}\text{-SAT}$ or $\varphi_i \in \text{SAT}$.

In order to define the class BH(level), we consider the following parameterized problem.

BH(level)-SAT
Instance: A positive integer k and a sequence $(\varphi_1, \ldots, \varphi_k)$ of propositional formulas.
Parameter: k.
Question: $(\varphi_1, \ldots, \varphi_k) \in \text{BH}_k\text{-SAT}$?

The parameterized complexity class BH(level) then consists of all parameterized problems that can be fpt-reduced to the problem BH(level)-SAT. In other words, the class BH(level) consists of all parameterized problems Q for which there exists an fpt-reduction that reduces each instance (x, k) of Q to an instance of some problem in the $f(k)$-th level of the Boolean Hierarchy, for some computable function f. In the remainder of this section, we show that the classes $\text{FPT}^{\text{NP}}[\text{few}]$ and BH(level) coincide.

Theorem 7.5. $\text{FPT}^{\text{NP}}[\text{few}] = \text{BH(level)}$

Proof. The result follows directly from Lemmas 7.6 and 7.7.

We begin by showing that $\text{BH(level)} \subseteq \text{FPT}^{\text{NP}}[\text{few}]$.

Lemma 7.6. *Let Q be a parameterized problem that is contained in* BH*(level). Then there exists an algorithm A that decides Q in fpt-time using at most $f(k)$ SAT queries, where k is the parameter value and f is some computable function.*

Proof. We construct an algorithm that decides whether $(x, k) \in Q$. Since $Q \in$ BH(level), we know that there exists an fpt-reduction R that reduces any instance (x, k) of Q to an instance $R(x, k) = (x', k')$ of BH(level)-SAT. We know that $x' = (\varphi_1, \ldots, \varphi_{k'})$, and that $k' \leq g(k)$ for some computable function g. The algorithm, given an instance (x, k), firstly computes (x', k'). Then, for each $i \in [k']$, it decides whether φ_i is satisfiable by a single SAT query. Since (x', k') corresponds to a Boolean combination of statements concerning the satisfiability of the formulas φ_i, the algorithm can then decide in fpt-time whether $(x', k') \in \text{BH(level)-SAT}$.

Next, we show the converse inclusion, that is, we show that FPT^{NP}[few] $\subseteq$ BH(level).

Lemma 7.7. *Let Q be a parameterized problem that is in FPT^{NP}[few], i.e., there exists an algorithm A that decides Q in fpt-time using at most $g(k)$ SAT queries, where k is the parameter value and g is some computable function. Then there exists an fpt-reduction that reduces an instance (x, k) of Q to an instance (x', k') of BH(level)-SAT, where $k' \leq 2^{g(k)+1}$.*

Proof. We use the algorithm A to construct an fpt-reduction from Q to BH(level)-SAT. We will use the known fact that a disjunction of m SAT-UNSAT instances can be reduced to a single instance of BH$_{2m}$-SAT [35]. Let (x, k) be an instance of Q. We may assume without loss of generality that A makes exactly $g(k)$ SAT queries on any input (x, k). Consider the set $B = \mathbb{B}^{g(k)}$. We interpret each sequence $\bar{b} = (b_1, \ldots, b_{g(k)}) \in B$ as a sequence of answers to the SAT queries made by A; a 0 corresponds to the answer of the SAT query being "unsatisfiable" and a 1 corresponds to the answer being "satisfiable." For each $\bar{b} \in B$, we simulate the algorithm A on input (x, k) by using the answer specified by b_i to the i-th SAT query. Let us write $A_{\bar{b}}(x, k)$ to denote the simulation of A on input (x, k) where the answers to the SAT queries are specified by $\bar{b}$. By performing this simulation for each $\bar{b} \in B$, we can determine in fpt-time the set $B' \subseteq B$ of sequences $\bar{b}$ such that $A_{\bar{b}}(x, k)$ accepts.

We know that A accepts (x, k) if and only if the "correct" sequence of answers is contained in B', in other words, A accepts (x, k) if and only if there exists some $\bar{b} = (b_1, \ldots, b_{g(k)}) \in B'$ such that for each b_i it holds that if $b_i = 0$ then ψ_i is unsatisfiable, and if $b_i = 1$ then ψ_i is satisfiable, where ψ_i denotes the formula used for the i-th SAT query made by $A_{\bar{b}}(x, k)$. For each $\bar{b} \in B'$, we construct an instance $I(\bar{b}) = (\varphi_1, \varphi_0)$ of SAT-UNSAT that is a yes-instance if and only if the above condition holds for sequence $\bar{b}$, as follows. Let $(\psi_1, \ldots, \psi_{g(k)})$ be the propositional formulas that $A_{\bar{b}}(x, k)$ uses for the SAT queries, i.e., ψ_i corresponds to the formula used for the i-th SAT query of $A_{\bar{b}}(x, k)$. We may assume without loss of generality that the formulas ψ_i are variable disjoint, i.e., for each $i, i' \in [g(k)]$ with $i < i'$, it holds that $\text{Var}(\psi_i) \cap \text{Var}(\psi_{i'}) = \emptyset$. We construct the instance (φ_1, φ_0) as follows:

$$C_1 = \{ i \in [g(k)] : b_i = 1 \};$$
$$\varphi_1 = \bigwedge_{j \in C_1} \psi_j;$$
$$C_0 = \{ i \in [g(k)] : b_i = 0 \}; \text{ and}$$
$$\varphi_0 = \bigvee_{j \in C_0} \psi_j.$$

It is straightforward to verify that $I(\bar{b}) \in$ SAT-UNSAT if and only if $\bar{b}$ corresponds to the "correct" sequence of answers for the SAT queries made by A, i.e., for each b_i with $b_i = 0$ it holds that ψ_i is unsatisfiable, and for each b_i with $b_i = 1$ it holds that ψ_i is satisfiable.

We constructed ℓ instances $I(\overline{b_1}), \ldots, I(\overline{b_\ell})$ of SAT-UNSAT, for some $\ell \leq 2^{g(k)}$, such that the algorithm A accepts the instance (x, k), and thus $(x, k) \in Q$, if and only if there exists some $i \in [\ell]$ such that $I(\overline{b_\ell}) \in$ SAT-UNSAT. In other words, we reduced our original instance (x, k) of Q to a disjunction of $\ell \leq 2^{g(k)}$ instances of SAT-UNSAT. We know that such a disjunction can be reduced to an instance of $BH_{2\ell}$-SAT [35]. This completes our fpt-reduction from Q to BH(level).

The bound of $2^{g(k)+1}$ in the proof of Lemma 7.7 can be improved to a bound of $2^{g(k)}$ using the "mind change technique" [14, 133, 194]. For our purposes any bound depending only on k suffices.

7.2.2 Satisfiability Problems Complete for FPT^{NP}[few]

We consider to parameterized problems that based on two notions of maximal models for propositional formulas. Using the characterization of $\text{FPT}^{\text{NP}}[\text{few}]$ as a parameterized variant of the Boolean Hierarchy, we show that these two problems are $\text{FPT}^{\text{NP}}[\text{few}]$-complete.

We begin by defining the two notions of maximal models. Let φ be a propositional formula, and let $X \subseteq \text{Var}(\varphi)$ be a subset of variables of φ. Moreover, fix an arbitrary linear ordering on X. We say that an assignment $\alpha : \text{Var}(\varphi) \to \mathbb{B}$ is *an X-maximal model of* φ if α satisfies φ and there exists no assignment α' that satisfies φ and that sets more variables in X to true than α. We say that an assignment $\alpha : \text{Var}(\varphi) \to \mathbb{B}$ is *the lexicographically X-maximal model of* φ if α satisfies φ and there exists no assignment α' that satisfies φ and that is lexicographically strictly larger than α, when restricted to the variables in X.

Consider the following three parameterized problems.

ODD-LOCAL-MAX-MODEL
Instance: A propositional formula φ, and a subset $X \subseteq \text{Var}(\varphi)$ of variables.
Parameter: $|X|$.
Question: Do the X-maximal models of φ set an odd number of variables in X to true?

ODD-LOCAL-LEX-MAX-MODEL
Instance: A propositional formula φ, and a subset $X \subseteq \text{Var}(\varphi)$ of variables.
Parameter: $|X|$.
Question: Does the lexicographically X-maximal model of φ set an odd number of variables in X to true?

LOCAL-MAX-MODEL
Instance: A satisfiable propositional formula φ, a subset $X \subseteq \text{Var}(\varphi)$ of variables, and a variable $w \in X$.
Parameter: $|X|$.
Question: Is there a model of φ that sets a maximal number of variables in X to true (among all models of φ) and that sets w to true?

We show that these three problems are $\text{FPT}^{\text{NP}}[\text{few}]$-complete. Membership in $\text{FPT}^{\text{NP}}[\text{few}]$ can be shown straightforwardly. To show $\text{FPT}^{\text{NP}}[\text{few}]$-hardness, we consider the following auxiliary problem.

> BOUNDED-SAT-UNSAT-DISJUNCTION
> *Instance:* A family $(\varphi_i, \varphi_i')_{i \in [k]}$ of pairs of propositional formulas.
> *Parameter:* k.
> *Question:* Is there some $\ell \in [k]$ such that $(\varphi_\ell, \varphi_\ell') \in$ SAT-UNSAT?

We begin by showing that BOUNDED-SAT-UNSAT-DISJUNCTION is FPT^{NP}[few]-complete.

Proposition 7.8. BOUNDED-SAT-UNSAT-DISJUNCTION *is* FPT^{NP}[few]-*complete.*

Proof. Membership in FPT^{NP}[few] can be shown routinely. Hardness for FPT^{NP}[few] follows directly from the fact that every instance $(\varphi_1, \ldots, \varphi_k)$ of BH(level)-SAT can be expressed as a disjunction of $f(k)$ instances of SAT-UNSAT [35], for some computable function f.

Then, we can show FPT^{NP}[few]-hardness for ODD-LOCAL-MAX-MODEL and ODD-LOCAL-LEX-MAX-MODEL by providing an fpt-reduction from BOUNDED-SAT-UNSAT-DISJUNCTION.

Proposition 7.9. ODD-LOCAL-MAX-MODEL *is* FPT^{NP}[few]-*complete.*

Proof. Membership in FPT^{NP}[few] can be shown routinely. We show hardness by giving an fpt-reduction from BOUNDED-SAT-UNSAT-DISJUNCTION. Let $(\varphi_i, \varphi_i')_{i \in [k]}$ be an instance of BOUNDED-SAT-UNSAT-DISJUNCTION. We assume without loss of generality that the formulas φ_i and φ_i' are all variable-disjoint. We construct an instance (ψ, Z) of ODD-LOCAL-MAX-MODEL as follows. We consider the following disjoint sets of propositional variables:

$$Y = \bigcup_{i \in [k]} (\text{Var}(\varphi_i) \cup \text{Var}(\varphi_i')),$$
$$X = \{ x_i, x_i' : i \in [k] \},$$
$$Y = \{ y_i, y_i' : i \in [k] \}, \text{ and}$$
$$W = \{w\}.$$

We let $Z = X \cup Y \cup W$.

We then define the formula ψ to be the conjunction of the following propositional formulas. Firstly, we ensure that whenever some x_i is true, then φ_i must be satisfied, and whenever some x_i' is true, then φ_i' must be satisfied. We do so by means of the following formula:

$$\bigwedge_{i \in [k]} ((x_i \to \varphi_i) \wedge (x_i' \to \varphi_i')).$$

Then, we ensure that the variables y_i and y_i' get the same truth value as the variables x_i and x_i' (respectively):

$$\bigwedge_{i \in [k]} ((x_i \leftrightarrow y_i) \wedge (x_i' \leftrightarrow y_i')).$$

Finally, we ensure that w can only be true if there is some $i \in [k]$ such that x_i is true and x_i' is false:

$$w \leftrightarrow \bigvee_{i \in [k]} (x_i \wedge \neg x_i').$$

The satisfying assignment of ψ that sets as many variables in Z to true as possible satisfies as many of the formulas φ_i and φ_i' as possible. Moreover, the number of variables in Z that are set to true is odd if and only if w is satisfied. This is the case for the Z-maximal model if and only if there is some $i \in [k]$ where φ_i is satisfied and φ_i' is not satisfied. Because this model is Z-maximal, we know that φ_i' is unsatisfiable. Therefore $(\varphi_i, \varphi_i') \in$ SAT-UNSAT. Conversely, if $(\varphi_i, \varphi_i') \in$ SAT-UNSAT for some $i \in [k]$, we get that the Z-maximal model satisfies w.

Thus $(\varphi_i, \varphi_i')_{i \in [k]} \in$ ODD-LOCAL-MAX-MODEL if and only if $(\psi, Z) \in$ ODD-LOCAL-MAX-MODEL.

To show FPT^{NP}[few]-hardness for ODD-LOCAL-LEX-MAX-MODEL, we can also use the reduction that we used in the FPT^{NP}[few]-hardness proof in Proposition 7.9.

Proposition 7.10. ODD-LOCAL-LEX-MAX-MODEL *is* FPT^{NP}[few]-*complete.*

Proof. Membership in FPT^{NP}[few] can be shown routinely. Hardness follows from the fpt-reduction in the proof of Proposition 7.9, which is also an fpt-reduction from BOUNDED-SAT-UNSAT-DISJUNCTION to ODD-LOCAL-LEX-MAX-MODEL, for an arbitrary linear ordering on the variables in Z where the variable w is ordered last.

Similarly, to show FPT^{NP}[few]-hardness for LOCAL-MAX-MODEL, we can also use the reduction that we used in the FPT^{NP}[few]-hardness proof in Proposition 7.9.

Proposition 7.11. LOCAL-MAX-MODEL *is* FPT^{NP}[few]-*complete.*

Proof. Membership in FPT^{NP}[few] can be shown routinely. To show hardness, we can straightforwardly modify the fpt-reduction in the proof of Proposition 7.9, to an fpt-reduction from BOUNDED-SAT-UNSAT-DISJUNCTION to LOCAL-MAX-MODEL, by taking (ψ, Z, w) as the constructed instance of LOCAL-MAX-MODEL.

7.3 Lower Bounds on the Number of Oracle Queries

In this section, we show how hardness for the parameterized complexity class FPT^{NP}[few] (for a parameterized problem Q) can be used to show a lower bound on the number of SAT queries made by any fpt-algorithm that solves Q. In particular, we show that any FPT^{NP}[few]-hard problem Q cannot be solved by an fpt-algorithm that queries a SAT oracle only $O(1)$ times, unless the Polynomial Hierarchy collapses. Moreover, we preview a result from Chap. 14 that hardness for the parameterized complexity classes A[2], $\Sigma_2^p[k*]$ and $\Sigma_2^p[*k, t]$ can be used to show that problems cannot be solved in fpt-time using $f(k)$ SAT queries.

We begin with showing the $\omega(1)$ lower bound based on FPT^{NP}[few]-hardness.

Proposition 7.12. *Let Q be any $FPT^{NP}[few]$-hard parameterized problem. Then Q is not solvable by an fpt-algorithm that uses only $O(1)$ SAT queries, unless the Polynomial Hierarchy collapses.*

Proof. Assume that Q is solvable by an fpt-algorithm that uses only c SAT queries, where c is a constant. We will show that the PH collapses. Since Q is BH(level)-hard, we know that there exists an fpt-reduction R_1 from BH(level)-SAT to Q. Then, by Lemma 7.7, there exists an fpt-reduction R_2 from Q to BH(level)-SAT, that reduces any instance (x', k') of Q to an instance (x'', k'') of BH(level)-SAT, where $k'' \leq 2^{c+1}$. Then, the composition R of R_1 and R_2 is an fpt-reduction from BH(level)-SAT to itself such that any instance (x, k) of BH(level)-SAT is reduced to an equivalent instance (x'', k'') of BH(level)-SAT, where $k'' \leq m = 2^{c+1}$. We can straightforwardly modify this reduction to always produce an instance (x'', m) of BH(level)-SAT, by adding trivial instances of SAT to the sequence x''.

We now show that the Boolean Hierarchy collapses to the m-th level, where $m = 2^{c+1}$. Let y be an instance of BH_{m+1}-SAT. We can then see the reduction R as a polynomial-time reduction from BH_{m+1}-SAT to BH_m-SAT: the fpt-reduction R runs in time $f(k) \cdot n^{O(1)}$, and since $k = m + 1$ is a constant, the factor $f(k)$ is constant. From this we can conclude that $BH_m = BH_{m+1}$. Thus, the BH collapses, and consequently the PH collapses [41, 126]. $\qquad \square$

In Chap. 14, we show that we can rule out that a parameterized problem Q is solvable in fpt-time using only $f(k)$ SAT queries (for some computable function f) by showing that Q is hard for A[2], $\Sigma_2^p[k*]$ or $\Sigma_2^p[*k, t]$ —under various complexity-theoretic assumptions. For instance, for the case of A[2], we show that if any A[2]-hard problem is in $FPT^{NP}[few]$, then there exists a subexponential-time Turing reduction from $QSAT_2(3DNF)$ to SAT, that is, a Turing reduction that runs in time $2^{o(n)}$, where n is the number of variables (Corollary 14.7). For more details, we refer to Chap. 14.

7.4 Bounded Optimization Problems

In Sect. 7.2, we showed that the class $FPT^{NP}[few]$ can be seen as a parameterized variant of the Boolean Hierarchy. In this section, we consider another alternative characterization of the class $FPT^{NP}[few]$ that is based on bounded optimization problems. In particular, this characterization is based on the problem of maximizing the output of a non-deterministic Turing machine that outputs positive integers represented by binary strings of length at most $f(k)$, where k is the parameter value and f is some computable function. Moreover, we illustrate this characterization by showing $FPT^{NP}[few]$-completeness for yet another parameterized problem.

The characterization in terms of bounded optimization problems is similar to results of Spakowski for the class Θ_2^p [184]. We use the following notation. Let $w \in \mathbb{B}^*$. By $|w|_1$ we denote the number of 1's occurring in w. We characterize $FPT^{NP}[few]$ as follows.

Theorem 7.13. A parameterized problem Q is in $\text{FPT}^{\text{NP}}[\text{few}]$ if and only if there is an NTM $\mathbb{M}$ with output tape such that there exist a polynomial p and computable functions f and g such that for every instance (x, k) of Q:

1. Every computation path of $\mathbb{M}(x, k)$ has length at most $f(k) \cdot p(|x|)$.
2. The length of the output of every computation path of $\mathbb{M}(x, k)$ is bounded by $g(k)$.
3. Every two paths ρ_1 and ρ_2 of $\mathbb{M}(x, k)$ have the same acceptance behavior whenever they have the same number of 1's in the output.
4. It holds that $(x, k) \in Q$ if and only if $\mathbb{M}$ accepts (x, k) on $\rho_{\max}$, where $\rho_{\max}$ is a computation path with the maximum number of 1's in the output.

Proof. ($\Rightarrow$) Let Q be an arbitrary language in $\text{FPT}^{\text{NP}}[\text{few}]$. Then there exists a deterministic Turing machine $\mathbb{M}_1$, with an oracle $C \in \text{NP}$, that decides Q. Without loss of generality, we assume that $\mathbb{M}_1$ queries exactly $f(k)$ strings to the oracle for every input (x, k). Because $C \in \text{NP}$, there exists some $D \in \text{P}$ and a polynomial r such that:

$$z \in C \text{ if and only if there exists some } y \text{ with } |y| \leq r(|z|) \text{ such that } (x, y) \in D.$$

We construct an NTM $\mathbb{M}_2$ that operates as follows on input (x, k):

Step 1: Non-deterministically guess $b_1, b_2, \ldots, b_{f(k)} \in \mathbb{B}$.
Step 2: On each path ρ with guessed $b_1, b_2, \ldots, b_{f(k)}$, construct the oracle queries $q_1, q_2, \ldots, q_{f(k)} \in \Sigma^*$ by simulating $\mathbb{M}_1^C(x, k)$, where the answers to the queries of $\mathbb{M}_1$ to C are substituted by $b_1, b_2, \ldots, b_{f(k)}$—take "yes" as an answer to the i-th query if and only if $b_i = 1$.
Step 3: Successively, for each i with $b_i = 1$, non-deterministically guess a string y_i with $|y_i| \leq r(|q_i|)$. Verify that each (q_i, y_i) is in D. Output the string "0" and reject on the current path ρ if at least one such test fails. Otherwise, continue as follows on ρ.
Step 4: Output the string $w = 11 \ldots 1$, where $|w| = \text{number}(b_1 \ldots b_{f(k)}) + 1$. We know that $|w| = 2^{O(f(k))}$.
Step 5: Accept on ρ if and only if the computation of $\mathbb{M}_1^C(x, k)$ simulated in Step 2 was accepting.

Clearly, the machine $\mathbb{M}_2$ runs in fpt-time. Let $\rho_{\max}$ be any path reaching Step 4 with lexicographically maximum $b_1 \ldots b_{f(k)}$ among all paths reaching Step 4. It is straightforward to see that:

- The bits $b_1 \ldots b_{f(k)}$ guessed on $\rho_{\max}$ represent the correct oracle answers to the queries made by $\mathbb{M}_1^C$ on input (x, k).
- The output string w for $\rho_{\max}$ has a maximum number of 1's among the output strings of all paths of $\mathbb{M}_2$.

Therefore, $\mathbb{M}_1^C$ accepts (x, k) if and only if $\mathbb{M}_2(x, k)$ accepts in Step 5 on path $\rho_{\max}$.

($\Leftarrow$) Let $\mathbb{M}_1$ be a suitable non-deterministic Turing machine with output tape. Hence, $(x, k) \in Q$ if and only if $\mathbb{M}_1$ accepts (x, k) on ρ_{max}, where ρ_{max} is a computation path of $\mathbb{M}_1(x, k)$ with a maximum number of 1's in the output. We have to show that $Q \in \text{FPT}^{NP}[\text{few}]$.

We sketch an $\text{FPT}^{NP}[\text{few}]$ algorithm A that decides Q. We know that there exists a computable function g such that the length of the output w_ρ of any computation path ρ is bounded by $g(k)$, i.e., $|w_\rho| \le g(k)$. Let O be the following decision problem. Instances of O are triples $(\mathbb{M}, \ell, b)$, where $\mathbb{M}$ is an NTM, ℓ is a positive integer, and $b \in \mathbb{B}^*$. Any instance $(\mathbb{M}, \ell, 1)$ is a yes-instance if and only if $\mathbb{M}$ has an accepting computation path ρ of $\mathbb{M}_1$ where $|w_\rho|_1 = \ell$, and any instance $(\mathbb{M}, \ell, 0)$ is a yes-instance if and only if $\mathbb{M}$ has a rejecting computation path ρ with $|w_\rho|_1 = \ell$. Clearly, O is a language in NP. For each $\ell \in [g(k)]$, the algorithm A queries $(\mathbb{M}_1, \ell, b)$ for all $b \in \mathbb{B}$. Then, the algorithm A determines ℓ_{max}, where:

$$\ell_{max} = \max\{\ell : \ell \in [g(k)], \text{ for some } b \in \mathbb{B}: (\mathbb{M}_1, \ell, b) \in O\}.$$

Finally, the algorithm accepts if and only if $(\mathbb{M}_1, \ell_{max}, 1) \in O$. It is straightforward to verify that A correctly decides whether $\mathbb{M}_1$ accepts (x, k) on ρ_{max}. Also, A is an fpt-algorithm that queries the language O at most $2g(k)$ times.

Next, we consider another parameterized problem that we show to be $\text{FPT}^{NP}[\text{few}]$-complete. We use the characterization of $\text{FPT}^{NP}[\text{few}]$ in terms of bounded optimization problems to show $\text{FPT}^{NP}[\text{few}]$-hardness for this problem. This is a proof "from first principles," i.e., a proof that does not rely on a reduction from another $\text{FPT}^{NP}[\text{few}]$-complete problem.

Let φ be a propositional formula, and let $X \subseteq \text{Var}(\varphi)$ be a subset of variables occurring in φ. We define $\max^1(\varphi, X)$ as follows:

$$\max^1(\varphi, X) = \max\{|\{x \in X : \alpha(x) = 1\}| : (\alpha : \text{Var}(\varphi) \to \mathbb{B}), \varphi[\alpha] = 1\}.$$

That is, $\max^1(\varphi, X)$ is the maximum number of variables in X set to true by any assignment that satisfies φ. Now, consider the following parameterized problem.

LOCAL-MAX-MODEL-COMPARISON

Instance: Two satisfiable propositional formulas φ_1 and φ_2, a positive integer k, and a subset $X \subseteq \text{Var}(\varphi_1) \cap \text{Var}(\varphi_2)$ of k variables.

Parameter: k.

Question: $\max^1(\varphi_1, X) = \max^1(\varphi_2, X)$?

We show that this problem is $\text{FPT}^{NP}[\text{few}]$-complete.

Proposition 7.14. LOCAL-MAX-MODEL-COMPARISON *is* $\text{FPT}^{NP}[\text{few}]$-*complete.*

Proof. We firstly show membership in $\text{FPT}^{NP}[\text{few}]$. Let $(\varphi_1, \varphi_2, k, X)$ be an instance of LOCAL-MAX-MODEL-COMPARISON. Then the following problem O is in NP:

$$O = \{(\varphi, m) : \varphi \text{ is a propositional formula, t.e. } (\alpha : \text{Var}(\varphi) \to \mathbb{B})$$
$$\text{s.t. } \varphi[\alpha] = 1, \text{ and } \alpha \text{ sets exactly } m \text{ variables in } X \text{ to } 1 \quad\}.$$

We sketch an fpt-algorithm A that decides LOCAL-MAX-MODEL-COMPARISON and that makes $2k$ queries to the oracle O. First, A asks whether $(\varphi_i, m) \in O$ for all $i \in \{1, 2\}$ and all $m \in [k]$. From the answers to these queries, it is straightforward to determine whether $\max^1(\varphi_1, X) = \max^1(\varphi_2, X)$.

Next, we show that LOCAL-MAX-MODEL-COMPARISON is $\text{FPT}^{\text{NP}}[\text{few}]$-hard. Let Q be an arbitrary problem in $\text{FPT}^{\text{NP}}[\text{few}]$. We give an fpt-reduction from Q to LOCAL-MAX-MODEL-COMPARISON. By Theorem 7.13, we know that there exists an NTM $\mathbb{M}$ with output tape such that there exists a polynomial p and computable functions f and g such that for every $(x, k) \in \Sigma^* \times \mathbb{N}$: (1) every computation path of $\mathbb{M}(x, k)$ has length at most $f(k) \cdot p(|x|)$, (2) the length of the output of every computation path of $\mathbb{M}(x, k)$ is bounded by $g(k)$, (3) every two paths ρ_1 and ρ_2 of $\mathbb{M}(x, k)$ have the same acceptance behavior whenever they have the same number of 1's in the output, and (4) $(x, k) \in Q$ if and only if $\mathbb{M}$ accepts (x, k) on $\rho_{\max}$. Assume without loss of generality that for every input (x, k) there is at least one (accepting or non-accepting) computation path of $\mathbb{M}(x, k)$ with at least one 1 in the output.

Let (x, k) be an instance of Q. We now construct two propositional formulas $\varphi_1(x, k)$ and $\varphi_2(x, k)$ as follows. In order to do so, we will make use of the well-known proof of the Cook-Levin Theorem [54, 144], in which a non-deterministic Turing machine $\mathbb{M}'$ together with an input $(x, k) \in \Sigma^* \times \mathbb{N}$ is transformed into a formula $\varphi_{\mathbb{M}'(x,k)}$ propositional atoms of the form:

- *state*(q, i), representing whether $\mathbb{M}(x, k)$ is in state q at time step i;
- *tape*(σ, i, ℓ, m), representing whether in the execution of $\mathbb{M}(x, k)$ the ℓ-th cell of the m-th tape contains symbol σ at time step i;
- *head*(i, ℓ, m), representing whether in the execution of $\mathbb{M}(x, k)$ the head of the m-th tape is at position ℓ at time step i.

Furthermore, whenever $\mathbb{M}'$ halts within $f(k) \cdot p(|x|)$ steps when given an input (x, k), for some polynomial p and some computable function f, then the formula $\varphi_{\mathbb{M}(x,k)}$ is bounded in size by $f'(k) \cdot p'(|x|)$, for some polynomial p' and some computable function f'. Moreover, we know that the output tape of $\mathbb{M}$ can only contain 0's and 1's. Therefore, we introduce the following additional variables:

- *output*(i, ℓ), representing whether in the execution of $\mathbb{M}(x, k)$ the ℓ-th cell of the output tape contains a 1 at time step i.

Observe that we only need the variables *output*(i, ℓ) for $\ell \in [g(k)]$.

Now, we construct the formulas $\varphi_1(x, k)$ and $\varphi_2(x, k)$. Assume without loss of generality that every computation path of $\mathbb{M}(x, k)$ is of length exactly $f(k) \cdot p(|x|) = r$. The formula $\varphi_1(x, k)$ contains the variables described above, and it satisfiable if and only if the instantiation of the variables corresponds to a *valid (accepting or non-accepting)* computation path of $\mathbb{M}(x, k)$. The formula $\varphi_2(x, k)$ also contains the variables described above, and it satisfiable if and only if either (1) all variables are set to 0, or (2) the instantiation of the variables corresponds to a *valid and accepting* computation path of $\mathbb{M}(x, k)$. It is straightforward to verify that both $\varphi_1(x, k)$

and $\varphi_2(x, k)$ are satisfiable. Furthermore, the set $X \subseteq \mathrm{Var}(\varphi_1(x, k)) \cap \mathrm{Var}(\varphi_2(x, k))$ consists of the variables *output*(r, ℓ), for $\ell \in [g(k)]$. Clearly, $|X| = g(k)$.

By construction of $\varphi_1(x, k)$ and $\varphi_2(x, k)$, it is straightforward to verify that $\max^1(\varphi_1(x, k), X)) \geq \max^1(\varphi_2(x, k), X))$ (since every satisfying assignment for $\varphi_2(x, k)$ that corresponds to a computation path is also a satisfying assignment for $\varphi_1(x, k)$). It is also straightforward to verify that $\max^1(\varphi_1(x, k), X)) = \max^1(\varphi_2(x, k), X))$ if and only if $\mathbb{M}(x, k)$ accepts on $\rho_{\max}$. Therefore we know that $(x, k) \in Q$ if and only if $(\varphi_1(x, k), \varphi_2(x, k), X, g(k)) \in$ LOCAL-MAX-MODEL-COMPARISON.

7.5 Witness-Producing SAT Oracles

In this section, we consider an extension of the class $\mathrm{FPT}^{\mathrm{NP}}[\text{few}]$ based on a more powerful oracle model, where the SAT oracles return a satisfying assignment for yes-answers. It can be argued that this extended oracle model more closely resembles the setting of algorithms that invoke a SAT solver, because SAT solvers in practice also return a satisfying assignment when given a satisfying formula as input. We investigate the additional solving power that this more powerful oracle model yields. We show that for the case of decision problems, exactly the same set of problems can be solved in fpt-time using $f(k)$ SAT queries in both oracle models. On the other hand, we show that for the case of search problems, the oracles that return satisfying assignments lead to more solving power.

Because our analysis of the additional solving power yielded by the witness oracle model also applies to several classical complexity classes (based on polynomial-time algorithms that query a SAT oracle), we consider various notions both from a classical complexity perspective as well as from a parameterized complexity perspective.

Search Problems

We begin with defining the concept of *search problems*. A *(classical) search problem* is a set $Q \subseteq \Sigma^* \times \Sigma^*$ of pairs of strings. Intuitively, each such pair $(x, y) \in Q$ specifies an answer y to an input x. We say that an algorithm *solves* Q if on any input $x \in \Sigma^*$, it outputs a string $y \in \Sigma^*$ such that $(x, y) \in Q$ if such a string y exists, and outputs "none" otherwise. Similarly, a *parameterized search problem* is a set $Q \subseteq \Sigma^* \times \Sigma^* \times \mathbb{N}$ of triples (x, y, k). Intuitively, each such triple (x, y, k) specifies an answer y to an input (x, k). *solves* Q if on any input $(x, k) \in \Sigma^* \times \mathbb{N}$, it outputs a string $y \in \Sigma^*$ such that $(x, y, k) \in Q$ if such a string y exists, and outputs "none" otherwise. When speaking about inputs (x, k) of Q, we let k denote the parameter value.

Witness Oracles

Next, we formally define the oracle model based on witness oracles. Let O be a search problem, i.e., a set $O \in \Sigma^* \times \Sigma^*$ of pairs of strings. A Turing machine with access to a *witness O oracle* is a Turing machine $\mathbb{M}$ that has a dedicated oracle tape and dedicated states q_{query}, q_{yes} and q_{no}. Whenever $\mathbb{M}$ is in the state q_{query} it transitions into the state q_{yes} if the oracle tape contains a string x such that there exists some $y \in \Sigma^*$ such that $(x, y) \in O$, and in addition the contents of the oracle tape are replaced by (the encoding of) such a string y; it transitions into the state q_{no} if there exists no y such that $C(x, y)$. Such transitions are called *oracle queries*.

Throughout this section, we will also consider SAT as a search problem. That is, in this case we let:

$$SAT = \{ (\varphi, \alpha) : \varphi \text{ is a propositional formula}, \alpha : \text{Var}(\varphi) \to \mathbb{B}, \varphi[\alpha] = 1 \}.$$

Whether we use SAT to denote this search problem or the usual decision problem will be clear from the context. For instance, whenever we write "witness SAT oracle", we denote the search problem. Moreover, if we want to emphasize that we use the decision problem SAT as an oracle, we write "decision SAT oracle."

When speaking about a witness NP oracle, we mean a search problem O for which the following problem is in NP:

$$\{ x : x \in \Sigma^*, \text{ there is some } y \text{ with } |y| = |x|^{O(1)} \text{ such that } (x, y) \in O \}.$$

Parallel Oracle Access

We also consider some complexity classes that are based on Turing machines that have parallel access to an oracle. Let O be a decision problem. A Turing machine $\mathbb{M}$ with *parallel access to a decision O oracle* is a Turing machine with a dedicated oracle tape, and dedicated states q_{query} and q_{done}. The machine $\mathbb{M}$ can write several strings $x_1, \ldots, x_u$ to the oracle tape, separated by a designated symbol \$. Then, once during the computation, $\mathbb{M}$ can enter the state q_{query}, after which it does not proceed according to the transition relation, but instead it transitions into the state q_{done}. Moreover, when this special transition occurs, the contents of the oracle tape are replaced by a bitstring $b_1 \ldots b_u$, where $b_i = 1$ if and only if $x_i \in C$, for all $i \in [u]$, where $x_1\$ \ldots \$x_u$ is the content of the oracle tape. In this case, we say that the machine makes u oracle queries.

In a similar way, we define Turing machines with *access to a witness O oracle*, for search problems O. The only difference is that the contents of the oracle tape are not replaced by a bitstring representing whether $x_i \in O$, but by a sequence of answers $a_1\$ \ldots \$a_u$, where a_i encodes "none" if there exists no y such that $(x_i, y) \in O$; and otherwise, if there exists some y such that $(x_i, y) \in O$, the answer a_i encodes such a string y.

Complexity Classes

Finally, before we start the investigation of the additional power yielded by the witness oracle model, we consider various (parameterized) complexity classes containing decision and search problems.

In Sect. 7.1, we considered the class Θ_2^p, consisting of those decision problems that are solved by a polynomial-time algorithm that queries a decision SAT oracle $O(\log n)$ times. This class coincides with the class $P_{||}^{NP}$, consisting of all decision problems that are decided by a polynomial-time algorithm that has access to a decision SAT oracle [33, 121, 133]. Similarly, we consider the class $F\Theta_2^p$ of search problems that are solved by a polynomial-time algorithm that queries a decision SAT oracle $O(\log n)$ times. Also, the class $FP_{||}^{NP}$ consists of all search problems that are solved by a polynomial-time algorithm that has parallel access to a SAT oracle. Analogously, we define the classes $F\Theta_2^p[\text{wit}]$ and $FP_{||}^{NP}[\text{wit}]$, based on algorithms with (parallel) access to witness SAT oracles. (The complexity class $F\Theta_2^p[\text{wit}]$ coincides with the class $FNP//OptP[\log]$, that is defined as the set of all search problems that can be solved by a non-deterministic polynomial-time Turing machine that receives as advice the answer to one "NP optimization" computation [48, 134].) In Sect. 7.1, we also considered the class Δ_2^p, consisting of those decision problems that are solved by a polynomial-time algorithm with access to a decision SAT oracle. Similarly to the definitions above, we define the complexity classes $F\Delta_2^p$ and $F\Delta_2^p[\text{wit}]$ of search problems.

Finally, we consider several classes of parameterized search problems. We consider the complexity class $FPT^{NP}[\text{few,wit}]$ of parameterized decision problems that are decided by an fpt-algorithm that queries a witness SAT oracle at most $f(k)$ times, where k denotes the parameter value and f is some computable function. Similarly, the class $FFPT^{NP}[\text{few,wit}]$ consists of all parameterized search problems that are solved by an fpt-algorithm that queries a witness SAT oracle at most $f(k)$ times. The classes $FPT_{||}^{NP}[\text{few,wit}]$ and $FFPT_{||}^{NP}[\text{few,wit}]$ consist of all parameterized decision and search problems (respectively), that are solved by an fpt-algorithm with parallel access to a witness SAT oracle, that makes at most $f(k)$ queries for any input (x, k), for some computable function f.

7.5.1 Comparing the Oracle Models for Decision Problems

We firstly compare the two different oracle models (decision vs. witness SAT oracles) in the setting of decision problems. We begin by showing that for polynomial-time algorithms without any further bounds on the number of oracle queries, the two models are equally powerful. Then, we show a similar result for polynomial-time algorithms that can query the oracle a logarithmic number of times. Finally, we show how these results can be adapted to the parameterized setting. In particular, we show that for fixed-parameter tractable algorithms that can query the oracle $f(k)$ times, the two oracle models are equally powerful.

We begin with showing that for the case of Δ_2^p, witness oracles are not more powerful than decision oracles.

Proposition 7.15. $\Delta_2^p[\text{wit}] = \Delta_2^p$.

Proof (sketch). By the self-reducibility of SAT, one can query a decision oracle a linear number of times to extract a satisfying assignment for any satisfying propositional formula φ. Therefore, any polynomial-time algorithm with access to a witness SAT oracle can straightforwardly be modified to work also with a decision SAT oracle.

Next, we turn our attention to the case of Θ_2^p. The following proof is based on a result by Spakowski [184, Proposition 3.2.3]. The following result was actually already known from the literature [137, Corollary 6.3.5]. However, since we want to apply a modification of the proof to the case of $\text{FPT}^{\text{NP}}[\text{few}]$, we give a proof of the statement.

Proposition 7.16. $\Theta_2^p[\text{wit}] = \Theta_2^p$.

Proof (sketch). Clearly $\Theta_2^p \subseteq \Theta_2^p[\text{wit}]$. We show that $\Theta_2^p[\text{wit}] \subseteq \Theta_2^p$. Let $Q \in \Theta_2^p[\text{wit}]$ be an arbitrary problem. We show that $Q \in \Theta_2^p$. Let A be the algorithm that decides for any instance x of Q with $|x| = n$ whether $x \in Q$ in polynomial time, by querying a witness NP oracle at most $O(\log n)$ times. We may assume without loss of generality that M queries the oracle exactly $z(n) = O(\log n)$ times on any input of length n.

We construct an NTM M that has an output tape, and that has the following property: $x \in Q$ if and only if M accepts x on some computation path with a maximum number of 1's in the output. Here the maximum is taken over all possible (accepting or rejecting) computation paths of M.

We let M implement the following guess-and-check algorithm. Let x be an instance of Q with $|x| = n$. For $i \in [z(n)]$, the algorithm guesses some $b_i \in \mathbb{B}$. Furthermore, for each $i \in [z(n)]$ such that $b_i = 1$, the algorithm guesses a (polynomial-length) witness w_i corresponding to the "yes-answer" b_i. Then, we let the algorithm simulate A where the bits $b_1, \ldots, b_{z(n)}$ are used for the oracle answers, and where the witnesses w_i are used for the witnesses corresponding to the "yes-answers." Then, for each $i \in [z(n)]$ the algorithm verifies whether w_i is indeed a correct witness that justifies the "yes-answer" for the i-th query made to the NP oracle. If there is any incorrect witness w_i, the algorithm outputs the binary string 0 of length 1, and rejects. Otherwise, it continues. Then, the algorithm outputs the string 1^{w+1}, where w is the number whose binary representation is the string $b_1 \ldots b_{z(n)}$. Since $z(n) = O(\log n)$, we know that w is polynomial in n. Finally, the algorithm accepts the input x if and only if the simulation of A with the guessed bits b_i and the guessed witnesses w_i accepted. It is straightforward to verify that M satisfies the required property.

Next, we develop an algorithm witnessing that $Q \in \Theta_2^p$, i.e., a polynomial-time algorithm that decides for any input x of Q with $|x| = n$ whether $x \in Q$ by making $O(\log n)$ queries to a decision NP oracle. We use two different NP oracles. The first oracle decides, given a non-determinisic TM $\mathbb{M}$, an input x, and an integer m, whether there exists a computation path of $\mathbb{M}$ on input x with m 1's in the output. The second oracle decides, given a non-determinisic TM $\mathbb{M}$, an input x, and an integer m, whether there exists an accepting computation path of $\mathbb{M}$ on input x with m 1's in the output.

Let x be an instance of Q with $|x| = n$. We know that $x \in Q$ if and only if there is some accepting computation path of $\mathbb{M}$ on input x with a maximum number of 1's in the output. Moreover, we know that any computation path of $\mathbb{M}$ on input x has at most $2^{z(n)+1} = n^{O(1)}$ 1's in the output. With binary search, we can determine the maximum number w_{max} of 1's in any computation path of N on input x with $O(\log n)$ queries to the first NP oracle. Subsequently, we can use one query to the second NP oracle to decide whether there exists an accepting computation path of N with input x that has w_{max} 1's in the output, and therefore whether $x \in Q$. This concludes our proof that $Q \in \Theta_2^p$.

Finally, we show how the proof of Proposition 7.16 can be modified to show for the case of $FPT^{NP}[few]$ that witness oracles do not yield more solving power.

Proposition 7.17. $FPT^{NP}[few, wit] = FPT^{NP}[few]$.

Proof (sketch). We show that $FPT^{NP}[few,wit] \subseteq FPT^{NP}[few]$. Let Q be a problem in $FPT^{NP}[few,wit]$, that is, there is an fpt-algorithm A for Q that queries a witness NP oracle $f(k)$ times. We transform this algorithm into an fpt-algorithm A' for Q that queries a decision NP oracle $f(k)$ times. The general idea of this transformation is entirely similar to the construction used in the proof of Proposition 7.16. It is readily verified that applying this construction to this case yields the required fpt-algorithm A'.

An interesting topic for future research is the relation between the classes $FPT^{NP}[few]$ and $FPT_{||}^{NP}[few,wit]$. It is unclear if and how the argument why the class $FPT^{NP}[few]$ does not change when defined based on parallel oracle access (as sketched in the beginning of Sect. 7.2) can be extended to answer this question. Similarly, the relation between the classes $\Theta_2^p[wit]$ and $P_{||}^{NP}[wit]$ remains open.

7.5.2 Comparing the Oracle Models for Search Problems

We then turn our attention to the setting of search problems. We show that in this setting, the witness oracle model is in fact more powerful than the decision oracle model (except in the case of $F\Delta_2^p$).

We begin with the case of $F\Delta_2^p$.

Proposition 7.18. $F\Delta_2^p[wit] = F\Delta_2^p$.

Proof. The proof of Proposition 7.15 also works to show this result.

Then, for the case of $F\Theta_2^p$ and $FFPT^{NP}[few]$, we will make use of a result by Gottlob and Fermüller [99, Theorem 5.4]. This result states that the search problem SAT cannot be solved by a polynomial-time algorithm that queries a decision NP oracle $O(\log n)$ times (unless $P = NP$). This directly gives us the following result. The inclusion $F\Theta_2^p \subseteq F\Theta_2^p[wit]$ is trivial, and is likely to be strict.

Observation 7.19. $F\Theta_2^p \subsetneq F\Theta_2^p[wit]$, *unless $P = NP$.*

Proof. This follows directly from a result by Gottlob and Fermüller [99, Theorem 5.4].

Similarly, the inclusion $FFPT^{NP}[few] \subseteq FFPT^{NP}[few,wit]$ is trivial, and is likely to be strict for the same reason.

Observation 7.20. $FFPT^{NP}[few] \subsetneq FFPT^{NP}[few, wit]$, *unless $P = NP$.*

Proof. This follows directly from a result by Gottlob and Fermüller [99, Theorem 5.4].

Moreover, in Section 10.2, we will consider a search problem that is in $FFPT^{NP}$ [few,wit], but not in $FFPT^{NP}[few]$, unless $W[P] = FPT$ (see Proposition 10.18).

Finally, we argue that in the setting of search problems in fact only the last SAT query needs to return a satisfying assignment. We firstly show this for the case of $F\Theta_2^p[wit]$. Again, the result for the case of $F\Theta_2^p[wit]$ was already known from the literature [137, Lemma 6.3.4]. Since we want to apply a modification of the proof to the case of $FFPT^{NP}[few]$, we nevertheless indicate how the statement can be proved.

Proposition 7.21. *Let $Q \in F\Theta_2^p[wit]$ be an arbitrary search problem. Then there exists a polynomial-time algorithm with access to a decision NP oracle O_1 and a witness NP oracle O_2, that solves Q by making $O(\log n)$ queries to the decision oracle O_1 and subsequently a single query to the witness oracle O_2, where n denotes the input size.*

Proof (idea). Let A be an polynomial-time algorithm that solves Q by making logarithmically many queries to a witness NP oracle. Construct an algorithm similar to the algorithm constructed in the proof of Proposition 7.16, i.e. an algorithm that makes logarithmically many queries to a decision NP oracle, but that replaces the (single) last query to the decision NP oracle by a query to the witness NP oracle in order to compute some output of $Q(x)$.

This proof can be modified to show a similar result for the case of $FFPT^{NP}[few,wit]$.

Proposition 7.22. *Let $Q \in FFPT^{NP}[few, wit]$ be an arbitrary parameterized search problem. Then there exists an fpt-algorithm with access to a decision NP oracle O_1 and a witness NP oracle O_2, that solves Q by making $f(k)$ queries to the decision oracle O_1 and subsequently a single query to the witness oracle O_2.*

Proof (sketch). The proof of Proposition 7.21 can be straightforwardly modified to show this statement.

Similarly to the case for decision problems, interesting topics for future research are the relation between the classes $\mathrm{FFPT}^{\mathrm{NP}}[\text{few,wit}]$ and $\mathrm{FFPT}_{||}^{\mathrm{NP}}[\text{few,wit}]$ and the relation between the classes $\mathrm{F\Theta}_2^{\mathrm{p}}[\text{wit}]$ and $\mathrm{FP}_{||}^{\mathrm{NP}}[\text{wit}]$. It has been shown that $\mathrm{F\Theta}_2^{\mathrm{p}} \neq \mathrm{FP}_{||}^{\mathrm{NP}}$, unless the ETH fails [125]—that is, unless there is some $2^{o(n)}$-time algorithm for 3SAT, where n denotes the number of variables. It would be interesting to investigate whether the techniques used to show this can be extended to separate $\mathrm{FFPT}^{\mathrm{NP}}[\text{few,wit}]$ from $\mathrm{FFPT}_{||}^{\mathrm{NP}}[\text{few,wit}]$ or to separate $\mathrm{F\Theta}_2^{\mathrm{p}}[\text{wit}]$ from $\mathrm{FP}_{||}^{\mathrm{NP}}[\text{wit}]$.

Summary

In this chapter, we considered several ways of generalizing (many-to-one) fpt-reducibility to SAT to the notion of fpt-time Turing reductions to SAT—that is, fpt-algorithms that have access to an NP oracle. These different variants are based on different bounds on the number of oracle queries that the algorithms can make. The variant that we studied in most detail leads to the parameterized complexity class $\mathrm{FPT}^{\mathrm{NP}}[\text{few}]$, where the number of queries is allowed to depend only on the parameter value (and not on the input size). We showed that this class can alternatively be seen as a parameterized variant of the Boolean Hierarchy. Moreover, we showed how hardness results for various parameterized complexity classes can be employed to obtain lower bounds on the number of NP oracle queries made by any fpt-algorithm that solves a particular problem. Finally, we briefly considered an extension of the class $\mathrm{FPT}^{\mathrm{NP}}[\text{few}]$, where the oracles return a witness in case of a yes-answer.

Notes

The results in Sects. 7.2.1 and 7.3 were shown in a paper that appeared in the proceedings of COMSOC 2014 [77] and in the proceedings of AAMAS 2015 [78].

In previous work [77, 78, 109, 111, 112], the class $\mathrm{FPT}^{\mathrm{NP}}[\text{few}]$ appeared under the name $\mathrm{FPT}^{\mathrm{NP}[f(k)]}$.

Applying the Theory

Chapter 8
Problems in Knowledge Representation and Reasoning

In Chaps. 4–7, we developed parameterized complexity tools to adequately characterize the computational complexity of parameterized variants of problems that
lie at the second level of the Polynomial Hierarchy (or higher). We are particularly
interested in using these tools (as well as other parameterized complexity tools) to
discriminate parameterized problems that admit an fpt-reduction to SAT from problems that do not. In Chaps. 8–13, we will apply these tools to concrete problems
from various domains of computer science and artificial intelligence.

For an overview of all parameterized problems that we consider in this thesis
(grouped by their computational complexity), we refer to the Index of Parameterized
Problems on p. 397.

In this chapter, we begin by investigating the parameterized complexity of several
problems from the area of Knowledge Representation and Reasoning.

Outline of This Chapter

In Sect. 8.1, we give an overview of the parameterized complexity results that we
found in Chaps. 4 and 6 for the consistency problem for disjunctive answer set
programming. This problem served as a running example throughout Chap. 6. For
this problem, there are several parameterizations that admit an fpt-reduction to SAT,
and several other parameterizations that do not (the latter are complete for $\Sigma_2^p[k*]$
and $\Sigma_2^p[*k, \mathrm{P}]$).

Then, in Sect. 8.2, we consider several parameterizations of the problem of abductive reasoning over propositional theories. In particular, we consider two types
of parameters that measure the distance to a tractable base class. One of these types
leads to fpt-reductions to SAT (for two tractable base classes). We show that the other
type of parameters leads to problems that do not admit fpt-reductions to SAT (these
problems are complete for $\Sigma_2^p[*k, 1]$ and $\Sigma_2^p[*k, \mathrm{P}]$).

Finally, in Sect. 8.3, we consider a parameterized problem that captures a robust
notion of satisfiability in the setting of constraint satisfaction. We show that this
problem is $\Pi_2^p[k*]$-complete.

© Springer-Verlag GmbH Germany, part of Springer Nature 2019
R. de Haan: Parameterized Complexity in the Polynomial Hierarchy, LNCS 11880,
https://doi.org/10.1007/978-3-662-60670-4_8

Table 8.1 Parameterized complexity results for ASP-CONSISTENCY.

Problem	Parameterized complexity	
ASP-CONS(norm.bd-size)	para-NP-complete	(Proposition 5.2, [82])
ASP-CONS(#cont.atoms)	para-co-NP-complete	(Proposition 5.3)
ASP-CONS(#cont.rules)	$\Sigma_2^p[k*]$-complete	(Theorem 6.3)
ASP-CONS(#disj.rules)	$\Sigma_2^p[*k, P]$-complete	(Theorem 6.24)
ASP-CONS(#non-dual-normal.rules)	$\Sigma_2^p[*k, P]$-complete	(Theorem 6.26)
ASP-CONS(max.atom.occ.)	para-Σ_2^p-complete	(Proposition 5.4)

8.1 Disjunctive Answer Set Programming

We begin by briefly reviewing the parameterized complexity results for the consistency problem for disjunctive answer set programming that we discussed in Chaps. 4–6. For a detailed definition of this problem, we refer to Sect. 5.1.

The problem ASP-CONSISTENCY is Σ_2^p-complete [71]. The different parameterizations that we consider for this problem are complete for various parameterized complexity classes between para-NP and para-co-NP, on the one hand, and para-Σ_2^p, on the other hand. An overview of these results is provided in Table 8.1. (For an overview of the definition of the parameterized problems referred to in this table, we refer to Table 5.1 on p. 74.)

We hope that these parameterized complexity results can help guide engineering efforts for practical *answer set solvers* (i.e., algorithms to solve reasoning problems based on answer set programming). Many answer set solvers already employ SAT solving techniques, e.g., Cmodels [92], ASSAT [147], and Clasp [89]. Work has also been done on translations from ASP to SAT, both for classes of programs for which reasoning problems are within NP or co-NP [16, 79, 124, 147] and for classes of programs with reasoning problems beyond NP and co-NP [124, 143, 146].

8.2 Abductive Reasoning

Next, we analyze the computational complexity of two additional parameterizations of the propositional abduction problem. We already considered two parameterized variants of this problem in Chap. 4 (for a definition of the problem of propositional abduction and these parameterizations, we refer to Sect. 4.2.2). The problem is Σ_2^p-complete in general [72], but it is only NP-complete when restricted to Horn theories [181] or to Krom (2CNF) theories [168, Lemma 61]. The parameters that we considered for this problem in Sect. 4.2.2 capture a distance measure to these tractable base classes, based on the notion of (strong) backdoors. For these parameters, the problem admits an fpt-reduction to SAT (see the work of Pfandler, Rümmele and Szeider [173], and Propositions 4.5 and 4.6).

In this section, we consider two more parameterizations for the propositional abduction problem, that also capture a distance measure to the tractable base classes of Horn and Krom theories. The parameterizations considered by Pfandler et al. and in Sect. 4.2.2 count the (minimum) number of variables that need to be eliminated to end up in the base class. The parameters that we consider in this section count the number of clauses that need to be eliminated to end up in the base class.

In particular, we consider the following two parameterized problems.

ABDUCTION(#non-Horn-clauses):
Input: An abduction instance $\mathcal{P} = (V, H, M, T)$, and a positive integer m.
Parameter: The number of clauses of T that are not Horn clauses.
Question: Does there exist a solution S of $\mathcal{P}$ of size at most m?

ABDUCTION(#non-Kron-clauses):
Input: An abduction instance $\mathcal{P} = (V, H, M, T)$, and a positive integer m.
Parameter: The number of clauses of T that are not Krom clauses.
Question: Does there exist a solution S of $\mathcal{P}$ of size at most m?

We show that these parameterized problems are complete for different levels of the $\Sigma_2^p[*k, t]$ hierarchy. We begin by proving $\Sigma_2^p[*k, \mathrm{P}]$-completeness for the problem ABDUCTION(#non-Horn-clauses). It will be instructive to firstly consider an alternative NP-hardness proof for the problem ABDUCTION restricted to Horn theories. We give a reduction directly from the canonical NP-complete problem 3SAT. The $\Sigma_2^p[*k, \mathrm{P}]$-hardness proof for ABDUCTION(#non-Horn-clauses) that we give below is based on a similar idea as the one behind this NP-hardness proof.

Proposition 8.1. ABDUCTION *is NP-hard, even when restricted to instances where the theory is a Horn formula.*

Proof. We show NP-hardness by giving a polynomial-time reduction from 3SAT. Let $\varphi = c_1 \wedge \cdots \wedge c_m$ be a 3CNF formula with $\mathrm{Var}(\varphi) = \{x_1, \ldots, x_n\}$. We construct an instance $(\mathcal{P}, m)$ of ABDUCTION, where $\mathcal{P} = (V, H, M, T)$ and where the theory T is a Horn formula. We let:

$X = \{x_i : i \in [n]\};$

$X' = \{x_i' : i \in [n]\};$

$Y = \{y_j : j \in [m]\};$

$V = X \cup X' \cup Y;$

$H = X \cup X';$

$M = Y;$

$T = \{(\neg x_i \vee \neg x_i') : i \in [n]\} \cup$

$\qquad \{(\sigma(l_1^j) \rightarrow y_j), (\sigma(l_1^j) \rightarrow y_j), (\sigma(l_1^j) \rightarrow y_j) : j \in [m], c_j = (l_1^j \vee l_2^j \vee l_3^j)\};$

$m = |X|,$

where the function $\sigma : \text{Lit}(X) \to X \cup X'$ is defined by letting $\sigma(x) = x$ and $\sigma(\neg x) = x'$, for all $x \in X$. Clearly, T is a Horn formula. We claim that φ is satisfiable if and only if $\mathcal{P}$ has a solution of size at most m.

($\Rightarrow$) Assume that φ is satisfiable, i.e., that there exists a truth assignment $\alpha : X \to \mathbb{B}$ that satisfies φ. We construct the solution $S = \{ x \in X : \alpha(x) = 1 \} \cup \{ x' : x \in X, \alpha(x) = 0 \} \subseteq H$, which has exactly m elements. We know that $T \cup S$ is satisfiable, because the following truth assignment $\beta : V \to \mathbb{B}$ satisfies $T \cup S$. For each $x \in X$, we let $\beta(x) = 1$ if and only if $\alpha(x) = 1$ and $\beta(x') = 1$ if and only if $\alpha(x) = 0$, and for each $y \in Y$ we let $\beta(y) = 1$. We show that $T \cup S \models M$. Let $y_j \in M$ be an arbitrary manifestation. We show that $T \cup S \models y_j$. Since α satisfies c_j, we know that alpha satisfies some literal l_j^ℓ. Therefore, $\sigma(l_j^\ell) \in S$, and thus since $(\sigma(l_j^\ell) \to y_j) \in T$, we know that $T \cup S \models y_j$. Since y_j was arbitrary, we can conclude that $T \cup S \models M$. Therefore, S is a solution of $\mathcal{P}$.

($\Leftarrow$) Assume that $\mathcal{P}$ has a solution $S \subseteq H$ of size at most m. Since $(\neg x \vee \neg x') \in T$ for each $x \in X$, and $T \cup S$ is consistent, we know that for no $x \in X$ it holds that both $x \in S$ and $x' \in S$. We then define the truth assignment $\alpha : X \to \mathbb{B}$ by letting $\alpha(x) = 1$ if and only if $x \in S$, for each $x \in X$. We show that α satisfies φ. Let c_j be an arbitrary clause of φ. We know that $T \cup S \models M$, and so in particular $T \cup S \models y_j$. This can only be the case if $z \in S$ and $(z \to y_j) \in T$. We distinguish two cases: either (i) $z = x$ for some $x \in X$, or (ii) $z = x'$ for some $x \in X$. In case (i), we know that $x \in c_j$. Moreover, we know that $x \in S$ and therefore that $\alpha(x) = 1$. Thus α satisfies c_j. In case (ii), we know that $\neg x \in c_j$. Moreover, we know that $x' \in S$, and thus that $x \notin S$, and therefore that $\alpha(x) = 0$. Thus α satisfies c_j. Since c_j was arbitrary, we can conclude that α satisfies φ.

Before we can show that ABDUCTION(#non-Horn-clauses) is $\Sigma_2^p[*k, \text{P}]$-complete, we need to consider the following observation.

Observation 8.2. *Let φ be a CNF formula consisting of k non-Horn clauses $c_1, \dots, c_k$, and m Horn clauses $c_1', \dots, c_m'$. Let φ' denote the Horn part of φ, i.e., $\varphi' = \{ c_1', \dots, c_m' \}$. Then φ is satisfiable if and only if for some $L \in \{ \{ l_1, \dots, l_k \} : l_i \in c_i \}$ the formula $\varphi' \wedge \bigwedge L$ is satisfiable.*

Proof. ($\Rightarrow$) Assume that φ is satisfiable, i.e., that there is a truth assignment $\alpha : \text{Var}(\varphi) \to \mathbb{B}$ that satisfies all clauses of φ. So in particular, α satisfies all clauses in φ'. Then, also, for each $i \in [k]$, α satisfies some literal $l_i \in c_i$. In other words α satisfies $\varphi' \wedge \bigwedge L$, where $L = \{ l_1, \dots, l_k \}$.

($\Leftarrow$) Assume that there is some $L \in \{ \{ l_1, \dots, l_k \} : l_i \in c_i \}$ such that $\varphi' \wedge \bigwedge L$ is satisfiable. Then there exists an assignment $\alpha : \text{Var}(\varphi) \to \mathbb{B}$ that satisfies all clauses in φ' and that satisfies $\bigwedge L$. It suffices to show that α satisfies all clauses $c_1, \dots, c_k$. This follows directly from the fact that α satisfies L and that L contains a literal $l_i \in c_i$ for each $i \in [k]$.

We can now establish $\Sigma_2^p[*k, \text{P}]$-completeness for ABDUCTION(#non-Horn-clauses). The hardness proof is based on an idea that is similar to the one behind the above NP-hardness proof, and uses the normalization result for $\Sigma_2^p[*k, \text{P}]$ of Proposition 6.23 (in Sect. 6.3.2).

Proposition 8.3. ABDUCTION(*#non-Horn-clauses*) is $\Sigma_2^P[*k, P]$-*complete.*

Proof. Firstly, we show that ABDUCTION(#non-Horn-clauses) is $\Sigma_2^P[*k, P]$-hard by giving an fpt-reduction from the problem $\Sigma_2^P[*k]$-WSAT(CIRC). Let (C, k) be an instance of $\Sigma_2^P[*k]$-WSAT(CIRC), where $C = \exists X.\forall Y.\mathcal{C}$ is a quantified Boolean circuit, where $X = \{x_1, \ldots, x_n\}$ and $Y = \{y_1, \ldots, y_m\}$. We construct an instance $(\mathcal{P}, m)$ of ABDUCTION(#non-Horn-clauses), where $\mathcal{P} = (V, H, M, T)$ is an abduction instance whose theory T is a CNF formula with k non-Horn clauses. By Proposition 6.23, we may assume without loss of generality that C is in negation normal form and that the only negation nodes in C have input nodes in X as input. Moreover, we may assume without loss of generality that for each truth assignment $\alpha : X \to \mathbb{B}$ there exists a truth assignment $\beta : Y \to \mathbb{B}$ of weight k such that $\alpha \cup \beta$ satisfies $\mathcal{C}$. We let:

$$
\begin{aligned}
X^0 &= \{ x^0 : x \in X \}; \\
X^1 &= \{ x^1 : x \in X \}; \\
W &= \{ w_g : g \text{ is an internal node of } \mathcal{C} \}; \\
V &= X^0 \cup X^1 \cup Y \cup W \cup \{z\} \cup \\
&\quad \{ y_j^i : i \in [k], j \in [m] \}; \\
H &= X^0 \cup X^1; \\
m &= |X|; \\
M &= \{z\}; \\
T &= \{ (y_1^i \vee \cdots \vee y_m^i) : i \in [k] \} \cup \\
&\quad \{ (y_j^i \to y_j) : i \in [k], j \in [m] \} \cup \\
&\quad \{ (\neg y_j^i \vee \neg y_j^{i'}) : i, i' \in [k], i < i', j \in [m] \} \cup \\
&\quad \{ (\neg x^0 \vee \neg x^1) : x \in X \} \cup \\
&\quad T_{\mathcal{C}},
\end{aligned}
$$

where we construct the set $T_{\mathcal{C}}$ as follows. Firstly, we define the following mapping σ from nodes of $\mathcal{C}$ to variables in V. For each non-negated input node $x \in X$, we let $\sigma(x) = x^1$. For each negated input node $\neg x$, for $x \in X$, we let $\sigma(\neg x) = x^0$. For each input node $y_j \in Y$, we let $\sigma(y) = y_j$. For each internal node g, we let $\sigma(g) = w_g$. Then, for each internal conjunction node g with inputs $g_1, \ldots, g_u$, we add the Horn clause $((\sigma(g_1) \wedge \cdots \wedge \sigma(g_u)) \to \sigma(g))$ to $T_{\mathcal{C}}$. For each internal disjunction node g with inputs $g_1, \ldots, g_u$, we add the Horn clauses $(\sigma(g_1) \to \sigma(g)), \ldots, (\sigma(g_u) \to \sigma(g))$ to $T_{\mathcal{C}}$. In addition, we add the Horn clause $(\sigma(g_o) \to z)$ to $T_{\mathcal{C}}$, where g_o is the output node of $\mathcal{C}$. Clearly, T is a CNF formula that contains k non-Horn clauses.

Intuitively, for each $x \in X$, the variable x^1 represents setting x to true and the variable x^0 represents setting x to false. Then, each subset $S \subseteq H$ of size m that satisfies T corresponds to a truth assignment to the input nodes in X, because T contains a clause $(\neg x^0 \vee \neg x^1)$ for each $x \in X$. Moreover, each assignment that satisfies T must set exactly k variables y_j to true. Finally, the clauses in T_C ensure that the variable z must be set to true for each assignment that corresponds to a satisfying assignment of C.

We now show that $(C, k) \in \Sigma_2^p[*k]$-WSAT(CIRC) if and only if $(\mathcal{P}, m) \in$ ABDUCTION(#non-Horn-clauses).

($\Rightarrow$) Assume that $(C, k) \in \Sigma_2^p[*k]$-WSAT(CIRC), i.e., that there exists a truth assignment $\alpha : X \to \mathbb{B}$ such that for all truth assignments $\beta : Y \to \mathbb{B}$ of weight k the assignment $\alpha \cup \beta$ satisfies C. We show that $(\mathcal{P}, m) \in$ ABDUCTION. We construct the solution $S \subseteq H$ by letting $S = \{ x^1 : x \in X, \alpha(x) = 1 \} \cup \{ x^0 : x \in X, \alpha(x) = 0 \}$. Clearly, $|S| = m$. Since $C[\alpha]$ is satisfiable, and since $C[\alpha]$ is monotone, we know that $T \cup S$ is satisfiable; setting all variables in $Y \cup W \cup \{z\}$ to true (and setting the variables y_j^i appropriately) satisfies $T \cup S$. Next, we show that $T \cup S \models M$. Let $\gamma : V \to \mathbb{B}$ be an arbitrary truth assignment that satisfies $T \cup S$. We then know that $\gamma(x^1) = 1$ if and only if $\alpha(x) = 1$ and $\gamma(x^0) = 1$ if and only if $\alpha(x) = 0$. Moreover, we know that γ sets exactly k variables y_j to true. Then, since $\alpha \cup \beta$ satisfies C, for each assignment β of weight k, we know that the clauses in T_C ensure that γ sets z to true. Therefore, since γ was arbitrary, we know that $T \cup S \models \{z\}$. Thus $(\mathcal{P}, m) \in$ ABDUCTION.

($\Leftarrow$) Conversely, assume that $(\mathcal{P}, m) \in$ ABDUCTION, i.e., that there exists a solution $S \subseteq H$ of size at most m such that $T \cup S$ is satisfiable and $T \cup S \models M$. We may assume without loss of generality that $|S| = m$. We show that $(C, k) \in \Sigma_2^p[*k]$-WSAT(CIRC). We define the truth assignment $\alpha : X \to \mathbb{B}$ as follows. For each $x \in X$, we let $\alpha(x) = 1$ if and only if $x^1 \in S$. We claim that for any truth assignment $\beta : Y \to \mathbb{B}$ of weight k it holds that $\alpha \cup \beta$ satisfies C. Let $\beta : Y \to \mathbb{B}$ be an arbitrary truth assignment. Let $\{y_{\ell_1}, \ldots, y_{\ell_k}\} = \{ y_j : j \in [m], \beta(y_j) = 1 \}$. We construct the partial assignment $\gamma : V \to \mathbb{B}$ as follows. We let $\gamma(x^1) = 1$ if and only if $x^1 \in S$ and $\gamma(x^0) = 1$ if and only if $x^0 \in S$, for each $x \in X$. Moreover, we let $\gamma(y_j) = \beta(y_j)$, and $\gamma(y_j^i) = 1$ if and only if $j = \ell_i$, for each $j \in [m]$. Since $T \cup S \models z$, we know that $(T \cup S)[\gamma] \models z$. Moreover, we know that $(T \cup S)[\gamma]$ is satisfiable. By a straightforward inductive argument, we then know that for each internal node g of C it holds that $\alpha \cup \beta$ sets g to true if and only if $(T \cup S)[\gamma] \models w_g$. Then, since $(w_o \to z) \in T$ (where o is the output node of C) and since $(T \cup S)[\gamma] \models z$, we know that $\alpha \cup \beta$ satisfies C. Since β was arbitrary, we can conclude that $(C, k) \in \Sigma_2^p[*k]$-WSAT(CIRC).

Next, we show membership in $\Sigma_2^p[*k, P]$. Let $(\mathcal{P}, m)$ be an instance of ABDUCTION(#non-Horn-clauses), where $\mathcal{P} = (V, H, M, T)$ is an abduction instance whose theory T contains k non-Horn clauses. Let $V = \{v_1, \ldots, v_u\}$ and $H = \{h_1, \ldots, h_n\}$. Let $c_1, \ldots, c_k$ be the non-Horn clauses of T, with $c_i = l_i^1 \vee \cdots \vee l_i^{\ell_i}$ for each $i \in [k]$. Also, let T' denote the set of Horn clauses in T. Moreover, let $\ell = \sum_{i \in [k]} \ell_i$. We sketch an algorithm that takes as input three bitstrings: one string $\bar{x} = x_1 \ldots x_n$, of length n, one string $\bar{y} = y_1 \ldots y_u$, of length u, and one string $\bar{z} z_1 \ldots z_\ell$ of length ℓ containing exactly k 1's. Moreover, we consider the string $\bar{z}$ as the concatenation of the strings $\bar{z}_1, \ldots, \bar{z}_k$, where $\bar{z}_i = z_{i,1} \ldots z_{i,\ell_i}$. The algorithm checks (1) whether $\bar{x}$ contains at most m 1's, it checks (2) whether the truth assignment $\alpha : V \to \mathbb{B}$ defined by $\alpha(v_j) = 1$ if and only if $y_j = 1$ satisfies $T \cup S$, where the set $S \subseteq H$ is given by $S = \{ h_i \in H : x_i = 1 \}$, and it checks (3) if it is the case that if (3a) each string $\bar{z}_i$ contains exactly one 1, then (3b) it holds that $L_{\bar{z}} \cup T' \cup S \models M$, where $L_{\bar{z}} = \{ l_b \in c_i : b \in [\ell_i], z_{i,b} = 1 \}$. Since T' is a Horn formula and $L_{\bar{z}}$ and S consist of unit clauses, we can choose this algorithm so that it runs in polynomial time.

Now, it is straightforward to verify that $(\mathcal{P}, m) \in$ ABDUCTION if and only if there exists some strings $\bar{x}$ and $\bar{y}$ such that for all strings $\bar{z}$ containing k 1's the algorithm accepts. Since the algorithm runs in polynomial time, we can construct a quantified Boolean circuit $C = \exists X. \exists Y. \forall Z. C$ in polynomial time with the property that the algorithm accepts for some strings $\bar{x}$, $\bar{y}$ and for all suitable strings $\bar{z}$ if and only if there is a truth assignment $\alpha : X \cup Y \to \mathbb{B}$ such that for all truth assignments $\beta : Z \to \mathbb{B}$ of weight k the assignment $\alpha \cup \beta$ satisfies C. In other words, $(C, k) \in \Sigma_2^p[*k]$-WSAT if and only if $(\mathcal{P}, m) \in$ ABDUCTION.

Next, we turn to showing $\Sigma_2^p[*k, 1]$-completeness for ABDUCTION(#non-Krom-clauses). In order to show $\Sigma_2^p[*k, 1]$-hardness, we firstly show that we can restrict our attention to instances of the problem $\Pi_2^p[*k]$-WSAT(2CNF) of a particular type.

Lemma 8.4. *Let (φ, k) be an instance of $\Pi_2^p[*k]$-WSAT(2CNF), where $\varphi = \forall X. \exists Y. \psi$. In fpt-time, we can construct an equivalent instance (φ', k') of $\Pi_2^p[*k]$-WSAT where $\varphi' = \forall X'. \exists Y'. \psi'$, and where ψ' is a CNF formula that has exactly $k'' = k + 2$ clauses of size more than 2, with the property that for each truth assignment $\gamma : X' \cup Y' \to \mathbb{B}$ that sets more than k' variables in Y' to true, the formula $\psi'[\gamma]$ is false.*

Proof. Let (φ, k) be an instance of $\Pi_2^p[*k]$-WSAT(2CNF), where $\varphi = \forall X. \exists Y. \psi$ and where $Y = \{y_1, \ldots, y_m\}$. We construct the following instance (φ', k') with $\varphi' = \forall X. \exists Y'. \psi'$ as follows. We let $Y' = Y \cup Y' \cup W$, where

$$Y' = \{ y_j^i : i \in [k], j \in [m] \},$$

and where

$$W = \{ w_{j,j'}^i : i \in [0, k+1], j, j' \in [0, m+1], j \le j' \}.$$

Intuitively, the variables y_j^i encode which variables in Y are set to true, and the variables $w_{j,j'}^i$ encode "gaps" between the true variables in Y. If a variable y_j^i is true, we will require that y_j is true as well. Moreover, if a variable $w_{j,j'}^i$ is true, we will require that y_j^{i-1} and $y_{j'}^i$ are true, and that $y_{j''}$ is false, for all $j'' \in [j+1, j'-1]$ (for the cases where $i = 0$ or $i = k+1$, the requirements are slightly different). Finally, we will have non-binary clauses to ensure that for each $i \in [0, k+1]$, there is some gap variable $w_{j,j'}^i$ that is set to true.

Concretely, we let ψ' consist of the following clauses:

$$\psi' = \psi \cup$$

$$\{ (\neg y^i_j \vee \neg y^i_{j'}) : i \in [k], j, j' \in [m], j < j' \} \cup \tag{8.1}$$

$$\{ (\neg w^i_{j_1, j'_1} \vee \neg w^i_{j_2, j'_2}) : i \in [0, k], j_1, j'_1 \in [m+1], j_1 \le j'_1,$$

$$j_2, j'_2 \in [m+1], j_2 \le j'_2, (j_1, j'_1) \ne (j_2, j'_2) \} \cup \tag{8.2}$$

$$\{ (y^i_j \to y_j) : i \in [k], j \in [m] \} \cup \tag{8.3}$$

$$\{ (y^i_j \to \neg y^{i'}_{j'}) : i, i' \in [k], i < i', j, j' \in [m], j' \le j \} \cup \tag{8.4}$$

$$\{ (w^i_{j,j'} \to y^{i-1}_j), (w^i_{j,j'} \to y^i_{j'}) : i \in [2, k], j, j' \in [m+1], j \le j' \} \cup \tag{8.5}$$

$$\{ (w^i_{j,j'} \to \neg y_{j''}) : i \in [0, k+1], j'' \in [j+1, j'-1] \} \cup \tag{8.6}$$

$$\{ (w^0_{j,j'} \to y^1_{j'}) : j, j' \in [m+1], j \le j' \} \cup \tag{8.7}$$

$$\{ (\neg w^0_{j,j'}) : j, j' \in [m+1], j \le j' \} \cup \tag{8.8}$$

$$\{ (w^{k+1}_{j,j'} \to y^k_j) : j, j' \in [m+1], j \le j' \} \cup \tag{8.9}$$

$$\{ (\neg w^{k+1}_{j,j'}) : j, j' \in [0, m+1], j \le j' \} \cup \tag{8.10}$$

$$\{ \bigvee \{ w^i_{j,j'} : j, j' \in [0, m+1], j \le j' \} : i \in [0, k+1] \}. \tag{8.11}$$

Finally, we let $k' = 3k + 2$. Clearly, ψ' has k'' clauses of size more than 2. We claim that $(\psi, k) \in \Pi^p_2[*k]$-WSAT if and only if $(\psi', k') \in \Pi^p_2[*k]$-WSAT.

($\Rightarrow$) Assume that $(\psi, k) \in \Pi^p_2[*k]$-WSAT. We show that $(\psi', k') \in \Pi^p_2[*k]$-WSAT. Let $\alpha : X \to \mathbb{B}$ be an arbitrary assignment. We show that there exists an assignment $\beta : Y' \to \mathbb{B}$ of weight k' such that $\psi'[\alpha \cup \beta]$ is true. Since $(\psi, k) \in \Pi^p_2[*k]$-WSAT, we know that there exists an assignment $\beta' : Y \to \mathbb{B}$ of weight k such that $\psi[\alpha \cup \beta']$ is true. Let $\{ \ell_1, \dots, \ell_k \} = \{ \ell \in [m] : \beta'(y_\ell) = 1 \}$ be the indices of variables y_i that β' sets to true, such that $\ell_1 < \cdots < \ell_k$. We construct the assignment $\beta : Y' \to \mathbb{B}$ as follows. For each $y \in Y$, we let $\beta(y) = \beta'(y)$. For each $i \in [k]$ and each $j \in [m]$, we let $\beta(y^i_j) = 1$ if and only if $j = \ell_i$. Moreover, for each $j, j' \in [m+1]$ with $j \le j'$, we let $\beta(w^0_{j,j'}) = 1$ if and only if $j = 0$ and $j' = \ell_1$. For each $j, j' \in [m+1]$ with $j \le j'$, we let $\beta(w^{k+1}_{j,j'}) = 1$ if and only if $j = \ell_k$ and $j' = m + 1$. Finally, for each $i \in [2, k]$ and each $j, j' \in [m+1]$ with $j \le j'$, we let $\beta(w^i_{j,j'}) = 1$ if and only if $j = \ell_{i-1}$ and $j' = \ell_i$. Then β has weight exactly $k' = 3k + 2$. Also, it holds that $\psi'[\alpha \cup \beta]$ is true. Therefore, since α was arbitrary, we know that $(\psi', k') \in \Pi^p_2[*k]$-WSAT.

($\Leftarrow$) Conversely, assume that $(\psi', k') \in \Pi^p_2[*k]$-WSAT. We show that then also $(\psi, k) \in \Pi^p_2[*k]$-WSAT. Let $\alpha : X \to \mathbb{B}$ be an arbitrary truth assignment. Then, since $(\psi', k') \in \Pi^p_2[*k]$-WSAT, we know that there exists a truth assignment $\beta : Y' \to \mathbb{B}$ of weight $k' = 3k + 2$ such that $\psi'[\alpha \cup \beta]$ is true. We claim that the truth assignment $\beta' : Y \to \mathbb{B}$ defined by $\beta(y) = \beta'(y)$ for each $y \in Y$, i.e., the restriction of β to Y, has weight k. Firstly, we know that β' sets exactly $k + 2$ variables in W to true, by clauses (8.2) and (8.11). Then, by clauses (8.1), (8.4–8.5) and (8.7–8.10), we know that β' sets exactly k variables in Y' to true. Thus, by clauses (8.4), β' sets exactly k variables in Y to true as well. Moreover, since β satisfies ψ and ψ contains only variables in $X \cup Y$, we know that $\alpha \cup \beta'$ satisfies ψ. Since α was arbitrary, we know that $(\psi, k) \in \Pi_2^p[*k]$-WSAT.

Moreover, we claim that for each truth assignment $\gamma : X \cup Y' \to \mathbb{B}$ that sets more than k' variables in Y' to true, the formula $\psi'[\gamma]$ is false. Let γ be a truth assignment that satisfies ψ'. We know that γ sets exactly $k + 2$ variables in W to true, by clauses (8.2) and (8.11). Then, by clauses (8.1), (8.4–8.5) and (8.7–8.10), we know that γ sets exactly k variables in Y' to true. Then, by clauses (8.3), γ sets at least k variables in Y to true as well. By clauses (8.6), we know that γ sets exactly k variables in Y to true. Thus, γ sets exactly k' variables in Y' to true.

Then, in order to show membership in $\Sigma_2^p[*k, 1]$, we use the following observation, that is similar to Observation 8.2.

Observation 8.5. *Let φ be a CNF formula consisting of k non-binary clauses $c_1, \ldots, c_k$, and m binary clauses $c_1', \ldots, c_m'$. Let φ' denote the binary part of φ, i.e., $\varphi' = \{c_1', \ldots, c_m'\}$. Then φ is satisfiable if and only if for some $L \in \{\{l_1, \ldots, l_k\} : l_i \in c_i\}$ the formula $\varphi' \wedge \bigwedge L$ is satisfiable.*

Proof. The proof is entirely similar to the proof of Observation 8.2.

We are now ready to show $\Sigma_2^p[*k, 1]$-completeness for ABDUCTION-(#non-Krom-clauses).

Proposition 8.6. ABDUCTION(*#non-krom-clauses*) *is $\Sigma_2^p[*k, 1]$-complete.*

Proof. We begin by showing that ABDUCTION(#non-Krom-clauses) is $\Sigma_2^p[*k, 1]$-hard. We do so by giving an fpt-reduction from the problem $\Sigma_2^p[*k]$-WSAT(2DNF). Let (φ, k) be an instance of $\Sigma_2^p[*k]$-WSAT(2DNF), where $\varphi = \exists X.\forall Y.\psi$, $X = \{x_1, \ldots, x_n\}$, $Y = \{y_1, \ldots, y_m\}$. By Lemma 8.4, we can transform (φ, k) in fpt-time into an equivalent instance (φ', k') with $\varphi = \exists X'.\forall Y'.\psi'$, for which holds that any truth assignment that sets more than k' variables in Y' to true satisfies ψ', under the condition that ψ' contains k' terms of size more than 2. Therefore, we will assume without loss of generality that our instance (φ, k) satisfies these properties, i.e., we assume that any truth assignment $\gamma : X \cup Y \to \mathbb{B}$ that sets more than k variables in Y to true satisfies ψ, and that ψ contains k terms of size more than 2. We may also assume without loss of generality that for any truth assignment $\alpha : X \to \mathbb{B}$, the formula $\psi[\alpha]$ is falsifiable. We will construct an equivalent instance $(\mathcal{P}, m)$ of ABDUCTION(#non-Krom-clauses), where the theory T of the abduction instance $\mathcal{P} = (V, H, M, T)$ contains exactly $k' = 2k$ non-binary clauses. We let:

$$X^0 = \{x^0 : x \in X\};$$
$$X^1 = \{x^1 : x \in X\};$$
$$V = X^0 \cup X^1 \cup X \cup Y \cup \{z\} \cup$$
$$\{y_j^i : i \in [k], j \in [m]\};$$
$$H = X^0 \cup X^1;$$
$$m = |X|;$$
$$M = \{z\}; \text{ and}$$
$$T = \{(y_1^i \vee \cdots \vee y_m^i \vee z) : i \in [k]\} \cup$$
$$\{(y_j^i \to y_j) : i \in [k], j \in [m]\} \cup$$
$$\{(\neg y_j^i \vee \neg y_j^{i'}) : i, i' \in [k], i < i', j \in [m]\} \cup$$
$$\{(\neg x^0 \vee \neg x^1) : x \in X\} \cup$$
$$\{(x^0 \to \neg x), (x^1 \to x) : x \in X\} \cup$$
$$\{\neg \delta : \delta \in \psi\}.$$

Clearly, T is a CNF formula that contains $2k$ non-binary clauses. We claim that $(\varphi, k) \in \Sigma_2^p[*k]$-WSAT if and only if $(\mathcal{P}, m) \in$ ABDUCTION(#non-Kron-clauses).

($\Rightarrow$) Assume that there exists a truth assignment $\alpha : X \to \mathbb{B}$ such that for all truth assignments $\beta : Y \to \mathbb{B}$ of weight k it holds that $\psi[\alpha \cup \beta]$ is true. We show that the set $X' = \{x^0 : x \in X, \alpha(x) = 0\} \cup \{x^1 : x \in X, \alpha(x) = 1\}$ is a solution of $\mathcal{P}$ of size $m = |X|$. Since $\psi[\alpha]$ is falsifiable, we can satisfy $T \cup X'$ by setting all variables y_j^i to false (and choosing appropriate truth values for the remaining variables). Thus, it suffices to show that $T \cup X' \models z$. Assume that there exists a truth assignment γ that simultaneously satisfies $T \cup X'$ and $\neg z$. Then there must exist distinct indices $i_1, \ldots, i_k \in [k]$ such that $\gamma(y_{i_\ell}^\ell) = 1$ for all $\ell \in [k]$. We distinguish two cases: either (i) γ restricted to Y has weight more than k or (ii) γ restricted to Y has weight exactly k. In either case, we know that γ does not satisfy the clauses $\{\neg \delta : \delta \in \psi\}$, and thus does not satisfy T, which is a contradiction with our previous assumption. Therefore, we can assume that there exists no γ that satisfies both $T \cup X'$ and sets z to false. Thus $T \cup X' \models z$.

($\Leftarrow$) Assume that there exists a solution $X' \subseteq H$ of size at most m such that $T \cup X'$ is satisfiable and such that $T \cup X' \models z$. Consider the truth assignment $\alpha : X \to \mathbb{B}$ defined by letting $\alpha(x) = 1$ if and only if $x^1 \in X'$. We show that for any truth assignment $\beta : Y \to \mathbb{B}$ of weight k it holds that $\psi[\alpha \cup \beta]$ is true. Let $\beta : Y \to \mathbb{B}$ be an arbitrary truth assignment. Moreover, let $\gamma : V \to \mathbb{B}$ be defined as follows. For each $x \in X$ we let $\gamma(x) = \gamma(x^1) = \alpha(x) = 1 - \gamma(x^0)$. Since $T \cup X'$ is satisfiable, we then know that α sets all variables in X' to true. For each $y \in Y$, we let $\gamma(y) = \beta(y)$. Let $\{y_{i_1}, \ldots, y_{i_k}\} = \{y \in Y : \beta(y) = 1\}$. Then, we let $\gamma(y_j^\ell) = 1$ if and only if $i_\ell = j$. Finally, we let $\gamma(z) = 0$. We know that $T \cup X' \models z$. We also know that γ sets all the variables in X' to true. Moreover, it is straightforward to verify that $\gamma \models T \setminus \{\neg \delta : \delta \in \psi\}$. Therefore, it must hold that $\gamma \not\models \{\neg \delta : \delta \in \psi\}$, and thus that $\gamma \models \psi$. Since γ restricted to the variables in ψ coincides with $\alpha \cup \beta$, we can conclude that $\psi[\alpha \cup \beta]$ is true.

Next, we show that ABDUCTION(#non-Kron-clauses) is in $\Sigma_2^P[*k, 1]$ by giving an fpt-reduction to the problem $\Sigma_2^P[*k]$-WSAT($\Gamma_{1,u}$) for some constant u. Let $(\mathcal{P}, m)$ be an instance of ABDUCTION(#non-Kron-clauses) where $\mathcal{P} = (V, H, M, T)$ and where T is a 2CNF formula consisting of n binary clauses $c_1', \ldots, c_m'$ and k non-binary clauses $c_1, \ldots, c_k$. We will construct a Boolean circuit $C(X, Y)$ of weft 1 over the variables in $X \cup Y$, in such a way that $(\mathcal{P}, m) \in$ ABDUCTION(#non-Kron-clauses) if and only if there exists a truth assignment $\alpha : X \to \mathbb{B}$ such that for every truth assignment $\beta : Y \to \mathbb{B}$ of weight k it holds that $C[\alpha \cup \beta]$ is true.

The idea of this reduction is the following. We will use existential variables to encode a guess for a solution $S \subseteq H$ of size at most m. Then, we will use existential variables to encode a guess for the truth assignment that is consistent with S and that satisfies T, which we will ensure using a set of clauses of bounded size. Then, we will use universal variables to encode the condition that all assignments satisfying one literal in each of the clauses $c_1, \ldots, c_k$ are inconsistent with S and $\bigvee_{m \in M} \neg m$, which we will ensure using a DNF formula with terms of bounded size.

Firstly, we introduce the existential variables $X = X_S \cup V$, where

$$X_S = \{ x_h^j : j \in [m], h \in H \} \cup \{ x_h : h \in H \}.$$

Then we construct the CNF formula χ (with clauses of unbounded size) as follows:

$$\begin{aligned}
\chi = &\{ \{ x_h^j : h \in H \} : j \in [m] \} \\
&\cup \{ (\neg x_h^j \vee \neg x_{h'}^j) : j \in [m], h \in H, h' \in H, h \neq h' \} \\
&\cup \{ (x_h^j \to x_h) : j \in [m], h \in H \} \\
&\cup \{ (x_h \to h) : h \in H \} \\
&\cup T.
\end{aligned}$$

The formula χ is satisfiable if and only if there exists a subset $S \subseteq H$ of size at most m such that $T \cup S$ is satisfiable.

Then, in polynomial time we can transform χ into a 3CNF formula χ' with $\text{Var}(\chi') = X \cup X'$ that is satisfiable if and only if χ is satisfiable. Moreover, for any truth assignment $\alpha : X \to \mathbb{B}$ it is the case that $\chi[\alpha]$ is true if and only if $\chi'[\alpha]$ is satisfiable. This can be done using the standard Tseitin transformation [188].

Next, we will introduce the set of universal variables Y where:

$$Y = \{ y_l^i : i \in [k], l \in c_i \} \cup \{ y_m : m \in M \}.$$

We will construct a DNF formula φ with clauses of bounded size that encodes whether all assignments satisfying (at least) one literal in each of the clauses $c_1, \ldots, c_k$ are inconsistent with S and $\bigvee_{m \in M} \neg m$. We will define $\varphi = \varphi_1 \vee \varphi_2 \vee \varphi_3$, where we will define the subformulas φ_1, φ_2 and φ_3 below.

We define the DNF formula φ_1 by letting:

$$\varphi_1 = \{\, (y_l^i \wedge y_{l'}^i) : i \in [k], l \in c_i, l' \in c_i, l \neq l'\,\},$$

and we define the DNF formula φ_2 by letting:

$$\varphi_2 = \{\, (y_m \wedge y_{m'}) : m, m' \in M, m \neq m'\,\}.$$

We will define the formula φ_3, that encodes whether S and the set $L = \{l_i \in c_i : i \in [k]$, the variable $y_{l_i}^i$ is set to true $\}$ are inconsistent with the binary part T' of T and $\neg m$, where m is the manifestation for which holds that y_m is set to true. We may assume without loss of generality that $T' \wedge \neg m$ is consistent. If it were the case that $T' \models m$, we would be able to detect this in polynomial time, and we could remove m from M. We may also assume without loss of generality that T' is consistent.

We claim that T' is inconsistent with $S \cup L \cup \{\neg m\}$ if and only if either (i) there are some literals $l, l' \in S \cup L$. such that there is a path from l to $\overline{l'}$ in the implication graph of T', or (ii) there is some literal $l \in S \cup L$ such that there is a path from l to m in the implication graph of T'. Assume that neither (i) nor (ii) is the case. Then we can satisfy T' and $S \cup L \cup \{\neg m\}$ simultaneously by setting all literals in $S \cup L \cup \{\neg m\}$ to true, as well all literals l' for which there is a path from some literal $l \in S \cup L$ to l' in the implication graph of T' (and setting the remaining variables appropriately). Conversely, if either (i) or (ii) holds, then T' is inconsistent with $S \cup L \cup \{\neg m\}$.

We now define φ_3 as follows:

$$
\begin{aligned}
\varphi_3 = \{\, &(x_h \wedge x_{h'}) : h \in H, h' \in H, \\
&\text{there is a path from } h \text{ to } \overline{h'} \text{ in the implication graph of } T'\,\} \\
\cup \{\, &(x_h \wedge y_l^i) : i \in [k], h \in H, l \in c_i, \\
&\text{there is a path from } h \text{ to } \overline{l} \text{ in the implication graph of } T'\,\} \\
\cup \{\, &(y_l^i \wedge y_{l'}^{i'}) : i, i' \in [k], i \neq i', l \in c_i, l' \in c_{i'}, \\
&\text{there is a path from } l \text{ to } \overline{l'} \text{ in the implication graph of } T'\,\} \\
\cup \{\, &(x_h \wedge y_m) : h \in H, m \in M, \\
&\text{there is a path from } h \text{ to } m \text{ in the implication graph of } T'\,\} \\
\cup \{\, &(y_l^i \wedge y_m) : i \in [k], l \in c_i, m \in M, \\
&\text{there is a path from } l \text{ to } m \text{ in the implication graph of } T'\,\}.
\end{aligned}
$$

Finally, we let:

$$C = \exists X \cup X'.\forall Y.(\chi' \wedge \varphi).$$

The (quantified) circuit C has at most one gate of unbounded arity in each root-to-leaf path, and thus has weft 1.

To show the correctness of our reduction, we prove that $(\mathcal{P}, m) \in \text{ABDUCTION}$ (#non-kron-clauses) if and only if $(C, k) \in \Sigma_2^p[*k]\text{-WSAT}$.

($\Rightarrow$) Assume that $(\mathcal{P}, m) \in \text{ABDUCTION}(\text{#non-kron-clauses})$, i.e., that there exists a solution $S \subseteq H$ of size at most m such that $T \cup S$ is consistent and $T \cup S \models M$.

Let $S = \{s_1, \ldots, s_{m'}\}$, with $m' \leq m$. Then define the truth assignment $\alpha : X_S \to \mathbb{B}$ by letting $\alpha(x_h) = 1$ if and only if $h \in S$, and $\alpha(x_h^i) = 1$ if and only if either $h = s_i$ or $m' < i \leq m$ and $h = s_{m'}$. Moreover, since $T \cup S$ is consistent, we know there exists a truth assignment $\alpha : V \to \mathbb{B}$ that satisfies $T \cup S$. Then, $\alpha \cup \alpha'$ satisfies χ, and so we can extend $\alpha \cup \alpha'$ to a truth assignment $\alpha'' : X \cup X' \to \mathbb{B}$ that satisfies χ'. We show that for any truth assignment $\beta : Y \to \mathbb{B}$ of weight k it holds that $\varphi[\alpha'' \cup \beta]$ is true. Let β be an arbitrary such truth assignment. We distinguish two cases: either (i) β satisfies exactly one y_i^j for each $j \in [k]$, and exactly one y_m, for $m \in M$, or (ii) this is not the case. In case (ii), β satisfies $\varphi_1 \vee \varphi_2$, and thus $\alpha'' \cup \beta$ satisfies φ. In case (i), let m be the unique manifestation $m \in M$ for which holds that $\beta(y_m) = 1$. Moreover, let $L = \{l_1, \ldots, l_k\}$ be the (unique) set of literals for which holds that $\beta(y_{l_i}^i) = 1$ for all $i \in [k]$. Since S is a solution for $\mathcal{P}$, we know that $T \cup S \models m$. Then, we also know that $T' \cup S \cup L \cup \{\neg m\}$ is inconsistent. Since T' contains only binary clauses, this can only be the case if there is an appropriate path in the implication graph of T' corresponding to a term in φ_3, satisfied by β. Thus $\alpha'' \cup \beta$ satisfies φ.

($\Leftarrow$) Conversely, assume that $(C, k) \in \Sigma_2^p[*k]$-WSAT, i.e., that there exists an assignment $\alpha : X \cup X' \to \mathbb{B}$ such that for all assignments $\beta : Y \to \mathbb{B}$ of weight k it holds that $(\psi' \wedge \varphi)[\alpha \cup \beta]$ is true. We show that $(\mathcal{P}, m) \in$ ABDUCTION(#non-Kron-clauses). We define the set $S \subseteq H$ by letting $S = \{h \in H : \alpha(x_h) = 1\}$. We claim that S is a solution of the abduction instance $\mathcal{P}$ of size at most m. Since χ' contains only variables in $X \cup X'$, we know that α satisfies χ'. Since any assignment satisfies χ' if and only if it satisfies χ, and by construction of χ any satisfying assignment sets at most m variables x_h, for $h \in H$, to true, we know that $|S| \leq m$. Moreover, since α satisfies χ', we know that $T \cup S$ is consistent. All that remains to show is that $T \cup S \cup \{m\}$ is inconsistent, for each $m \in M$. Let $m \in M$ be an arbitrary manifestation. We show that $T \cup S \cup \{m\}$ is unsatisfiable. By Observation 8.5, we know that $T \cup S \cup \{m\}$ is unsatisfiable if and only if $L \cup T' \cup S \cup \{m\}$ is unsatisfiable for each $L \in \{\{l_1, \ldots, l_k\} : l_i \in c_i\}$, where T' is the subset of T consisting of all Krom clauses of T. Let $L = \{l_1, \ldots, l_k\}$ be an arbitrary such set. We show that $L \cup T' \cup S \cup \{m\}$ is unsatisfiable. We construct the truth assignment $\beta : Y \to \mathbb{B}$ as follows. For each $\ell \in [k]$ we let $\beta(y_l^\ell) = 1$ if and only if $l = l_\ell$, and for each $m' \in M$ we let $\beta(y_{m'}) = 1$ if and only if $m' = m$. Clearly, β has weight k. Therefore, we know that $\alpha \cup \beta$ satisfies φ. Also, $\alpha \cup \beta$ falsifies φ_1 and φ_2. Thus, $\alpha \cup \beta$ must satisfy φ_3. However, by construction of φ_3, then there must exist a path in the implication graph of T' that witnesses that $L \cup T' \cup S \cup \{m\}$ is unsatisfiable. From this we can conclude that $T \cup S \cup \{m\}$ is unsatisfiable for each $m \in M$, and thus that $T \cup S \models M$. Hence, S is a solution of $\mathcal{P}$ of size at most m, and thus $(\mathcal{P}, m) \in$ ABDUCTION(#non-Krom-clauses).

8.3 Robust Constraint Satisfaction

The final example of a problem that arises in the area of Knowledge Representation and Reasoning that we consider, is the problem of robust constraint satisfaction.

Various robust constraint satisfaction problems were considered by Gottlob [98] and Abramsky, Gottlob and Kolaitis [3]. These problems are concerned with the question of whether, for an instance of the constraint satisfaction problem, every partial assignment of a particular size can be extended to a full solution. We consider a natural parameterization of this problem (where the parameter is the size of the partial assignment), and we show that this parameterized problem is complete for $\Sigma_2^p[k*]$.

We begin with formally defining the concept of robust satisfiability for CSP instances. For a definition of basic notions from the area of constraint satisfaction, we refer to Sect. 4.4. Let $\mathcal{I}$ be a CSP instance, and let $\alpha : V \to D$ be an assignment to some subset $V \subseteq \text{Var}(\mathcal{I})$ of variables of $\mathcal{I}$. We say that α *violates* a constraint $C = ((v_1, \ldots, v_r), R)$ of $\mathcal{I}$ if there is no extension β of α to the variables in $\text{Var}(\mathcal{I})$ such that $(\beta(v_1), \ldots, \beta(v_r)) \in R$. Let k be a positive integer. We say that a CSP instance $\mathcal{I}$ is *k-robustly satisfiable* if for each instantiation $\alpha : V \to D$ defined on some subset $V \subseteq \text{Var}(\mathcal{I})$ of k variables (i.e., $|V| = k$) that does not violate any constraint of $\mathcal{I}$, it holds that α can be extended to a solution for the CSP instance $\mathcal{I}$.

We consider the following parameterized problem.

ROBUST-CSP-SAT
Instance: A CSP instance $\mathcal{I}$, and a positive integer k.
Parameter: k.
Question: Is $\mathcal{I}$ k-robustly satisfiable?

We show that this problem is $\Pi_2^p[k*]$-complete. In order to show $\Pi_2^p[k*]$-hardness, we firstly prove the following technical lemma.

Lemma 8.7. *Let (φ, k) be an instance of $\Sigma_2^p[k*]$-WSAT. In polynomial time, we can construct an instance (φ', k) of $\Sigma_2^p[k*]$-WSAT with $\varphi' = \exists X.\forall Y.\psi$, such that:*

- $(\varphi', k) \in \Sigma_2^p[k*]$-WSAT *if and only if* $(\varphi, k) \in \Sigma_2^p[k*]$-WSAT*; and*
- *for any assignment $\alpha : X \to \mathbb{B}$ that has weight $m \neq k$, it holds that $\forall Y.\psi[\alpha]$ is true.*

Proof. Let (φ, k) be an instance of $\Sigma_2^p[k*]$-WSAT, where $\varphi = \exists X.\forall Y.\psi$, and where $X = \{x_1, \ldots, x_n\}$. We construct an instance (φ', k), with $\varphi' = \exists X.\forall Y \cup Z.\psi'$. We define:

$$Z = \{z_j^i : i \in [k], j \in [n]\}.$$

Intuitively, one can think of the variables z_j^i as being positioned in a matrix with n rows and k columns: the variable z_j^i is placed in the j-th row and the i-th column. We will use this matrix to verify whether exactly k variables in X are set to true.

We define ψ' as follows:

$$\psi' = (\psi_{\text{corr}}^{X,Z} \wedge \psi_{\text{row}}^{Z} \wedge \psi_{\text{col}}^{Z}) \to \psi;$$

$$\psi_{\text{corr}}^{X,Z} = \bigwedge_{j \in [n]} \left(\bigwedge_{i \in [k]} (z_j^i \to x_j) \wedge (x_j \to \bigvee_{i \in [k]} z_j^i) \right);$$

$$\psi_{\text{row}}^{Z} = \bigwedge_{j \in [n]} \bigwedge_{i,i' \in [k], i < i'} (\neg z_j^i \vee \neg z_j^{i'}); \text{ and}$$

$$\psi_{\text{col}}^{Z} = \bigwedge_{i \in [k]} \left(\bigwedge_{j,j' \in [n], j < j'} (\neg z_j^i \vee \neg z_{j'}^i) \wedge \bigvee_{j \in [n]} z_j^i \right).$$

Intuitively, the formula $\psi_{\text{corr}}^{X,Z}$ ensures that exactly those x_j are set to true for which there exists some z_j^i that is set to true. The formula ψ_{row}^{Z} ensures that in each row of the matrix filled with variables z_j^i, there is at most one variable set to true. The formula ψ_{col}^{Z} ensures that in each column there is exactly one variable set to true.

We show that $(\varphi, k) \in \Sigma_2^p[k*]$-WSAT if and only if $(\varphi', k') \in \Sigma_2^p[k*]$-WSAT.

($\Rightarrow$) Let $\alpha : X \to \mathbb{B}$ be an assignment of weight k such that $\forall Y.\psi[\alpha]$ is true. We show that $\forall Y \cup Z.\psi'[\alpha]$ is true. Let $\beta : Y \cup Z \to \mathbb{B}$ be an arbitrary truth assignment. We distinguish two cases: either (i) for each $i \in [k]$ there is a unique $j_i \in [n]$ such that $\beta(z_{j_i}^i) = 1$ and $\alpha(x_{j_i}) = 1$, or (ii) this is not the case. In case (i), it is straightforward to verify that $\alpha \cup \beta$ satisfies $(\psi_{\text{corr}}^{X,Z} \wedge \psi_{\text{row}}^{Z} \wedge \psi_{\text{col}}^{Z})$. Then, since $\forall Y.\psi[\alpha]$ is true, we know that $\alpha \cup \beta$ satisfies ψ, and thus that $\alpha \cup \beta$ satisfies ψ'. In case (ii), we know that $\alpha \cup \beta$ does not satisfy $(\psi_{\text{corr}}^{X,Z} \wedge \psi_{\text{row}}^{Z} \wedge \psi_{\text{col}}^{Z})$, and thus that $\alpha \cup \beta$ satisfies ψ'. Therefore, $\forall Y \cup Z.\psi'[\alpha]$ is true.

($\Leftarrow$) Assume that there exists an assignment $\alpha : X \to \mathbb{B}$ of weight k such that $\forall Y \cup Z.\psi'[\alpha]$ is true. We show that $\forall Y.\psi[\alpha]$ is true. Let $\beta : Y \to \mathbb{B}$ be an arbitrary truth assignment. Let $\{ x_j : j \in [n], \alpha(x_j) = 1 \} = \{x_{j_1}, \ldots, x_{j_k}\}$. We construct the assignment $\gamma : Z \to \mathbb{B}$ by letting $\gamma(z_j^i) = 1$ if and only if $j = j_i$. We know that $\alpha \cup \beta \cup \gamma$ satisfies ψ'. Clearly, $\alpha \cup \gamma$ satisfies $(\psi_{\text{corr}}^{X,Z} \wedge \psi_{\text{row}}^{Z} \wedge \psi_{\text{col}}^{Z})$. Hence, $\alpha \cup \beta$ must satisfy ψ. Since β was arbitrary, we know that $\forall Y.\psi[\alpha]$ is true.

Next, it is straightforward to verify that for any assignment $\alpha : X \to \mathbb{B}$ that has weight $m \neq k'$, it holds that for no assignment $\gamma : Z \to \mathbb{B}$ it is the case that $\alpha \cup \gamma$ satisfies $(\psi_{\text{corr}}^{X,Z} \wedge \psi_{\text{row}}^{Z} \wedge \psi_{\text{col}}^{Z})$, and thus that $\forall Y.\psi[\alpha]$ is true.

We are now ready to show $\Pi_2^p[k*]$-completeness for ROBUST-CSP-SAT.

Proposition 8.8. ROBUST-CSP-SAT *is* $\Pi_2^p[k*]$-*complete. Moreover,* $\Pi_2^p[k*]$-*hardness holds even when the domain size* $|D|$ *is restricted to 2.*

Proof. We begin by showing membership in $\Pi_2^p[k*]$. We do so by giving an fpt-reduction from ROBUST-CSP-SAT to $\Pi_2^p[k*]$-WSAT. Let $(\mathcal{I}, k)$ be an instance of ROBUST-CSP-SAT, where $\mathcal{I}$ is a CSP instance, where $\text{Var}(\mathcal{I}) = \{v_1, \ldots, v_n\}$, where $D = \{d_1, \ldots, d_m\}$ is the domain of $\mathcal{I}$, and where k is a positive integer. We construct an instance (φ, k) of $\Pi_2^p[k*]$-WSAT. For the formula φ, we use proposi-tional variables $Z = \{ z_j^i : i \in [n], j \in [m] \}$ and $Y = \{ y_j^i : i \in [n], j \in [m] \}$. Intu-itively, the variables z_j^i will represent an arbitrary assignment α that assigns values to k variables in $\mathrm{Var}(\mathcal{I})$. Any variable z_j^i represents that variable v_i gets assigned value d_j. The variables y_j^i will represent the solution β that extends the arbitrary assignment α. Similarly, any variable y_j^i represents that variable v_i gets assigned value d_j.

We then let $\varphi = \forall Z.\exists Y.\psi$ with $\psi = (\psi_{\mathrm{proper}}^Z \wedge \neg\psi_{\mathrm{violate}}^Z) \rightarrow (\psi_{\mathrm{corr}}^{Y,Z} \wedge \psi_{\mathrm{proper}}^Y \wedge \bigwedge_{C \in \mathcal{I}} \psi_C^Y)$. We will describe the subformulas of φ below, as well as the intuition behind them.

We start with the formula ψ_{proper}^Z. This formula represents whether for each vari-able v_i at most one value is chosen for the assignment α. We let:

$$\psi_{\mathrm{proper}}^Z = \bigwedge_{i \in [n]} \bigwedge_{j,j' \in [m], j < j'} (\neg z_j^i \vee \neg z_{j'}^i).$$

Next, we consider the formula $\psi_{\mathrm{violate}}^Z$. This subformula encodes whether the as-signment α violates some constraint $C \in \mathcal{I}$. We let:

$$\psi_{\mathrm{violate}}^Z = \bigvee_{C=(S,R) \in \mathcal{I}} \bigwedge_{\overline{d} \in R} \bigvee_{z \in \psi^{\overline{d},C}} z,$$

where we define the set $\psi^{\overline{d},C} \subseteq Z$ as follows. Let $C = ((v_{i_1}, \ldots, v_{i_r}), R) \in \mathcal{I}$ and $\overline{d} = (d_{j_1}, \ldots, d_{j_r}) \in R$. Then we let:

$$\psi^{\overline{d},C} = \{ z_j^{i_\ell} : \ell \in [r], j \neq j_\ell \}.$$

Intuitively, the set $\psi^{\overline{d},C}$ contains the variables z_j^i that represent those variable as-signments in α that prevent that β satisfies C by assigning $\mathrm{Var}(C)$ to $\overline{d}$.

Then, the formula ψ_{proper}^Y ensures that for each variable v_i exactly one value d_j is chosen in β. We define:

$$\psi_{\mathrm{proper}}^Y = \bigwedge_{i \in [n]} \left(\bigvee_{j \in [m]} y_j^i \wedge \bigwedge_{j,j' \in [m], j < j'} (\neg y_j^i \vee \neg y_{j'}^i) \right).$$

Next, the formula $\psi_{\mathrm{corr}}^{Y,Z}$ ensures that β is indeed an extension of α. We define:

$$\psi_{\mathrm{corr}}^{Y,Z} = \bigwedge_{i \in [n]} \bigwedge_{j \in [m]} (z_j^i \rightarrow y_j^i).$$

Finally, for each $C \in \mathcal{I}$, the formula ψ_C^Y represents whether β satisfies C. Let $C = ((v_{i_1}, \ldots, v_{i_r}), R) \in \mathcal{I}$. We define:

$$\psi_C^Y = \bigvee_{(d_{j_1}, \ldots, d_{j_r}) \in R} \bigwedge_{\ell \in [r]} y_{j_\ell}^{i_\ell}.$$

We now argue that $(\mathcal{I}, k) \in \text{ROBUST-CSP-SAT}$ if and only if $(\varphi, k) \in \Pi_2^P[k*]\text{-WSAT}$.

$(\Rightarrow)$ Assume that $\mathcal{I}$ is k-robustly satisfiable. We show that $(\varphi, k) \in \Pi_2^P[k*]\text{-WSAT}$. Let $\alpha : Z \to \mathbb{B}$ be an arbitrary assignment of weight k. If $\alpha(z_j^i) = \alpha(z_{j'}^i) = 1$ for some $i \in [n]$ and some $j, j' \in [m]$ with $j < j'$, then α (and any extension of it) satisfies $\neg\psi_{\text{proper}}^Z$. Therefore, $\exists Y.\psi[\alpha]$ is true. We can thus restrict our attention to the case where for each $i \in [n]$, there is at most one $j_i \in [m]$ such that $\alpha(z_{j_i}^i) = 1$. Define the subset $V \subseteq \text{Var}(\mathcal{I})$ and the instantiation $\mu : V \to D$ by letting $v_i \in V$ and $\mu(v_i) = d_j$ if and only if $\alpha(z_j^i) = 1$. Clearly, μ assigns values to k different variables.

We distinguish two cases: either (i) μ violates some constraint $c \in C$, or (ii) μ violates no constraint in C. In case (i), it is straightforward to verify that α satisfies ψ_{violate}^Z. Therefore, $\exists Y.\psi[\alpha]$ is true. Next, consider case (ii). Since $\mathcal{I}$ is k-robustly satisfiable, we know that there exists some complete instantiation $\nu : \text{Var}(\mathcal{I}) \to D$ that extends μ, and that satisfies all constraints $C \in \mathcal{I}$. Now, define the assignment $\beta : Y \to \mathbb{B}$ by letting $\beta(y_j^i) = 1$ if and only if $\nu(v_i) = d_j$. It is straightforward to verify that $\alpha \cup \beta$ satisfies $\psi_{\text{corr}}^{Y,Z}$, and that β satisfies ψ_{proper}^Y and ψ_C^Y for all $C \in \mathcal{I}$. Therefore, $\alpha \cup \beta$ satisfies ψ. This concludes our proof that $(\varphi, k) \in \Pi_2^P[k*]\text{-WSAT}$.

$(\Leftarrow)$ Assume that $(\varphi, k) \in \Pi_2^P[k*]\text{-WSAT}$. We show that $\mathcal{I}$ is k-robustly satisfiable. Let $V \subseteq \text{Var}(\mathcal{I})$ be an arbitrary subset of k variables, and let $\mu : V \to D$ be an arbitrary instantiation that does not violate any constraint $C \in \mathcal{I}$. Define the assignment $\alpha : Z \to \mathbb{B}$ by letting $\alpha(z_j^i) = 1$ if and only if $v_i \in V$ and $\mu(v_i) = d_j$. Clearly, the assignment α has weight k. Therefore, there must exist an assignment $\beta : Y \to \mathbb{B}$ such that $\alpha \cup \beta$ satisfies ψ. It is straightforward to verify that α satisfies $\psi_{\text{proper}}^Z \wedge \neg\psi_{\text{violate}}^Z$. Therefore, $\alpha \cup \beta$ must satisfy $\psi_{\text{corr}}^{Y,Z} \wedge \psi_{\text{proper}}^Y$ and ψ_C^Y for all $C \in \mathcal{I}$. Since $\alpha \cup \beta$ satisfies ψ_{proper}^Y, we know that for each $i \in [n]$, there is a unique $j_i \in [m]$ such that $\beta(y_{j_i}^i) = 1$. Define the complete instantiation $\nu : V \to D$ by letting $\nu(v_i) = d_{j_i}$. It is straightforward to verify that ν extends μ, since $\alpha \cup \beta$ satisfies $\psi_{\text{corr}}^{Y,Z}$. Also, since $\alpha \cup \beta$ satisfies ψ_C^Y for all $C \in \mathcal{I}$, it follows that ν satisfies each $C \in \mathcal{I}$. Therefore, ν is a solution of the CSP instance $\mathcal{I}$. This concludes our proof that $\mathcal{I}$ is k-robustly satisfiable.

Next, we show that ROBUST-CSP-SAT is $\Pi_2^P[k*]$-hard. We do so by giving an fpt-reduction from $\Pi_2^P[k*]\text{-WSAT}(\text{3CNF})$ to ROBUST-CSP-SAT. Let (φ, k) be an instance of $\Pi_2^P[k*]\text{-WSAT}(\text{3CNF})$, with $\varphi = \forall X.\exists Y.\psi$, and $\psi = c_1 \wedge \cdots \wedge c_u$. By Lemma 8.7, we may assume without loss of generality that for any assignment $\alpha : X \to \mathbb{B}$ of weight $m \neq k$, we have that $\exists Y.\psi[\alpha]$ is false. We construct an instance $(\mathcal{I}, k)$ of ROBUST-CSP-SAT as follows. We define the set $\text{Var}(\mathcal{I})$ of variables by $\text{Var}(\mathcal{I}) = X \cup Y'$, where $Y' = \{y^i : y \in Y, i \in [2k+1]\}$, and we let $D = \mathbb{B}$. We will define the set $\mathcal{I}$ of constraints below, by representing each clause c_i of ψ as a set of clauses whose length is bounded by $f(k)$, for some fixed function f.

The intuition behind the construction of $\mathcal{I}$ is the following. We replace each variable $y \in Y$, by $2k + 1$ copies y^i of it. Assigning a variable $y \in Y$ to a value $b \in \mathbb{B}$ will then correspond to assigning a majority of variables y^i to b, i.e., assigning at least $k + 1$ variables y^i to b. In order to encode this transformation in the constraints of $\mathcal{I}$, intuitively, we will replace each occurrence of a variable y by the conjunction:

$$\psi_y = \bigwedge_{i_1,\ldots,i_{k+1} \in [2k+1], i_1 < \cdots < i_{k+1}} (y^{i_1} \vee \cdots \vee y^{i_{k+1}}),$$

and replace each occurrence of a literal $\neg y$ by a similar conjunction. We will then multiply the resulting formula out into CNF. Note that whenever a majority of variables y^i is set to $b \in \mathbb{B}$, then the formula ψ_y will also evaluate to b.

In the construction of C, we will directly encode the CNF formula that is a result of the transformation described above. For each literal $l = y \in Y$, let l^i denote y^i, and for each literal $l = \neg y$ with $y \in Y$, let l^i denote $\neg y^i$. For each literal l over the variables $X \cup Y$, we define a set $\sigma(l)$ of clauses:

$$\sigma(l) = \begin{cases} \{ (l^{i_1} \vee \cdots \vee l^{i_{k+1}}) : i_1, \ldots, i_{k+1} \in [2k+1], i_1 < \cdots < i_{k+1} \} \\ \qquad\qquad\qquad\qquad\qquad\qquad\qquad \text{if } l \text{ is a literal over } Y; \\ \{l\} \qquad\qquad\qquad\qquad\qquad\qquad\quad \text{if } l \text{ is a literal over } X. \end{cases}$$

Note that for each literal l, it holds that $|\sigma(l)| \leq g(k) = \binom{2k+1}{k+1}$. Next, for each clause $c_i = l_1^i \vee l_2^i \vee l_3^i$ of ψ, we introduce to $\mathcal{I}$ a set $\sigma(c_i)$ of clauses:

$$\sigma(c_i) = \{ d_1 \vee d_2 \vee d_3 : d_1 \in \sigma(l_1^i), d_2 \in \sigma(l_2^i), d_3 \in \sigma(l_3^i) \}.$$

Note that $|\sigma(c_i)| \leq g(k)^3$. Formally, we let $\mathcal{I}$ be the set of constraints corresponding to the set $\bigcup_{i \in [u]} \sigma(c_i)$ of clauses. Since each such clause is of length at most $3(k + 1)$, representing a clause by means of a constraint can be done by specifying $\leq 2^{3(k+1)} - 1$ tuples, i.e., all tuples satisfying the clause. Therefore, the instance $(\mathcal{I}, k)$ can be constructed in fpt-time.

We now argue that $(\varphi, k) \in \Pi_2^P[k*]$-WSAT(3CNF) if and only if $(\mathcal{I}, k) \in$ ROBUST-CSP-SAT.

($\Rightarrow$) Assume that $(\varphi, k) \in \Pi_2^P[k*]$-WSAT(3CNF). We show that $\mathcal{I}$ is k-robustly satisfiable. Let $\mu : \text{Var}(\mathcal{I}) \to D$ be an arbitrary partial assignment with $|\text{Dom}(\mu)| = k$ that does not violate any constraint in $\mathcal{I}$. We know that $|\text{Dom}(\mu) \cap X| \leq k$, and in particular that $|\{ x \in \text{Dom}(\mu) \cap X : \mu(x) = 1 \}| = m \leq k$. Now define the assignment $\alpha : X \to \mathbb{B}$ as follows. For any $x \in X$, if $x \in \text{Dom}(\mu)$, then let $\alpha(x) = \mu(x)$. Also, for $k - m$ variables $x' \in X \backslash \text{Dom}(\mu)$, we let $\alpha(x') = 1$. For all other variables $x' \in X \backslash \text{Dom}(\mu)$, we let $\alpha(x') = 0$. Then α has weight k. Therefore, there must exist an assignment $\beta : Y \to \mathbb{B}$ such that $\psi[\alpha \cup \beta]$ is true. Now, define the assignment $\nu : \text{Var}(\mathcal{I}) \to D$ extending μ as follows. For each $z \in \text{Dom}(\mu)$, we let $\nu(z) =$

$\mu(z)$. For each $x \in X \backslash \text{Dom}(\mu)$, we let $\nu(x) = \alpha(x)$. For each $y^i \in Y' \backslash \text{Dom}(\mu)$, we let $\nu(y^i) = \beta(y)$. It is straightforward to verify that for each $y \in Y$, $\beta(y) = b \in \mathbb{B}$ if and only if ν sets at least $k + 1$ variables y^i to b. Using this fact, and the fact that $\alpha \cup \beta$ satisfies each clause c_i of ψ, it is straightforward to verify that ν satisfies each constraint of $\mathcal{I}$, and therefore that ν is a solution of the CSP instance $\mathcal{I}$. This concludes our proof that $(\mathcal{I}, k) \in$ROBUST-CSP-SAT.

($\Leftarrow$) Now, assume that (Z, D, C) is k-robustly satisfiable. We show that $(\varphi, k) \in \Pi_2^P[k*]$-WSAT(3CNF). Let $\alpha : X \to \mathbb{B}$ be an arbitrary assignment of weight k. Now define the partial assignment $\mu : Z \to D$ by letting $x \in \text{Dom}(\mu)$ and $\mu(x) = \alpha(x)$ if and only if $\alpha(x) = 1$, for all $x \in X$. Clearly, $|\text{Dom}(\mu)| = k$, and μ does not violate any constraints of C. Therefore, we know that there exists an extension $\nu : Z \to D$ of μ that is a solution for the CSP instance (Z, D, C). For all $x \in X \backslash \text{Dom}(\mu)$ it holds that $\nu(x) = 0$. If this weren't the case, this would violate our assumption that for all assignments $\alpha' : X \to \mathbb{B}$ with weight $m \neq k$ we have that $\exists Y.\psi[\alpha']$ is false. Therefore, we know that ν coincides with α on the variables X. Now, we define the assignment $\beta : Y \to \mathbb{B}$ by letting $\beta(y) = 1$ if and only if for at least $k + 1$ different y^i it holds that $\nu(y^i) = 1$.

We then verify that $\psi[\alpha \cup \beta]$ is true. Let c_i be a clause of ψ. We know that ν satisfies all clauses in $\sigma(c_i)$. Assume that $\alpha \cup \beta$ does not satisfy $c_i = l_1^i \vee l_2^i \vee l_3^i$. Then, $\alpha \cup \beta$ sets all literals l_1^i, l_2^i and l_3^i to false. Therefore, for any literal l_j^i over Y, ν sets at least $k + 1$ copies of l_j^i to false. From this, we can conclude that there is some clause in $\sigma(c_i)$ that is set to false by ν, which is a contradiction. Therefore, we know that $\psi[\alpha \cup \beta]$ is true. This concludes our proof that $(\varphi, k) \in \Pi_2^P[k*]$-WSAT(3CNF).

Notes

The results in Sects. 8.1 and 8.3 appeared in a paper in the proceedings of KR 2014 [116, 117]. The results in Sect. 8.2 were shown in an unpublished manuscript [110].

Chapter 9
Model Checking for Temporal Logics

In Chap. 8, we started to apply the parameterized complexity tools that we developed in Chaps. 4–7 to concrete problems from various domains of computer science and artificial intelligence. In this chapter, we continue this investigation by analyzing the parameterized complexity of various parameterized problems related to temporal reasoning. In particular, we study several parameterized variants of the model checking problem for various (fragments of) temporal logics.

Temporal logic model checking is an important problem with applications in key areas of computer science and engineering, among others in the verification of software and hardware systems (see, e.g., [11, 51, 53, 83]). The problem consists of checking whether a model, given in the form of a labelled relational structure (a *Kripke structure*), satisfies a temporal property, given as a logic formula. Underlining the importance of temporal logic model checking, the ACM 2007 Turing Award was given for foundational research on the topic [50]. Indispensable for the state-of-the-art in solving this problem in industrial-size settings is the algorithmic technique of symbolic model checking using SAT solvers (called *bounded model checking*), where the SAT solvers are employed to find counterexamples [18, 20, 21, 52].

The approach of bounded model checking generally works well in cases where the Kripke structure is large, but the temporal logic specification is small. Since the framework of parameterized complexity is able to distinguish an additional measure of the input, that can be much smaller than the input size, a parameterized complexity approach would be especially suited for a theoretical complexity analysis. However, previous parameterized complexity analyses have not been able to fill the gap. First of all, existing parameterized complexity analyses [60, 85, 95, 151, 161] have only considered the problem for settings where the Kripke structure is spelled-out explicitly (or consists of a small number of explicitly spelled-out components), which is highly impractical in many cases. In fact, the so-called state explosion problem is a major obstacle for developing practically useful techniques [49]. For this reason, the Kripke structures are often described *symbolically*, for instance using propositional formulas, which allows for exponentially more succinct encodings of the structures. Secondly, whereas parameterized complexity analysis is traditionally focused on fixed-parameter tractability for positive results, the technique of bounded

R. de Haan: Parameterized Complexity in the Polynomial Hierarchy, LNCS 11880, https://doi.org/10.1007/978-3-662-60670-4_9

model checking revolves around encoding the problem as an instance of SAT. There-fore, the standard parameterized complexity analysis is bound to concentrate on very restrictive cases in order to obtain fixed-parameter tractability, unaware of some of the more liberal settings where bounded model checking can be applied.

In this chapter, we provide a parameterized complexity analysis that reveals the possibilities and limits of the technique of bounded model checking. More specif-ically, we analyze the complexity of the model checking problem for fragments of various temporal logics, where we take the size of the temporal logic formula as parameter. In our formalization of the problem, the Kripke structures are repre-sented symbolically (and can thus be of size exponential in the size of their descrip-tion). Moreover, our complexity analysis focuses on whether the problems admit fpt-reductions to SAT or not.

As a by-product of our investigation, we introduce the parameterized complexity class PH(level), that is another parameterized variant of the Polynomial Hierarchy. This class can be seen as an analogue of the parameterized variant BH(level) of the Boolean Hierarchy, that we considered in Chap. 7. It is based on the satisfiability problem of quantified Boolean formulas parameterized by the number of quantifier alternations.

Outline of This Chapter

We begin in Sect. 9.1 by introducing the syntax and semantics of three of the most widespread temporal logics (LTL, CTL and CTL*). These linear-time and branching-time propositional modal logics are the temporal logics that we consider in this chapter. Moreover, for each of these logics, we also consider the fragments where several temporal operators (namely, U and/or X) are disallowed. Additionally, we briefly review known (parameterized) complexity results for their model checking problems.

Then, in Sect. 9.2, we consider another formalization of the model checking prob-lem, where the Kripke structures are represented symbolically. We argue that without restrictions on the Kripke structures, in this setting the problem is PSPACE-hard even for temporal logic formulas of constant size (Proposition 9.2). However, for the set-ting where the *recurrence diameter* of the Kripke structures (the size of the largest loop-free path) is required to be bounded by a polynomial of the input size, we identify a logic fragment whose model checking problem admits an fpt-reduction to SAT. In particular, we interpret a known result for bounded model checking for the fragment of LTL without U and X operators as membership in para-co-NP (Proposition 9.3).

In Sect. 9.3, we introduce the new parameterized complexity class PH(level). We show that this class can also be characterized by means of an parameterized first-order logic model checking problem, as well as by means of alternating Turing machines that alternate between existential and universal configurations only a small number of times (depending only on the parameter). Moreover, we briefly relate this class to other parameterized variants of the Polynomial Hierarchy.

Finally, in Sect. 9.4, we develop the parameterized complexity results that indicate that for all other fragments of the temporal logics that we consider, the symbolic model checking problem does not admit an fpt-reduction to SAT. More precisely, we

Table 9.1 Parameterized complexity results for the problem SYMBOLIC- MC*[$\mathcal{L}$] for the different (fragments of) logics $\mathcal{L}$. In this problem, the recurrence diameter of the structure is polynomially bounded. The problem SYMBOLIC- MC[$\mathcal{L}$], where the recurrence diameter is unbounded, is para-PSPACE-complete in all cases.

Logic $\mathcal{L}$	LTL	CTL	CTL*
$\mathcal{L}$	para-PSPACE-c (Theorems 9.6, 9.7)	PH(level)-c (Theorem 9.9)	para-PSPACE-c (Theorems 9.6, 9.7)
$\mathcal{L}\backslash X$	para-PSPACE-c (Theorem 9.7)	PH(level)-c (Theorem 9.9)	para-PSPACE-c (Theorem 9.7)
$\mathcal{L}\backslash U$	para-PSPACE-c (Theorem 9.6)	PH(level)-c (Theorem 9.9)	para-PSPACE-c (Theorem 9.6)
$\mathcal{L}\backslash U,X$	para-co-NP-c (Proposition 9.3)	PH(level)-c (Theorem 9.9)	PH(level)-c (Theorem 9.10)

give a complete parameterized complexity classification of the problem of checking whether a given Kripke structure, that is specified symbolically using a propositional formula and whose recurrence diameter is polynomial in the size of the propositional formula, satisfies a given temporal logic specification, parameterized by the size of the temporal logic formula.

- We extend the para-co-NP-membership result for the logic LTL where both operators U and X are disallowed to a para-co-NP-completeness result (Proposition 9.3 in Sect. 9.2).
- We show that the problem is para-PSPACE-complete for LTL (and so also for its generalization CTL*) when at least one of the operators U and X is allowed (Theorems 9.6 and 9.7).
- We show that in all remaining cases (all fragments of CTL, and the fragment of CTL* without the operators U and X) the problem is complete for PH(level) (Theorems 9.9 and 9.10).

In short, we show that the only case (for the fragments of temporal logics that we consider) where the technique of bounded model checking can be applied is the fragment of LTL without the operators U and X. An overview of the parameterized complexity results that we develop in this chapter can be found in Table 9.1.

Related Work

Computational complexity analysis has been a central aspect in the study of temporal logic model checking problems, and naturally these problems have been analyzed from a parameterized complexity point of view. For instance, LTL model checking parameterized by the size of the logic formula features as a textbook example for fixed-parameter tractability [85]. For the temporal logic CTL, parameterized complexity has also been used to study the problems of model checking and satisfiability [60, 95, 151, 161]. As the SAT encoding techniques used for bounded LTL model checking result in an incomplete solving method in general, limits on the cases in

Fig. 9.1 An example Kripke
structure $\mathcal{M}_1$ for the
set $P = \{p_1, p_2\}$ of
propositions.

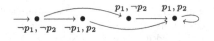

which this particular encoding can be used as a complete solving method have been
studied [32, 52, 139].

9.1   The (Parameterized) Complexity of Model Checking

In this section, we give a brief overview of the definition of the temporal logics LTL,
CTL and CTL* that we consider in this chapter. Moreover, we survey some well-
known (parameterized) complexity results for the model checking problem for these
logics.

9.1.1   Temporal Logics

We begin with defining the semantical structures for all temporal logics. In the
remainder of the chapter, we let P be a finite set of propositions. A *Kripke structure*
is a tuple $\mathcal{M} = (S, R, V, s_0)$, where S is a finite set of *states*, where $R \subseteq S \times S$ is a
binary relation on the set of states called the *transition relation*, where $V : S \to 2^P$ is
a *valuation function* that assigns each state to a set of propositions, and where $s_0 \in S$
is the *initial state*. An example of a Kripke structure is given in Fig. 9.1. We say that
a finite sequence $s_1 \ldots s_\ell$ of states $s_i \in S$ is a *finite path* in $\mathcal{M}$ if $(s_i, s_{i+1}) \in R$ for
each $i \in [\ell - 1]$. Similarly, we say that an infinite sequence $s_1 s_2 s_3 \ldots$ of states $s_i \in S$
is an *infinite path* in $\mathcal{M}$ if $(s_i, s_{i+1}) \in R$ for each $i \geq 1$.

Now, we can define the syntax of the logic LTL. LTL formulas over the set P of
atomic propositions are formed according to the following grammar (here p ranges
over P), given by:

$$\varphi ::= p \mid \neg\varphi \mid (\varphi \wedge \varphi) \mid X\varphi \mid F\varphi \mid (\varphi U\varphi).$$

We consider the usual abbreviations, such as $\varphi_1 \vee \varphi_2 = \neg(\neg\varphi_1 \wedge \neg\varphi_2)$. In addition,
we let the abbreviation $G\varphi$ denote $\neg F\neg\varphi$. Intuitively, the formula $X\varphi$ expresses that φ
is true in the next (time) step, $F\varphi$ expresses that φ becomes true at some point in the
future, $G\varphi$ expresses that φ is true at all times from now on, and $\varphi_1 U\varphi_2$ expresses
that φ_2 becomes true at some point in time, and until then the formula φ_1 is true at all
points. Formally, the semantics of LTL formulas are defined for Kripke structures,
using the notion of (infinite) paths. Let $\mathcal{M} = (S, R, V, s_0)$ be a Kripke structure,
and $\bar{s}_1 = s_1 s_2 s_3 \ldots$ be a path in $\mathcal{M}$. Moreover, let $\bar{s}_i = s_i s_{i+1} s_{i+2} \ldots$ for each $i \geq 2$.
Truth of LTL formulas φ on paths $\bar{s}$ (denoted $\bar{s} \models \varphi$) is defined inductively as follows:

$$\bar{s}_i \models p \qquad \text{if } p \in V(s_i)$$
$$\bar{s}_i \models \varphi_1 \wedge \varphi_2 \text{ if } \bar{s}_i \models \varphi_1 \text{ and } \bar{s}_i \models \varphi_2$$
$$\bar{s}_i \models \neg\varphi \qquad \text{if } \bar{s}_i \not\models \varphi$$
$$\bar{s}_i \models X\varphi \qquad \text{if } \bar{s}_{i+1} \models \varphi$$
$$\bar{s}_i \models F\varphi \qquad \text{if for some } j \geq 0, \bar{s}_{i+j} \models \varphi$$
$$\bar{s}_i \models \varphi_1 U\varphi_2 \text{ if there is some } j \geq 0 \text{ such that } \bar{s}_{i+j} \models \varphi_2$$
$$\text{and } \bar{s}_{i+j'} \models \varphi \text{ for each } j' \in [0, j-1]$$

Then, we say that an LTL formula φ is true in the Kripke structure $\mathcal{M}$ (denoted $\mathcal{M} \models \varphi$) if for all infinite paths $\bar{s}$ starting in s_0 it holds that $\bar{s} \models \varphi$. For instance, considering the example $\mathcal{M}_1$ from Fig. 9.1, it holds that $\mathcal{M}_1 \models FGp_2$.

Next, we can define the syntax of the logic CTL*, which consists of two different types of formulas: state formulas and path formulas. When we refer to CTL* formulas without specifying the type, we refer to state formulas. Given the set P of atomic propositions, the syntax of CTL* formulas is defined by the following grammar (here Φ denotes CTL* *state formulas*, φ denotes CTL* *path formulas*, and p ranges over P), given by:

$$\Phi ::= p \mid \neg\Phi \mid (\Phi \wedge \Phi) \mid \exists\varphi.$$

$$\varphi ::= \Phi \mid \neg\varphi \mid (\varphi \wedge \varphi) \mid X\varphi \mid F\varphi \mid (\varphi U\varphi).$$

Again, we consider the usual abbreviations, such as $\varphi_1 \vee \varphi_2 = \neg(\neg\varphi_1 \wedge \neg\varphi_2)$, for state formulas as well as for path formulas. Moreover, we let the abbreviation $G\varphi$ denote $\neg F\neg\varphi$, and we let the abbreviation $\forall\varphi$ denote $\neg\exists\neg\varphi$. Path formulas have the same intended meaning as LTL formulas. State formulas, in addition, allow explicit quantification over paths, which is not possible in LTL.

Formally, the semantics of CTL* formulas are defined inductively as follows. Let $\mathcal{M} = (S, R, V, s_0)$ be a Kripke structure, $s \in S$ be a state in $\mathcal{M}$ and $\bar{s}_1 = s_1 s_2 s_3 \ldots$ be a path in $\mathcal{M}$. Again, let $\bar{s}_i = s_i s_{i+1} s_{i+2} \ldots$ for each $i \geq 2$. The truth of CTL* state formulas Φ on states s (denoted $s \models \Phi$) is defined as follows:

$$s \models p \qquad \text{if } p \in V(s)$$
$$s \models \Phi_1 \wedge \Phi_2 \text{ if } s \models \Phi_1 \text{ and } s \models \Phi_2$$
$$s \models \neg\Phi \qquad \text{if } s \not\models \Phi$$
$$s \models \exists\varphi \qquad \text{if there is some path } \bar{s} \text{ in } \mathcal{M} \text{ starting in } s \text{ such that } \bar{s} \models \varphi$$

The truth of CTL* path formulas φ on paths $\bar{s}$ (denoted $\bar{s} \models \varphi$) is defined as follows:

$$\bar{s}_i \models \Phi \qquad \text{if } s_i \models \Phi$$
$$\bar{s}_i \models \varphi_1 \wedge \varphi_2 \text{ if } \bar{s}_i \models \varphi_1 \text{ and } \bar{s}_i \models \varphi_2$$
$$\bar{s}_i \models \neg\varphi \qquad \text{if } \bar{s}_i \not\models \varphi$$
$$\bar{s}_i \models X\varphi \qquad \text{if } \bar{s}_{i+1} \models \varphi$$
$$\bar{s}_i \models F\varphi \qquad \text{if for some } j \geq 0, \bar{s}_{i+j} \models \varphi$$
$$\bar{s}_i \models \varphi_1 U\varphi_2 \quad \text{if there is some } j \geq 0 \text{ such that } \bar{s}_{i+j} \models \varphi_2$$
$$\text{and } \bar{s}_{i+j'} \models \varphi \text{ for each } j' \in [0, j-1]$$

Then, we say that a CTL* formula Φ is true in the Kripke structure $\mathcal{M}$ (denoted $\mathcal{M} \models \Phi$) if $s_0 \models \Phi$. For example, again taking the structure $\mathcal{M}_1$, it holds that $\mathcal{M}_1 \models \exists(Xp_1 \wedge \forall GXXp_2)$.

Next, the syntax of the logic CTL is defined similarly to the syntax of CTL*. Only the grammar for path formulas φ differs, namely:

$$\varphi ::= X\Phi \mid F\Phi \mid (\Phi U\Phi).$$

In particular, this means that every CTL state formula, (CTL formula for short) is also a CTL* formula. The semantics for CTL formulas is defined as for their CTL* counterparts. Moreover, we say that a CTL formula Φ is true in the Kripke structure $\mathcal{M}$ (denoted $\mathcal{M} \models \Phi$) if $s_0 \models \Phi$.

For each of the logics $\mathcal{L} \in \{\text{LTL, CTL, CTL*}\}$, we consider the fragments $\mathcal{L}\backslash X, \mathcal{L}\backslash U$ and $\mathcal{L}\backslash U,X$. In the fragment $\mathcal{L}\backslash X$, the X-operator is disallowed. Similarly, in the fragment $\mathcal{L}\backslash U$, the U-operator is disallowed. In the fragment $\mathcal{L}\backslash U,X$, neither the X-operator nor the U-operator is allowed. Note that the logic LTL$\backslash$X is also known as UTL, and the logic LTL$\backslash$U,X is also known as UTL$\backslash$X (see, e.g., [139]).

9.1.2 (Parameterized) Complexity Results

Next, we review some known (parameterized) complexity results for the model checking problem of the different temporal logics. Formally, we consider the problem MC[$\mathcal{L}$], for each of the temporal logics $\mathcal{L}$, where the input is a Kripke structure $\mathcal{M}$ and an $\mathcal{L}$ formula φ, and the question is to decide whether $\mathcal{M} \models \varphi$. Note that in this problem the Kripke structure $\mathcal{M}$ is given explicitly in the input.

MC[$\mathcal{L}$]
Input: A Kripke structure $\mathcal{M}$, and an $\mathcal{L}$ formula φ.
Question: $\mathcal{M} \models \varphi$?

We will also consider this computational task as a parameterized problem, where the parameter is the size of the logic formula. We will use the same name for the parameterized problem (it is clear from the context which problem we refer to). It is well known that the problems MC[LTL] and MC[CTL*] are PSPACE-complete, and that the problem MC[CTL] is polynomial-time solvable (see, e.g., [11]). It is also well known that the problems MC[LTL] and MC[CTL*] are fixed-parameter tractable when parameterized by the size of the logic formula (see, e.g., [11, 85]).

9.2 Symbolically Represented Kripke Structures

A challenge occurring in practical verification settings is that the Kripke structures are too large to handle. Therefore, these Kripke structures are often not written down explicitly, but rather represented symbolically by encoding them succinctly using propositional formulas. In this section, we consider a (parameterized) decision problem that can be used to model the task of symbolic model checking. We show that this problem is PSPACE-hard for all (fragments of) temporal logics that we consider, even when restricted to formulas of constant size.

Moreover, we consider a restriction of the problem that admits an fpt-reduction to SAT. In particular, we interpret a known result for bounded model checking for the fragment of LTL without U and X operators as membership in para-co-NP, for the setting where the recurrence diameter of the Kripke structures (the size of the largest loop-free path) is bounded by a polynomial of the input size. Additionally, we extend this to a para-co-NP-completeness result.

9.2.1 PSPACE-hardness for Symbolic Model Checking

We begin by defining how Kripke structures can be represented symbolically using propositional formulas. Let $P = \{p_1, \ldots, p_m\}$ be a finite set of propositional variables. A *symbolically represented Kripke structure* over P is a tuple $\mathcal{M} = (\varphi_R, \alpha_0)$, where $\varphi_R(x_1, \ldots, x_m, x_1', \ldots, x_m')$ is a propositional formula over the variables $x_1, \ldots, x_m, x_1', \ldots, x_m'$, and where $\alpha_0 \in \mathbb{B}^m$ is a truth assignment to the variables in P. The Kripke structure associated with $\mathcal{M}$ is (S, R, V, α_0), where $S = \mathbb{B}^m$ consists of all truth assignments to P, where $(\alpha, \alpha') \in R$ if and only if $\varphi_R[\alpha, \alpha']$ is true, and where $V(\alpha) = \{ p_i : \alpha(p_i) = 1 \}$.

Example 9.1. Let $P = \{p_1, p_2\}$. The Kripke structure $\mathcal{M}_1$ from Fig. 9.1 can be symbolically represented by (φ_R, α_0), where $\varphi_R(x_1, x_2, x_1', x_2') = [(\neg x_1 \wedge \neg x_2) \rightarrow (\neg x_1' \leftrightarrow x_2')] \wedge [(\neg x_1 \leftrightarrow x_2) \rightarrow (x_1' \wedge x_2')] \wedge [(x_1 \wedge x_2) \rightarrow (x_1' \wedge x_2')]$, and $\alpha_0 = (0, 0)$. ⊣

We can now consider the symbolic variant SYMBOLIC- MC[$\mathcal{L}$] of the model checking problem, for each of the temporal logics $\mathcal{L}$.

SYMBOLIC- MC[$\mathcal{L}$]
Input: A symbolically represented Kripke structure $\mathcal{M}$, and an $\mathcal{L}$ formula φ.
Question: $\mathcal{M} \models \varphi$?

Similarly to the case of MC[$\mathcal{L}$], we will also consider SYMBOLIC- MC[$\mathcal{L}$] as a parameterized problem, where the parameter is $|\varphi|$. Interestingly, for the logics

LTL and CTL*, the complexity of the model checking problem does not change when Kripke structures are represented symbolically: SYMBOLIC- MC[LTL] and SYMBOLIC- MC[CTL*] are PSPACE-complete (see [141]). However, for the logic CTL, the complexity of the problem does show an increase. In fact, the problem is already PSPACE-hard for very simple formulas.

Proposition 9.2. SYMBOLIC- MC[LTL] *is* PSPACE-*hard even when restricted to the case where* $\varphi = Gp$. SYMBOLIC- MC[CTL] *and* SYMBOLIC- MC[CTL*] *are* PSPACE-*hard even when restricted to the case where* $\varphi = \forall Gp$.

Proof. We give a polynomial-time reduction from QSAT. Let $\varphi = \exists x_1.\forall x_2 \ldots \exists x_{m-1}.$ $\forall x_m.\psi$ be a quantified Boolean formula. We construct a symbolically represented Kripke structure $\mathcal{M}$ as follows. We consider the following set of variables:

$$Z = \{\, x_i, y_i : i \in [m] \,\} \cup \{d, t\}.$$

The initial state α_0 is the all-zeroes assignment to Z.

We construct the formula φ_R representing the transition relation of $\mathcal{M}$. We let:

$$\varphi_R(Z, Z') = \varphi_{R,0}(Z, Z') \vee \bigwedge_{j \in [4]} \varphi_{R,j}(Z, Z'),$$

where we define the formulas $\varphi_{R,j}(Z, Z')$ below. The intuition behind the construction of $\mathcal{M}$ is that any path will correspond to a strategy for choosing the valuation of the existentially quantified variables. We use the variables y_i to indicate which variables x_i have already been assigned a value. In fact, we ensure that in every reachable state, the variables y_i that are set to true are a consecutive sequence $y_1, \ldots, y_i$ for some $i \in [m]$. We use the following formula $\varphi_{R,1}(Z, Z')$ to do this:

$$\varphi_{R,1}(Z, Z') = \bigwedge_{i \in [m-1]} \neg y_i' \rightarrow \neg y_{i+1}'.$$

Moreover, we ensure for any transition from state α to state α', that α and α' differ on at most one variable y_i, using the following formula $\varphi_{R,2}(Z, Z')$:

$$\varphi_{R,2}(Z, Z') = \bigwedge_{i \in [m]} [\neg(y_i \leftrightarrow y_i') \rightarrow \bigwedge_{i' \in [i+1,m]} (y_{i'} \leftrightarrow y_{i'}')].$$

Furthermore, below we will use the following auxiliary formulas, that ensure that for any transition, the number of variables y_i that are true strictly increases (if not all y_i are set to true) or decreases (if not all y_i are set to false), respectively:

$$\varphi_{y\text{-incr}}(Z, Z') = (\neg y_1 \rightarrow y_1') \wedge \bigwedge_{i \in [m-1]} (y_i \wedge \neg y_{i+1}) \rightarrow y_{i+1}',$$

$$\varphi_{y\text{-decr}}(Z, Z') = (y_m \rightarrow \neg y_m') \wedge \bigwedge_{i \in [m-1]} (y_i \wedge \neg y_{i+1}) \rightarrow \neg y_i'.$$

Fig. 9.2 Gadgets for the proof of Proposition 9.2. The labels on the relations indicate what part of $\varphi_{R,4}$ is used to encode the relations.

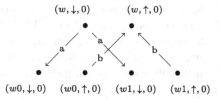

(a) Gadget for words w of odd length $< m$.

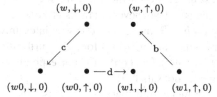

(b) Gadget for words w of even length $< m$.

$$(w,\downarrow,0) \quad\quad (w,\uparrow,0)$$
$$\bullet \longrightarrow{e}\longrightarrow \bullet$$

(c) Gadget for words w of length m that satisfy ψ.

$$(w,\downarrow,0) \quad\quad (w,\downarrow,1) \quad\quad (w,\uparrow,0)$$
$$\bullet \longrightarrow{f}\longrightarrow \bullet \longrightarrow{g}\longrightarrow \bullet$$

(d) Gadget for words w of length m that do not satisfy ψ.

Next, we ensure that in all reachable states, whenever y_i is false, x_i also has to be false. We do so using the following formula $\varphi_{R,3}(Z, Z')$:

$$\varphi_{R,3}(Z, Z') = \bigwedge_{i\in[m-1]} \neg y_i' \to \neg x_i'.$$

Because of the above restrictions, we can restrict our attention to states α for which holds (1) that $y_1, \ldots, y_i$ are true, for some $i \in [m]$, and all remaining variables y_j are false, and (2) that all variables x_j for $j \in [i+1, m]$ are false. We will denote these states by tuples (w, e, t), where $w \in \mathbb{B}^i, e \in \{\uparrow, \downarrow\}$ and $t \in \mathbb{B}$. A tuple (w, d, t) with $|w| = i$ denotes the state α that sets $y_1, \ldots, y_i$ to true, sets $x_1, \ldots, x_i$ according to w, sets d to true if and only if $e = \uparrow$, and sets the variable t according to the value in the tuple.

The idea behind how we continue constructing φ_R is that we piece together all possible instantiations of the gadgets in Fig. 9.2. This results in a large directed acyclic graph containing states (w, e, t), with the property that any path that visits a state $(w, \downarrow, 0)$ ultimately also visits the state $(w, \uparrow, 0)$. This property allows us to use the gadgets in the following way. The gadget for a word w of odd length $i < m$ enforces that whenever a path visits the state $(w, \downarrow, 0)$, it must also visit the state $(wb, \downarrow, 0)$ for some $b \in \mathbb{B}$. Intuitively, this simulates existential quantifiers. This property allows us to use the gadgets in the following way. The gadget for a word w of even length $i < m$ enforces that whenever a path visits the state $(w, \downarrow, 0)$, it must also visit both states $(wb, \downarrow, 0)$ for $b \in \mathbb{B}$. Intuitively, this simulates universal quantifiers. Moreover, the gadgets for words w of length m enforce that on the way from $(w, \downarrow, 0)$ to $(w, \uparrow, 0)$ the state $(w, \downarrow, 1)$ is visited if and only if w corresponds to a truth assignment to the variables in X that does not satisfy ψ.

We make sure that φ_R encodes exactly the transitions from $\alpha_1 = (w_1, e_1, t_1)$ to $\alpha_2 = (w_2, e_2, t_2)$ from the gadgets described above by means of the following (sub)formulas of φ_R. We distinguish seven cases. The labels on the arrows in Fig. 9.2 indicate which case applies to which relation in the gadgets.

(a) The string w_1 is of odd length less than m and $e_1 = \downarrow$ and $t_1 = 0$. We ensure that $w_2 = w_1 b$ for some $b \in \mathbb{B}$, that $e_2 = \downarrow$ and that $t_2 = 0$.

(b) It holds that $e_1 = \uparrow$, $t_1 = 0$ and the string w_1 either (i) is of even length less than m or (ii) is of odd length less than m and ends with 1. We ensure that w_2 is the string w_1 without the last symbol, that $e_2 = \uparrow$, and that $t_2 = 0$.

(c) The string w_1 is of even length less than m and $e_1 = \downarrow$ and $t_1 = 0$. We ensure that $w_2 = w_1 0$, that $e_2 = \downarrow$ and that $t_2 = 0$.

(d) It holds that $e_1 = \uparrow$, $t_1 = 0$ and the string w_1 is of odd length less than m and ends with 0. We ensure that w_2 is the string w_1 where the last symbol is replaced by a 1, that $e_2 = \downarrow$, and that $t_2 = 0$.

(e) The string w_1 is of length m, $e_1 = \downarrow$ and $t_1 = 0$. Moreover, w_1 satisfies ψ. We ensure that $w_2 = w_1$, that $e_2 = \uparrow$ and that $t_2 = 0$.

(f) The string w_1 is of length m, $e_1 = \downarrow$ and $t_1 = 0$. Moreover, w_1 does not satisfy ψ. We ensure that $w_2 = w_1$, that $e_2 = \downarrow$ and that $t_2 = 1$.

(g) The string w_1 is of length m, $e_1 = \downarrow$ and $t_1 = 1$. We ensure that $w_2 = w_1$, that $e_2 = \uparrow$ and that $t_2 = 0$.

Formally, we construct the formula $\varphi_{R,4}(Z, Z')$ as follows:

$$\varphi_{R,4}(Z, Z') =$$

$$\bigwedge_{\substack{i \in [m] \\ i \text{ odd}}} [[\neg y_m \wedge d \wedge \neg t \wedge y_i \wedge \neg y_{i+1}] \rightarrow [d' \wedge \neg t' \wedge y'_{i+1} \wedge \bigwedge_{i' \in [i]} (x_i \leftrightarrow x'_i)]] \quad \text{(a)};$$

$$\wedge \bigwedge_{\substack{i \in [m] \\ i \text{ even}}} [[\neg y_m \wedge \neg d \wedge \neg t \wedge y_i \wedge \neg y_{i+1}] \rightarrow$$

$$[\neg d' \wedge \neg t' \wedge \neg y'_i \wedge \bigwedge_{i' \in [i-1]} (x_i \leftrightarrow x'_i)]] \quad \text{(b.i)};$$

$$\wedge \bigwedge_{\substack{i \in [m] \\ i \text{ odd}}} [[\neg y_m \wedge \neg d \wedge \neg t \wedge y_i \wedge \neg y_{i+1} \wedge x_i] \rightarrow$$

$$[\neg d' \wedge \neg t' \wedge \neg y'_i \wedge \bigwedge_{i' \in [i-1]} (x_i \leftrightarrow x'_i)]] \quad \text{(b.ii)};$$

$$\wedge \bigwedge_{\substack{i \in [m] \\ i \text{ even}}} [[\neg y_m \wedge d \wedge \neg t \wedge y_i \neg y_{i+1}] \rightarrow$$

$$[d' \wedge \neg t' \wedge y'_{i+1} \wedge \neg x'_{i+1} \wedge \bigwedge_{i' \in [i]} (x_i \leftrightarrow x'_i)]] \quad \text{(c)};$$

$$\wedge \bigwedge_{\substack{i \in [m] \\ i \text{ odd}}} [[\neg y_m \wedge \neg d \wedge \neg t \wedge y_i \wedge \neg y_{i+1} \wedge \neg x_i] \rightarrow$$

$$[d' \wedge \neg t' \wedge y'_i \wedge \neg y'_{i+1} \wedge x'_i \wedge \bigwedge_{i' \in [i-1]} (x_i \leftrightarrow x'_i)] \quad \text{(d)};$$

$$\wedge [d \wedge \neg t \wedge \psi \wedge \bigwedge_{i \in [m]} y_i] \rightarrow [\neg d' \wedge \neg t' \wedge \bigwedge_{i \in [m]} (y'_i \wedge (x_i \leftrightarrow x'_i))] \quad \text{(e)};$$

$$\wedge [d \wedge \neg t \wedge \neg \psi \wedge \bigwedge_{i \in [m]} y_i] \rightarrow [d' \wedge t' \wedge \bigwedge_{i \in [m]} (y'_i \wedge (x_i \leftrightarrow x'_i))] \quad \text{(f)};$$

$$\wedge [d \wedge t \wedge \neg \psi \wedge \bigwedge_{i \in [m]} y_i] \rightarrow [\neg d' \wedge \neg t' \wedge \bigwedge_{i \in [m]} (y'_i \wedge (x_i \leftrightarrow x'_i))] \quad \text{(g)};$$

Finally, we make sure that the state $(\epsilon, \uparrow, 0)$ has a self-loop, by means of the following formula $\varphi_{R,0}$:

$$\varphi_{R,0}(Z, Z') = \bigwedge_{z \in Z \setminus \{d\}} \neg z \wedge \bigwedge_{z' \in Z' \setminus \{d'\}} \neg z'.$$

We let the temporal logic formula whose truth is to be checked be Gt in the case of LTL, and $\forall Gt$ in the case of CTL or CTL* (these formulas are equivalent).

We claim that φ has a QBF model if and only if $\mathcal{M} \models Gt$, where $\mathcal{M}$ is specified by (φ_R, α_0). This holds because there is a correspondence between QBF models for φ and paths in $\mathcal{M}$ that satisfy the proposition t in each state. Each such path in $\mathcal{M}$ can be transformed into a QBF model for φ by removing the direction of the arrows, removing self-loops, merging states $(w, \downarrow, 0)$ and $(w, \uparrow, 0)$ into a single truth assignment corresponding to the word w. Because such a path does not visit any state where t is true, the leaves of the resulting tree satisfy ψ, and therefore the resulting tree is a QBF model for φ. Vice versa, each QBF model can be used similarly to obtain a path in $\mathcal{M}$ that satisfies Gt.

9.2.2 An Fpt-Reduction to SAT for LTL\U,X

The result of Proposition 9.2 seems to indicate that the model checking problem for the temporal logics LTL, CTL and CTL* is intractable when Kripke structures are represented symbolically, even when the logic formulas are extremely simple. However, in the literature further restrictions have been identified that allow the problem to be solved by means of an encoding into SAT, which allows the use of practically very efficient SAT solving methods. In the hardness proof of Proposition 9.2, the Kripke structure has only a single path, which contains exponentially many different states. Intuitively, such exponential-length paths may be the cause of PSPACE-hardness. To circumvent this source of hardness, and to go towards the mentioned setting where the problem can be solved by means of a SAT encoding, we need to restrict the recurrence diameter. The *recurrence diameter* $rd(\mathcal{M})$ of a Kripke structure $\mathcal{M}$ is the length of the longest simple (non-repeating) path in $\mathcal{M}$. We consider the following variant of SYMBOLIC- MC[$\mathcal{L}$], where the recurrence diameter of the Kripke structures is restricted.[1]

SYMBOLIC- MC*[$\mathcal{L}$]
Input: A symbolically represented Kripke structure $\mathcal{M}$, $rd(\mathcal{M})$ in unary, and an $\mathcal{L}$ formula φ.
Parameter: $\vert\varphi\vert$.
Question: $\mathcal{M} \models \varphi$?

This restricted setting has been studied by Kroening et al. [139]. In particular, they showed that the model checking problem for LTL\U,X allows an encoding into SAT that is linear in $rd(\mathcal{M})$, even when the Kripke structure $\mathcal{M}$ is represented symbolically, and can thus be of exponential size. Using the result of Kroening et al., we obtain para-co-NP-completeness.

Proposition 9.3. SYMBOLIC- MC*[LTL\U, X] *is* para-co-NP-*complete.*

Proof (sketch). Kroening et al. [139] use the technique of bounded model checking [18, 21, 52], where SAT solvers are used to find a 'lasso-shaped' path in a Kripke structure that satisfies an LTL formula φ. They show that for LTL\U,X formulas, the largest possible length of such lasso-shaped paths that needs to be considered (also called the *completeness threshold*) is linear in $rd(\mathcal{M})$. However, the completeness threshold depends linearly on the size of a particular type of generalized Büchi automaton expressing φ, which in general is exponential in the size of φ. Therefore, this SAT encoding does not run in polynomial time, but it does run in fixed-parameter tractable time when the size of φ is the parameter. Their encoding of the problem of finding a counterexample into SAT can be seen as an encoding of the model checking problem into UNSAT.

We show para-co-NP-hardness by showing that the problem SYMBOLIC- MC* [LTL\U, X] is co-NP-hard already for formulas of constant size. We do so by a reduction from UNSAT. Let ψ be a propositional formula over the variables $x_1, \ldots, x_n$. We construct an instance of SYMBOLIC- MC*[LTL\U, X] as follows. We consider the set $P = \{y_0, y_1, x_1, \ldots, x_n\}$ of propositional variables. We then construct the symbolically represented Kripke structure $\mathcal{M}$ given by (φ_R, α_0) as follows. This structure $\mathcal{M}$ is depicted in Fig. 9.3. We let $\alpha_0 = \overline{0}$, i.e., the all-zeroes assignment. Then we define φ_R as follows:

---

[1] An equivalent way of phrasing the problem is to require that the recurrence diameter of the Kripke model $\mathcal{M}$ is polynomial in the size of its description (φ_R, α_0).

Fig. 9.3 The Kripke structure in the proof of Proposition 9.3. (Only the reachable part is depicted.)

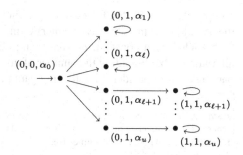

$$\varphi_{R,4}(y_0, y_1, \overline{x}, y_0', y_1', \overline{x}') =$$

$$(\neg y_1 \rightarrow (\neg y_0' \wedge y_1')) \qquad (a);$$

$$\wedge \; (y_1 \rightarrow \bigwedge_{i \in [n]} (x_i \leftrightarrow x_i')) \qquad (b);$$

$$\wedge \; ((y_1 \wedge \psi(x_1, \ldots, x_n) \rightarrow (y_0' \wedge y_1')) \qquad (c);$$

$$\wedge \; ((y_1 \wedge \neg\psi(x_1, \ldots, x_n) \rightarrow (\neg y_0' \wedge y_1')) \; (d).$$

The transition relation given by φ_R allows a transition from α_0 to the state $(0, 1, \alpha)$ for any truth assignment α to the variables $x_1, \ldots, x_n$. Then, if this assignment α satisfies ψ, a transition is allowed to the looping state $(1, 1, \alpha)$. Otherwise, if α does not satisfy ψ, the only transition from state $(0, 1, \alpha)$ is to itself. Finally, we define the LTL formula to be $\varphi = G\neg y_0$.

Moreover, $rd(\mathcal{M}) = 2$, and the LTL formula φ is of constant size, and contains only the temporal operator G. It is straightforward to verify that $\mathcal{M} \models \varphi$ if and only if ψ is unsatisfiable.

In the remainder of this chapter, we will give parameterized complexity results that give evidence that this is the only case in this setting where such an fpt-reduction to SAT is possible. In order to do so, we first make a little digression to introduce a new parameterized complexity class, that can be seen as a parameterized variant of the Polynomial Hierarchy.

9.3 Another Parameterized Variant of the Polynomial Hierarchy

In order to completely characterize the parameterized complexity of the problem SYMBOLIC-MC*[$\mathcal{L}$]—that we defined in Sect. 9.2—for all (fragments of) logics $\mathcal{L}$, we need to introduce a parameterized complexity class, that is another parameterized variant of the Polynomial Hierarchy.

For each level of the PH, the number of quantifier alternations is bounded by a constant. If we allow an unbounded number of quantifier alternations, we get the complexity class PSPACE (see, e.g., [7, Theorem 5.10]). Parameterized complexity theory allows a middle way: neither letting the number of quantifier alternations be bounded by a constant, nor removing all bounds on the number of quantifier alternations, but bounding the number of quantifier alternations by a function of the parameter. We consider the following parameterized problem QSAT(level).

QSAT(level)
Instance: A quantified Boolean formula $\varphi = \exists X_1 \forall X_2 \exists X_3 \ldots Q_k X_k \psi$, where Q_k is a universal quantifier if k is even and an existential quantifier if k is odd, and where ψ is quantifier-free.
Parameter: k.
Question: Is φ true?

We define the parameterized complexity class PH(level) to be the class of all parameterized problems that can be fpt-reduced to QSAT(level).

9.3.1 Alternative Characterizations

Next, we show an alternative characterization of the class PH(level) using Alternating Turing machines (ATMs). We will use this characterization below to show membership in PH(level). For the sake of convenience, we begin by briefly repeating some properties of alternating Turing machines. For more details, we refer to Sect. 2.2.

The states of an ATM are partitioned into *existential* and *universal states*. Intuitively, if the ATM $\mathbb{M}$ is in an existential state, it accepts if there is some successor state that accepts, and if $\mathbb{M}$ is in a universal state, it accepts if all successor states accept. We say that $\mathbb{M}$ is ℓ-alternating, for $\ell \geq 0$, if for each input x, for each run of $\mathbb{M}$ on x, and for each computation path in this run, there are at most ℓ transitions from an existential state to a universal state, or vice versa. The class PH(level) consists of all problems that can be solved in fixed-parameter tractable time by an ATM whose number of alternations is bounded by a function of the parameter.

Proposition 9.4. *Let Q be a parameterized problem. Then $Q \in$ PH(level) if and only if there exist a computable function $f : \mathbb{N} \to \mathbb{N}$ and an ATM $\mathbb{M}$ such that: (1) $\mathbb{M}$ solves Q in fixed-parameter tractable time, and (2) for each slice Q_k of Q, $\mathbb{M}$ is $f(k)$-alternating.*

Proof. First of all, we observe that the class of parameterized problems that can be solved in fpt-time by an $f(k)$-alternating ATM is closed under fpt-reductions. Next, we describe how an instance $\varphi = \exists X_1 \forall X_2 \ldots \forall X_k \psi$ of the problem QSAT(level) can be solved by a (fixed) ATM $\mathbb{M}$ that is k-alternating. Using non-determinism in its first existential phase, $\mathbb{M}$ guesses truth values for the variables in X_1. Then, using non-determinism in the subsequent universal phase, $\mathbb{M}$ guesses truth values for the variables in X_2. Similarly, using alternating existential and universal phases, $\mathbb{M}$ guesses truth values for all variables in all other X_i. Then, using deterministic computation, $\mathbb{M}$ verifies whether the guessed truth values satisfy ψ, and accepts the input if and only if ψ is satisfied. Clearly, φ is true if and only if $\mathbb{M}$ accepts φ.

For the other direction, let Q be an arbitrary parameterized problem that is decided in fpt-time by an ATM $\mathbb{M}$ that is $f(k)$-alternating. We assume without loss of generality that $f(k)$ is even. We give an fpt-reduction from Q to QSAT(level), using standard ideas from the proof of the Cook-Levin Theorem. We assume without loss of generality that for each non-deterministic step that $\mathbb{M}$ takes, there are only two possible transitions. If this were not the case, we could straightforwardly transform $\mathbb{M}$ so that it satisfies this property. Let y be an arbitrary instance for Q, with $|y| = n$. We know that $\mathbb{M}$ runs in fpt-time, i.e., in time $g(k)n^c$, for some computable function g and some constant c. Therefore, we know in particular that in each phase (existential or universal), $\mathbb{M}$ makes at most $g(k)n^c$ non-deterministic (binary) choices. We introduce $f(k)$ sets $X_1, \ldots, X_{f(k)}$, where each set X_i contains $g(k)n^c$ propositional variables $x_{i,1}, \ldots, x_{i,g(k)n^c}$. The interpretation of these variables is that variable $x_{i,j}$ specifies which transition to take in the j-th non-deterministic step in the i-th phase of the computation. Using (the truth values of) these variables in $X_1, \ldots, X_{f(k)}$, we can then in fpt-time decide whether $\mathbb{M}$ ends up in an accepting state using these transitions (when given input y). Therefore, we can in fpt-time construct a Boolean circuit C over the variables in $X_1, \ldots, X_{f(k)}$ that captures this simulation procedure. Then, the quantified Boolean circuit $\exists X_1 \forall X_2 \ldots \forall X_{f(k)} C$ is true if and only if $y \in Q$. Finally, we can easily transform this to a quantified Boolean formula of the right form, for instance by using a standard Tseitin transformation to transform C into an equivalent universally quantified DNF formula $\exists Z \psi$. The result of the reduction is then the quantified Boolean formula $\exists X_1 \forall X_2 \ldots \forall X_{f(k)} \forall Y \psi$, which is an instance of QSAT(level), and which is true if and only if $y \in Q$.

As a direct consequence of this definition, we get that the class PH(level) is closed under fpt-reductions. Next, to further illustrate the robustnest of the class PH(level), we characterize this class using first-order logic model checking (which has also been used to characterize the classes of the well-known W-hierarchy and the A-hierarchy; see, e.g., [85]). Consider the problem MC[FO], where the input consists of a relational structure $\mathcal{A}$, and a first-order formula $\varphi = \exists x_{1,1}, \ldots, x_{1,\ell_1}.\forall x_{2,1}, \ldots, x_{2,\ell_1}.\exists x_{3,1}, \ldots, x_{3,\ell_1} \ldots Q_k x_{k,1}, \ldots, x_{k,\ell_k}.\psi$ in prenex form, where $Q_k = \forall$ if k is even and $Q_k = \exists$ if k is odd. The question is whether $\mathcal{A} \models \varphi$. The problem MC[FO] is PH(level)-complete when parameterized by $(k-1)$.[2] We denote this parameterized problem by MC[FO](quant.alt.).

Proposition 9.5. MC[FO](quant.alt.) *is* PH(level)-*complete.*

Proof. We fpt-reduce QSAT(level) and MC[FO](quant.alt.) to each other. The reduction from QSAT(level) to MC[FO](quant.alt.) is very straightforward. We construct a relational structure $\mathcal{A}$ with two elements 0, 1 in its domain, and two unary predicates T and F, where $T^{\mathcal{A}} = \{1\}$ and $F^{\mathcal{A}} = \{0\}$. Then we transform the quantified Boolean formula φ to a first-order formula φ' by transforming each positive literal x to

---

[2]The problem MC[FO] is also PH(level)-complete when parameterized by k. We use the parameter $(k-1)$ because it corresponds to the number of quantifier alternations. The parameter k corresponds to the number of quantifier blocks.

the first-order atom $T(x)$ and transforming each negative literal $\neg x$ to the atom $F(x)$. It is easy to verify the correctness of this reduction.

For the other direction, we describe the fpt-reduction from MC[FO](quant.alt.) to QSAT(level). Let $\varphi = \exists x_{1,1}, \ldots, x_{1,\ell_1}.\forall x_{2,1}, \ldots, x_{2,\ell_1}.\exists x_{3,1}, \ldots, x_{3,\ell_1} \cdots Q_k x_{k,1}, \ldots, x_{k,\ell_k}.\psi$, together with the relational structure $\mathcal{A}$ with domain A, be an instance of MC[FO](quant.alt.) (we assume without loss of generality that k is even). We replace each first-order variable $x_{i,j}$ with $|A|$ propositional variables $x_{i,j}^a$, for $a \in A$. Let X_i' denote the set of propositional replacement variables $x_{i,j}^a$ for $j \in [\ell_i]$. Then, for each i, we construct a formula χ_i that ensures that for each $x_{i,j}$ with $j \in [\ell_i]$ there is exactly one $x_{i,j}^a$ that is true. Next, we transform ψ into a propositional formula ψ' as follows. Each occurrence of an atom $R(x_{i_1,j_1}, \ldots, x_{i_r,j_r})$ in ψ, where $R^{\mathcal{A}} = \{(a_{1,1}, \ldots, a_{1,r}), \ldots, (a_{\ell,1}, \ldots, a_{\ell,r})\}$, we replace by the disjunction $\bigvee_{i \in [\ell]}(x_{i_1,j_1}^{a_{i,1}} \wedge \cdots \wedge x_{i_r,j_r}^{a_{i,r}})$. Finally, we construct the propositional formula $\psi'' = \chi_1 \wedge (\chi_2 \rightarrow (\chi_3 \wedge (\chi_4 \rightarrow \cdots (\chi_k \rightarrow \psi')) \cdots))$. The final result of the reduction is the quantified Boolean formula $\exists X_1'.\forall X_2'.\exists X_3' \ldots \forall X_k'.\psi''$. The correctness of this reduction can be verified straightforwardly.

9.3.2 Relation to Other Parameterized Variants of the PH

In the previous chapters of this thesis, we considered more parameterized variants of the Polynomial Hierarchy. We briefly consider how the class PH(level) relates to these classes. Firstly, for each $i \geq 1$, the parameterized complexity classes para-Σ_i^p and para-Π_i^p are contained in the class PH(level). So in particular, the classes $\Sigma_2^p[k*]$ and $\Sigma_2^p[*k, t]$ are also contained in PH(level). Moreover, PH(level) is contained in para-PSPACE. These inclusions are all strict, unless the PH collapses.

Another parameterized variant of the PH that has been studied is the A-hierarchy, containing the parameterized complexity classes A[t] for each $t \geq 1$. Remember that each class A[t] is defined as the class of all problems that can be fpt-reduced to MC[FO], restricted to first-order formulas φ (in prenex form) with a quantifier prefix that starts with an existential quantifier and that contains t quantifier alternations, parameterized by the size of φ. From this definition, it directly follows that A[t] is contained in PH(level), for each $t \geq 1$. The A-hierarchy also contains the parameterized classes AW[*] $\subseteq$ AW[SAT] $\subseteq$ AW[P], each of which contains the classes A[t], for each $t \geq 1$. These classes are also contained in PH(level). Moreover, the inclusion of all these classes in PH(level) is strict, unless P = NP.

9.4 Completeness for PH(level) and para-PSPACE

In this section, we provide a complete parameterized complexity classification for the problem SYMBOLIC- MC*[$\mathcal{L}$]. We already considered the case for $\mathcal{L} = \text{LTL}\backslash\text{U,X}$ in Sect. 9.2.2, which was shown to be para-co-NP-complete. We give (negative) parameterized complexity results for the other cases. An overview of the results can be found in Table 9.1 on page 183. Firstly, we show that for the case of LTL, allowing at least one of the temporal operators U or X leads to para-PSPACE-completeness.

Theorem 9.6. SYMBOLIC- MC*[LTL\U] *is* para-PSPACE-*complete.*

Proof. Membership follows from the PSPACE-membership of SYMBOLIC- MC[LTL]. We show hardness by showing that the problem is already PSPACE-hard for a constant parameter value. We do so by giving a reduction from QSAT. Let $\varphi_0 = \forall x_1.\exists x_2\ldots Q_n x_n.\psi$ be a quantified Boolean formula. We may assume without loss of generality that $(n \bmod 4) = 1$, and thus that $Q_n = \forall$. We construct a Kripke structure $\mathcal{M}$ symbolically represented by (φ_R, α_0), whose reachability diameter is polynomial in the size of φ_0, and an LTL formula φ that does not contain the U operator, in such a way that φ_0 is true if and only if $\mathcal{M} \not\models \neg\varphi$. (So technically, we are reducing to the co-problem of SYMBOLIC- MC*[LTL\U]. Since PSPACE is closed under complement, this suffices to show PSPACE-hardness.)

The idea is to construct a full binary tree (of exponential size), with bidirectional transitions between each parent and child, and to label the nodes of this tree in such a way that a constant-size LTL formula can be used to force paths to be a traversal of this tree corresponding to a QBF model of the formula φ_0. The idea of using LTL formulas to force paths to be traversals of exponential-size binary trees was already mentioned by [139]. We construct the Kripke structure $\mathcal{M}$ as depicted in Fig. 9.4.

We first show how to construct $\mathcal{M} = (\varphi_R, \alpha_0)$. Remember that $P = \{x_1, \ldots, x_n, y_1, \ldots, y_n, a_1, a_l, a_r, a'_1, a'_l, a'_r, e_1, e_2, e'_1, e'_2, f, g\}$. We let α_0 be the assignment that sets only the propositional variables a_1, e'_2 to true, and all other propositional variables to false. Then, we define $\varphi_R(\overline{P}^*, \overline{P})$ to be the conjunction of the following subformulas. The first conjunct

$$(g \leftrightarrow e'_1 \wedge \bigwedge_{p\in P\setminus\{e'_1,g\}} \neg p)$$

ensures that g is only true in the sink state, and the second conjunct

$$(g \rightarrow (a_1 \wedge e'_2 \wedge \bigwedge_{p\in P\setminus\{a_1,e'_2\}} \neg p))$$

ensures that the sink state is only reachable from the initial state. Next, we make sure that the states in the tree have the correct truth values for the propositional variables $y_1, \ldots, y_n$, i.e., that for each node in the i-th level of the tree, exactly the variables $y_1, \ldots, y_i$ are true. This is ensured by the following conjuncts:

$$\bigwedge_{i\in[2,n]} (y_i \rightarrow y_{i-1}),$$

$$\bigwedge_{i\in[n-1]} (\neg y_i \rightarrow \neg y_{i+1}), \text{ and}$$

$$\bigwedge_{i\in[n-1]} (\neg(y_i \wedge y_{i+1} \wedge \neg y_i^* \wedge \neg y_{i+1}^*) \wedge \neg(y_i^* \wedge y_{i+1}^* \wedge \neg y_i \wedge \neg y_{i+1})).$$

We ensure that the propositional variable f is true exactly in the leafs of the tree, using the conjunct:

$$(f \leftrightarrow \bigwedge_{i \in [n]} y_i).$$

Then, we ensure that each node at the i-th level of the tree corresponds to a partial truth assignment to the variables $x_1, \ldots, x_i$ that agrees with its parent node on the variables $x_1, \ldots, x_{i-1}$. We do so by means of the following conjuncts:

$$\bigwedge_{i \in [n]} (\neg y_i \rightarrow \neg x_i), \text{ and}$$
$$\bigwedge_{i \in [n]} ((y_i \leftrightarrow y_i^\star) \rightarrow (x_i \leftrightarrow x_i^\star)).$$

Finally, we enforce the intended truth values for the propositional variables in $A = \{a_1, a_l, a_r, a_1', a_l', a_r'\}$ and in $E = \{e_1, e_2, e_1', e_2'\}$. In order to do so, we introduce two auxiliary formulas $\varphi_{\text{l-ch}}$ and $\varphi_{\text{r-ch}}$, that encode whether a node in the tree is a left or a right child:

$$\varphi_{\text{l-ch}}(\overline{p}) = \bigvee_{i \in [n]} (y_i \wedge \neg y_{i+1} \wedge \neg x_i) \vee (y_n \wedge \neg x_n), \text{ and}$$
$$\varphi_{\text{r-ch}}(\overline{p}) = \bigvee_{i \in [n]} (y_i \wedge \neg y_{i+1} \wedge x_i) \vee (y_n \wedge x_n).$$

Moreover, we introduce another auxiliary formula $\varphi_{\text{down}}(\overline{p}^\star, \overline{p})$, that encodes whether the transition goes down the tree:

$$\varphi_{\text{down}}(\overline{p}^\star, \overline{p}) = \bigvee_{i \in [n]} (\neg y_i \wedge y_i^\star).$$

Using these auxiliary formulas, we construct the following conjuncts of φ_R, that enforce the intended interpretation of the propositional variables in A and E:

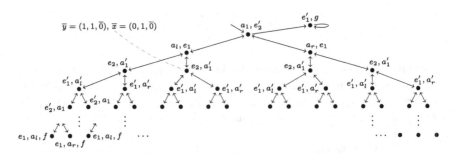

Fig. 9.4 The Kripke structure in the proof of Proposition 9.6. (Only the reachable part is depicted.)

$$(\neg g \wedge \varphi_{\text{down}}(\overline{p}^\star, \overline{p}) \wedge a_1^\star \wedge \varphi_{\text{l-ch}}(\overline{p})) \to (a_l \wedge e_1),$$
$$(\neg g \wedge \varphi_{\text{down}}(\overline{p}^\star, \overline{p}) \wedge a_1^\star \wedge \varphi_{\text{r-ch}}(\overline{p})) \to (a_r \wedge e_1),$$
$$(\neg g \wedge \varphi_{\text{down}}(\overline{p}^\star, \overline{p}) \wedge (a_1')^\star \wedge \varphi_{\text{l-ch}}(\overline{p})) \to (a_l' \wedge e_1'),$$
$$(\neg g \wedge \varphi_{\text{down}}(\overline{p}^\star, \overline{p}) \wedge (a_1')^\star \wedge \varphi_{\text{r-ch}}(\overline{p})) \to (a_r' \wedge e_1'),$$
$$(\neg g \wedge \varphi_{\text{down}}(\overline{p}^\star, \overline{p}) \wedge e_1^\star) \to (e_2 \wedge a_1'),$$
$$(\neg g \wedge \varphi_{\text{down}}(\overline{p}^\star, \overline{p}) \wedge (e_1')^\star) \to (e_2' \wedge a_1),$$
$$(\neg g \wedge \neg\varphi_{\text{down}}(\overline{p}^\star, \overline{p}) \wedge a_1^\star \wedge \varphi_{\text{l-ch}}(\overline{p})) \to (a_l' \wedge e_1'),$$
$$(\neg g \wedge \neg\varphi_{\text{down}}(\overline{p}^\star, \overline{p}) \wedge a_1^\star \wedge \varphi_{\text{r-ch}}(\overline{p})) \to (a_r' \wedge e_1'),$$
$$(\neg g \wedge \neg\varphi_{\text{down}}(\overline{p}^\star, \overline{p}) \wedge (a_1')^\star \wedge \varphi_{\text{l-ch}}(\overline{p})) \to (a_l \wedge e_1),$$
$$(\neg g \wedge \neg\varphi_{\text{down}}(\overline{p}^\star, \overline{p}) \wedge (a_1')^\star \wedge \varphi_{\text{r-ch}}(\overline{p})) \to (a_r \wedge e_1),$$
$$(\neg g \wedge \neg\varphi_{\text{down}}(\overline{p}^\star, \overline{p}) \wedge e_1^\star \to (a_1 \wedge e_2'),$$
$$(\neg g \wedge \neg\varphi_{\text{down}}(\overline{p}^\star, \overline{p}) \wedge (e_1')^\star \to (a_1' \wedge e_2),$$
$$\bigwedge_{c \in A} (c \to \bigwedge_{c' \in A \setminus \{c\}} \neg c'), \text{ and}$$
$$\bigwedge_{c \in E} (c \to \bigwedge_{c' \in E \setminus \{c\}} \neg c').$$

This concludes our construction of the Kripke structure $\mathcal{M}$ as depicted in Fig. 9.4.

It is straightforward to check that the recurrence diameter $rd(\mathcal{M})$ of $\mathcal{M}$ is bounded by $2n$ as the longest simple path in $\mathcal{M}$ is from some leaf in the tree to another leaf.

More concretely, the intuition behind the construction of $\mathcal{M}$ is as follows. Every transition from the i-th level to the $(i + 1)$-th level (where the root is at the 0-th level) corresponds to assigning a truth value to the variable x_{i+1}. We use variables $\overline{x} = (x_1, \ldots, x_n)$ to keep track of the truth assignment in the current position of the tree, and variables $\overline{y} = (y_1, \ldots, y_n)$ to keep track of what level in the tree the current position is (at level i, exactly the variables $y_1, \ldots, y_i$ are set to true). At the even levels i, we use the variables a_1, a_l, a_r (and a_1', a_l', a_r') to ensure that (in a single path) both possible truth assignments to the (universally quantified) variable x_{i+1} are used. At the odd levels i, we use the variables e_1, e_2 (and e_1', e_2') to ensure that one of both possible truth assignments to the (existentially quantified) variable x_{i+1} is used. We need the copies $a_1', e_1', \ldots$ to be able to enforce the intended (downward and upward) traversal of the tree. Then, the variable f is used to signal that a leaf has been reached, and the variable g is used to signal that the path is in the sink state.

Next, we give a detailed specification of the LTL formula φ, that enforces a traversal of the tree $\mathcal{M}$ corresponding to a QBF model of the formula φ_0. The LTL formula φ consists of a conjunction of several subformulas. The first conjunct of φ is Xa_l, which ensures that the path starts by going down to the left child of the root of the tree. The next conjuncts are $G[(a_1 \wedge Xe_1 \wedge \neg Xf) \to XXe_2]$ and $G[(a_1' \wedge Xe_1' \wedge \neg Xf) \to XXe_2']$, which ensure that after a transition corresponding to setting a universal variable to some value, the path goes further down the tree (if possible). The conjuncts $G[(e_1 \wedge Xe_2) \to XXa_1']$ and $G[(e_1' \wedge Xe_2') \to XXa_1]$ ensure that after setting an existential variable to some value, the path goes further down the tree by setting the next universal variable to 0. The conjunct $G[(a_1 \wedge Xe_1 \wedge Xf) \to XXa_1]$ ensures that after reaching a leaf of the tree, the path goes back up. The conjuncts $G[(a_l \wedge Xa_1) \to XXa_r]$ and $G[(a_l' \wedge Xa_1') \to XXa_r']$ ensure that after the path goes back up a transition corresponding to setting a universal variable x_i to 0, the path continues (downwards) with the transition corresponding to setting the variable x_i to 1. The conjuncts $G[(a_r \wedge Xa_1) \rightarrow XXe'_1]$ and $G[(a'_r \wedge Xa'_1) \rightarrow XXe_1]$ ensure that after the path goes back up a transition corresponding to setting a universal variable x_i to 0, the path continues (upwards) by going back up on the transition corresponding to setting variable x_{i-1}. The conjuncts $G[(e_2 \wedge Xe_1) \rightarrow XXa_1]$ and $G[(e'_2 \wedge Xe'_1 \wedge X\neg g) \rightarrow XXa'_1]$ ensure that after the path goes back up a transition corresponding to setting an existential variable x_i to some value, the path continues (upwards) by going back up on the transition corresponding to setting variable x_{i-1} (if possible).

Finally, we need to ensure that this tree traversal corresponds to a QBF model for the formula φ_0, i.e., that all total assignments that appear along the path satisfy the matrix ψ of φ_0. We do so by adding a last conjunct to φ. However, to keep the LTL formula φ of constant size, we need to introduce a propositional variable that abbreviates the truth of ψ. By Lemma 9.8, we may assume without loss of generality that there is a propositional variable z_ψ in P that in each state is set to 1 if and only if this state sets the formula ψ to true. We then let the final conjunct of φ be $G[f \rightarrow z_\psi]$. Clearly, the size of φ is constant.

We can then show that φ_0 is true if and only if $\mathcal{M} \not\models \neg\varphi$. By construction of $\mathcal{M}$ and φ, all paths starting in the initial state of $\mathcal{M}$ that satisfy φ naturally correspond to a QBF model of φ_0, and all QBF models of φ_0 correspond to such a path. Assume that φ_0 is true. Then there exists a QBF model of φ_0. Then there exists a path satisfying φ, and thus $\mathcal{M} \not\models \neg\varphi$. Conversely, assume that $\mathcal{M} \not\models \neg\varphi$. Then there exists a path that satisfies φ. Therefore, there exists a QBF model of φ_0, and thus φ_0 is true.

We continue with showing para-PSPACE-hardness for the fragment of LTL where the U-operator is allowed.

Theorem 9.7. SYMBOLIC- MC*[LTL\X] *is* para-PSPACE-*complete*.

Proof. Membership follows from the PSPACE-membership of SYMBOLIC- MC[LTL]. We show para-PSPACE-hardness by modifying the reduction in the proof of Theorem 9.6. The idea is to simulate the X operator using the U operator. Given an instance of QSAT, we construct the Kripke structure $\mathcal{M}$ and the LTL formula φ as in the proof of Theorem 9.6. Then, we modify $\mathcal{M}$ and φ as follows. Firstly, we add a fresh variable x_0 to the set of propositions P, we ensure that x_0 is false in the initial state α_0, and we modify φ_R so that in each transition, the variable x_0 swaps truth values. Then, it is straightforward to see that any LTL formula of the form $X\varphi'$ is equivalent to the LTL formula $(x_0 \rightarrow x_0U(\neg x_0 \wedge \varphi')) \wedge (\neg x_0 \rightarrow \neg x_0U(x_0 \wedge \varphi'))$, on structures where x_0 shows this alternating behavior. Using this equivalence, we can recursively replace all occurrences of the X operator in the LTL formula φ. This leads to an exponential blow-up in the size of φ, but since φ is of constant size, this blow-up is permissible.

Next, we show that for the case of CTL, the problem is complete for PH(level), even when both temporal operators U and X are disallowed. In order to establish this result, we need the following technical lemma.

Lemma 9.8. *Given a symbolically represented Kripke structure $\mathcal{M}$ given by (φ_R, α_0) over the set P of propositional variables, and a propositional formula ψ over P, we can construct in polynomial time a Kripke structure $\mathcal{M}'$ given by (φ'_R, α'_0) over the set $P \cup \{z\}$ of variables (where $z \notin P$) such that:*

- *there exists an isomorphism ρ between the states in the reachable part of $\mathcal{M}$ and the states in the reachable part of $\mathcal{M}'$ that respects the initial states and the transition relations,*
- *each state s in the reachable part of $\mathcal{M}$ agrees with $\rho(s)$ on the variables in P, and*
- *for each state s in the reachable part of $\mathcal{M}$ it holds that $\rho(s) \models z$ if and only if $s \models \psi$.*

Proof. Intuitively, the required Kripke structure $\mathcal{M}'$ can be constructed by adding the variable z to the set P of propositions, and modifying the formula φ_R specifying the transition relation and the initial state α_0 appropriately. In the new initial state α'_0, the variable z gets the truth value 1 if and only if $\alpha_0 \models \psi$. Moreover, the transition relation specified by φ'_R ensures that in any reached state α, the variable z gets the truth value 1 if and only if $\alpha \models \psi$.

Concretely, we define $\mathcal{M}' = (\varphi'_R, \alpha'_0)$ over the set $P \cup \{z\}$ as follows. We let $\alpha'_0(p) = \alpha_0(p)$ for all $p \in P$, and we let $\alpha'_0(z) = 1$ if and only if $\alpha_0 \models \psi$. Then, we define φ'_R by letting:

$$\varphi'_R(\overline{x}, z, \overline{x}', z') = \varphi_R(\overline{x}, \overline{x}') \wedge (z' \leftrightarrow \psi(\overline{x})).$$

The isomorphism ρ can then be constructed as follows. For each state α in $\mathcal{M}$, $\rho(\alpha) = \alpha \cup \{z \mapsto 1\}$ if $\alpha \models \psi$, and $\rho(\alpha) = \alpha \cup \{z \mapsto 0\}$ if $\alpha \not\models \psi$.

We can now show PH(level)-completeness for the case of CTL\U,X.

Theorem 9.9. Symbolic- MC*[CTL] *is* PH*(level)-complete. Moreover, hardness already holds for* Symbolic- MC*[CTL\U, X].

Proof. In order to show hardness, we give an fpt-reduction from QSAT(level). Let $\varphi = \exists X_1 \forall X_2 \ldots Q_k X_k \psi$ be an instance of QSAT(level). We construct a Kripke structure $\mathcal{M}$ over a set P of propositional variables represented symbolically by (φ_R, α_0), with polynomial recurrence diameter, and a CTL formula Φ such that φ is true if and only if $\mathcal{M} \models \Phi$.

The idea is to let $\mathcal{M}$ consist of a (directed) tree of exponential size, as depicted in Fig. 9.5. The tree consists of k levels (where the root is at the 0-th level). All nodes on the i-th level are labelled with proposition l_i. Moreover, each node is associated with a truth assignment over the variables in $X = \bigcup_{i \in [k]} X_i$. For each node n at the i-th level (for $i \in [0, k-1]$) with corresponding truth assignment α_n, and for each truth assignment α to the variables in X_{i+1}, there is a child node of n (at the $(i+1)$-th level) whose corresponding assignment agrees with α on the variables in X_{i+1}. Also, the truth assignment corresponding to each child of n agrees with α_n on the variables

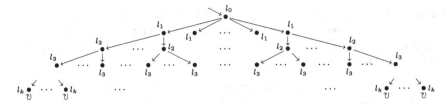

Fig. 9.5 The Kripke structure in the proof of Theorem 9.9. (Only the reachable part is depicted.)

in $X_1, \ldots, X_i$. Moreover, by Lemma 9.8, we may assume without loss of generality that there is a propositional variable z_ψ in P that in each state is set to 1 if and only if this state sets the propositional formula ψ (over X) to true.

We show how to construct the Kripke structure $\mathcal{M} = (\varphi_R, \alpha_0)$. Remember that $P = X_1 \cup \cdots \cup X_k \cup \{l_0, l_1, \ldots, l_k\}$ (for the sake of simplicity, we leave treatment of the propositional variable z_ψ to the technique discussed in the proof of Lemma 9.8). We let α_0 be the truth assignment that sets only the propositional variable l_0 to true, and all other propositional variables to false. Then, we define $\varphi_R(\overline{p}, \overline{p}')$ as the conjunction of several subformulas. The first conjuncts ensure that in each level of the tree, the propositional variables l_i get the right truth value:

$$\bigwedge_{i \in [0, k-1]} l_i \to l'_{i+1}, \quad (l_k \to l'_k) \quad \text{and} \quad \bigwedge_{i, i' \in [k], i < i'} \neg(l'_i \wedge l'_{i'}).$$

The following conjunct ensures that the partial truth assignment of a node at the i-th level of the tree agrees with its parent on all variables in $X_1, \ldots, X_{i-1}$.

$$\bigwedge_{i \in [n]} (l'_i \to \bigwedge_{x \in X_1 \cup \cdots \cup X_{i-1}} (x \leftrightarrow x')).$$

This concludes our construction of the Kripke structure $\mathcal{M}$ as depicted in Fig. 9.5. Clearly, the longest simple path in $\mathcal{M}$ is a root-to-leaf path, which has length k.

Then, using this structure $\mathcal{M}$, we can express the quantified Boolean formula φ in CTL as follows. We define $\Phi = \exists F(l_1 \wedge \forall F(l_2 \wedge \exists F(l_3 \wedge \cdots Q_k F(l_k \wedge z_\psi) \cdots)))$. By construction of Φ, we get that those subtrees of $\mathcal{M}$ that naturally correspond to witnesses for the truth of this CTL formula Φ exactly correspond to the QBF models for φ. From this, we directly get that φ is true if and only if $\mathcal{M} \models \Phi$.

In order to prove membership in PH(level), we show that SYMBOLIC- MC*[CTL] can be decided in fpt-time by an ATM $\mathbb{M}$ that is $f(k)$-alternating. The algorithm implemented by $\mathbb{M}$ takes a different approach than the well-known dynamic programming algorithm for CTL model checking for explicitly encoded Kripke structures (see, e.g., [11, Sect. 6.4.1]). Since symbolically represented Kripke structures can be of size exponential in the input, this bottom-up algorithm would require exponential time. Instead, we employ a top-down approach, using (existential and universal) non-determinism to quantify over the possibly exponential number of states.

We consider the function CTL-MC, given in pseudo-code in Algorithm 9.1, which takes as input the Kripke structure $\mathcal{M}$ in form of its representation (φ_R, α_0), a state α in $\mathcal{M}$, a CTL formula Φ and the recurrence diameter $rd(\mathcal{M})$ of $\mathcal{M}$ (in unary), and outputs 1 if and only if α makes Φ true. The algorithm only needs to check for paths of length at most $m = rd(\mathcal{M})$ in the case of the U operator, because any path longer than m must cycle. Note that in this algorithm, we omit the case for the operator F, as any CTL formula $\exists F \Phi$ is equivalent to $\exists T U \Phi$. It is readily verified that this algorithm correctly computes whether $\mathcal{M}, \alpha \models \Phi$. Therefore, $\mathcal{M} \models \Phi$ if and only if CTL-MC $(\mathcal{M}, \alpha_0, \Phi, m)$ returns 1, where m is the unary encoding of $rd(\mathcal{M})$.

Algorithm 9.1. Recursive CTL model checking using bounded alternation.

```
 1 function CTL-MC (M, α, Φ, m):
 2     switch Φ do
 3         case p ∈ P: return α(p);
 4         case ¬Φ₁: return not CTL-MC (M, α, Φ₁, m);
 5         case Φ₁ ∧ Φ₂: return CTL-MC (M, α, Φ₁, m) and CTL-MC (M, α, Φ₂, m);
 6         case ∃XΦ₁:
 7             (existentially) pick a state α' in M;              /* guess next state */
 8             if φR(α, α') is false then return 0;              /* check transition */
 9             return CTL-MC (M, α', Φ₁, m);                     /* recurse */
10         end
11         case ∃Φ₁UΦ₂:
12             (existentially) pick some m' ≤ m;                 /* guess path length */
13             (existentially) pick states α₁, ..., αₘ' in M;       /* guess path */
14             (universally) pick some j ∈ [m' − 1];            /* cover all states */
15             if φR(αⱼ, αⱼ₊₁) is false then return 0;           /* check transition */
16             if CTL-MC (M, αⱼ, Φ₁, m) = 0 then return 0;        /* recurse */
17             return CTL-MC (M, αₘ', Φ₂, m);                   /* recurse */
18         end
19     endsw
20 end
```

It remains to verify that the algorithm CTL-MC can be implemented in fpt-time by an $f(k)$-alternating ATM $\mathbb{M}$. We can construct $\mathbb{M}$ in such a way that the existential guesses are done using the existential non-deterministic states of $\mathbb{M}$, and the universal guesses by the universal non-deterministic states. Note that the recursive call in the case for $\neg \Phi_1$ is preceded by a negation, so the existential and universal non-determinism swaps within this recursive call. The recursion depth of the algorithm is bounded by $|\Phi| = k$, since each recursive call strictly decreases the size of the CTL formula used. Moreover, in each call of the function CTL-MC, at most two recursive calls are made (not counting recursive calls at deeper levels of recursion). Therefore, the running time of $\mathbb{M}$ is bounded by $2^k \text{poly}(n)$, where n is the input size. Also, since in each call of the function at most two alternations between existential and universal non-determinism are used (again, not counting at deeper levels of recursion), we know that $\mathbb{M}$ is 2^k-alternating. (This bound on the number of alternations needed can be improved with a more careful analysis and some optimizations to the algorithm.)

Finally, we complete the parameterized complexity classification of the problem SYMBOLIC- MC* by showing membership in PH(level) for the case of CTL*\U,X.

Theorem 9.10. SYMBOLIC- MC*[CTL*\U, X] *is* PH(level)-*complete.*

Proof. Hardness for PH(level) follows from Theorem 9.9. We show membership in PH(level), by describing an algorithm A to solve the problem that can be implemented by an ATM $\mathbb{M}$ that runs in fpt-time and that is $f(k)$-alternating. The algorithm works similarly to Algorithm 9.1, described in the proof of Theorem 9.9, and recursively decides the truth of a CTL* formula in a state. The difference with Algorithm 9.1 is that it does not look only at the outermost temporal operators of the CTL* formula in a recursive step, but considers possibly larger subformulas in each recursive step. Let $\exists\varphi$ be a CTL* formula, and let s be a state in $\mathcal{M}$. The algorithm A then considers all maximal subformulas $\psi_1, \ldots, \psi_\ell$ of φ that are CTL* state formulas as atomic propositions $p_1, \ldots, p_\ell$, turning the formula φ into an LTL formula. Since φ does not contain the operators U and X, we know that in order to check the existence of an infinite path satisfying φ, it suffices to look for lasso-shaped paths of bounded length (linear in $rd(\mathcal{M})$ and exponential in the size of φ), i.e., a finite path followed by a finite cycle [139]. The algorithm A then uses (existential) non-determinism to guess such a lasso-shaped path π, and to guess for each state which of the propositions $p_1, \ldots, p_\ell$ are true, and verifies that π witnesses truth of $\exists\varphi$. Then, in order to ensure that it correctly determines whether $\exists\varphi$ is true, the algorithm needs to verify that it guessed the right truth values for $p_1, \ldots, p_\ell$ in π. It does so by recursively determining, for each state s' in the lasso-shaped path π, and each p_i, whether ψ_i is true in s' if and only if it guessed p_i to be true in s'. (In order to ensure that in each level of recursion there are only a constant number of recursive calls, like Algorithm 9.1, the algorithm A uses universal non-determinism iterate over each p_i and each s'.) The algorithm then reports that $\exists\varphi$ is true in s if and only if (1) the guesses for π and the truth values of $p_1, \ldots, p_\ell$ together form a correct witness for truth of $\exists\varphi$, and (2) for each p_i and each s' it holds that p_i was guessed to be true in s' if and only if ψ_i is in fact true in s'. The recursive cases for CTL* formulas where the outermost operator is not temporal are analogous to Algorithm 9.1. Like Algorithm 9.1, the algorithm runs in fpt-time and is 2^k-alternating.

Notes

The results in this chapter appeared in a paper in the proceedings of KR 2016 [115].

Chapter 10
Problems Related to Propositional Satisfiability

In this chapter, we continue the parameterized complexity investigation of problems at higher levels of the Polynomial Hierarchy using the parameterized complexity tools that we developed in Chaps. 4–7. We look at several parameterized variants of problems related to propositional satisfiability.

In particular, we consider several parameterized variants of the problems of minimizing implicants of DNF formulas and minimizing DNF formulas. Moreover, we consider two parameterized problems related to finding inconsistent subsets of propositional knowledge bases and identifying subsets of propositional knowledge bases whose deletion restores consistency of the knowledge base.

Outline of This Chapter

We begin in Sect. 10.1 by considering several parameterizations of minimization problems related to DNF formulas and implicants of DNF formulas. We firstly consider the problem of deciding, given a DNF formula φ and an implicant C of φ, whether there exists an implicant $C' \subseteq C$ whose size is at most a given upper bound. We consider this problem parameterized (1) by the size of the given implicant, (2) by the upper bound, and (3) by the difference between the size of the given implicant and the given upper bound. For the first parameterization, we show that the problem is para-co-NP-complete, and for the latter two, we show that the problem is $\Sigma_2^p[k*]$-complete.

Then, we consider the problem of deciding, given a DNF formula φ, whether there exists an equivalent subformula of φ whose size is at most a given upper bound. As parameters, we consider (1) the given upper bound and (2) the difference between the size of φ and the given upper bound. For the first parameter, we show membership in $\Sigma_2^p[k*]$ and in $\mathrm{FPT}^{\mathrm{NP}}[\text{few}]$, and we show hardness for para-co-NP. For the second parameter, we show $\Sigma_2^p[k*]$-completeness.

Finally, in Sect. 10.2, we investigate two parameterized problems related to finding inconsistent subsets of sets of propositional formulas, and finding subsets that can be deleted to repair the inconsistency. Firstly, we consider the problem of deciding, given

© Springer-Verlag GmbH Germany, part of Springer Nature 2019
R. de Haan: Parameterized Complexity in the Polynomial Hierarchy, LNCS 11880,
https://doi.org/10.1007/978-3-662-60670-4_10

a set Φ of propopositional formulas, whether the minimum size of any inconsistent subset of Φ is both odd and bounded by a given upper bound k. The parameter in this problem is the upper bound k. We show that this problem is $\Sigma_2^p[k*]$-complete. For the other problem that we consider, one is given a set Φ of propositional formulas, and the task is to compute a subset $\Phi' \subseteq \Phi$ of minimum size (smaller than a given upper bound) such that $\Phi \setminus \Phi'$ is satisfiable, if such a set exists. The parameter is the upper bound on the subsets Φ'. We show that this problem can be solved in fpt-time using $O(k)$ witness SAT oracle queries. Moreover, we show that a decision variant of this problem is $\text{FPT}^{\text{NP}}[\text{few}]$-complete.

10.1 Minimization of DNF Formulas and Implicants

In this section, we analyze the parameterized complexity of various problems related to minimizing DNF formulas and implicants of DNF formulas.

10.1.1 Minimizing Implicants

We begin with the problems related to minimizing implicants of DNF formulas. Let φ be a DNF formula. We say that a set C of literals is an *implicant of* φ if all assignments that satisfy $\bigwedge_{l \in C} l$ also satisfy φ.

The following decision problem—that is related to the question of whether a given implicant of a DNF formula can be reduced in size— is Σ_2^p-complete [190].

SHORTEST-IMPLICANT-CORE
Instance: A DNF formula φ, an implicant C of φ of size n, and an integer m.
Question: Does there exist an implicant $C' \subseteq C$ of φ of size m?

We consider three parameterizations of this problem:

- SHORTEST-IMPLICANT-CORE(implicant size), where the parameter $k = n$ is the size of the given implicant;
- SHORTEST-IMPLICANT-CORE(core size), where the parameter $k = m$ is the size of the minimized implicant; and
- SHORTEST-IMPLICANT-CORE(reduction size), where the parameter $k = n - m$ is the difference in size between the original implicant and the minimized implicant.

We firstly show that the first parameterized problem admits an fpt-reduction to SAT. Concretely, we show that the problem is para-co-NP-complete.

Proposition 10.1. SHORTEST-IMPLICANT-CORE(implicant size) *is* para-co-NP-*complete.*

Proof. To show para-co-NP-hardness, we give a polynomial-time reduction from UNSAT to the slice of SHORTEST-IMPLICANT-CORE(implicant size) for the parameter value 1. Let φ be an instance of UNSAT consisting of a CNF formula. Let $x \notin \text{Var}(\varphi)$

be a fresh variable. We construct an instance $(\psi, \{x\})$ of SHORTEST-IMPLICANT-CORE(implicant size) as follows. We let $\psi = (\neg\varphi \vee x)$. We know that there is exactly one $C' \subsetneq C$, namely $C' = \emptyset$. It is straightforward to verify that $\emptyset$ is an implicant of ψ if and only if $\neg\varphi$ is valid, or equivalently, if and only if φ is unsatisfiable.

To show para-co-NP-membership, we give an fpt-reduction from SHORTEST-IMPLICANT-CORE(implicant size) to UNSAT. Let (ψ, C) be an instance of SHORTEST-IMPLICANT-CORE(implicant size). For each $D \subsetneq C$, we let ψ^D be a copy of ψ where each $x \in \mathrm{Var}(\psi)$ is replaced by a copy x^D of x. Furthermore, for each $D \subsetneq C$, we define the set $\sigma(D) = \{ x^D : x \in C' \}$ containing a copy x^D for each $x \in C'$. We construct an instance φ of UNSAT by letting $\varphi = \bigwedge_{D \subsetneq C}(\bigwedge \sigma(D) \wedge \neg\psi^{C'})$. Clearly, φ can be constructed in time $2^k \cdot \|\psi\|$. It is straightforward to verify that φ is satisfiable if and only if no $D \subsetneq C$ is an implicant of ψ.

Next, we analyze the parameterized complexity of the problem SHORTEST-IMPLICANT-CORE(core size). We show that the problem is $\Sigma_2^p[k*]$-complete. In order to prove $\Sigma_2^p[k*]$-hardness of SHORTEST-IMPLICANT-CORE(core size), we need the following technical lemma. We omit its straightforward proof.

Lemma 10.2. *Let (φ, k) be an instance of $\Sigma_2^p[k*]$-WSAT. In polynomial time, we can construct an equivalent instance (φ', k) of $\Sigma_2^p[k*]$-WSAT with $\varphi' = \exists X.\forall Y.\psi$, such that for every assignment $\alpha : X \to \mathbb{B}$ that has weight $m \neq k$, it holds that $\forall Y.\psi[\alpha]$ is true.*

We are now in a position to show $\Sigma_2^p[k*]$-completeness for the problem SHORTEST-IMPLICANT-CORE(core size).

Proposition 10.3. SHORTEST-IMPLICANT-CORE(core size) *is $\Sigma_2^p[k*]$-complete.*

Proof (sketch). To show hardness, we give an fpt-reduction from $\Sigma_2^p[k*]$-WSAT(DNF) to SHORTEST-IMPLICANT-CORE(core size). Intuitively, the choice for some $C' \subseteq C$ with $|C'| = k$ corresponds directly to the choice of some assignment $\alpha : X \to \mathbb{B}$ of weight k. Both involve a choice between $\binom{n}{k}$ candidates, and in both cases verifying whether the chosen candidate witnesses that the instance is a yes-instance involves solving a co-NP-complete problem. Any implicant C' forces those variables x that are included in C' to be set to true (and the other variables are not forced to take any truth value). However, by Lemma 10.2, any assignment that sets more than k variables x to true will trivially satisfy ψ. Therefore, the only relevant assignment is the assignment that sets only those x to true that are forced to be true by some C' of length k, and hence the choice for such a C' corresponds exactly to the choice for some assignment α of weight k. To verify whether some C' of length k is an implicant of the formula φ is equivalent to checking whether the formula $\bigwedge_{c \in C'} c \wedge \varphi$ is valid, which in turn is equivalent to checking whether a formula $\forall Y.\psi[\alpha]$ is true, for some assignment α.

Let (φ, k) be an instance of $\Sigma_2^p[k*]$-WSAT(DNF), with $\varphi = \exists X. \forall Y. \psi$. By Lemma 10.2, we may assume without loss of generality that for any assignment $\alpha :$ $X \to \mathbb{B}$ of weight $m \neq k$, $\forall Y. \psi[\alpha]$ is true. We may also assume without loss of generality that $|X| > k$; if this were not the case, (φ, k) would trivially be a no-instance. We construct an instance (φ', C, k) of SHORTEST-IMPLICANT-CORE(core size) by letting $\mathrm{Var}(\varphi') = X \cup Y$, $C = \bigwedge_{x \in X} x$, and $\varphi' = \psi$. Clearly, φ' is a Boolean formula in DNF. Also, consider the assignment $\alpha : X \to \mathbb{B}$ where $\alpha(x) = 1$ for all $x \in X$. We know that $\forall Y. \psi[\alpha]$ is true, since α has weight more than k. Therefore C is an implicant of φ'. We omit a detailed proof of correctness for this reduction.

To show membership in $\Sigma_2^p[k*]$, we give an fpt-reduction from SHORTEST-IMPLICANT-CORE(core size) to $\Sigma_2^p[k*]$-WSAT. This reduction uses exactly the same similarity between the two problems, i.e., the fact that assignments of weight k correspond exactly to implicants of length k, and that verifying whether this choice witnesses that the instance is a yes-instance in both cases involves checking validity of a propositional formula. We describe the reduction, and omit a detailed proof of correctness. Let (φ, C, k) be an instance of SHORTEST-IMPLICANT-CORE(core size), where $C = \{c_1, \ldots, c_n\}$. We construct an instance (φ', k) of $\Sigma_2^p[k*]$-WSAT, where $\varphi' = \exists X. \forall Y. \psi$, by defining $X = \{x_1, \ldots, x_n\}$, $Y = \mathrm{Var}(\varphi)$, $\psi = \psi_{\mathrm{corr}}^{X,Y} \to \varphi$, and $\psi_{\mathrm{corr}}^{X,Y} = \bigwedge_{i \in [n]} (x_i \to c_i)$.

Finally, we show $\Sigma_2^p[k*]$-completeness for SHORTEST-IMPLICANT-CORE (reduction size).

Proposition 10.4. SHORTEST-IMPLICANT-CORE(reduction size) *is* $\Sigma_2^p[k*]$*-complete.*

Proof (sketch). As an auxiliary problem, we consider the parameterized problem $\Sigma_2^p[k*]$-WSAT^{n-k}, which is a variant of $\Sigma_2^p[k*]$-WSAT. Given an input consisting of a QBF $\varphi = \exists X. \forall Y. \psi$ with $|X| = n$ and an integer k, the problem is to decide whether there exists an assignment α to X with weight $n - k$ such that $\forall Y. \psi[\alpha]$ is true. The parameter for this problem is k. We claim that this problem has the following properties. We omit the straightforward proof of these claims.

Claim 1. $\Sigma_2^p[k*]$-WSAT^{n-k} is $\Sigma_2^p[k*]$-complete.

Claim 2. Let (φ, k) be an instance of $\Sigma_2^p[k*]$-WSAT^{n-k}. In polynomial time, we can construct an equivalent instance (φ', k) of $\Sigma_2^p[k*]$-WSAT^{n-k} with $\varphi' = \exists X. \forall Y. \psi$, such that for any assignment $\alpha : X \to \mathbb{B}$ that has weight $m \neq (|X| - k)$, it holds that $\forall Y. \psi[\alpha]$ is true.

Using these claims, both membership and hardness for $\Sigma_2^p[k*]$ follow straightforwardly using arguments similar to the $\Sigma_2^p[k*]$-completeness proof of SHORTEST-IMPLICANT-CORE(core size). The fpt-reductions in the proof of Proposition 10.3 show that SHORTEST-IMPLICANT-CORE(reduction size) fpt-reduces to and from $\Sigma_2^p[k*]$-WSAT^{n-k}.

10.1.2 Minimizing DNF Formulas

Next, we turn our attention to the problems related to minimizing DNF formulas.

The following decision problem—that is related to the question of whether a given DNF can be reduced in size—is Σ_2^p-complete [190]. We say that a DNF formula φ is a *term-wise subformula* of another DNF formula φ' if for all terms $t \in \varphi$ there exists a term $t' \in \varphi'$ such that $t \subseteq t'$.

DNF-MINIMIZATION

Instance: A DNF formula φ of size n, and an integer m.
Question: Does there exist a term-wise subformula φ' of φ of size m such that $\varphi \equiv \varphi'$?

We consider two parameterizations of this problem:

- DNF-MINIMIZATION(reduction size), where the parameter $k = n - m$ is the difference in size between the original formula φ and the minimized formula φ'; and
- DNF-MINIMIZATION(core size), where the parameter $k = m$ is the size of the minimized formula φ'.

We firstly show that the problem DNF-MINIMIZATION(reduction size) is $\Sigma_2^p[k*]$-complete.

Proposition 10.5. DNF-MINIMIZATION(reduction size) *is* $\Sigma_2^p[k*]$-*complete.*

Proof (sketch). To show $\Sigma_2^p[k*]$-hardness, we use the reduction from the literature that is used to show Σ_2^p-hardness for DNF-MINIMIZATION. The polynomial-time reduction from SHORTEST-IMPLICANT-CORE to DNF-MINIMIZATION given by Umans [190, Theorem 2.2] is an fpt-reduction from SHORTEST-IMPLICANT-CORE(reduction size) to DNF-MINIMIZATION(reduction size).

To show membership in $\Sigma_2^p[k*]$, we describe an algorithm A that solves the problem and that can be implemented by an $\Sigma_2^p[k*]$-machine. Let (φ, k) be an instance of DNF-MINIMIZATION(reduction size). The algorithm A firstly guesses k literal occurrences in φ that are to be removed, resulting in the DNF formula φ'. This can be done in the existential phase using $f(k) \log n$ non-deterministic steps. Then, in the universal phase, the algorithm A verifies that $\varphi \equiv \varphi'$. This can be done using polynomially many non-deterministic steps.

We then turn to analyzing the parameterized complexity of DNF-MINIMIZATION (core size). We begin with an easy lower bound.

Proposition 10.6. DNF-MINIMIZATION(core size) *is* para-co-NP-*hard.*

Proof. We show that the slice of DNF-MINIMIZATION(core size) for the parameter value 2 is co-NP-hard, by giving a polynomial time reduction from UNSAT. Let ψ be an instance of UNSAT. Assume without loss of generality that ψ is in CNF. We construct an instance $(\varphi, 2)$ of DNF-MINIMIZATION(core size) by letting $\varphi = \neg \psi \vee (x_1 \wedge x_2 \wedge x_3)$, for fresh variables $x_1, x_2, x_3 \notin \text{Var}(\psi)$. We show that ψ is unsatisfiable if and only there exists a DNF formula φ' of size 2 that is equivalent to φ.

($\Rightarrow$) Assume that ψ is unsatisfiable. Then $\neg\psi$ is valid, and therefore so is φ. Let $y \in \text{Var}(\psi)$ be some variable that occurs both positively and negatively in ψ. Then $\varphi' = y \vee \neg y$ is a DNF formula of size 2 that is equivalent to φ.

($\Leftarrow$) Assume that ψ is satisfiable. Then there exists an assignment $\alpha : \text{Var}(\psi) \to \mathbb{B}$ such that $\psi[\alpha] = 1$. Then the variables x_1, x_2, x_3 are all relevant in φ. The assignment $\alpha_1 = \alpha \cup \{x_2 \mapsto 1, x_3 \mapsto 1\}$ witnesses that x_1 is relevant in φ, for instance. Therefore, by Lemma 10.8, any DNF formula equivalent to φ must contain all variables x_1, x_2, x_3, and thus must be of size at least 3.

Next, we give an upper bound on the complexity of DNF-MINIMIZATION(core size), by showing $\Sigma_2^p[k*]$-membership.

Proposition 10.7. DNF-MINIMIZATION(core size) *is in* $\Sigma_2^p[k*]$.

Proof. We give an fpt-reduction to $\Sigma_2^p[k*]$-WSAT. Let (φ, k) be an instance of DNF-MINIMIZATION(core size), where φ is a DNF formula. Let $X = \text{Var}(\varphi)$. We construct an instance (φ', k') of $\Sigma_2^p[k*]$-WSAT as follows. We introduce the set $Y = \{y_{i,x} : i \in [k], x \in X\}$ of variables. Intuitively, the variables $y_{i,x}$ represent a choice of (at most) k variables $x \in X$ to be used in the minimized DNF formula. Let $\psi_1, \dots, \psi_b$ be an enumeration of all possible DNF formulas of size $\leq k$ on the (fresh) variables $z_1, \dots, z_k$. By straightforward counting, we know that $b \leq f(k)$, for some function $f = 2^{O(k \log k)}$. Moreover, for each $j \in [b]$, let $\psi_j = t_{j,1} \vee \cdots \vee t_{j,w_j}$, and for each $\ell \in [w_j]$, let $t_{j,\ell} = l_{j,\ell,1} \wedge \cdots \wedge l_{j,\ell,v_{j,\ell}}$. We introduce another set $U = \{u_j : j \in [b]\}$ of variables. Intuitively, these variables will be used to select the shape ψ_j of the minimized DNF formula. We then perform our construction by letting $\varphi' = \exists Y. \exists U. \forall X. \varphi''$, where $\varphi'' = \varphi_{\text{proper}}^Y \wedge \varphi_{\text{one}}^U \wedge \bigwedge_{j \in [b]} (u_j \to (\varphi \leftrightarrow \bigvee_{\ell \in [w_j]} \varphi_{\text{sat}}^{j,\ell}))$. Here, the formula $\varphi_{\text{proper}}^Y$ ensures that for each $i \in [k]$ there is exactly one $x_i \in X$ such that y_{i,x_i} is true, and that the x_i are all distinct. It consists of clauses $\bigvee_{x \in X} y_{i,x}$, $(\neg y_{i,x} \vee \neg y_{i',x})$ and $(\neg y_{i,x} \vee \neg y_{i,x'})$, for each $i, i' \in [k]$ with $i < i'$ and each $x, x' \in X$ such that $x \neq x'$. The formula φ_{one}^U ensures there is exactly one $j \in [b]$ such that u_j is true, and consists of the clause $\bigvee_{j \in [b]} u_j$ and clauses $(\neg u_j \vee \neg u_{j'})$ for each $j, j' \in [u]$ with $j < j'$. For each $j \in [b]$, the assignment to the variables in Y represents a DNF formula χ_j that is obtained by taking the ψ_j and replacing each z_i in ψ_j by the unique x_i for which y_{i,x_i} is true. Next, the formulas $\varphi_{\text{sat}}^{j,\ell} = \bigwedge_{t \in [v_{j,\ell}]} \varphi_{\text{sat}}^{j,\ell,t}$ encode whether the ℓ-th term of χ_j is satisfied by the assignment to the variables in X. For this, we let $\varphi^{j,\ell,t} = \bigwedge_{x \in X} (y_{m,x} \to x)$ if $l_{j,\ell,t} = z_m$ for some $m \in [k]$, and we let $\varphi^{j,\ell,t} = \bigwedge_{x \in X} (y_{m,x} \to \neg x)$ if $l_{j,\ell,t} = \neg z_m$ for some $m \in [k]$. Finally, we let $k' = k + 1$. We show that $(\varphi, k) \in$ DNF-MINIMIZATION(core size) if and only if $(\varphi', k') \in \Sigma_2^p[k*]$-WSAT.

($\Rightarrow$) Assume that there exists a DNF χ of size $\leq k$ such that $\varphi \equiv \chi$. By Lemma 10.8, we may assume without loss of generality that $\text{Var}(\chi) \subseteq \text{Var}(\varphi)$. Then clearly there exists some DNF formula ψ_j over the variables $z_1, \dots, z_k$ and some distinct variables $x_1, \dots, x_k \in X$ such that $\chi = \psi_j[z_1 \mapsto x_1, \dots, z_k \mapsto x_k]$. We construct the assignment $\alpha : Y \cup U \to \mathbb{B}$ of weight k' as follows. We let $\alpha(y_{i,x}) = 1$ if and only if $x = x_i$, and we let $\alpha(u_{j'}) = 1$ if and only if $j' = j$.

It is straightforward to verify that α satisfies the formulas $\varphi^Y_{\text{proper}}$ and φ^U_{one}. We show that $\forall X.(\bigwedge_{j\in[b]}(u_j \to (\varphi \leftrightarrow \bigvee_{\ell\in[w_j]} \varphi^{j,\ell}_{\text{sat}})))[\alpha]$ is true. Let $\beta : X \to \mathbb{B}$ be an arbitrary truth assignment. Clearly, for each $j' \in [u]$ such that $j' \neq j$, the implication is satisfied, because $\alpha(u_{j'}) = 0$. We show that $(\varphi \leftrightarrow \bigvee_{\ell\in[w_j]} \varphi^{j,\ell}_{\text{sat}})[\alpha \cup \beta]$ is true. If $\varphi[\beta]$ is true, then since $\chi \equiv \varphi$, we know that some term $t_{j,\ell}[z_1 \mapsto x_1, \ldots, z_k \mapsto x_k]$ of χ is satisfied. It is straightforward to verify that $\varphi^{j,\ell}_{\text{sat}}$ is satisfied then as well. Conversely, if $\varphi[\beta]$ is not true, then an analogous argument shows that for no $\ell \in [w_j]$ the formula $\varphi^{j,\ell}_{\text{sat}}$ is satisfied. This concludes our proof that $(\varphi', k') \in \Sigma^p_2[k*]$-WSAT.

($\Leftarrow$) Assume that there exists some assignment $\alpha : Y \cup U \to \mathbb{B}$ of weight k such that $\forall X.\varphi''$ is true. Clearly, for each $i \in [k]$, α there must be exactly one $x_i \in X$ such that α sets y_{i,x_i} to true, and there must be exactly one $j \in [b]$ such that α sets u_j to true, since otherwise the formulas $\varphi^Y_{\text{proper}}$ and φ^U_{one} would not be satisfied. Consider the formula χ that is obtained from ψ_j by replacing the variables $z_1, \ldots, z_k$ by the variables $x_1, \ldots, x_k$, respectively. We know that the size of χ is at most k. We show that $\varphi \equiv \chi$.

Let $\beta : X \to \mathbb{B}$ be an arbitrary truth assignment. We show that $\varphi[\beta] = \chi[\beta]$. Since $\alpha(u_j) = 1$, we know that $\alpha \cup \beta$ must satisfy the formula $(\varphi \leftrightarrow \bigvee_{\ell\in[w_j]} \varphi^{j,\ell}_{\text{sat}})$. Assume that β satisfies χ, i.e., β satisfies some term $t_{j,\ell}[z_1 \mapsto x_1, \ldots, z_k \mapsto x_k]$ of χ. It is straightforward to verify that then $\varphi^{j,\ell}_{\text{sat}}[\alpha \cup \beta]$ is true. Then also $\varphi[\alpha \cup \beta] = \varphi[\beta]$ is true. Conversely, assume that β satisfies φ. Then we know that $\bigvee_{\ell\in[w_j]} \varphi^{j,\ell}_{\text{sat}}[\alpha \cup \beta]$ is true, i.e., there exists some $\ell \in [w_j]$ such that $\varphi^{j,\ell}_{\text{sat}}[\alpha \cup \beta]$ is true. It is then straightforward to verify that the term $t_{j,\ell}[z_1 \mapsto x_1, \ldots, z_k \mapsto x_k]$ is satisfied by β. This concludes our proof that $(\varphi, k) \in$ DNF-MINIMIZATION(core size). □

Finally, we show that the problem DNF-MINIMIZATION(core size) is in $\text{FPT}^{\text{NP}}[\text{few}]$. In particular, we show that it can be solved with an fpt-algorithm that uses at most $\lceil \log_2 k \rceil + 1$ queries to a SAT oracle. Moreover, this algorithm works even for the case where equivalent DNF formulas that are not term-wise subformulas of φ are also accepted.

In order to show this, we will consider the notion of relevant variables, and establish several lemmas that help us to describe and analyze the algorithm. Let φ be a DNF formula and let $x \in \text{Var}(\varphi)$ be a variable occurring in φ. We call x *relevant* in φ if there exists some assignment $\alpha : \text{Var}(\varphi)\setminus\{x\} \to \mathbb{B}$ such that $\varphi[\alpha \cup \{x \mapsto 0\}] \neq \varphi[\alpha \cup \{x \mapsto 1\}]$. We begin with the following lemma, that we state without proof.

Lemma 10.8. *Let φ be a DNF formula and let φ' be a DNF formula of minimal size that it is equivalent to φ. Then for every variable $x \in \text{Var}(\varphi)$ it holds that $x \in \text{Var}(\varphi')$ if and only if x is relevant in φ.*

Proof. Assume that $x \in \text{Var}(\varphi)$ is relevant in φ. We show that $x \in \text{Var}(\varphi')$. Let $X = \text{Var}(\varphi) \cup \text{Var}(\varphi')$. Then there exists some assignment $\alpha : X\setminus\{x\} \to \mathbb{B}$ such that $\varphi[\alpha_1] \neq \varphi[\alpha_2]$, where $\alpha_1 = \alpha \cup \{x \mapsto 0\}$ and $\alpha_2 = \alpha \cup \{x \mapsto 1\}$. Assume that $x \notin \text{Var}(\varphi')$. Then $\varphi'[\alpha_1] = \varphi'[\alpha_2]$ because φ_1 and φ_2 coincide on $\text{Var}(\varphi')$. This is a contradiction with the assumption that φ and φ' are equivalent. Therefore, $x \in \text{Var}(\varphi')$.

Conversely, assume that $x \in \mathrm{Var}(\varphi)$ is not relevant in φ. We show that $x \notin \mathrm{Var}(\varphi')$. By definition we know that for each assignment $\alpha : X \backslash \{x\} \to \mathbb{B}$ it holds that $\varphi[\alpha_1] = \varphi[\alpha_2]$, where $\alpha_1 = \alpha \cup \{x \mapsto 0\}$ and $\alpha_2 = \alpha \cup \{x \mapsto 1\}$. Assume that $x \in \mathrm{Var}(\varphi')$. Then φ' is equivalent to the DNF formula $\varphi'[x \mapsto 0]$, which is strictly smaller than φ'. This contradicts minimality of φ'. Therefore, $x \notin \mathrm{Var}(\varphi')$. $\qquad \square$

The algorithm for DNF-MINIMIZATION(core size) that we will construct below uses a SAT oracle to answer the question of whether for a DNF formula φ there exist (at least) some given number of variables that are relevant in φ. We show how to encode this problem into SAT (in polynomial time).

Lemma 10.9. *Given a DNF formula φ and a positive integer m (given in unary), deciding whether there are at least m variables that are relevant in φ is in NP.*

Proof. We describe a guess-and-check algorithm that decides the problem. The algorithm first guesses m distinct variables occurring in φ, and for each guessed variable x the algorithm guesses an assignment α_x to the remaining variables $\mathrm{Var}(\varphi) \backslash \{x\}$. Then, the algorithm verifies whether the guessed variables are really relevant by checking that, under α_x, assigning different values to x changes the outcome of the Boolean function represented by φ, i.e., $\varphi[\alpha_x \cup \{x \mapsto 0\}] \neq \varphi[\alpha_x \cup \{x \mapsto 1\}]$. It is straightforward to construct a SAT instance ψ that implements this guess-and-check procedure. Moreover, from any assignment that satisfies ψ it is easy to extract the relevant variables. $\qquad \square$

Before we can show $\mathrm{FPT}^{\mathrm{NP}}[\mathrm{few}]$-membership for DNF-MINIMIZATION(core size), we consider one more technical lemma.

Lemma 10.10. *Let $x_1, \ldots, x_k$ be propositional variables. There are $2^{O(k \log k)}$ different DNF formulas ψ over the variables $x_1, \ldots, x_k$ that are of size k.*

Proof. Each suitable DNF formula $\psi = t_1 \vee \cdots \vee t_\ell$ can be formed by writing down a sequence $\sigma = (l_1, \ldots, l_k)$ of literals l_i over $x_1, \ldots, x_k$, and splitting this sequence into terms, i.e., choosing integers $d_1, \ldots, d_{\ell+1} \in [k+1]$ with $d_1 = 1$ and $d_1 < \cdots < d_{\ell+1}$ such that $t_i = \{l_{d_i}, \ldots, l_{d_{i+1}-1}\}$ for each $i \in [\ell]$. To see that there are $2^{O(k \log k)}$ formulas ψ, it suffices to see that there are $O(k^k)$ sequences σ, and $O(2^k)$ choices for the integers d_i. $\qquad \square$

Proposition 10.11. DNF-MINIMIZATION(core size) *can be solved by an fpt-algorithm that uses $\lceil \log k \rceil + 1$ queries to a SAT oracle. Moreover, the first $\lceil \log k \rceil$ queries to the oracle are of size $O(k^2 n^2)$, and the last query is of size $2^{O(k \log k)} \cdot n$, where n is the input size.*

Proof. The algorithm given in pseudo-code in Algorithm 10.1 solves the problem DNF-MINIMIZATION(core size) in the required time bounds. By Lemma 10.8, we know that any minimal equivalent formula of φ must contain all and only the variables that are relevant in φ. By Proposition 7.17, we may assume without loss of

---

Algorithm 10.1: Solving the problem DNF-Minimization(core size) in fpt-time using $\lceil \log k \rceil + 1$ queries to a SAT oracle.

---

input : an instance (φ, k) of DNF-Minimization(core size)
output: YES iff $(\varphi, k) \in$ DNF-Minimization(core size)

```
1  rvars ← ∅ ;                              // relevant variables in φ
2  i ← 0; j ← k + 2 ;                        // bounds on # of rvars
3  while i + 1 < j do            // logarithmic search for the # of rvars
4  │  ℓ ← ⌈(i + j)/2⌉ ;
5  │  check using SAT oracle if there are at least ℓ relevant variables in φ ;
6  │  if the SAT oracle returns a model M then
7  │  │  rvars ← the ℓ relevant variables encoded by the model M ;
8  │  end
9  │  else break;
10 end
11 if |rvars| > k then
12 │  return "no" ;                 // too many rvars for any DNF of size ≤ k
13 else
14 │  foreach DNF formula ψ of size k over rvars do        // 2^O(k log k) of these
15 │  │  construct a formula φψ that is unsatisfiable iff ψ ≡ φ;
   │  │                        // the formulas φψ must be variable disjoint
16 │  end
17 │  query the SAT oracle whether ⋀ψ φψ is satisfiable ;
18 │  if the SAT oracle returns "yes" then
19 │  │  return "no" ;              // no candidate ψ is equivalent to φ
20 │  else
21 │  │  return "yes" ;             // some candidate ψ is equivalent to φ
22 │  end
23 end
```

---

generality that the SAT oracle is a witness oracle. The algorithm firstly determines how many variables are relevant in φ. By Lemma 10.9, we know that this can be done with a binary search using $\lceil \log k \rceil$ SAT queries. If there are more than k relevant variables, the algorithm rejects. Otherwise, the algorithm will have computed the set rvars of relevant variables. Next, with a single oracle query, it checks whether there exists some equivalent DNF formula ψ of size at most k over the variables in rvars. By Lemma 10.10, we know that there are $2^{O(k \log k)}$ different DNF formulas ψ of size at most k over the variables in rvars. Verifying whether a particular DNF formula ψ is equivalent to the original formula φ can be done by checking whether the formula $\varphi_\psi = (\psi \wedge \neg\varphi) \vee (\neg\psi \wedge \varphi)$ is unsatisfiable. Verifying whether there exists some suitable DNF formula ψ that is equivalent to φ can be done by making variable-disjoint copies of all φ_ψ and checking whether the conjunction of these copies is unsatisfiable.

The algorithm used to show membership in $\text{FPT}^{\text{NP}}[\text{few}]$ can be modified straightforwardly to return a DNF formula ψ of size at most k that is equivalent to an input φ if such a formula ψ exists. The algorithm would need to search for this ψ that is equivalent to φ, for which it would need an additional $O(k \log k)$ SAT queries.

Corollary 10.12. *There is an algorithm A that (1) when given a DNF formula φ and a positive integer k, computes a DNF formula ψ of size at most k such that $\psi \equiv \varphi$, if such a ψ exists, and returns "none" otherwise, and (2) runs in time $f(k)n^c$ and queries a witness SAT oracle at most $O(k \log k)$ times, for some computable function f and some constant c, where n denotes the input size.*

10.2   Inconsistency Repair

Finally, we analyze two parameterized problems that are related to the task of finding inconsistent subsets of sets of propositional formulas, and the task of finding subsets that can be deleted to repair the inconsistency. We begin with the following parameterized problem, that is related to the former task.

ODD- BOUNDED- INCONSISTENT- SET
Instance: An inconsistent set Φ of propositional formulas, and a positive integer k.
Parameter: k.
Question: Is the minimum size of an inconsistent subset Φ' of Φ both odd and at most k?

We show that this problem is $\Sigma_2^p[k*]$-hard. In order to prove $\Sigma_2^p[k*]$-hardness, we need the following technical lemma, that is similar to Lemma 10.2. We omit its straightforward proof.

Lemma 10.13. *Let (φ, k) be an instance of $\Sigma_2^p[k*]$-WSAT. In polynomial time, we can construct an equivalent instance (φ', k) of $\Sigma_2^p[k*]$-WSAT with $\varphi' = \exists X.\forall Y.\psi$, such that for every assignment $\alpha : X \to \mathbb{B}$ that has weight $m \neq k$, it holds that $\forall Y.\psi[\alpha]$ is false.*

Using this lemma, we can show the $\Sigma_2^p[k*]$-hardness result.

Proposition 10.14. ODD- BOUNDED- INCONSISTENT- SET *is $\Sigma_2^p[k*]$-hard.*

Proof. We show $\Sigma_2^p[k*]$-hardness by giving an fpt-reduction from $\Sigma_2^p[k*]$-WSAT. Let (φ, k) be an instance of $\Sigma_2^p[k*]$-WSAT, where $\varphi = \exists X.\forall Y.\psi$ and $X = \{x_1, \ldots, x_n\}$. By Lemma 10.13, we may assume without loss of generality that for any truth assignment $\alpha : X \to \mathbb{B}$ of weight $m \neq k$ it holds that $\forall Y.\psi[\alpha]$ is false. We construct an instance (Φ, k') of ODD- BOUNDED- INCONSISTENT- SET as follows. We introduce fresh variables $z_1, \ldots, z_n$. We let $k' = 2k + 1$ and:

$$\Phi = \{z_1, \ldots, z_n, (z_1 \to x_1), \ldots, (z_n \to x_n), \neg\psi\}.$$

We show that $(\varphi, k) \in \Sigma_2^p[k*]$-WSAT if and only if $(\Phi, k') \in$ ODD- BOUNDED- INCONSISTENT- SET.

($\Rightarrow$) Suppose that $(\varphi, k) \in \Sigma_2^p[k*]$-WSAT. This means that there is some truth assignment $\alpha : X \to \mathbb{B}$ of weight k such that $\forall Y.\psi[\alpha]$ is true. Consider the following set $\Phi' \subseteq \Phi$:

$$\Phi' = \{\neg\psi\} \cup \{ z_i, (z_i \to x_i) : i \in [n], \alpha(x_i) = 1 \}.$$

Since $\forall Y.\psi[\alpha]$ is true, we know that Φ' is unsatisfiable. Moreover, since for all truth assignments $\alpha' : X \to \mathbb{B}$ of weight $m < k$ it holds that $\forall Y.\psi[\alpha]$ is false, we can conclude that there is no smaller set $\Phi'' \subseteq \Phi$ that is unsatisfiable. Then, since $|\Phi'| \le k'$ and $|\Phi'|$ is odd, we can conclude that $(\Phi, k') \in$ ODD- BOUNDED- INCONSISTENT- SET.

($\Leftarrow$) Conversely, suppose that $(\Phi, k') \in$ ODD- BOUNDED- INCONSISTENT- SET. This means that there is some subset $\Phi' \subseteq \Phi$ of odd size at most k that is unsatisfiable, and that there is no unsatisfiable subset $\Phi'' \subseteq \Phi$ of smaller size. Clearly $\neg\psi \in \Phi'$, since otherwise Φ' would be satisfiable. Also, since Φ' is an unsatisfiable subset of minimal size, we know that for any $i \in [n]$, it holds that $z_i \in \Phi$ if and only if $(z_i \to x_i) \in \Phi$. Construct the truth assignment $\alpha : X \to \mathbb{B}$ as follows. For each $i \in [n]$, we let $\alpha(x_i) = 1$ if and only if $z_i \in \Phi$. Then, because $|\Phi'| \le k'$, we know that α has weight at most k. Moreover, since for each $\alpha' : X \to \mathbb{B}$ of weight $m < k$ it holds that $\forall Y.\psi[\alpha]$ is false, we know that α must be of weight exactly k. If this were not the case, Φ' could not be unsatisfiable. Then, by construction of α, we know that $\forall Y.\psi[\alpha]$ is true. Therefore, $(\varphi, k) \in \Sigma_2^p[k*]$-WSAT.

Next, we consider a parameterized problem that is related to repairing inconsistencies in a propositional knowledge base. Let Φ be a set of propositional formulas. We say that a subset $\Phi' \subseteq \Phi$ is a *repair set of* Φ if $\Phi \setminus \Phi'$ is consistent. We denote the minimum size of any repair set of Φ by the *minimum repair size of* Φ. Now consider the following parameterized decision problem.

ODD- BOUNDED- REPAIR- SET
Instance: A set Φ of propositional formulas, and a positive integer k.
Parameter: k.
Question: Is the minimum repair size of Φ both odd and at most k?

We show that the problem can be solved in fpt-time using $f(k)$ queries to a SAT oracle.

Proposition 10.15. ODD- BOUNDED- REPAIR- SET *is* FPT^{NP}[few]*-complete. Moreover,* FPT^{NP}[few]*-hardness holds even for the setting where all propositional formulas are single clauses.*

Proof. To show membership in FPT^{NP}[few], we describe an algorithm A that solves ODD- BOUNDED- REPAIR- SET, that runs in fpt-time, and that queries an NP oracle at most $k + 1$ times, for some computable function f. Let (Φ, k) be an instance of ODD- BOUNDED- REPAIR- SET. The following problem O is in NP:

$O = \{ (\Phi, \ell) : \Phi$ is a set of propositional formulas,
    there is some $\Phi' \subseteq \Phi$ such that $\Phi \setminus \Phi'$ is satisfiable, and $|\Phi'| \le \ell \}$.

Then, by querying an O oracle $k + 1$ times—namely using the queries $(\Phi, 0)$, $(\Phi, 1)$, $(\Phi, 2)$, ..., (Φ, k)—the algorithm A can determine the minimum repair size of Φ if it is at most k—namely, the smallest ℓ such that $(\Phi, \ell) \in O$. If $(\Phi, k) \notin O$, the algorithm A rejects the input. Otherwise, the algorithms A accepts the input (Φ, k) if and only if the minimum repair size ℓ is odd. It is straightforward to verify that the algorithm correctly decides whether $(\Phi, k) \in$ ODD- BOUNDED- REPAIR- SET, and runs in fpt-time.

Next, we show $\mathrm{FPT}^{\mathrm{NP}}$[few]-hardness by giving an fpt-reduction from ODD- LOCAL- MAX- MODEL. Let (φ, X) be an instance of ODD- LOCAL- MAX- MODEL, where $X = \{x_1, \ldots, x_k\}$. We may assume without loss of generality that φ is a CNF formula, i.e., that $\varphi = c_1 \wedge \cdots \wedge c_u$, where each c_j is a clause. If this were not the case, we could transform φ to a CNF formula φ' (such that for each assignment α to X, it holds that $\varphi[\alpha]$ is satisfiable if and only if $\varphi'[\alpha]$ is satisfiable), using the standard Tseitin transformation [188]. Moreover, we may assume without loss of generality that k is even.

We construct an instance (Φ, k) of ODD- BOUNDED- REPAIR- SET as follows. Let $m = 2k$. We introduce the set $Z = \{z_{i,j} : i \in [m], j \in [u]\}$ of fresh variables. We let:

$$\Phi = \{x_1, \ldots, x_k\} \cup \{(c_j \vee z_{i,j}), (c_j \vee \overline{z_{i,j}}) : i \in [m], j \in [u]\}.$$

We show that $(\Phi, k) \in$ ODD- BOUNDED- REPAIR- SET if and only if $(\varphi, X) \in$ ODD- LOCAL- MAX- MODEL.

$(\Rightarrow)$ Suppose that $(\Phi, k) \in$ ODD- BOUNDED- REPAIR- SET. This means that there is a set $\Phi' \subseteq \Phi$ of odd size such that $\Phi \backslash \Phi'$ is consistent, and for each $\Phi'' \subseteq \Phi$ with $|\Phi''| < |\Phi'|$ it holds that $\Phi \backslash \Phi''$ is inconsistent. Since $m > k$, we know that $\Phi \backslash \Phi'$ contains some formulas of the form $(c_j \vee z_{i,j})$ and of the form $(c_j \vee \overline{z_{i,j}})$, for each $j \in [u]$. Therefore, we know that $\Phi \backslash \Phi' \models \varphi$. We show that Φ' contains no formulas of the form $(c_j \vee z_{i,j})$ or of the form $(c_j \vee \overline{z_{i,j}})$. Suppose that this is not the case, i.e., that $(c_j \vee z_{i,j}) \in \Phi'$ for some $j \in [m]$; the case for $(c_j \vee \overline{z_{i,j}})$ is entirely analogous. Then $\Phi'' = \Phi' \backslash \{(c_j \vee z_{i,j})\} \subsetneq \Phi'$ has the property that $\Phi \backslash \Phi''$ is consistent. This is a contradiction, so we can conclude that Φ' contains no formulas of the form $(c_j \vee z_{i,j})$ or of the form $(c_j \vee \overline{z_{i,j}})$. In other words, $\Phi' = \{x_{i_1}, \ldots, x_{i_u}\}$, for some $i_1, \ldots, i_u \in [k]$ with $i_1 < \cdots < i_u$. Since $\Phi \backslash \Phi'$ is consistent, we know that there is a model of φ that sets exactly the variables in $X \backslash \Phi'$ to true. Moreover, there is no model of φ that sets more variables in X to true. Since k is even and $|\Phi'|$ is odd, we know that the X-maximal models of φ set an odd number of variables in X to true. In other words, $(\varphi, X) \in$ ODD- LOCAL- MAX- MODEL.

$(\Leftarrow)$ Conversely, suppose that $(\varphi, X) \in$ ODD- LOCAL- MAX- MODEL. This means that the X-maximal models of φ set an odd number of variables in X to true. Let α be an arbitrary X-maximal model of φ. Consider the set $\Phi' \subseteq \Phi$ that is defined as follows:

$$\Phi' = \{x_i : i \in [k], \alpha(x_i) = 0\}.$$

Clearly, $\Phi \backslash \Phi'$ is consistent. We show that for any $\Phi'' \subseteq \Phi$ with $|\Phi''| < |\Phi'|$ it holds that $\Phi \backslash \Phi''$ is inconsistent. Suppose that the contrary would be true, i.e., that there is some $\Phi'' \subseteq \Phi$ with $|\Phi''| < |\Phi'|$ such that $\Phi \backslash \Phi''$ is consistent. Then, by construction of Φ, there must be some model α' of φ that sets all the variables in $X \cap \Phi''$ to true. Moreover, it holds that $|X \cap \Phi''| < |X \cap \Phi'|$, so α' sets more variables in X to true than α. This is a contradiction with the fact that α is an X-maximal model of φ. Therefore, we can conclude that Φ' is a repair set of Φ of minimum size. In other words, $(\Phi, k) \in$ ODD- BOUNDED- REPAIR- SET.

The algorithm described in the membership part of the proof of Proposition 10.15 (when given access to a witness SAT oracle) can directly be used to compute minimum size repair sets of size at most k.

Corollary 10.16. *There is an algorithm A that (1) when given a set Φ of propositional formulas and a positive integer k, computes a minimum size repair set of Φ of size at most k, if this exists, and returns "none" otherwise, and (2) runs in time $f(k)n^c$ and queries a* witness *SAT oracle at most k times, for some computable function f and some constant c, where n denotes the input size.*

Moreover, we know that the number of SAT queries that any fpt-algorithm needs to make for computing minimum size repair sets of size at most k must depend on the value of k.

Proposition 10.17. *There is no algorithm A that (1) when given a set Φ of clauses and a positive integer k, computes a minimum size repair set of Φ of size at most k, if this exists, and returns "none" otherwise, and (2) runs in time $f(k)n^c$ and queries a* witness *SAT oracle at most c times, for some computable function f and some constant c, where n denotes the input size, unless the PH collapses.*

Proof (sketch). Suppose that such an algorithm A exists. We can then show that the problem ODD- BOUNDED- REPAIR- SET can be solved in fpt-time using $O(1)$ decision SAT oracle queries, using an argument similar to the one used in the proof of Proposition 7.17. Then, by Proposition 7.12, we can conclude that the PH collapses.

Finally, we show that computing minimum size repair sets of size at most k cannot be done in fpt-time using a small number of queries to a decision SAT oracle.

Proposition 10.18. *There is no algorithm A that (1) when given a set Φ of clauses and a positive integer k, computes a minimum size repair set of Φ of size at most k, if this exists, and returns "none" otherwise, and (2) runs in time $f(k)n^c$ and queries a decision SAT oracle at most $f(k) + O(\log n)$ times, for some computable function f and some constant c, where n denotes the input size, unless W[P] = FPT.*

Proof. Suppose that a suitable algorithm A exists. We show that W[P] = FPT, by providing an fpt-algorithm for the problem WSAT(CIRC). Let (C, k) be an instance of WSAT(CIRC), where C is a Boolean circuit with input nodes $X = \{x_1, \ldots, x_n\}$. We may assume without loss of generality that for each truth assignment $\alpha : X \rightarrow$

$\mathbb{B}$ of weight $m \neq k$ it holds that $C[\alpha]$ is false. If this were not the case, one can straightforwardly transform C into a circuit C' that satisfies this property such that for each truth assignment $\alpha : X \to \mathbb{B}$ of weight k it holds that $C[\alpha] = C'[\alpha]$.

We firstly transform C into a CNF formula φ that satisfies the property that for each truth assignment $\alpha : X \to \mathbb{B}$ it holds that $C[\alpha]$ is true if and only if $\varphi[\alpha]$ is satisfiable. This can be done straightforwardly using the standard Tseitin transformation [188]. Let $c_1, \ldots, c_u$ be the clauses of φ.

We then construct a set Φ of clauses as follows. Let $m = 2k$. We introduce the set $Z = \{ z_{i,j} : i \in [m], j \in [u] \}$ of fresh variables. Then we let:

$$\Phi = \{\neg x_1, \ldots, \neg x_n\} \cup \{ (c_j \vee z_{i,j}), (c_j \vee \overline{z_{i,j}}) : i \in [m], j \in [u] \}.$$

By using an argument that is similar to the one in the proof of Proposition 10.15, one can show that repair sets Φ' of Φ of size k are in a one-to-one correspondence with satisfying assignments of C of weight k.

Next, we simulate the algorithm A with input (Φ, k), without querying the SAT oracle. Instead, we iterate over all possible answers that the oracle gives. Since the oracle is queried at most $f(k) + O(\log n)$ times, and the oracle gives binary answers (it is a decision oracle), we know that there are at most $2^{f(k)} n^c$ possible sequences of answers, for some constant c. For each such sequence $\overline{b}$ of answers, we simulate the algorithm A. If the simulation of A on (Φ, k) using $\overline{b}$ returns "none", we continue with the next sequence $\overline{b}'$. If the simulation returns some repair set Φ' of size at most k, we do the following. Firstly, we check whether $|\Phi'| = k$ and $\Phi' \subseteq X$. If this is not the case, we continue with next sequence $\overline{b}'$. If this is the case, we construct the truth assignment $\alpha : X \to \mathbb{B}$ of weight k corresponding to Φ'. We then check whether $C[\alpha]$ is true. If this is the case, we know that $(C, k) \in \text{WSAT}(\text{CIRC})$, and we are done. If this is not the case, we continue with the next sequence $\overline{b}'$.

Finally, if we iterated over all sequences $\overline{b}$ of answers for the oracle queries, and we did not find a satisfying truth assignment for C of weight k, we can conclude that $(C, k) \notin \text{WSAT}(\text{CIRC})$. This is true because if it would be the case that $(C, k) \in \text{WSAT}(\text{CIRC})$, then there is some minimum size repair set Φ' of Φ of size k (from which a suitable truth assignment α can be constructed), and thus there is some sequence $\overline{b}$ of answers to the oracle queries for which the simulation of A would have resulted in such a set Φ'. Since we found no such set Φ', we know that it cannot be the case that $(C, k) \in \text{WSAT}(\text{CIRC})$, and we thus know that $(C, k) \notin \text{WSAT}(\text{CIRC})$.

It is readily verified that the algorithm to solve $\text{WSAT}(\text{CIRC})$ runs in fixed-parameter tractable time. Thus $W[P] = \text{FPT}$.

Notes

The results in Sect. 10.1 appeared in a paper in the proceedings of SAT 2014 [112].

Chapter 11
Problems in Judgment Aggregation

In this chapter, we investigate various parameterized variants of problems that arise in the area of judgment aggregation in computational social choice—this is a continuation of our parameterized complexity investigation of problems at higher levels of the Polynomial Hierarchy using the parameterized complexity tools that we developed in Chaps. 4–7. In particular, we study the problem of agenda safety for the majority rule in judgment aggregation, as well as the problem of computing outcomes for the Kemeny judgment aggregation procedure.

Overview of This Chapter
We begin in Sect. 11.1 with introducing some basic notions from the area of judgment aggregation. In particular, we introduce the two judgment aggregation procedures that play a role in this chapter: the majority rule and the Kemeny rule. Moreover, we introduce two formal frameworks that have been used in the literature to study judgment aggregation: a formula-based framework and a constraint-based framework.

Then, in Sect. 11.2, we investigate the parameterized complexity of the problem of deciding whether an agenda is safe for the majority judgment aggregation rule. This problem is Σ_2^p-complete in general. The parameterizations that we consider for this problem range from simple syntactic parameters (such as the number of formulas in the agenda or the maximum size of any formula in the agenda) to more intricate parameters (such as the treewidth of several different graphs that capture the interaction between formulas in the agenda). An overview of all parameterized variants that we consider, together with the parameterized complexity results that we obtain, can be found in Table 11.1.

Finally, in Sect. 11.3, we provide a parameterized complexity analysis of the problem of computing an outcome for the Kemeny judgment aggregation procedure, a Θ_2^p-complete problem. We study this problem both in the setting of formula-based and in the setting of constraint-based judgment aggregation (these are two formal judgment aggregation frameworks). We consider all possible combinations of a number of parameters, including the number of issues in the agenda, the maximum size of

© Springer-Verlag GmbH Germany, part of Springer Nature 2019
R. de Haan: Parameterized Complexity in the Polynomial Hierarchy, LNCS 11880,
https://doi.org/10.1007/978-3-662-60670-4_11

Table 11.1 Complexity results for the different parameterized variants of the agenda safety problem for the majority rule.

Parameter	Complexity	
maximum formula size (ℓ)	para-Π_2^p-complete	(Proposition 11.5)
maximum variable degree (d)	para-Π_2^p-complete	(Proposition 11.5)
$\ell + d$	para-Π_2^p-complete	(Proposition 11.5)
agenda size	in $\mathrm{FPT}^{NP}[\text{few}]$, i.e., solvable in fpt-time with $f(k)$ SAT queries, where $f(k) = 2^{O(k)}$ (Proposition 11.6) and $f(k) = \Omega(\log k)$ (Proposition 11.9)	
counterexample size	$\Pi_2^p[k*]$-hard	(Proposition 11.15)
formula primal treewidth	fixed-parameter tractable	(Proposition 11.10)
clausal primal treewidth	para-Π_2^p-complete	(Proposition 11.11)
formula incidence treewidth	para-Π_2^p-complete	(Proposition 11.12)
clausal incidence treewidth	para-Π_2^p-complete	(Proposition 11.13)

Table 11.2 Parameterized complexity results for FB-OUTCOME-KEMENY.

Parameters	Parameterized complexity result	
c, n, m	in FPT	(Proposition 11.18)
h, p	in $\mathrm{FPT}^{NP}[\text{few}]$	(Proposition 11.16)
n	in $\mathrm{FPT}^{NP}[\text{few}]$	(Proposition 11.17)
h, n, m, p	$\mathrm{FPT}^{NP}[\text{few}]$-hard	(Proposition 11.23)
c, h, n, p	$\mathrm{FPT}^{NP}[\text{few}]$-hard	(Proposition 11.24)
c, h, m, p	$\mathrm{FPT}^{NP}[\text{few}]$-hard	(Proposition 11.25)
c, h, m	para-Θ_2^p-hard	(Corollary 11.21)
c, m, p	para-Θ_2^p-hard	(Proposition 11.22)

formulas in the agenda, and the number of individuals that are involved in the judgment aggregation scenario. The parameterized complexity results that we obtain for the formula-based judgment aggregation framework are summarized in Table 11.2, and the results for the constraint-based framework are summarized in Table 11.3 (in both tables, a star denotes an arbitrary choice for the subset).

Table 11.3 Parameterized complexity results for CB-OUTCOME-KEMENY.

Parameters	Parameterized complexity result	
c	in FPT	(Proposition 11.27)
n	in FPT	(Proposition 11.26)
h	in XP	(Proposition 11.28)
h, p	W[SAT]-hard	(Proposition 11.29)
p	para-Θ_2^p-hard	(Proposition 11.30)

Fig. 11.1 The outcome of the majority vote by the three friends.

	easily accessible	beautiful	suitable
friend 1	no	yes	no
friend 2	yes	no	no
friend 3	yes	yes	yes
majority	yes	yes	no

11.1 Judgment Aggregation

Judgment aggregation studies how a group of individuals can make collective judgments on a logically related set of issues, and is situated in the interdisciplinary field of computational social choice [30]. Various issues that arise in the area of judgment aggregation play a role in the following example, known as the *discursive dilemma* (or *doctrinal paradox*). Suppose three friends want to go for a scenic run on the weekend, and they are choosing the location for their run. None of them has access to a car, and they only want to go to beautiful locations. Consequently, a location is suitable if and only if (1) it is easily accessible by public transport or by bike and (2) it is beautiful. Further, suppose the three friends have a possible location in mind, and they want to reach a decision by a simple majority vote, and suppose that the results of the vote are as in Fig. 11.1.

Clearly, the majority outcome of the vote is not consistent with the friends' definition of a suitable location for their run. The group opinion is that the location is easily accessible by public transport or bike and that it is beautiful, but that the location is not suitable.

Judgment aggregation offers various ways around this awkward situation. A first way of avoiding this dilemma is to identify the settings in which such an inconsistent majority outcome could possibly occur, and to use majority vote only in cases where this is guaranteed not to happen. This is known as *agenda safety* (for the majority rule)—a set of logically connected issues, or an *agenda*, is called *safe* if for no possible way of casting votes, the outcome of a majority vote results in an inconsistent group opinion.

A second method of solving the issue of the doctrinal paradox is to use a different way of forming a group opinion than the simple majority vote. There are plenty of *judgment aggregation procedures* that guarantee a logically consistent collective opinion, for every possible combination of individual opinions. One of the best-known judgment aggregation procedures that ensures a consistent outcome is the *Kemeny rule*. This procedure selects those consistent opinions that minimize the cumulative distance to the individual opinions—here every yes-no disagreement adds 1 to the distance (for more details, see Sects. 11.1.1 and 11.1.2). That is, the Kemeny rule might appoint multiple candidates for a group opinion, and to decide between them an additional tie-breaking mechanism would be needed.

In the remainder of this chapter, we study the parameterized complexity of various problems related to deciding the safety of an agenda, and to computing outcomes for the Kemeny judgment aggregation procedure. Before we are ready to do this, we need to introduce two formal frameworks for modeling the setting of judgment aggregation: the framework of *formula-based judgment aggregation* (as used by, e.g., [62, 73, 75, 104, 142, 148, 149]) and the framework of *constraint-based judgment aggregation* (as used by, e.g., [102, 103]). We describe these two frameworks in the following sections.

11.1.1 Formula-Based Judgment Aggregation

We begin with explaining the framework of formula-based judgment aggregation. For more details, we refer to textbooks and overview articles on the topic [30, 104, 148, 149].

In this framework, a set of logically related issues is modeled by an agenda. An *agenda* is a finite, nonempty set Φ of formulas that does not contain any doubly-negated formulas and that is closed under complementation, i.e., $\Phi = \{\varphi_1, \ldots, \varphi_n, \neg\varphi_1, \ldots, \neg\varphi_n\}$, where for each $i \in [n]$, the formula φ_i does not have negation as its outermost logical connective. Moreover, if $\Phi = \{\varphi_1, \ldots, \varphi_n, \neg\varphi_1, \ldots, \neg\varphi_n\}$ is an agenda, then we let $[\Phi] = \{\varphi_1, \ldots, \varphi_n\}$ denote the *pre-agenda* associated to the agenda Φ. We denote the bitsize of the agenda Φ by $\|\Phi\| = \sum_{\varphi \in \Phi} |\varphi|$.

(Individual and group) opinions are modeled by judgment sets. A *judgment set* J for an agenda Φ is a subset $J \subseteq \Phi$. We call a judgment set J *complete* if for each $\varphi \in [\Phi]$ it holds that $\varphi \in J$ or $\neg\varphi \in J$; and we call it *consistent* if there exists a truth assignment that makes all formulas in J true. It is natural to require that every feasible (individual or group) opinion is consistent.

We associate with each agenda Φ an integrity constraint Γ, that can be used to further restrict the set of feasible opinions. Such an *integrity constraint* consists of a single propositional formula. In the remainder of the paper, if no integrity constraint is specified, we implicitly assume that $\Gamma = \top$. We say that a judgment set J is Γ-*consistent* if there exists a truth assignment that simultaneously makes all formulas in J and Γ true. Let $\mathcal{J}(\Phi, \Gamma)$ denote the set of all complete and Γ-consistent subsets of Φ. We say that finite sequences $\boldsymbol{J} \in \mathcal{J}(\Phi, \Gamma)^+$ of complete and Γ-consistent judgment sets are *profiles*, and where convenient we equate a profile $\boldsymbol{J} = (J_1, \ldots, J_p)$ with the multiset $\{J_1, \ldots, J_p\}$.[1]

---

[1] By a slight abuse of notation, we use the same brackets for sets and multisets.

A *judgment aggregation procedure* (or *rule*) for the agenda Φ and the integrity constraint Γ is a function F that takes as input a profile $\boldsymbol{J} \in \mathcal{J}(\Phi, \Gamma)^+$, and that produces a non-empty set of non-empty judgment sets, i.e., it produces an element in $2^{2^{\Phi} \setminus \{\emptyset\}} \setminus \{\emptyset\}$. We call a judgment aggregation procedure F *resolute* if for any profile $\boldsymbol{J}$ it returns a singleton, i.e., $|F(\boldsymbol{J})| = 1$; otherwise, we call F *irresolute*.

An example of a resolute judgment aggregation procedure is the *majority rule* Majority, where Majority$(\boldsymbol{J}) = \{J^*\}$ and where $\varphi \in J^*$ if and only if φ occurs in the majority of judgment sets in $\boldsymbol{J}$, for all $\varphi \in [\Phi]$ (in case of a tie between φ and $\neg\varphi$, for $\varphi \in [\Phi]$, we arbitrarily let $\varphi \in J^*$). We call a judgment aggregation procedure F *complete* and Γ-*consistent*, if J is complete and Γ-consistent, respectively, for every $\boldsymbol{J} \in \mathcal{J}(\Phi, \Gamma)^+$ and every $J \in F(\boldsymbol{J})$.

As we have seen earlier in Sect. 11.1, the procedure Majority is not consistent. Consider the agenda Φ with $[\Phi] = \{p, q, p \wedge q\}$, and the profile $\boldsymbol{J} = (J_1, J_2, J_3)$, where $J_1 = \{\neg p, q, \neg(p \wedge q)\}$, $J_2 = \{p, \neg q, \neg(p \wedge q)\}$, and $J_3 = \{p, q, (p \wedge q)\}$. The unique outcome $\{p, q, \neg(p \wedge q)\}$ in Majority$(\boldsymbol{J})$ is inconsistent.

The *Kemeny judgment aggregation procedure* is based on a notion of distance. This distance is based on the Hamming distance $d(J, J') = |\{ \varphi \in [\Phi] : \varphi \in (J \setminus J') \cup (J' \setminus J) \}|$ between two complete judgment sets J, J'. Intuitively, the Hamming distance $d(J, J')$ counts the number of issues on which two judgment sets disagree. Let J be a single Γ-consistent and complete judgment set, and let $(J_1, \dots, J_p) = \boldsymbol{J} \in \mathcal{J}(\Phi, \Gamma)^+$ be a profile. We define the distance between J and $\boldsymbol{J}$ to be Dist$(J, \boldsymbol{J}) = \sum_{i \in [p]} d(J, J_i)$. Then, we let the outcome Kemeny$_{\Phi,\Gamma}(\boldsymbol{J})$ of the Kemeny rule be the set of those $J^* \in \mathcal{J}(\Phi, \Gamma)$ for which there is no $J \in \mathcal{J}(\Phi, \Gamma)$ such that Dist$(J, \boldsymbol{J}) < $ Dist$(J^*, \boldsymbol{J})$. If Φ and Γ are clear from the context, we often write Kemeny$(\boldsymbol{J})$ to denote Kemeny$_{\Phi,\Gamma}(\boldsymbol{J})$. Intuitively, the Kemeny rule selects those complete and Γ-consistent judgment sets that minimize the cumulative Hamming distance to the judgment sets in the profile. The Kemeny rule is irresolute, complete and Γ-consistent.

11.1.2 Constraint-Based Judgment Aggregation

We continue with explaining the framework of constraint-based judgment aggregation. For more details, we refer to the work of Grandi and Endriss [102, 103].

In this framework, a set of logically related issues is modeled by a number of propositional variables that are connected by an integrity constraint. Let $\mathcal{I} = \{x_1, \dots, x_n\}$ be a finite set of *issues* (in the form of propositional variables). Intuitively, these issues are the topics about which the individuals want to combine their judgments. A truth assignment $\alpha : \mathcal{I} \to \mathbb{B}$ is called a *ballot*, and represents an opinion that individuals and the group can have. We will also denote ballots α by a binary vector $(b_1, \dots, b_n) \in \mathbb{B}^n$, where $b_i = \alpha(x_i)$ for each $i \in [n]$. Moreover, we say that $(p_1, \dots, p_n) \in \{0, 1, \star\}^n$ is a *partial ballot*, and that $(p_1, \dots, p_n)$ *agrees with* a ballot $(b_1, \dots, b_n)$ if $p_i = b_i$ whenever $p_i \neq \star$, for all $i \in [n]$.

Additionally, we consider an *integrity constraint* Γ, that can be used to restrict the set of feasible opinions (for both the individuals and the group). The integrity constraint Γ is a satisfiable propositional formula on the variables $x_1, \ldots, x_n$. We define the set $\mathcal{R}(\mathcal{I}, \Gamma)$ of *rational ballots* to be the ballots (for $\mathcal{I}$) that satisfy the integrity constraint Γ. Rational ballots in the constraint-based judgment aggregation framework correspond to complete and Γ-consistent judgment sets in the formula-based judgment aggregation framework. We say that finite sequences $r \in \mathcal{R}(\mathcal{I}, \Gamma)^+$ of rational ballots are *profiles*, and where convenient we equate a profile $r = (r_1, \ldots, r_p)$ with the (multi)set $\{r_1, \ldots, r_p\}$.

A *judgment aggregation procedure* (or *rule*), for the set $\mathcal{I}$ of issues and the integrity constraint Γ, is a function F that takes as input a profile $r \in \mathcal{R}(\mathcal{I}, \Gamma)^+$, and that produces a non-empty set of ballots. We call a judgment aggregation procedure F *rational* (or *consistent*), if r is rational for every $r \in \mathcal{R}(\mathcal{I}, \Gamma)^+$ and every $r \in F(r)$.

As an example of a judgment aggregation procedure we consider the *majority rule* Majority, where Majority$(r) = \{(b_1, \ldots, b_n)\}$ and where each b_i agrees with the majority of the i-th bits in the ballots in r (in case of a tie, we arbitrarily let $b_i = 1$). To see that Majority is not rational, consider the set $\mathcal{I} = \{x_1, x_2, x_3\}$ of issues, the integrity constraint $\Gamma = x_3 \leftrightarrow (x_1 \rightarrow x_2)$, and the profile $r = (r_1, r_2, r_3)$, where $r_1 = (1, 1, 1)$, $r_2 = (1, 0, 0)$, and $r_3 = (0, 0, 1)$. The unique outcome $(1, 0, 1)$ in Majority(r) is not rational.

The *Kemeny aggregation procedure* is defined for the constraint-based judgment aggregation framework as follows. Similarly to the case for formula-based judgment aggregation, the Kemeny rule is based on the Hamming distance $d(r, r') = |\{i \in [n] : b_i \neq b_i'\}|$, between two rational ballots $r = (b_1, \ldots, b_n)$ and $r' = (b_1', \ldots, b_n')$ for the set $\mathcal{I}$ of issues and the integrity constraint Γ. Let r be a single ballot, and let $(r_1, \ldots, r_p) = r \in \mathcal{R}(\mathcal{I}, \Gamma)^+$ be a profile. We define the distance between r and r to be $\text{Dist}(r, r) = \sum_{i \in [p]} d(r, r_i)$. Then, we let the outcome Kemeny$_{\mathcal{I}, \Gamma}(r)$ of the Kemeny rule be the set of those ballots $r^* \in \mathcal{R}(\mathcal{I}, \Gamma)$ for which there is no $r \in \mathcal{R}(\mathcal{I}, \Gamma)$ such that $\text{Dist}(r, r) < \text{Dist}(r^*, r)$. If $\mathcal{I}$ and Γ are clear from the context, we often write Kemeny(r) to denote Kemeny$_{\mathcal{I}, \Gamma}(r)$. The Kemeny rule is irresolute and rational.

One can transform each agenda (together with an integrity constraint) in the formula-based judgment aggregation framework to an equivalent set of issues together with an integrity constraint in the constraint-based judgment aggregation framework. In the worst case, this translation leads to a blow-up in size. Vice versa, translating from the constraint-based framework to the formula-based framework is possible with only a polynomial size increase. However, this latter translation is not possible in polynomial time (unless P = NP). For more details on these issues, we refer to the work of Endriss, Grandi, De Haan, and Lang [74].

11.2 Agenda Safety for the Majority Rule

In this section, we will study the parameterized complexity of the problem of deciding whether a given agenda Φ is safe for the majority rule. We will only study this problem in the framework of formula-based judgment aggregation. Moreover, we will study this problem only in the setting where the integrity constraint is trivial, that is, where $\Gamma = \top$. In the remainder of this section, we will omit any mention of the integrity constraint Γ. We will consider a number of parameterizations for this problem.

Remember that an agenda Φ is *safe* for the majority rule if for any possible profile J for the agenda Φ, the majority outcome J^*, where Majority(J) = $\{J^*\}$, is consistent. Safety for the majority rule can be characterized using the following property. An agenda Φ satisfies the *median property (MP)* if every inconsistent subset of Φ has itself an inconsistent subset of size at most 2. An agenda Φ is safe for the majority rule if and only if Φ satisfies the median property [75, 165].

As an illustration of this characterization of agenda safety for the majority rule using the median property, consider the example of an agenda Φ that is not safe for the majority rule—that we discussed in Sect. 11.1—where $[\Phi] = \{p, q, p \wedge q\}$. We have that Φ does not satisfy the median property, because it contains the subset $\{p, q, \neg(p \wedge q)\} \subseteq \Phi$ that is inconsistent, but that itself contains no inconsistent subset of size 2.

Concretely, we study the following decision problem MAJORITY-SAFETY. This problem is Π_2^p-complete [75].

MAJORITY-SAFETY
Instance: An agenda Φ.
Question: Is Φ safe for the majority rule?

The parameterizations that we consider for this problem are both simple syntactic parameters that count the number of formulas or the maximum formula size, for instance, and more intricate parameters such as the treewidth of various graphs that model the structure of agendas. An overview of all parameterized variants that we consider, together with the parameterized complexity results that we obtain, can be found in Table 11.1.

11.2.1 CNF Formulas

Before we start with the parameterized complexity investigation of the problem MAJORITY-SAFETY, we show that we can restrict our attention to agendas containing only formulas in CNF. In particular, we show how to transform any agenda Φ to an agenda Φ', containing only formulas in CNF (and their negations), that is safe if and only if Φ is safe. Moreover, the formulas in Φ' are in one-to-one correspondence with the formulas in Φ, and any conjunction of a subset of Φ' is satisfiable if and only if the corresponding subset of Φ is satisfiable. For this, we will need the following lemma, whose proof is based on the well-known Tseitin transformation [188].

Lemma 11.1. *Let φ be a propositional formula. We can construct a CNF formula φ' such that $\mathrm{Var}(\varphi') \supseteq \mathrm{Var}(\varphi)$ and for each truth assignment $\alpha : \mathrm{Var}(\varphi) \to \mathbb{B}$ it holds that α satisfies φ if and only if there exists an assignment $\beta : (\mathrm{Var}(\varphi')\backslash\mathrm{Var}(\varphi)) \to \mathbb{B}$ such that the assignment $\alpha \cup \beta$ satisfies φ'.*

Proof. Assume without loss of generality that φ contains only the connectives $\wedge$ and $\neg$. Let $\mathrm{Sub}(\varphi)$ denote the set of all subformulas of φ. We let $\mathrm{Var}(\varphi') = \mathrm{Var}(\varphi) \cup \{ z_\chi : \chi \in \mathrm{Sub}(\varphi) \}$, where each z_χ is a fresh variable. We then define φ' to be the formula $\chi_\varphi \wedge \bigwedge_{\chi \in \mathrm{Sub}(\varphi)} \sigma(\chi)$, where we define the formulas $\sigma(\chi)$, for each $\chi \in \mathrm{Sub}(\varphi)$ as follows. If $\chi = l$ is a literal, we let $\sigma(\chi) = (z_l \to l) \wedge (l \to z_l)$; if $\chi = \neg\chi'$, we let $\sigma(\chi) = (z_\chi \to \neg z_{\chi'}) \wedge (z_{\chi'} \to \neg z_\chi)$; and if $\chi = \chi_1 \wedge \chi_2$, we let $\sigma(\chi) = (z_\chi \to z_{\chi_1}) \wedge (z_\chi \to z_{\chi_2}) \wedge (\neg z_{\chi_1} \vee \neg z_{\chi_2} \to \neg z_\chi)$. Let $\alpha : \mathrm{Var}(\varphi) \to \mathbb{B}$ be an arbitrary truth assignment. We claim that α satisfies φ if and only if there exists an assignment $\beta : (\mathrm{Var}(\varphi')\backslash\mathrm{Var}(\varphi)) \to \mathbb{B}$ such that $\alpha \cup \beta$ satisfies φ'. Define the assignment β' as follows. For each $\chi \in \mathrm{Sub}(\varphi)$, we let $\beta'(z_\chi) = 1$ if and only if α satisfies χ. Clearly, if α satisfies φ, then $\alpha \cup \beta'$ satisfies φ'. Conversely, for any assignment $\beta : (\mathrm{Var}(\varphi')\backslash\mathrm{Var}(\varphi)) \to \mathbb{B}$ that does not coincide with β', clearly, the assignment $\alpha \cup \beta$ does not satisfy some clause of φ'. Moreover, if $\alpha \cup \beta'$ satisfies φ', then α satisfies φ. $\square$

We can now show how to transform any agenda Φ in polynomial time to an agenda Φ', containing only formulas in CNF (and their negations), that is safe if and only if Φ is safe.

Proposition 11.2. *Let Φ be an agenda with $[\Phi] = \{\varphi_1, \ldots, \varphi_n\}$. We can construct in polynomial time an agenda Φ' with $[\Phi'] = \{\varphi_1', \ldots, \varphi_n'\}$ such that each φ_i' is in CNF and any subset $\Psi = \{\varphi_{i_1}, \ldots, \varphi_{i_{m_1}}, \neg\varphi_{j_1}, \ldots, \neg\varphi_{j_{m_2}}\}$ of Φ is consistent if and only if $\Psi' = \{\varphi_{i_1}', \ldots, \varphi_{i_m}', \neg\varphi_{j_1}', \ldots, \neg\varphi_{j_{m_2}}'\}$ is consistent.*

Proof (Proof). Let Φ be an agenda with $[\Phi] = \{\varphi_1, \ldots, \varphi_n\}$. By Lemma 11.1, we can transform each φ_i in linear time to a CNF formula φ_i' such that $\mathrm{Var}(\varphi_i') \supseteq \mathrm{Var}(\varphi_i)$ and for each truth assignment $\alpha : \mathrm{Var}(\varphi_i) \to \mathbb{B}$ we have that α satisfies φ_i if and only if there exists an assignment $\beta : (\mathrm{Var}(\varphi_i')\backslash\mathrm{Var}(\varphi_i)) \to \mathbb{B}$ such that the assignment $\alpha \cup \beta$ satisfies φ_i'. Because we can introduce fresh variables for constructing each φ_i', we can assume without loss of generality that for each $i, i' \in [n]$ with $i < i'$ it is the case that $(\mathrm{Var}(\varphi_i')\backslash\mathrm{Var}(\varphi_i)) \cap (\mathrm{Var}(\varphi_{i'}')\backslash\mathrm{Var}(\varphi_{i'})) = \emptyset$. Let $\Psi = \{\varphi_{i_1}, \ldots, \varphi_{i_{m_1}}, \neg\varphi_{j_1}, \ldots, \neg\varphi_{j_{m_2}}\}$ be an arbitrary subset of Φ. We show that Ψ is consistent if and only if $\Psi' = \{\varphi_{i_1}', \ldots, \varphi_{i_{m_1}}', \neg\varphi_{j_1}', \ldots, \neg\varphi_{j_{m_2}}'\}$ is consistent.

($\Rightarrow$) Let $\alpha : \mathrm{Var}(\Psi) \to \mathbb{B}$ be an assignment that satisfies all formulas in Ψ. By construction of the formulas φ_i', by Lemma 11.1, and by the fact that for each $i, i' \in [n]$ with $i < i'$ it is the case that $(\mathrm{Var}(\varphi_i')\backslash\mathrm{Var}(\varphi_i)) \cap (\mathrm{Var}(\varphi_{i'}')\backslash\mathrm{Var}(\varphi_{i'})) = \emptyset$, we know that there exists an assignment $\beta : (\mathrm{Var}(\Psi')\backslash\mathrm{Var}(\Psi)) \to \mathbb{B}$ such that $\alpha \cup \beta$ satisfies all formulas in Ψ.

($\Leftarrow$) Conversely, assume that there exists an assignment $\alpha : \text{Var}(\Psi') \to \mathbb{B}$ that satisfies all formulas in Ψ'. Then, by construction of the formulas φ_i', we know that $\text{Var}(\Psi') \subseteq \text{Var}(\Psi)$. Now, by Lemma 11.1, we know that α satisfies all formulas in Ψ as well.

Intuitively, the above result shows that, using additional auxiliary variables, each agenda can be rewritten into another agenda that contains only formulas in CNF (or their negation) that are equivalent (with respect to satisfiability) to the formulas in the original agenda.

11.2.2 Syntactic Restrictions

We begin with considering a number of parameterizations for the problem MAJORITY-SAFETY that impose syntactic restrictions on the agenda, that is, these parameters measure various aspects that can be directly read off from the syntactic representation of the agenda. Concretely, we parameterize by the size of formulas $\varphi \in \Phi$, by the maximum number of times that any variable occurs in Φ (the degree of Φ), and by the number of formulas occurring in Φ. Formally, we consider the following parameterized problems:

- MAJORITY-SAFETY(formula-size), where the parameter is $\ell = \max\{ |\varphi| : \varphi \in \Phi \}$;
- MAJORITY-SAFETY(degree), where the parameter is the degree d of Φ;
- MAJORITY-SAFETY(degree + formula size), where the parameter is $\ell + d$; and
- MAJORITY-SAFETY(agenda-size), where the parameter is $|\Phi|$.

Here we define the *degree* of an agenda Φ to be the maximum number of times that any variable $x \in \text{Var}(\Phi)$ occurs in $[\Phi]$, i.e., $\max_{x \in \text{Var}(\Phi)}(\sum_{\varphi \in [\Phi]} \text{occ}(x, \varphi))$, where $\text{occ}(x, \varphi)$ denotes the number of times that x occurs in φ.

These parameters can be assumed to be small in many natural settings. The assumption that the size of formulas in an agenda is small corresponds to the expectation that the separate statements that the individuals are judging are in a sense atomic, and therefore of bounded size. The assumption that the degree of an agenda is small corresponds to the expectation that each proposition that occurs in the statements to be judged occurs only a small number of times. The assumption that the number of formulas in the agenda is small is based on the fact that the individuals need to form an opinion on all formulas in the agenda.

Agendas with Small Formulas and Small Degree

We start by showing that parameterizing by (the sum of) the maximum formula size and the degree of the agenda Φ does not decrease the complexity of deciding whether the agenda is safe, even when the pre-agenda associated to Φ contains only formulas in 2CNF $\cap$ HORN. Intuitively, these restrictions on the form and size of the formulas in the agenda do not rule out the complex interactions between the formulas in the agenda that involve many formulas simultaneously, and that give rise to the Π_2^p-hardness of the problem.

Proposition 11.3. MAJORITY-SAFETY(formula-size) *is* para-Π_2^p-*complete*.

Proof. Membership in para-Π_2^p follows from the Π_2^p-membership of MAJORITY-SAFETY. We show para-Π_2^p-hardness by giving a polynomial-time reduction from co-QSAT$_2$(3CNF) to the problem $\{x : (x, c) \in$ MAJORITY-SAFETY(formula-size) $\}$, where c is bounded by the size of formulas of the form $\neg((\neg x_1 \vee \neg x_2 \vee \neg x_3) \wedge \neg z)$. This reduction is a modified variant of a reduction given by Endriss et al. [75, Lemma 11]. Let $\varphi = \forall X.\exists Y.\psi$ be an instance of co-QSAT$_2$, where $\psi = c_1 \wedge \cdots \wedge c_m$ is in 3CNF, and where $X = \{x_1, \ldots, x_m\}$. We may assume without loss of generality that none of the c_i is a unit clause. We construct the agenda $\Phi = \{x_1, \neg x_1, \ldots, x_n, \neg x_n, (c_1 \wedge \neg z_1), \neg(c_1 \wedge \neg z_1), \ldots, (c_m \wedge \neg z_m), \neg(c_m \wedge \neg z_m)\}$, where $Z = \{z_1, \ldots, z_m\}$ is a set of fresh variables. We show that Φ satisfies the median property if and only if φ is true.

($\Rightarrow$) Suppose that φ is false, i.e., there exists some $\alpha : X \to \mathbb{B}$ such that $\forall Y.\neg\psi[\alpha]$ is true. Let $L = \{x_i : i \in [n], \alpha(x_i) = 1\} \cup \{\neg x_i : i \in [n], \alpha(x_i) = 0\}$. We know that α is the unique assignment to the variables in X that satisfies L. Now consider $\Phi' = L \cup \{(c_1 \wedge z_1), \ldots, (c_m \wedge z_m)\}$.

We firstly show that Φ' is inconsistent. We proceed indirectly and assume that Φ' is consistent, i.e., there exists an assignment $\beta : Y \cup Z \to \mathbb{B}$ such that $\alpha \cup \beta$ satisfies Φ'. Then $\alpha \cup \beta$ must satisfy each c_i. Therefore, β satisfies $\psi[\alpha]$, which contradicts our assumption that $\forall Y.\neg\psi[\alpha]$ is true. Therefore, we can conclude that Φ' is inconsistent.

Next, we show that each subset $\Phi'' \subseteq \Phi'$ of size 2 is consistent. Let $\Phi'' \subseteq \Phi'$ be an arbitrary subset of size 2. We distinguish three cases: either (i) $\Phi'' = \{l_i, l_j\}$ for some $i, j \in [n]$ with $i < j$; (ii) $\Phi'' = \{l_i, (c_j \wedge \neg z_j)\}$ for some $i \in [n]$ and some $j \in [m]$; or (iii) $\Phi'' = \{(c_i \wedge \neg z_i), (c_j \wedge \neg z_j)\}$ for some $i, j \in [m]$ with $i < j$. In case (i), clearly Φ'' is consistent. In case (ii) and (iii), Φ'' is consistent because c_i and c_j are not unit clauses.

($\Leftarrow$) Conversely, suppose that Φ does not satisfy the median property, i.e., there exists an inconsistent subset $\Phi' \subseteq \Phi$ that itself does not contain an inconsistent subset of size 2. We show that φ is false. Firstly, we show that $\Psi' = \Phi' \backslash \{\neg(c_1 \wedge \neg z_1), \ldots, \neg(c_m \wedge \neg z_m)\}$ is inconsistent. We proceed indirectly, and assume that Ψ' is consistent, i.e., there exists an assignment $\gamma : \text{Var}(\Psi') \to \mathbb{B}$ such that γ satisfies Ψ'. Now let $Z' = \{z_i : i \in [m], \neg(c_i \wedge \neg z_i) \in \Phi'\}$ and let $\gamma' : Z' \to \mathbb{B}$ be defined by letting $\gamma'(z) = 0$ for all $z \in Z'$. Since Ψ' contains no negated pairs of formulas, we know that $Z' \cap \text{Var}(\Psi') = \emptyset$. Then the assignment $\gamma \cup \gamma'$ satisfies Φ', since γ satisfies all $\psi \in \Psi'$ and γ' satisfies all $\varphi \in \Phi' \backslash \Psi'$. This is a contradiction with our assumption that Φ' is inconsistent, so we can conclude that Ψ' is inconsistent.

Now let the assignment $\alpha : X \to \mathbb{B}$ be defined as follows. For each $x \in X$, we let $\alpha(x) = 1$ if $x \in \Psi'$, we let $\alpha(x) = 0$ if $\neg x \in \Psi'$, and we (arbitrarily) define $\alpha(x) = 1$ otherwise. We now show that $\neg\exists Y.\psi[\alpha]$ is true. We proceed indirectly, and assume that there exists an assignment $\beta : Y \to \mathbb{B}$ such that $\psi[\alpha \cup \beta]$ is true.

Now consider the assignment $\gamma : Z \to \mathbb{B}$ such that $\gamma(z) = 0$ for all $z \in Z$. We claim that the assignment $\alpha \cup \beta \cup \gamma$ satisfies Ψ'. Let $\chi \in \Psi'$ be an arbitrary formula. We distinguish two cases: either (i) $\chi \in \{x_i, \neg x_i\}$ for some $i \in [n]$; or (ii) $\chi = (c_i \wedge \neg z_i)$ for some $i \in [m]$. In case (i), we know that α satisfies χ. For case (ii), we know that $\alpha \cup \beta$ satisfies c_i, since $\alpha \cup \beta$ satisfies ψ. Moreover, we know that γ satisfies $\neg z_i$. Therefore, $\alpha \cup \beta \cup \gamma$ satisfies χ. This is a contradiction with our previous conclusion that Ψ' is inconsistent, so we can conclude that $\neg \exists Y.\psi[\alpha]$ is true. From this, we know that $\forall X.\exists Y.\psi$ is false.

Next, using the following technical lemma and the reduction given in the proof of Proposition 11.3, we get para-Π_2^p-completeness of MAJORITY-SAFETY(degree + formula size). The hardness result holds even when we restrict the formulas to be in HORN $\cap$ 2CNF.

Lemma 11.4. *The problem* co-QSAT$_2$(3CNF) *is* Π_2^p-*hard even when restricted to instances* $\varphi = \forall X.\exists Y.\psi$ *where each* $x \in X$ *occurs at most 2 times in* ψ *and each* $y \in Y$ *occurs at most 3 times in* ψ.

Proof. Let $\varphi = \forall X.\exists Y.\psi$ be an instance of co-QSAT$_2$(3CNF). We construct in polynomial time an equivalent instance $\varphi' = \forall X'.\exists Y'.\psi'$ of co-QSAT$_2$(3CNF) such that each $x \in X'$ occurs at most 2 times in ψ' and each $y \in Y'$ occurs at most 3 times in ψ'.

Firstly, we construct an equivalent formula $\varphi_1 = \forall X.\exists Y_1.\psi_1$ such that each $x \in X_1$ occurs at most 2 times in ψ_1. We do this by repeatedly applying the following transformation. Let $z \in X$ be any variable that occurs $m > 3$ times in ψ. We create m copies $z_1, \ldots, z_m$ of z, that we add to the set Y of existentially quantified variables. We replace each occurrence of z in ψ by a distinct copy z_i. Finally, we ensure equivalence of ψ_1 and ψ by letting $\psi_1 = \psi \wedge \psi_{\text{equiv}}^z$, where we define ψ_{equiv}^z to be the conjunction of binary clauses $(z_i \to z_{i+1})$ for each $i \in [m-1]$, the binary clause $(z_m \to z_1)$, and the binary clauses $(z \to z_1)$ and $(z_1 \to z)$. Repeated application of this transformation results in a formula φ_1 that satisfies the required properties.

Then, we transform φ_1 into an equivalent formula $\varphi_2 = \forall X.\exists Y_2.\psi_2$ such that each $y \in Y_2$ occurs at most 3 times in ψ_2. Moreover, each $x \in X$ occurs as many times in ψ_2 as it did in ψ_1 (i.e., twice). We use a similar strategy as we did in the first phase: we repeatedly apply the following transformation. Let $y \in Y_1$ be any variable that occurs $m > 3$ times in ψ_1. We create m copies $y_1, \ldots, y_m$ of y, that we add to the set Y_1 of existentially quantified variables. Then we replace each occurrence of y in ψ by a distinct copy y_i. Finally, we ensure equivalence of ψ_2 and ψ_1 by letting $\psi_2 = \psi_{\text{equiv}}^y \wedge \psi_1$, where we define ψ_{equiv}^y to the conjunction of the binary clauses $(y_i \to y_{i+1})$ for all $i \in [m-1]$ and the binary clause $(y_m \to y_1)$. Again, repeated application of this transformation results in a formula φ_2 that satisfies the required properties.

We now show that MAJORITY-SAFETY(degree + formula size) is para-Π_2^p-hard, even when we restrict the formulas to be in HORN $\cap$ 2CNF.

Proposition 11.5. MAJORITY-SAFETY(degree + formula size) *is para-*Π_2^p-*hard even when restricted to agendas* Φ *such that all formulas* $\varphi \in [\Phi]$ *are in* HORN $\cap$ 2CNF.

Proof. We consider the reduction used to show Proposition 11.3. The agenda Φ that we constructed contains only formulas of the form x_i or their negation, and formulas of the form $(c_i \wedge \neg z_i)$, where c_i is a clause, or their negation. Clearly, the formulas x_i and $\neg x_i$ are (equivalent to formulas) in HORN $\cap$ 2CNF. It suffices to show that each formula $\varphi \in \Phi$ with $\varphi = (c_i \wedge \neg z_i)$ is equivalent to a formula $\varphi' \in$ HORN $\cap$ 2CNF. Let $c_i = (l_1^i \vee l_2^i \vee l_3^i)$. Observe that $(c_i \wedge \neg z_i) = ((l_1^i \vee l_2^i \vee l_3^i) \wedge \neg z_i) \equiv (l_1^i \vee \neg z_i) \wedge (l_2^i \vee \neg z_i) \wedge (l_3^i \vee \neg z_i)$. Thus, we can construct Φ in such a way that $[\Phi]$ contains only formulas in HORN $\cap$ 2CNF.

Agendas with Few Formulas

Next, we parameterize the problem by the number of formulas occurring in the agenda. We will show that the problem MAJORITY-SAFETY(agenda-size) can be solved by an fpt-algorithm that uses $f(k)$ SAT queries, where k denotes the parameter value. That is, the problem is in the class FPT^{NP}[few]. Intuitively, the fpt-algorithm that we construct will exploit the fact that the agenda only contains few formulas, by considering all possible inconsistent subsets of the agenda, and using a SAT solver to verify that these all have an inconsistent subset of size at most 2.

Proposition 11.6. *There exists an algorithm that decides* MAJORITY-SAFETY *(agenda-size) in fpt-time using at most* $2^{O(k)}$ *queries to a SAT oracle, where k is the parameter value. That is,* MAJORITY-SAFETY*(agenda-size) is in* FPT^{NP}[few].

Proof. The algorithm that we described in Sect. 4.3 solves the problem in fixed-parameter tractable time and uses at most $2^{O(k)}$ queries to a SAT oracle.

Moreover, we give evidence that this is the best that one can do, i.e., that there exists no fpt-algorithm that uses a significantly smaller number of SAT queries, under a common complexity-theoretic assumption. We show that MAJORITY-SAFETY(agenda-size) is complete for the class FPT^{NP}[few]. We start with identifying an easier hardness result, which we will then extend to hardness for the class FPT^{NP}[few].

Lemma 11.7. MAJORITY-SAFETY*(agenda-size) is* para-*co-DP-hard.*

Proof. We prove hardness for para-co-DP by giving a polynomial-time reduction from SAT-UNSAT to co-MAJORITY-SAFETY, such that the resulting instance is an agenda of constant size. Let (φ_1, φ_2) be an instance of SAT-UNSAT. We construct the agenda Φ with $[\Phi] = \{\psi_1, \psi_2, \psi_3\}$ by letting $\psi_1 = r_1 \wedge p_1 \wedge \varphi_1$, $\psi_2 = r_2 \wedge p_2$, and $\psi_3 = r_3 \wedge ((p_1 \wedge p_2) \rightarrow \varphi_2)$, where $\{r_1, r_2, r_3, p_1, p_2\}$ are distinct fresh variables not occurring in φ_1 nor in φ_2. We show that Φ does not satisfy the MP if and only if $(\varphi_1, \varphi_2) \in$ SAT-UNSAT.

($\Rightarrow$) Assume that Φ does not satisfy the MP. Then there exists a satisfiable complement-free subagenda $\Phi' \subseteq \Phi$ such that each subset $\Phi'' \subseteq \Phi'$ of size 2 is satisfiable. We distinguish several cases: either (i) $\Phi' = [\Phi] = \{\psi_1, \psi_2, \psi_3\}$, or (ii) the above case does not hold and Φ' contains ψ_1, or (iii) the above two cases do not hold.

We show that in case (i) we can conclude that $(\varphi_1, \varphi_2) \in$ SAT-UNSAT. By assumption, every subset $\Phi'' \subseteq \Phi$ of size 2 is satisfiable. Therefore, we can conclude that the formula ψ_1 is satisfiable. Hence, φ_1 is satisfiable. Next, we show that φ_2 is unsatisfiable. We proceed indirectly, and we assume that there exists some assignment $\alpha : \text{Var}(\varphi_2) \to \mathbb{B}$ that satisfies φ_2. We construct a satisfying assignment $\alpha' : \text{Var}(\Phi) \to \mathbb{B}$ for Φ, which leads to a contradiction. We let α' coincide with α on the variables in $\text{Var}(\varphi_2)$. Moreover, we know that there exists some satisfying assignment $\beta : \text{Var}(\varphi_1) \to \mathbb{B}$ for φ_1. We let α' coincide with β on the variables in $\text{Var}(\varphi_1)$. Finally, we let $\alpha'(x) = 1$ for each $x \in \{r_1, r_2, r_3, p_1, p_2\}$. Clearly, α' satisfies all formulas in Φ then. This leads to a contradiction with the fact that Φ is unsatisfiable, and therefore we can conclude that φ_2 is unsatisfiable.

Next, we show that case (ii) cannot occur. We know that $\psi_1 \in \Phi'$, and that each subset $\Phi'' \subseteq \Phi$ of size 2 is satisfiable. Therefore, we know that φ_1 is satisfiable. Let $\beta : \text{Var}(\varphi_1) \to \mathbb{B}$ be a satisfying assignment for φ_1. We extend the assignment β to an assignment $\beta' : \text{Var}(\Phi) \to \mathbb{B}$ that satisfies Φ'. We let $\beta'(r_1) = \beta'(p_1) = 1$. If $\psi_2 \in \Phi$, we let $\beta'(r_2) = \beta'(p_2) = 1$; otherwise, if $\neg\psi_2 \in \Phi$, we let $\beta'(r_2) = 0$. If $\psi_3 \in \Phi$, we let $\beta'(r_3) = 1$ and $\beta'(p_2) = 0$; otherwise, if $\neg\psi_3 \in \Phi$, we let $\beta'(r_3) = 0$. On the other variables, we let β' be defined arbitrarily. Since not both $\psi_2 \in \Phi$ and $\psi_3 \in \Phi$, we know that β' is well-defined. It is easy to verify that β' satisfies Φ', which is a contradiction with our assumption that Φ' is unsatisfiable. From this we can conclude that case (ii) cannot occur.

Finally, we show that case (iii) cannot occur either. We construct an assignment $\beta : \text{Var}(\Phi) \to \mathbb{B}$ that satisfies Φ'. We know that $\neg\psi_1 \in \Phi'$. Let $\beta(r_1) = \beta(p_1) = 0$. If $\psi_2 \in \Phi'$, we let $\beta(r_2) = \beta(p_2) = 1$; otherwise, if $\neg\psi_2 \in \Phi'$, we let $\beta(r_2) = 0$; If $\psi_3 \in \Phi'$, we let $\beta(r_3) = 1$; otherwise, if $\neg\psi_3 \in \Phi'$, we let $\beta(r_3) = 0$. It is easy to verify that β satisfies Ψ, which is a contradiction with our assumption that Φ' is unsatisfiable. From this we can conclude that case (iii) cannot occur.

($\Leftarrow$) Conversely, assume that φ_1 is satisfiable and that φ_2 is unsatisfiable. Then consider the complement-free subagenda $\Phi' \subseteq \Phi$ given by $\Phi' = [\Phi] = \{\psi_1, \psi_2, \psi_3\}$. Since $\psi_1, \psi_2 \models p_1 \wedge p_2$ and φ_2 is unsatisfiable, we get that Φ' is unsatisfiable. However, since φ_1 is satisfiable, we get that each subset of Φ' of size 2 is satisfiable. Therefore, Φ does not satisfy the MP. $\square$

Proposition 11.8. Majority-Safety(agenda-size) *is* $\text{FPT}^{\text{NP}}[\text{few}]$-*hard.*

Proof. We give an fpt-reduction from BH(level)-Sat to co-Majority-Safety (agenda-size). Since co-BH(level)-Sat is $\text{FPT}^{\text{NP}}[\text{few}]$-complete, this suffices. Without loss of generality, we assume that $k \geq 2$ is even. Let the sequence $(\varphi_1, \ldots, \varphi_k)$ specify an instance of BH(level)-Sat. We know that we can construct in polynomial time a sequence of formulas $(\varphi_1, \psi_1, \ldots, \varphi_\ell, \psi_\ell)$, where $\ell = k/2$, such that $(\varphi_1, \ldots, \varphi_k) \in \text{BH}_k\text{-Sat}$ if and only if for some $i \in [\ell]$ it holds that $(\chi_i, \psi_i) \in$ SAT-UNSAT [35].

Now, for each $i \in [\ell]$, we can use the reduction in the proof of Lemma 11.7 to construct in polynomial time an agenda Φ_i of constant size such that Φ_i does not satisfy the median property if and only if $(\chi_i, \psi_i) \in$ SAT-UNSAT. Moreover, we can ensure that the agendas Φ_i are variable-disjoint. We now construct the agenda $\Phi = \bigcup_{i \in [\ell]} \Phi_i$. Below, we show that Φ does not satisfy the median property if and only if $(\chi_i, \psi_i) \in$ SAT-UNSAT for some $i \in [\ell]$. We know this latter condition holds if and only if our original instance $(\varphi_1, \ldots, \varphi_k) \in$ BH$_k$-SAT. Moreover, since $|\Phi| = O(k)$, we obtain a correct fpt-reduction.

All that remains is to show that Φ does not satisfy the median property if and only if $(\chi_i, \psi_i) \in$ SAT-UNSAT for some $i \in [\ell]$.

($\Rightarrow$) Assume that Φ does not satisfy the median property. Then there exists a subset $\Phi' \subseteq \Phi$ that is unsatisfiable such that each $\Phi'' \subseteq \Phi'$ of size 2 is satisfiable. Moreover, we can assume Φ' to be minimal with this property. Since Φ is partitioned into the variable disjoint subsets Φ_i, and since Φ' is minimal, we know that $\Phi' \subseteq \Phi_i$, for some $i \in [\ell]$. Then Φ_i does not satisfy the median property, from which we can conclude that $(\chi_i, \psi_i) \in$ SAT-UNSAT.

($\Leftarrow$) Conversely, assume that $(\chi_i, \psi_i) \in$ SAT-UNSAT for some $i \in [\ell]$. Then by construction of Φ_i, we know that Φ_i does not satisfy the median property. Therefore, since $\Phi_i \subseteq \Phi$, we know that Φ does not satisfy the median property.

By Proposition 7.12, we know that MAJORITY-SAFETY(agenda-size) is not solvable in fixed-parameter tractable time using $O(1)$ queries to a SAT oracle, unless the Polynomial Hierarchy collapses. This lower bound holds for any FPT^{NP}[few]-hard problem. For the particular case of MAJORITY-SAFETY(agenda-size), we can improve this bound from $\omega(1)$ to $\Omega(\log k)$.

Proposition 11.9. MAJORITY-SAFETY(agenda-size) *is not solvable by an fpt-algorithm that uses* $o(\log k)$ *SAT queries, where k denotes the parameter value, unless the PH collapses.*

Proof. The proof is analogous to the proof of Proposition 7.12. Suppose that MAJORITY-SAFETY(agenda-size) is solvable by an algorithm that runs in fixed-parameter tractable time and that uses $h(k) = o(\log k)$ SAT queries. We show that the BH collapses, and thus that consequently the PH collapses. By the proof of Proposition 11.8, we know that BH(level)-SAT can be fpt-reduced to the problem MAJORITY-SAFETY(agenda-size) in such a way that the parameter value k increases at most linearly to $h'(k) = O(k)$. Moreover, the proof of Proposition 11.6 can be straightforwardly modified to show that MAJORITY-SAFETY(agenda-size) can be fpt-reduced to BH(level)-SAT in such a way that the resulting parameter value k' is bounded by a function $h''(k) = 2^{O(k)}$, where k is the original parameter value. We can now combine these fpt-reductions to obtain a polynomial-time reduction that witnesses the collapse of the BH. We know that there exists some integer ℓ such that $h''(h'(h(\ell))) = \ell' < \ell$. Applying the composing the fpt-reductions gives us a polynomial-time reduction from the problem BH$_\ell$-SAT to the problem BH$_{\ell'}$-SAT. Since $\ell' < \ell$, this shows that the BH collapses to the ℓ'-th level. Since a collapse of the BH implies a collapse of the PH [41, 126], we can conclude that MAJORITY-SAFETY(agenda-size) is not solvable by an fpt-algorithm that uses $o(\log k)$ SAT queries, unless the PH collapses.

11.2.3 Bounded Treewidth

In the previous section, we considered various parameters for the problem MAJORITY-SAFETY that capture structure in the problem input that can easily be read off from the syntactic representation of the problem input. A more intricate type of structure that the agenda Φ can exhibit is the way in which the formulas $\varphi \in \Phi$ interact with each other. As an extreme example, consider the case of an agenda Φ with $[\Phi] = \{\varphi_1, \ldots, \varphi_m\}$, where all formulas φ_i are variable-disjoint. Clearly, any minimal inconsistent subset of this agenda has size 1, and thus this agenda is safe for the majority rule. In less extreme cases, the formulas of the agenda are allowed to interact (i.e., to have variables in common), but their interaction is structured in a particular way. The type of structured interaction that we consider in this section is the extent to which various graphs representing the interaction between formulas of the agenda resemble a tree—this is captured by the treewidth of these graphs. Treewidth is commonly used in the parameterized complexity analysis of hard problems in various fields. Intuitively, one could think of agendas of bounded treewidth as agendas where the propositional variables are divided into a number of groups, each containing a small number of variables, where the interaction between such groups is tree-like.

Let Φ be an agenda with $[\Phi] = \{\varphi_1, \ldots, \varphi_m\}$, where each φ_i is a CNF formula. We define the following graphs that are intended to capture the interaction between formulas in Φ. The *formula primal graph* $\mathcal{G}^{\text{fp}}(\Phi)$ of Φ has as vertices the variables Var(Φ) occurring in the agenda, and two variables are connected by an edge if there exists a formula φ_i in which they both occur. The *formula incidence graph* $\mathcal{G}^{\text{fi}}(\Phi)$ of Φ is a bipartite graph whose vertices consist of (1) the variables Var(Φ) occurring in the agenda and (2) the formulas $\varphi_i \in [\Phi]$. A variable $x \in$ Var(Φ) is connected by an edge with a formula $\varphi_i \in [\Phi]$ if x occurs in φ_i, i.e., $x \in$ Var(φ_i). The *clausal primal graph* $\mathcal{G}^{\text{cp}}(\Phi)$ of Φ has as vertices the variables Var(Φ) occurring in the agenda, and two variables are connected by an edge if there exists a formula φ_i and a clause $c \in \varphi_i$ in which they both occur. The *clausal incidence graph* $\mathcal{G}^{\text{ci}}(\Phi)$ of Φ is a bipartite graph whose vertices consist of (1) the variables Var(Φ) occurring in the agenda and (2) the clauses c occurring in formulas $\varphi_i \in [\Phi]$. A variable $x \in$ Var(Φ) is connected by an edge with a clause c of the formula $\varphi_i \in [\Phi]$ if x occurs in c, i.e., $x \in$ Var(c).

We consider the following parameterized variants of the problem MAJORITY-SAFETY:

- MAJORITY-SAFETY(f-tw), where the parameter is the treewidth of the formula primal graph (the *formula primal treewidth*);
- MAJORITY-SAFETY(c-tw), where the parameter is the treewidth of the clausal primal graph (the *clausal primal treewidth*);
- MAJORITY-SAFETY(f-tw*) where the parameter is the treewidth of the formula incidence graph (the *formula incidence treewidth*); and

- MAJORITY-SAFETY(c-tw*) where the parameter is the treewidth of the clausal incidence graph (the *clausal incidence treewidth*).

We show that the presence of tree-like structure in only one of these four graphs leads to a reduction in the complexity of the problem MAJORITY-SAFETY. We show that MAJORITY-SAFETY(f-tw) is fixed-parameter tractable, and for the remaining three parameters, the problem is para-Π_2^p-complete.

We begin by showing that the problem is fixed-parameter tractable when parameterized by the formula primal treewidth.

Proposition 11.10. MAJORITY-SAFETY(f-tw) *is fixed-parameter tractable.*

Proof. We will use Courcelle's Theorem, which states that checking whether a relational structure $\mathcal{A}$ satisfies a monadic second-order logic (MSOL) sentence φ is fixed-parameter tractable, parameterized by the treewidth of the Gaifman graph of $\mathcal{A}$ plus the size of φ (cf. [56, 85]). The Gaifman graph of $\mathcal{A}$ has as vertices all elements in the universe of $\mathcal{A}$, and two elements a, b are connected with an edge if they occur together in some tuple in the interpretation $R^{\mathcal{A}}$ of some relation symbol R.

Let Φ be an instance of MAJORITY-SAFETY, where $[\Phi] = \{\varphi_1, \ldots, \varphi_m\}$ and each φ_i is a CNF formula, that has formula primal treewidth k. That is, there is a tree decomposition of the formula primal graph of Φ of width $k + 1$. We construct a relational structure $\mathcal{A} = (A, \cdot^{\mathcal{A}})$ and a (fixed) MSOL sentence φ, such that $\mathcal{A} \models \varphi$ if and only if $\Phi \in$ MAJORITY-SAFETY. We let $A = \Phi \cup \mathrm{Var}(\Phi) \cup \{c \in \varphi_i : i \in [m]\}$. Moreover, we introduce unary relation symbols F, V, C and binary relation symbols I^+, I^-, D. We let:

$$F^{\mathcal{A}} = \Phi;$$
$$V^{\mathcal{A}} = \mathrm{Var}(\Phi);$$
$$C^{\mathcal{A}} = \{c \in \varphi_i : i \in [m]\};$$
$$(I^+)^{\mathcal{A}} = \{(c, x) : c \in \varphi_i, i \in [m], x \text{ occurs pos. in } c\};$$
$$(I^-)^{\mathcal{A}} = \{(c, x) : c \in \varphi_i, i \in [m], x \text{ occurs neg. in } c\}; \text{ and}$$
$$D = \{(\varphi_i, c) : i \in [m], c \in \varphi_i\}.$$

We can transform a tree decomposition T of width $k + 1$ for the formula primal graph of Φ into a tree decomposition T' of the Gaifman graph of $\mathcal{A}$ of width $k + 3$. Because all variables occurring in any formula $\varphi_i \in \Phi$ form a clique in the formula primal graph, they must occur in some bag of T, we can extend this bag to a subtree where all edges between φ_i, all clauses $c \in \varphi_i$ and the variables in $\mathrm{Var}(\varphi_i)$ are covered as well. This can be done in such a way that T' has width $k + 3$.

We then use the following MSOL sentence φ (that does not depend on Φ), where:

$$\varphi = \neg \exists P_1 \subseteq F. \exists P_2 \subseteq F.$$
$$[\forall p \in P_1. \neg P_2(p) \wedge (|P_1 \cup P_2| \geq 3) \wedge$$
$$\neg \varphi_{\text{sat}}(P_1, P_2) \wedge \varphi_{\min}(P_1, P_2);$$
$$\varphi_{\text{sat}}(P_1, P_2) = \exists S. [\forall p \in P_1. \forall c. [C(c) \wedge D(p, c)] \rightarrow$$
$$[\exists s. (S(s) \wedge I^+(c, s)) \vee (\neg S(s) \wedge I^-(c, s))]] \wedge$$
$$[\forall p \in P_2. \exists c. [C(c) \wedge D(p, c) \wedge \forall s.$$
$$(I^+(c, s) \rightarrow \neg S(s)) \wedge (I^-(c, s) \rightarrow S(s))]]; \text{ and}$$
$$\varphi_{\min}(P_1, P_2) = \forall P_1' \subseteq P_1. \forall P_2' \subseteq P_2.$$
$$((P_1' \cup P_2') \subsetneq (P_1 \cup P_2)) \rightarrow \varphi_{\text{sat}}(P_1', P_2').$$

Here we use the abbreviation $\exists P \subseteq F. \psi$ to denote the formula $\exists P. \forall p(P(p) \rightarrow F(p)) \wedge \psi$. Moreover, we also use the abbreviation $(|P| \geq q)$ and $(P \subsetneq P')$ with the usual meaning.

Intuitively, the second-order quantification $\exists P_1$ guesses a subset of $[\Phi]$ and the second-order quantification $\exists P_2$ guesses a subset of $\{ \neg \varphi : \varphi \in [\Phi] \}$, such that $P_1 \cup P_2$ is a minimally unsatisfiable subset of Φ of cardinality ≥ 3. The formula $\neg \varphi_{\text{sat}}(P_1, P_2)$ enforces that $P_1 \cup P_2$ is unsatisfiable, and the formula $\varphi_{\min}$ encodes that it is minimally so, i.e., that all strict subsets of $P_1 \cup P_2$ are satisfiable.

It is readily verified that $\mathcal{A} \models \varphi$ if and only if $\Phi \in$ MAJORITY-SAFETY. Therefore, since the size of φ is constant and $\mathcal{A}$ has treewidth at most $k + 2$, we get that MAJORITY-SAFETY(f-tw) is fixed-parameter tractable by Courcelle's Theorem.

Next, we show that the problem is para-Π_2^p-complete when parameterized by the clausal primal treewidth.

Proposition 11.11. MAJORITY-SAFETY(c-tw) *is* para-Π_2^p-*complete.*

Proof. We show para-Π_2^p-hardness by showing that the problem is already Π_2^p-hard for constant values of the parameter. We do so by giving a reduction from co-QSAT$_2$(3CNF). This reduction is a modified variant of a reduction given by Endriss et al. [75, Lemma 11]. Let $\varphi = \forall X. \exists Y. \psi$ be an instance of co-QSAT$_2$, where $\psi = c_1 \wedge \cdots \wedge c_m$ is in 3CNF, and where $X = \{x_1, \ldots, x_m\}$. Moreover, for each $i \in [m]$, let c_i consist of the literals l_1^i, l_2^i and l_3^i. We may assume without loss of generality that none of the c_i is equivalent to a unit clause.

We construct the agenda Φ as follows. We introduce fresh variables z_j^i for $i \in [m]$ and $j \in [3]$. Let Z denote the set of all such variables z_j^i. Then, we let $[\Phi] = \{x_1, \ldots, x_n\} \cup \{ (z_1^i \vee \neg l_1^i) \wedge (z_2^i \vee \neg l_2^i) \wedge (z_3^i \vee \neg l_3^i) : i \in [m] \}$. It is straightforward to verify that the clausal primal graph of Φ is a tree, and thus that Φ has clausal primal treewidth 1. We show that Φ satisfies the median property if and only if φ is true.

($\Rightarrow$) Suppose that φ is false, i.e., there exists some $\alpha : X \rightarrow \mathbb{B}$ such that $\forall Y. \neg \psi[\alpha]$ is true. Let $L = \{ x_i : i \in [n], \alpha(x_i) = 1 \} \cup \{ \neg x_i : i \in [n], \alpha(x_i) = 0 \}$. We know that α is the unique assignment to the variables in X that satisfies L. Now consider $\Phi' = L \cup \{ \neg ((z_1^i \vee \neg l_1^i) \wedge (z_2^i \vee \neg l_2^i) \wedge (z_3^i \vee \neg l_3^i)) : i \in [m] \}$.

We firstly show that Φ' is inconsistent. We proceed indirectly and assume that Φ' is consistent, i.e., there exists an assignment $\beta : Y \cup Z \to \mathbb{B}$ such that $\alpha \cup \beta$ satisfies Φ'. Then $\alpha \cup \beta$ must satisfy each c_i, since $\neg((z_1^i \vee \neg l_1^i) \wedge (z_2^i \vee \neg l_2^i) \wedge (z_3^i \vee \neg l_3^i)) \models c_i$. Therefore, β satisfies $\psi[\alpha]$, which contradicts our assumption that $\forall Y. \neg \psi[\alpha]$ is true. Therefore, we can conclude that Φ' is inconsistent.

Next, we show that each subset $\Phi'' \subseteq \Phi'$ of size 2 is consistent. Let $\Phi'' \subseteq \Phi'$ be an arbitrary subset of size 2. We distinguish three cases: either (i) $\Phi'' = \{l_i, l_j\}$ for some $i, j \in [n]$ with $i < j$; (ii) $\Phi'' = \{l_i, \neg((z_1^j \vee \neg l_1^j) \wedge (z_2^j \vee \neg l_2^j) \wedge (z_3^j \vee \neg l_3^j))\}$ for some $i \in [n]$ and some $j \in [m]$; or (iii) $\Phi'' = \{\neg((z_1^i \vee \neg l_1^i) \wedge (z_2^i \vee \neg l_2^i) \wedge (z_3^i \vee \neg l_3^i)), \neg((z_1^j \vee \neg l_1^j) \wedge (z_2^j \vee \neg l_2^j) \wedge (z_3^j \vee \neg l_3^j))\}$ for some $i, j \in [m]$ with $i < j$. In case (i), clearly Φ'' is consistent. In case (ii) and (iii), Φ'' is consistent because c_i and c_j are not equivalent to unit clauses.

($\Leftarrow$) Conversely, suppose that Φ does not satisfy the median property, i.e., there exists an inconsistent subset $\Phi' \subseteq \Phi$ that itself does not contain an inconsistent subset of size 2. We show that φ is false. Firstly, we show that $\Psi' = \Phi' \setminus \{ (z_1^i \vee \neg l_1^i) \wedge (z_2^i \vee \neg l_2^i) \wedge (z_3^i \vee \neg l_3^i) : i \in [m] \}$ is inconsistent. We proceed indirectly, and assume that Ψ' is consistent, i.e., there exists an assignment $\gamma : \mathrm{Var}(\Psi') \to \mathbb{B}$ such that γ satisfies Ψ'. Now let $Z' = \{ z_1^i, z_2^i, z_3^i : i \in [m], (z_1^i \vee \neg l_1^i) \wedge (z_2^i \vee \neg l_2^i) \wedge (z_3^i \vee \neg l_3^i) \in \Phi' \}$ and let $\gamma' : Z' \to \mathbb{B}$ be defined by letting $\gamma'(z) = 1$ for all $z \in Z'$. Since Ψ' contains no negated pairs of formulas, we know that $Z' \cap \mathrm{Var}(\Psi') = \emptyset$. Then the assignment $\gamma \cup \gamma'$ satisfies Φ', since γ satisfies all $\psi \in \Psi'$ and γ' satisfies all $\varphi \in \Phi' \setminus \Psi'$. This is a contradiction with our assumption that Φ' is inconsistent, so we can conclude that Ψ' is inconsistent.

Now let the assignment $\alpha : X \to \mathbb{B}$ be defined as follows. For each $x \in X$, we let $\alpha(x) = 1$ if $x \in \Psi'$, we let $\alpha(x) = 0$ if $\neg x \in \Psi'$, and we (arbitrarily) define $\alpha(x) = 1$ otherwise. We now show that $\neg \exists Y. \psi[\alpha]$ is true. We proceed indirectly, and assume that there exists an assignment $\beta : Y \to \mathbb{B}$ such that $\psi[\alpha \cup \beta]$ is true. Consider the assignment $\gamma : Z \to \mathbb{B}$ such that $\gamma(z) = 0$ for all $z \in Z$. We claim that the assignment $\alpha \cup \beta \cup \gamma$ satisfies Ψ'. Let $\chi \in \Psi'$ be an arbitrary formula. We distinguish two cases: either (i) $\chi \in \{x_i, \neg x_i\}$ for some $i \in [n]$; or (ii) $\chi = \neg((z_1^i \vee \neg l_1^i) \wedge (z_2^i \vee \neg l_2^i) \wedge (z_3^i \vee \neg l_3^i))$ for some $i \in [m]$. In case (i), we know that α satisfies χ. For case (ii), we know that $\alpha \cup \beta$ satisfies c_i, since $\alpha \cup \beta$ satisfies ψ. Moreover, we know that γ sets each z_j^i to 0. Therefore, we know that $\alpha \cup \beta \cup \gamma$ satisfies χ. This is a contradiction with our previous conclusion that Ψ' is inconsistent, so we can conclude that $\neg \exists Y. \psi[\alpha]$ is true. From this, we know that $\forall X. \exists Y. \psi$ is false.

We show that bounding the formula incidence treewidth also does not improve the complexity of the problem MAJORITY-SAFETY. The problem is para-Π_2^p-complete when parameterized by the formula incidence treewidth.

Proposition 11.12. MAJORITY-SAFETY(f-tw*) *is para-Π_2^p-complete.*

Proof. We observe that the Π_2^p-hardness proof of MAJORITY-SAFETY given by Endriss, Grandi and Porello [75, Lemmas 22 and 24] shows that the problem

MAJORITY-SAFETY is already Π_2^P-hard for agendas with formula incidence treewidth 1. This implies that MAJORITY-SAFETY(f-tw*) is para-Π_2^P-hard.

Finally, when parameterized by the clausal incidence treewidth, the problem is also para-Π_2^P-complete.

Proposition 11.13. MAJORITY-SAFETY(c-tw*) *is* para-Π_2^P-*complete.*

Proof. The agenda Φ used in the construction in the proof of Proposition 11.11 also has clausal incidence treewidth 1. Therefore, para-Π_2^P-hardness also holds for this case.

11.2.4 Small Counterexamples

Next, we consider the following parameterized variant MAJORITY-SAFETY(c.e.-size) of the problem MAJORITY-SAFETY. The problem consists of deciding, given an agenda Φ, and an integer k, whether every inconsistent subset Φ' of Φ of size k has itself an inconsistent subset of size at most 2. The parameter is k.

Assuming that counterexamples to the MP are small in practice corresponds to the assumption that whenever several statements together imply another statement, this latter statement is already implied by a small number of the former statements. In other words, the interaction between statements is, in a sense, local.

This problem is also related to agenda safety for supermajority rules. A supermajority rule accepts any proposition in the agenda if and only if a certain supermajority of the individuals, specified by a threshold proportion $\frac{1}{2} < q \leq 1$, accepts the proposition. Such rules always produce consistent outcomes if the threshold is greater than $\frac{k-1}{k}$, where k is the size of the largest minimally inconsistent subagenda [61, 148].

Unfortunately, it turns out that this parameterization does not lead to a significant (practically exploitable) improvement in the computational complexity. In order to prove this, we will need the following technical lemma.

Lemma 11.14. *Let (φ, k) be an instance of $\Pi_2^P[k*]$-WSAT. In polynomial time, we can construct an equivalent instance (φ', k) of $\Pi_2^P[k*]$-WSAT such that: (1) for every assignment $\alpha : X \rightarrow \mathbb{B}$ of weight $m > k$, the formula $\exists Y.\psi[\alpha]$ is false; and (2) for every assignment $\alpha : X \rightarrow \mathbb{B}$ of weigth $m < k$, the formula $\exists Y.\psi[\alpha]$ is true.*

Proof. Let (φ, k) be an instance of $\Pi_2^P[k*]$-WSAT, with $\varphi = \forall X.\exists Y.\psi$. We construct the instance $\varphi' = \forall X.\exists Y \cup Z.\psi'$ as follows. We define the set Z of variables by letting $Z = \{z_{x,i} : x \in X, i \in [k]\}$. Intuitively, these variables keep track of how many variables in X are set to true. We define the formula $\psi' = \psi_{\text{proper}}^Z \wedge (\psi_{\text{few}}^Z \vee \psi)$, where $\psi_{\text{proper}}^Z = \bigwedge_{x \in X} \bigvee_{i \in [k]} z_{x,i} \wedge \bigwedge_{i \in [k]} \bigwedge_{x,x' \in X, x \neq x'} (\neg z_{x,i} \vee \neg z_{x',i}) \wedge \bigwedge_{x \in X} \bigwedge_{i,i' \in [k], i < i'} (\neg z_{x,i} \vee \neg z_{x,i'})$, and $\psi_{\text{few}}^Z = \bigvee_{i \in [k]} \bigwedge_{x \in X} \neg z_{x,i}$. The formula ψ_{proper}^Z enforces that for any $x \in X$ that is set to true, there must be some $i \in [k]$ such that $z_{x,i}$

is set to true as well. Moreover, it enforces that for each $x \in X$ there is at most one $i \in [k]$ such that $z_{x,i}$ is true, and for each $i \in [k]$, there is at most one $x \in X$ such that $z_{x,i}$ is true. The formula ψ_{few}^Z is true if and only if there exists some $i \in [k]$ such that $z_{x,i}$ is false for all $x \in X$.

It is now straightforward to verify that for each assignment $\alpha : X \rightarrow \mathbb{B}$ it holds that (i) if α has weight k, then $\exists Y \cup Z.\psi'[\alpha]$ is true if and only if $\exists Y.\psi[\alpha]$ is true, (ii) if α has weight less than k, then $\exists Y \cup Z.\psi'[\alpha]$ is always true, and (iii) if α has weight more than k, then $\exists Y \cup Z.\psi'[\alpha]$ is never true.

Using the previous lemma, we can now show that the problem MAJORITY-SAFETY (c.e.-size) is $\Pi_2^p[k*]$-hard. This result is very similar to Proposition 10.14 in Chap. 10.

Proposition 11.15. MAJORITY-SAFETY(c.e.-size) *is* $\Pi_2^p[k*]$-*hard.*

Proof. In order to show $\Pi_2^p[k*]$-hardness, we provide an fpt-reduction from $\Pi_2^p[k*]$-WSAT to MAJORITY-SAFETY(c.e.-size). Let (φ, k) be an instance of $\Pi_2^p[k*]$-WSAT, where $\varphi = \forall X.\exists Y.\psi$ is a quantified Boolean formula, $X = \{x_1, \ldots, x_n\}$, and k is a positive integer. We may assume without loss of generality that φ satisfies properties (1) and (2) described in Lemma 11.14. We define the agenda $\Phi = \{x_1, \neg x_1, \ldots, x_n, \neg x_n, (\psi \wedge z), \neg(\psi \wedge z)\}$, where z is a fresh variable. We show that for all assignments $\alpha : X \rightarrow \mathbb{B}$ of weight k it is the case that $\exists Y.\psi[\alpha]$ is true if and only if every inconsistent subset Φ' of Φ of size $k + 1$ has itself an inconsistent subset of size 2.

($\Rightarrow$) Assume that there exists an inconsistent subset Φ' of Φ of size $k + 1$ that has itself no inconsistent subset of size 2. It is straightforward to see that for no $\varphi \in \Phi$, Φ' contains both φ and $\sim\varphi$. If Φ' does not contain $(\psi \wedge z)$, we can easily satisfy Φ' by setting z to false and satisfying all literals in Φ'. Therefore, $(\psi \wedge z) \in \Phi'$. We show that Φ' contains exactly k positive literals x_j for some $j \in [m]$. We proceed indirectly, and assume the contrary, i.e., that Φ' contains at most $k - 1$ positive literals x_j for some $j \in [m]$. Let $L = \Phi' \cap X$. Consider the assignment $\alpha : X \rightarrow \mathbb{B}$ such that $\alpha(x) = 1$ if and only if $x \in \Phi$. Clearly, α has weight strictly less than k. Therefore, we know that there exists an assignment $\beta : Y \rightarrow \mathbb{B}$ such that $\alpha \cup \beta$ satisfies ψ. Additionally, consider the assignment $\gamma : \{z\} \rightarrow \mathbb{B}$ such that $\gamma(z) = 1$. Then $\alpha \cup \beta \cup \gamma$ satisfies Φ', which contradicts our assumption that Φ' is inconsistent. From this we can conclude that $|\Phi' \cap X| = k$.

Now, again consider the assignment $\alpha : X \rightarrow \mathbb{B}$ such that $\alpha(x) = 1$ if and only if $x \in \Phi$. Clearly, α has weight k. We show that the formula $\exists Y.\psi[\alpha]$ is false. We proceed indirectly, and assume that there exists an assignment $\beta : Y \rightarrow \mathbb{B}$ such that $\alpha \cup \beta$ satisfies ψ. Consider the assignment $\gamma : \{z\} \rightarrow \mathbb{B}$ such that $\gamma(z) = 1$. It is straightforward to verify that $\alpha \cup \beta \cup \gamma$ satisfies Φ', which contradicts our assumption that Φ' is inconsistent. Therefore, we conclude that $\exists Y.\psi[\alpha]$ is false, and thus that it is not the case that for all assignments $\alpha : X \rightarrow \mathbb{B}$ of weight k it is the case that $\exists Y.\psi[\alpha]$ is true.

($\Leftarrow$) Assume that there exists an assignment $\alpha : X \rightarrow \mathbb{B}$ of weight k such that $\neg\exists Y.\psi[\alpha]$ is true. Let $L = \{x_i : i \in [n], \alpha(x_i) = 1\}$. Consider the subagenda

$\Phi' = L \cup \{(\psi \wedge z)\}$. We show that Φ' is inconsistent. We proceed indirectly, and assume that there exists an assignment $\beta : X \cup Y \cup \{z\} \to \mathbb{B}$ that satisfies Φ'. Clearly, $\beta(x_i) = 1$ for all $x_i \in L$. We show that $\beta(x) = 0$ for all $x \in X \backslash L$. We proceed indirectly, and assume the contrary, i.e., $\beta(x) = 1$ for some $x \in X \backslash L$. Then the restriction of β to the variables in X has weight $m > k$. Therefore, since for all assignments $\beta' : X \to \{0, 1\}$ of weight strictly larger than k the formula $\exists Y . \psi[\beta']$ is false, we know that β does not satisfy ψ. From this we can conclude that $\beta(x) = 0$ for all $x \in X \backslash L$. We then know that the restriction $\beta|_X$ of β to the variables in X has weight k. Also, since $(\psi \wedge z) \in \Phi$, we know that β satisfies ψ. This is a contradiction with our assumption that $\neg \exists Y . \psi[\beta|_X]$ is true. Therefore, we know that β cannot exist, and thus that Φ' is inconsistent.

We now show that each subset Φ'' of Φ' of size 2 is consistent. Let $\Phi'' \subseteq \Phi'$ be an arbitrary subset of size 2. We distinguish two cases: either (i) $\Phi'' = \{x_i, x_j\}$ for some $i, j \in [n]$ with $i < j$, or (ii) $\Phi'' = \{x_i, (\psi \wedge z)\}$ for some $i \in [n]$. In case (i), clearly Φ'' is consistent. In case (ii), we get that Φ'' is consistent by the fact that for every assignment $\alpha : X \to \mathbb{B}$ of weight $m < k$ the formula $\exists Y . \psi[\alpha]$ is true. This completes our proof that Φ' does not satisfy the median property.

Intuitively, restricting attention to counterexamples of size k, still leaves a search space of $O(n^k)$ possible counterexamples (where n is the input size). Moreover, since there is no restriction on the agenda, searching this space for a counterexample (or verifying that no such counterexample exists) is computationally hard.

11.3 Computing Outcomes for the Kemeny Rule

In this section, we will study the parameterized complexity of computing an outcome of the Kemeny judgment aggregation procedure. We investigate this problem both for the formula-based judgment aggregation framework and for the constraint-based judgment aggregation framework. In both settings, this problem is Θ_2^p-complete [75, 102, 142].

We consider a number of natural parameters for this problem—capturing various aspects of the problem input that can reasonably be expected to be small in some applications—and we give a complete parameterized complexity classification for the problem of computing the outcome of the Kemeny rule, for every combination of these parameters. The parameters that we consider are:

- the number n of issues that the individuals (and the group) form an opinion on;
- the maximum size m of formulas used to represent the issues;
- the size c of the integrity constraint used to limit the set of feasible opinions;
- the number p of individuals; and
- the maximum (Hamming) distance h between any two individual opinions.

For the formula-based judgment aggregation framework, we will study the following formalization FB-OUTCOME-KEMENY of the problem of computing an outcome for the Kemeny rule. Any algorithm that solves FB-OUTCOME-KEMENY can

be used to construct some $J^* \in \text{Kemeny}(\boldsymbol{J})$, with polynomial overhead, by itera-
tively calling the algorithm and adding formulas to the set L. Moreover, multiple
outcomes $J_1^*, J_2^*, \ldots$ can be constructed by adding previously found outcomes as
the sets L_i.

FB-OUTCOME-KEMENY
Instance: An agenda Φ with an integrity constraint Γ, a profile $\boldsymbol{J} \in \mathcal{J}(\Phi, \Gamma)^+$ and
subsets $L, L_1, \ldots, L_u \subseteq \Phi$ of the agenda, with $u \geq 0$.
Question: Is there a judgment set $J^* \in \text{Kemeny}(\boldsymbol{J})$ such that $L \subseteq J^*$ and $L_i \not\subseteq J^*$ for each
$i \in [u]$?

The parameters that we consider for the problem FB-OUTCOME-KEMENY are de-
fined straightforwardly. For an instance $(\Phi, \Gamma, \boldsymbol{J}, L, L_1, \ldots, L_u)$ of FB-OUTCOME-
KEMENY with $\boldsymbol{J} = (J_1, \ldots, J_p)$, we let $n = \|[\Phi]\|$, $m = \max\{ |\varphi| : \varphi \in [\Phi]\}$, $c =$
$|\Gamma|$, $p = |\boldsymbol{J}|$, and $h = \max\{ d(J_i, J_{i'}) : i, i' \in [p] \}$.

For the constraint-based judgment aggregation framework, we will study the following problem formalization CB-OUTCOME-KEMENY. Similarly to algorithms for
FB-OUTCOME-KEMENY, algorithms that solve CB-OUTCOME-KEMENY can be used
to construct multiple outcomes subsequently.

CB-OUTCOME-KEMENY
Instance: A set $\mathcal{I}$ of issues with an integrity constraint Γ, a profile $\boldsymbol{r} \in \mathcal{R}(\mathcal{I}, \Gamma)^+$ and
partial ballots $l, l_1, \ldots, l_u$ (for $\mathcal{I}$), with $u \geq 0$.
Question: Is there a ballot $r^* \in \text{Kemeny}(\boldsymbol{r})$ such that l agrees with r^* and each l_i does not agree
with r^*?

We define the parameters that we consider for CB-OUTCOME-KEMENY as fol-
lows. For an instance $(\mathcal{I}, \Gamma, \boldsymbol{r}, l, l_1, \ldots, l_u)$ of CB-OUTCOME-KEMENY with $\boldsymbol{r} =$
$(r_1, \ldots, r_p)$, we let $n = |\mathcal{I}|$, $c = |\Gamma|$, $p = |\boldsymbol{r}|$, and $h = \max\{ d(r_i, r_{i'}) : i, i' \in [p] \}$.
We remark that the parameter m does not make sense in the constraint-based frame-
work, as issues are not represented by a logic formula. When needed, the parameter m
for CB-OUTCOME-KEMENY is defined by letting $m = 1$.

For the framework of formula-based judgment aggregation, we give a tight classification for each possible case. In particular, we show the following. When parameterized by any set of parameters that includes c, n and m, the problem is fixed-parameter
tractable (Proposition 11.18). Otherwise, when parameterized by any set of param-
eters that includes either n or both h and p, the problem is $\text{FPT}^{\text{NP}}[\text{few}]$-complete
(Propositions 11.16, 11.17, 11.23, 11.24 and 11.25). For all remaining cases, the
problem is para-Θ_2^{p}-complete (Corollary 11.21 and Proposition 11.22).

For the framework of constraint-based judgment aggregation, we show the following results. When parameterized by any set of parameters that includes either c
or n, the problem is fixed-parameter tractable (Propositions 11.26 and 11.27). Otherwise, when parameterized by any set of parameters that includes h, the problem is
W[SAT]-hard and is in XP (Propositions 11.28 and 11.29). For all remaining cases,
the problem is para-Θ_2^{p}-complete (Proposition 11.30).

The results for the formula-based judgment aggregation framework are summarized in Table 11.2, and the results for the constraint-based framework are summarized in Table 11.3.

The remainder of this section is organized as follows. We develop upper and lower bounds for the parameterized variants of FB-OUTCOME-KEMENY in Sects. 11.3.1, and 11.3.2, respectively, and we develop upper and lower bounds for the parameterized variants of CB-OUTCOME-KEMENY in Sects. 11.3.3, and 11.3.4, respectively. Then, in Sect. 11.3.5, we provide a short overview of the parameterize complexity of FB-OUTCOME-KEMENY and CB-OUTCOME-KEMENY for all possible combinations of the parameters that we considered for these two problems.

11.3.1 Upper Bounds for the Formula-Based Framework

We begin with showing upper bounds for FB-OUTCOME-KEMENY. When parameterized either (i) by both h and p or (ii) by n, the problem is in $\text{FPT}^{\text{NP}}[\text{few}]$.

Proposition 11.16. FB-OUTCOME-KEMENY *parameterized by h and p is in* $\text{FPT}^{\text{NP}}[\text{few}]$.

Proof. The main idea behind this proof is that with these parameters, we can derive a suitable upper bound on the minimum distance of any complete and Γ-consistent judgment set to the profile J, such that the usual binary search algorithm with access to an NP oracle only needs to make $O(\log h + \log p)$ oracle queries.

We describe an algorithm A that solves FB-OUTCOME-KEMENY with the required number of oracle queries. Let $(\Phi, \Gamma, J, L, L_1, \ldots, L_u)$ be an instance. The algorithm needs to determine the minimum distance $d(J, J)$ for any complete and Γ-consistent judgment set J to the profile J. Let d^* denote this minimum distance. An upper bound on d^* is given by $h(p - 1)$. This upper bound can be derived as follows. Take an arbitrary $J \in J$. Clearly $d(J, J) = 0$, and for every $J' \in J$ with $J \neq J'$ we know that $d(J, J') \leq h$. Therefore, $d(J, J) \leq h(p - 1)$. Since $J \in J$, we know that J is complete and Γ-consistent. Therefore, the minimum distance of any complete and Γ-consistent judgment set to the profile J is at most $h(p - 1)$.

The algorithm A firstly computes d^*. Since $d^* \leq h(p - 1)$, with binary search this can be done using at most $\lceil \log h(p - 1) \rceil = O(\log h + \log p)$ queries to an oracle—the oracle decides for a given value d_0 whether there exists a complete and Γ-consistent judgment set J with $d(J, J) \leq d_0$. Then, with a single additional oracle query, the algorithm A determines whether there exists a complete and Γ-consistent judgment set J^* with $d(J^*, J) = d^*$, $L \subseteq J^*$, and $L_j \not\subseteq J^*$ for each $j \in [u]$.

When parameterized by the number n of formulas in the pre-agenda, the number of possible judgment sets is bounded by a function of the parameter. This allows the problem to be solved in fixed-parameter tractable time, using a single query to an NP oracle for each judgment set to determine whether it it Γ-consistent.

Proposition 11.17. FB-OUTCOME-KEMENY *parameterized by n is in* $\text{FPT}^{\text{NP}}[\text{few}]$.

Proof. The main idea behind this proof is that the number of possible judgment sets is bounded by the parameter, that is, there are only 2^n possible complete judgment sets. We describe an algorithm A that solves the problem in fixed-parameter tractable time by querying an NP oracle at most 2^n times. Let $(\Phi, \Gamma, J, L, L_1, \ldots, L_u)$ be an instance. Firstly, the algorithm A enumerates all possible complete judgment sets $J_1, \ldots, J_{2^n} \subseteq \Phi$. Then, for each such set J_i, the algorithm uses the NP oracle to determine whether J_i is Γ-consistent. Each judgment set J_i that is not Γ-consistent, is discarded. This can be done straightforwardly using 2^n calls to the NP oracle—one for each set J_i. (The number of oracle calls that are needed can be improved to $O(n)$ by using binary search on the number of Γ-consistent sets J_i.)

Then, for each of the remaining (Γ-consistent) judgment sets J_i, the algorithm A computes the cumulative Hamming distance $d(J_i, J)$ to the profile J. This can be done in polynomial time. Then, those J_i for which this distance is not minimal—that is, those J_i for which there exists some $J_{i'}$ such that $d(J_{i'}, J) < d(J_i, J)$—are discarded as well. The remaining judgment sets J_i then are exactly the complete and Γ-consistent judgment sets with a minimum distance to the profile J.

Finally, the algorithm goes over each of these remaining sets J_i, and checks whether $L \subseteq J_i$ and $L_j \not\subseteq J_i$ for all $j \in [u]$. This can clearly be done in polynomial time. If this check succeeds for some J_i, the algorithm A accepts the input, and otherwise, the algorithm rejects the input.

When additionally parameterizing by c and m, Γ-consistency of the judgment sets can be decided in fixed-parameter tractable time, and thus the whole problem becomes fixed-parameter tractable.

Proposition 11.18. FB-OUTCOME-KEMENY *parameterized by c, n and m is fixed-parameter tractable.*

Proof. We describe an fpt-algorithm A that solves the problem. Let $(\Phi, \Gamma, J, L, L_1, \ldots, L_u)$ be an instance. The algorithm A works exactly in the same way as the algorithm in the proof of Proposition 11.17. The only difference is that in order to check whether a given judgment set J_i is Γ-consistent, it does not need to make an oracle query. Determining whether a given judgment set J_i is Γ-consistent can be done in a brute-force fashion (e.g., using truth tables) in time $2^{c+nm} \cdot |J_i|$, since there are at most $c + nm$ propositional variables involved. Therefore, the algorithm runs in fixed-parameter tractable time.

11.3.2 Lower Bounds for the Formula-Based Framework

Next, we turn to parameterized hardness results for the problem FB-OUTCOME-KEMENY. We begin with showing that the problem is para-Θ_2^p-hard even when parameterized by c, h and m. We will use the following lemma.

Lemma 11.19. *Let φ be a propositional formula on the variables $x_1, \ldots, x_n$. In polynomial time we can construct a propositional formula φ' with $\mathrm{Var}(\varphi') \supseteq \mathrm{Var}(\varphi) \cup \{z_1, \ldots, z_n\}$ such that for every truth assignment $\alpha : \mathrm{Var}(\varphi) \to \mathbb{B}$ it holds* *that (1) $\varphi[\alpha]$ is true if and only if $\varphi'[\alpha]$ is satisfiable, and (2) if α sets exactly i variables to true, then $\varphi'[\alpha] \models z_i$.*

Proof. Let φ be a propositional formula on the variables $x_1, \ldots, x_n$. We construct the formula φ' as follows. We introduce propositional variables $z_{i,j}$ and z_i for each $i \in [n]$ and each $j \in [i]$. Intuitively, the variables $z_{i,j}$ encodes whether among the variables $x_1, \ldots, x_i$ at least j variables are set to true, and the variables z_i encode whether among the variables $x_1, \ldots, x_n$ exactly i variables are set to true.

We let φ' be a conjunction of several formulas. The first conjunct of φ' is the original formula φ. Then, we add the following conjunct:

$$z_{1,1} \leftrightarrow x_1.$$

Moreover, for each $i \in [n]$ such that $i > 1$, we add:

$$\left(x_i \leftrightarrow \bigwedge_{j \in [i]} (z_{i,j} \leftrightarrow z_{i-1,j-1}) \right) \wedge \left(\neg x_i \leftrightarrow \bigwedge_{j \in [i]} (z_{i,j} \leftrightarrow z_{i-1,j}) \right),$$

where for any $i \in [n]$, $z_{i,0}$ abbreviates $\top$. Finally, for each $i \in [n]$, we add:

$$z_i \leftrightarrow (z_{n,i} \wedge \neg z_{n,i+1}),$$

where $z_{n,n+1}$ abbreviates $\bot$.

It is straightforward to verify that the formula φ' satisfies the required properties. $\square$

Proposition 11.20. FB-OUTCOME-KEMENY *parameterized by c and h is* para-Θ_2^p-*hard.*

Proof. We show that FB-OUTCOME-KEMENY is Θ_2^p-hard already for a constant value of the parameters, by giving a reduction from MAX-MODEL. Let (φ, w) be an instance of MAX-MODEL with $\text{Var}(\varphi) = \{x_1, \ldots, x_n\}$ and $w = x_1$. Without loss of generality, we may assume that there is a model α of φ that sets at least two variables x_i to true. By Lemma 11.19, we can construct a suitable formula $\varphi' = c_1 \wedge \cdots \wedge c_b$ with additional variables $z_1, \ldots, z_n$ that represent a lower bound on the number of variables among $x_1, \ldots, x_n$ that are true in models of φ.

We construct the agenda Φ by letting $[\Phi] = \{z_w, z_{\neg w}, z_1, \ldots, z_n\} \cup \{y_{w,i}, y_{\neg w,i} : i \in [n+1]\} \cup \{y_{i,j} : i \in [n], j \in [i]\} \cup \{\chi, \chi'\}$, where $z_w, z_{\neg w}$ and all $y_{w,i}, y_{\neg w,i}, y_{i,j}$ are fresh variables. We let $Y = \{y_{w,i}, y_{\neg w,i} : i \in [n+1]\} \cup \{y_{i,j} : i \in [n], j \in [i]\}$. Moreover, we let χ be such that $\chi \equiv \neg((\bigvee Y \wedge \bigvee ([\Phi] \backslash Y)) \vee ((z_w \leftrightarrow w \leftrightarrow \neg z_{\neg w}) \wedge \varphi'))$, and we define χ' such that $\chi' \equiv \chi$ (that is, we let χ' be a syntactic variant of χ).

Then, we construct the profile J as follows. We let $J = \{J_{w,i}, J_{\neg w,i} : i \in [n+1]\} \cup \{J_{i,j} : i \in [n], j \in [i]\}$. Each of the judgment sets in the profile includes exactly two formulas in $[\Phi]$. Consequently, the maximum Hamming distance between any two judgment sets in the profile is 4. For each $i \in [n+1]$, we let $\{y_{w,i}, z_w\} \subseteq J_{w,i}$ and $\{y_{\neg w,i}, z_{\neg w}\} \subseteq J_{\neg w,i}$. Moreover, for each $i \in [n]$ and each $j \in [i]$, we let $\{y_{i,j}, z_i\} \subseteq J_{i,j}$. It is straightforward to verify that each $J \in \boldsymbol{J}$ is consistent. Finally, we let $L = \{z_w\}$, $\Gamma = \top$, and $u = 0$.

In other words, all formulas in $[\Phi]$ are excluded in a majority of the judgment sets in the profile $\boldsymbol{J}$. However, some formulas in $[\Phi]$ are included in more judgment sets in the profile than others. The formulas z_w and $z_{\neg w}$ are both included in $n + 1$ sets. Each formula z_i (for $i \in [n]$) is included in exactly i sets. All formulas in Y are included in exactly one set. Finally, the formulas χ and χ' are included in none of the sets. Intuitively, the formulas that are included in more judgment sets in the profile are cheaper to include in any candidate outcome J^*.

The complete judgment set that minimizes the cumulative Hamming distance to the profile $\boldsymbol{J}$ is the set $J_0 = \{\neg\psi : \psi \in [\Phi]\}$ that includes no formulas in $[\Phi]$. However, this set is inconsistent, which is straightforward to verify using the definition of χ. It can be made consistent by adding two formulas ψ_1, ψ_2 from $[\Phi]$ (and removing their complements). The choice of ψ_1, ψ_2 that leads to a consistent judgment set with minimum distance to the profile is by letting $\psi_1 \in \{z_w, z_{\neg w}\}$ and letting $\psi_2 = z_\ell$, where ℓ is the maximum number of variables among $x_1, \ldots, x_n$ set to true in any model of φ. Moreover, whenever $\psi_1 = z_w$, the resulting judgment set is consistent if and only if there is a model of φ that sets ℓ variables among $x_1, \ldots, x_n$ to true, including the variable w. From this, we directly know that $(\varphi, w) \in$ MAX-MODEL if and only if $(\Phi, \Gamma, \boldsymbol{J}, L) \in$ FB-OUTCOME-KEMENY. This concludes our para-Θ_2^p-hardness proof.

This hardness result can straightforwardly be extended to the case where all formulas in the agenda are of constant size, by using the well-known Tseitin transformation [188], leading to the following corollary.

Corollary 11.21. FB-OUTCOME-KEMENY *parameterized by c, h and m is* para-Θ_2^p-*hard.*

Proof. We can modify the proof of Proposition 11.20 as follows. We replace the formula $\neg\chi$ (and its syntactic variant $\neg\chi'$) by a 3CNF formula that has the same effect. By using the standard Tseitin transformation [188], we can transform $\neg\chi$ into a 3CNF formula ψ such that for each truth assignment $\alpha : \text{Var}(\neg\chi) \to \mathbb{B}$ it holds that $\neg\chi[\alpha]$ is true if and only if $\psi[\alpha]$ is satisfiable. Moreover, we can do this in such a way that the variables in $\text{Var}(\psi)\backslash\text{Var}(\neg\chi)$ are fresh variables. Similarly, we transform $\neg\chi'$ into a 3CNF formula ψ'. Let $\psi = c_1 \wedge \cdots \wedge c_b$ and $\psi' = c'_1 \wedge \cdots \wedge c'_b$ (we can straightforwardly ensure that ψ and ψ' have the same number of clauses). Then, similarly to the proof of Proposition 11.20, we let $[\Phi] = \{z_w, z_{\neg w}, z_1, \ldots, z_n\} \cup \{y_{w,i}, y_{\neg w,i} : i \in [n+1]\} \cup \{y_{i,j} : i \in [n], j \in [i]\} \cup \{c_i, c'_i : i \in [b]\}$. That is, instead of adding χ and χ' to the agenda, we add the clauses of ψ and ψ' as separate formulas to the agenda.

In the proof of Proposition 11.20, we had that $\neg\chi, \neg\chi' \in J$ for all judgment sets $J \in \boldsymbol{J}$. Instead, we now ensure that for all $J \in \boldsymbol{J}$, we have $c_i, c'_i \in J$ for all $i \in [b]$. From this, it follows that the set Kemeny($\boldsymbol{J}$) of outcomes is in one-to-one correspondence with the set of outcomes in the proof of Proposition 11.20.

Moreover, the maximum Hamming distance between any two judgment sets in the profile J is 4.

The problem is also para-Θ_2^p-hard when parameterized by c, m and p.

Proposition 11.22. FB-OUTCOME-KEMENY *parameterized by c, m and p is para-Θ_2^p-hard.*

Proof. We firstly show para-Θ_2^p-hardness for the problem parameterized by c and p, by giving a reduction from MAX-MODEL that uses constant values of c and p. This reduction can be seen as a modification of the Θ_2^p-hardness proof for FB-OUTCOME-KEMENY given by Endriss and De Haan [76, Proposition 7 and Corollary 8].

Let (φ, w) be an instance of MAX-MODEL. We may assume without loss of generality that φ is satisfiable by some truth assignment that sets at least one variable in $\text{Var}(\varphi)$ to true. We construct an instance (Φ, Γ, J, L) of FB-OUTCOME-KEMENY as follows. Take an integer b such that $b > \frac{3}{2}|\text{Var}(\varphi)|$, e.g., $b = 3|\text{Var}(\varphi)| + 1$. Let $[\Phi] = \text{Var}(\varphi) \cup \{ z_{i,j} : i \in [b], j \in [3] \} \cup \{ \varphi'_i : i \in [b] \}$, where each of the formulas φ'_i is a syntactic variant of the following formula φ'. We define $\varphi' = (\bigvee_{j \in [3]} \bigwedge_{i \in [b]} z_{i,j}) \vee \varphi$. Intuitively, the formula φ' is true either if (i) all variables $z_{i,j}$ are set to true for some $j \in [3]$, or if (ii) φ is satisfied. Then we let $J = \{J_1, J_2, J_3\}$, where for each $j \in [3]$, we let J_j contain the formulas $z_{i,j}$ for all $i \in [b]$, all formulas in $\text{Var}(\varphi)$, all the formulas φ'_i, and no other formulas from $[\Phi]$. (For each $\varphi \in [\Phi]$, if $\varphi \notin J_j$, we let $\neg\varphi \in J_j$.) Clearly, the judgment sets J_1, J_2 and J_3 are all complete and consistent. Moreover, we let $\Gamma = \top$, and $L = \{w\}$. It is straightforward to verify that the parameters c and p have constant values.

We now argue that there is some $J^* \in \text{Kemeny}(J)$ with $L \subseteq J^*$ if and only if $(\varphi, w) \in \text{MAX-MODEL}$. To see this, we first observe that the only complete and consistent judgment sets J for which it holds that $d(J, J) < d(J_j, J)$ (for any $j \in [3]$) must satisfy that $J \models \varphi$. Moreover, among those judgment sets J for which $J \models \varphi$, the judgment sets that minimize the distance to the profile J satisfy that $z_{i,j} \notin J$ for all $i \in [b]$ and all $j \in [3]$, and $\varphi'_i \in J$ for all $i \in [b]$. Using these observations, we directly get that there is some $J^* \in \text{Kemeny}(J)$ with $L \subseteq J^*$ if and only if there is a model of φ that sets a maximal number of variables in $\text{Var}(\varphi)$ to true and that sets the variable w to true.

Then, to show that the problem is also para-Θ_2^p-hard when parameterized by c, m and p, we can modify the above reduction in a way that is entirely similar to the proof of Corollary 11.21, replacing the formulas φ'_i by the clauses of 3CNF formulas that have the same effect on the consistency of judgment sets as the formulas φ'_i. □

For all parameterizations that do not include all of the parameters c, n and m, the problem FB-OUTCOME-KEMENY is FPT^{NP}[few]-hard. We begin with the case where c can be unbounded; this proof can be extended straightforwardly to the other two cases.

Proposition 11.23. FB-OUTCOME-KEMENY *parameterized by h, n, m and p is* FPT^{NP}[few]-*hard.*

Proof. We show FPT^{NP}[few]-hardness by giving an fpt-reduction from LOCAL-MAX-MODEL. (This reduction from LOCAL-MAX-MODEL is very similar to the reduction from MAX-MODEL used in the proof of Proposition 11.22.) Let (φ, X, w) be an instance of LOCAL-MAX-MODEL, with $X = \{x_1, \ldots, x_k\}$. We construct an instance (Φ, Γ, J, L) as follows. Take an integer b such that $b > \frac{3}{2}|X|$, e.g., let $b = 3|X| + 1$. We let $[\Phi] = X \cup \{z_{i,j} : i \in [b], j \in [3]\}$. Moreover, we let $\Gamma = \varphi' = (\bigvee_{j \in [3]} \bigwedge_{i \in [b]} z_{i,j}) \vee \varphi$. Intuitively, the formula Γ is true either if (i) all variables $z_{i,j}$ are set to true for some $j \in [3]$, or if (ii) φ is satisfied. Then we let $J = \{J_1, J_2, J_3\}$, where for each $j \in [3]$, we let J_j contain the formulas $z_{i,j}$ for all $i \in [b]$, and all formulas in X, and no other formulas in $[\Phi]$. (For each $\varphi \in [\Phi]$, if $\varphi \notin J_j$, we let $\neg\varphi \in J_j$.) Clearly, the judgment sets J_1, J_2 and J_3 are all complete and Γ-consistent. Finally, we let $L = \{w\}$. It is easy to verify that $h = 2b = 6k + 2$ and $n = 3b + k = 10k + 3$, where $k = |X|$, and that m and p are constant. Therefore, all parameter values are bounded by a function of the original parameter k.

We now argue that there is some $J^* \in \text{Kemeny}(J)$ with $L \subseteq J^*$ if and only if $(\varphi, X, w) \in \text{LOCAL-MAX-MODEL}$. The argument for this conclusion is similar to the argument used in the proof of Proposition 11.22. We first observe that the only complete and consistent judgment sets J for which it holds that $d(J, J) < d(J_j, J)$ (for any $j \in [3]$) must satisfy that $J \models \varphi$. Moreover, among those judgment sets J for which $J \models \varphi$, the judgment sets that minimize the distance to the profile J satisfy that $z_{i,j} \notin J$ for all $i \in [b]$ and all $j \in [3]$. Using these observations, we directly get that there is some $J^* \in \text{Kemeny}(J)$ with $L \subseteq J^*$ if and only if there is a model of φ that sets a maximal number of variables in X to true and that sets the variable w to true.

Proposition 11.24. FB-OUTCOME-KEMENY *parameterized by* c, h, n *and* p *is* FPT^{NP}[few]-*hard.*

Proof. We can show FPT^{NP}[few]-hardness by modifying the reduction from LOCAL-MAX-MODEL used in the proof of Proposition 11.23. Rather than using the formula φ' as the integrity constraint Γ, we let $\Gamma = \top$, and we add b syntactic variants $\varphi'_1, \ldots, \varphi'_b$ of φ' (and their negations) to the agenda Φ—that is, the formulas φ'_i for $i \in [b]$ are all syntactically different from each other, but for each such formula φ'_i it holds that $\varphi' \equiv \varphi'_i$. The judgment sets J_1, J_2 and J_3 in the profile J all include each of these formulas φ'_i.

As a result, the parameter value h remains the same. The value of the parameter p remains a constant, and the value of the parameter n increases only by b, so it remains bounded by a function of the original parameter k.

It is straightforward to verify that there are enough syntactic variants of the formula φ' in all judgment sets in the profile that for any complete and consistent judgment set J^* that minimizes the distance to the profile, it must hold that $J^* \models \varphi'$. Therefore, we get that the modified reduction is a correct reduction from LOCAL-MAX-MODEL, and thus that the problem is FPT^{NP}[few]-hard.

Proposition 11.25. FB-OUTCOME-KEMENY *parameterized by* c, h, m *and* p *is* FPT^{NP}[few]-*hard.*

Proof. We show $\text{FPT}^{\text{NP}}[\text{few}]$-hardness by modifying the (already modified) reduction from LOCAL-MAX-MODEL given in the proof of Proposition 11.24. In this reduction, the agenda included a small number of formulas φ_i', that were each of unbounded size. By using the same trick that we used in the proof of Corollary 11.21, we can use the standard Tseitin transformation [188] to transform each of these formulas into a 3CNF formula φ_i'' that will have the same effect. Then, rather than including φ_i' in the agenda Φ, we include all clauses of the formula φ_i'' in the agenda Φ. Then, in the judgment sets J_1, J_2 and J_3 in the profile $\boldsymbol{J}$, we also include the clauses of φ_i'' instead of the single formula φ_i', for all $i \in [b]$.

As a result, the number n of formulas in Φ is not bounded by a function of the original parameter k anymore, but the maximum size m of any formula in the agenda Φ is now bounded by a constant. Using the arguments used in the proofs of Corollary 11.21 and Proposition 11.24, it is then straightforward to verify the correctness of this modified reduction.

11.3.3 Upper Bounds for the Constraint-Based Framework

We now turn to showing upper bounds for CB-OUTCOME-KEMENY. When parameterized by the number n of issues, the number of possible ballots is bounded by a function of the parameter. This allows the problem to be solved in fixed-parameter tractable time.

Proposition 11.26. CB-OUTCOME-KEMENY *parameterized by n is fixed-parameter tractable.*

Proof. The main idea behind this proof is that the number of possible ballots is bounded by the parameter, that is, there are only 2^n possible (rational) ballots. We describe an algorithm A that solves the problem in fixed-parameter tractable time. Let $(\mathcal{I}, \Gamma, \boldsymbol{r}, l, l_1, \ldots, l_u)$ be an instance. Firstly, the algorithm A enumerates all possible ballots $r_1, \ldots, r_{2^n} \in \mathbb{B}^n$. Then, for each such ballot r_i, the algorithm determines whether r_i is rational, by checking whether $\Gamma[r_i]$ is true. This can be done in polynomial time. Each irrational ballot is discarded.

Then, for each of the remaining (rational) ballots r_i, the algorithm A computes the cumulative Hamming distance $d(r_i, \boldsymbol{r})$ to the profile $\boldsymbol{r}$. This can also be done in polynomial time. Then, those r_i for which this distance is not minimal—that is, those r_i for which there exists some $r_{i'}$ such that $d(r_{i'}, \boldsymbol{r}) < d(r_i, \boldsymbol{r})$—are discarded as well. The remaining ballots r_i then are exactly those rational ballots with a minimum distance to the profile $\boldsymbol{r}$.

Finally, the algorithm goes over each of these remaining ballots r_i, and checks whether l agrees with r_i and whether for all $j \in [u]$, l_j does not agree with r_i. If this check succeeds for some r_i, the algorithm A accepts the input, and otherwise, the algorithm rejects the input.

Since the size c of the integrity constraint is an upper bound on the number of issues that play a non-trivial role in the problem, this fixed-parameter tractability result easily extends to the parameter c.

Proposition 11.27. CB-OUTCOME-KEMENY *parameterized by c is fixed-parameter tractable.*

Proof. Since $|\Gamma| = c$, we know that the number of propositional variables in Γ is also bounded by the parameter c. Take an instance $(\mathcal{I}, \Gamma, \boldsymbol{r}, l, l_1, \ldots, l_u)$. Then, let $\mathcal{I}' = \text{Var}(\Gamma) \subseteq \mathcal{I}$ be the subset of issues that are mentioned in the integrity constraint Γ. We know that any outcome $r^* \in \text{Kemeny}(\boldsymbol{r})$ agrees with the majority of ballots in $\boldsymbol{r}$ on every issue in $\mathcal{I} \backslash \mathcal{I}'$ (in case of a tie, either choice works). Therefore, all that remains is to determine whether there are suitable choices for the issues in $\mathcal{I}$ (to obtain some $r^* \in \text{Kemeny}(\boldsymbol{r})$ that agrees with l and does not agree with l_j for all $j \in [u]$). By Proposition 11.26, we know that this is fixed-parameter tractable in $|\mathcal{I}'|$. Since $|\mathcal{I}'| \leq c$, we get fixed-parameter tractability also for CB-OUTCOME-KEMENY parameterized by c.

Bounding the maximum Hamming distance h between any two ballots in the profile gives us membership in XP.

Proposition 11.28. CB-OUTCOME-KEMENY *parameterized by h is in XP.*

Proof. Let $(\mathcal{I}, \Gamma, \boldsymbol{r}, l, l_1, \ldots, l_u)$ be an instance, with $\boldsymbol{r} = (r_1, \ldots, r_p)$. We describe an algorithm to solve the problem in time $O(p \cdot n^h \cdot n^d)$, for some constant d. The main idea behind this algorithm is the fact that each ballot whose Hamming distance to every ballot in the profile is more than h is irrelevant.

Take a ballot r such that $d(r, r_i) > h$ for each $i \in [p]$. We show that there exists a rational ballot r' with $d(r', \boldsymbol{r}) < d(r, \boldsymbol{r})$. Take any ballot in the profile, e.g., $r' = r_1$. Clearly, r' is rational. Since $d(r, r_i) > h$ for each $i \in [p]$, we know that $d(r, \boldsymbol{r}) > hp$. On the other hand, for r' we know that $d(r', r_i) \leq h$ for each $i \in [p]$ (and $d(r', r_1) = 0$), so $d(r', \boldsymbol{r}) \leq h(p - 1)$. Therefore, $d(r', \boldsymbol{r}) < d(r, \boldsymbol{r})$.

We thus know that every rational ballot with minimum distance to the profile lies at Hamming distance at most h to some ballot r_i in the profile $\boldsymbol{r}$. The algorithm works as follows. It firstly enumerates all ballots with Hamming distance at most h to some $r_i \in \boldsymbol{r}$. This can be done in time $O(p \cdot n^h)$. Then, similarly to the algorithm in the proof of Proposition 11.26, it discards those ballots that are not rational, and subsequently discards those ballots that do not have minimum distance to the profile. Finally, it iterates over all remaining rational ballots with minimum distance to determine whether there is one among them that agrees with l and disagrees with each l_j.

11.3.4 Lower Bounds for the Constraint-Based Framework

Finally, we show parameterized hardness results for CB-OUTCOME-KEMENY. When parameterized by both h and p, the problem is W[SAT]-hard.

Proposition 11.29. CB-OUTCOME-KEMENY *parameterized by h and p is* W[SAT]-*hard.*

Proof. We give an fpt-reduction from the W[SAT]-complete problem MONO-TONE-WSAT(FORM). Let (φ, k) be an instance of MONOTONE-WSAT(FORM). We construct an instance $(\mathcal{I}, \Gamma, r, l)$ of CB-OUTCOME-KEMENY as follows. We let $\mathcal{I} = \mathrm{Var}(\varphi) \cup \{z\} \cup \{y_{i,j} : i \in [3], j \in [3k+3]\}$. Moreover, we let $\Gamma = (z \wedge \varphi) \vee (\neg z \wedge \bigvee_{i \in [3]}(\bigwedge_{j \in [3k+3]} y_{i,j}))$. We define $r = (r_1, r_2, r_3)$ as follows. For each r_i, we let $r_i(w) = 0$ for all $w \in \{z\} \cup \mathrm{Var}(\varphi)$. Moreover, for each r_i and each $y_{\ell,j}$, we let $r_i(y_{\ell,j}) = 1$ if and only if $\ell = i$. It is readily verified that r_1, r_2 and r_3 are all rational. Finally, we let l be the partial assignment for which $l(z) = 1$, and that is undefined on all remaining variables. This completes our construction. Clearly, $p = 3$. Moreover, $h = 6k + 6$.

By construction of Γ, the only ballots that are rational—and that can have a smaller distance to the profile r than the ballots r_1, r_2 and r_3—are those ballots r^* that satisfy $(z \wedge \varphi)$. The ballots r_1, r_2 and r_3 have distance $4(3k + 3) = 12k + 12$ to the profile r. Any ballot r^* that satisfies $(z \wedge \varphi)$ minimizes its distance to r by setting all variables $y_{i,j}$ to false. Any such ballot r^* has distance $3(3k + 3) + 3(w + 1) = 9k + 3w + 12$ to the profile r, where w is the number of variables among $\mathrm{Var}(\varphi)$ that it sets to true. Therefore, the distance of such a ballot r^* to the profile r is smaller than (or equal to) the distance of r_1, r_2 and r_3 to r if and only if $9k + 3w + 12 \leq 12k + 12$, which is the case if and only if $w \leq k$. From this we can conclude that there is some $r^* \in \mathrm{Kemeny}(r)$ that agrees with l if and only if $(\varphi, k) \in$ MONOTONE-WSAT(FORM).

Finally, the proof of Proposition 11.22 can be modified to work also for the problem CB-OUTCOME-KEMENY parameterized by p, showing para-Θ_2^p-hardness for this case.

Proposition 11.30. CB-OUTCOME-KEMENY *parameterized by p is* para-Θ_2^p-*hard.*

Proof. We modify the Θ_2^p-hardness reduction used in the proof of Proposition 11.22 to work also for the case of CB-OUTCOME-KEMENY for a constant value of the parameter p. Instead of adding the formulas φ_i' to the agenda Φ, as done in the proof of Proposition 11.22, we let $\Gamma = \varphi'$. The remaining formulas in the agenda Φ were all propositional variables, and thus we can transform the instance (Φ, Γ, J, L) that we constructed for FB-OUTCOME-KEMENY into an instance $(\mathcal{I}, \Gamma, r, l)$, where r and l are constructed entirely analogously to J and L. Clearly, $p = 3$. Moreover, by a similar argument to the one that is used in the proof of Proposition 11.22, we get that $(\mathcal{I}, \Gamma, r, l) \in$ CB-OUTCOME-KEMENY if and only if $(\varphi, w) \in$ MAX-MODEL.

Table 11.4 Parameterized complexity of FB-OUTCOME-KEMENY for different sets of parameters.

Complexity	When parameterized by
in FPT	$\{c, n, m\}, \{c, h, n, m\}, \{c, n, m, p\}, \{c, h, n, m, p\}$
FPT^{NP}[few]-complete	$\{n\}, \{c, n\}, \{h, n\}, \{h, p\}, \{n, m\}, \{n, p\}, \{c, h, n\}, \{c, h, p\}, \{c, n, p\}, \{h, n, m\}, \{h, n, p\}, \{h, m, p\}, \{n, m, p\}, \{c, h, n, p\}, \{c, h, m, p\}, \{h, n, m, p\}$
para-Θ_2^p-complete	$\emptyset, \{c\}, \{h\}, \{m\}, \{p\}, \{c, h\}, \{c, m\}, \{c, p\}, \{h, m\}, \{m, p\}, \{c, h, m\}, \{c, m, p\}$

Table 11.5 Parameterized complexity of CB-OUTCOME-KEMENY for different sets of parameters.

Complexity	When parameterized by
in FPT	$\{c\}, \{n\}, \{c, h\}, \{c, n\}, \{c, p\}, \{h, n\}, \{n, p\}, \{c, h, n\}, \{c, h, p\}, \{c, n, p\}, \{h, n, p\}, \{c, h, n, p\}$
in XP, W[SAT]-hard	$\{h\}, \{h, p\}$
para-Θ_2^p-complete	$\emptyset, \{p\}$

11.3.5   Overview

In Tables 11.4 and 11.5, we provide an overview of the parameterize complexity of FB-OUTCOME-KEMENY and CB-OUTCOME-KEMENY for all possible combinations of the parameters that we considered for these two problems.

Notes

The results in Sect. 11.2 were shown in a paper that appeared in the proceedings of COMSOC 2014 [77] and in the proceedings of AAMAS 2015 [78]. The results in Sect. 11.3 were shown in a paper that appeared in the proceedings of COMSOC 2016 [106] and in the proceedings of ECAI 2016 [107].

Chapter 12
Planning Problems

In this chapter, we continue our parameterized complexity investigation of problems at higher levels of the Polynomial Hierarchy by investigating various parameterized problems that arise in the area of planning. In particular, we study several parameterized variants of natural planning problems that involve uncertainty about the initial state, as well as a planning problem where one needs to satisfy as many "soft goals" as possible in addition to satisfying the "hard goal".

Overview of This Chapter
We begin in Sect. 12.1 with explaining the formal planning framework that we use in this chapter (SAS$^+$ planning). Then, in Sect. 12.2, we consider several parameterized problems that are related to planning with uncertainty in the initial state. We show that the parameterized problems that we consider are complete for para-NP, para-DP, $\Sigma_2^p[k*]$ and $\Sigma_2^p[*k, P]$, respectively.

Finally, in Sect. 12.3, we consider a parameterized problem that involves an optimization component. That is, in this problem a plan is sought that optimizes the number of "soft goals" that are satisfied in addition to the "hard goal." We show that this problem, parameterized by the number of soft goals, is FPT^{NP}[few]-complete.

12.1 SAS$^+$ Planning

We begin by describing the framework of SAS$^+$ planning, that we will use in this chapter (see, e.g., [10]). Let $V = \{v_1, \ldots, v_n\}$ be a finite set of *variables* over a finite *domain* D. Furthermore, let $D^+ = D \cup \{\mathbf{u}\}$, where $\mathbf{u}$ is a special *undefined* value not present in D. Then D^n is the set of *total states* and $(D^+)^n$ is the set of *partial states* over V and D. Intuitively, a state $(d_1, \ldots, d_n) \in D^n$ corresponds to an assignment that assigns to each variable $v_i \in V$ the value $d_i \in D$, and a partial state corresponds to a partial assignment that assigns a value to some variables $v_i \in V$. Clearly, $D^n \subseteq (D^+)^n$—that is, each total state is also a partial state. Let $(d_1, \ldots, d_n) = s \in (D^+)^n$ be a state. Then the value of a variable v_i in state s is denoted by $s[v_i] = d_i$.

© Springer-Verlag GmbH Germany, part of Springer Nature 2019
R. de Haan: Parameterized Complexity in the Polynomial Hierarchy, LNCS 11880,
https://doi.org/10.1007/978-3-662-60670-4_12

An *SAS$^+$ instance* is a tuple $\mathbb{P} = (V, D, A, I, G)$ where V is a set of variables, D is a domain, A is a set of *actions*, $I \in D^n$ is the *initial state* and $G \in (D^+)^n$ is the (partial) *goal state*. Each action $a \in A$ has a *precondition* $\mathsf{pre}(a) \in (D^+)^n$ and an *effect* $\mathsf{eff}(a) \in (D^+)^n$.

We will frequently use the convention that a variable has the value **u** in a precondition/effect unless a value is explicitly specified. Furthermore, by a slight abuse of notation, we denote actions and partial states such as preconditions, effects, and goals as follows. Let $a \in A$ be an action, and let $\{p_1, \ldots, p_m\} \subseteq V$ be the set of variables that are not assigned by $\mathsf{pre}(a)$ to the value **u**—that is, $\{ v \in V : \mathsf{pre}(a)[v] \neq \mathbf{u} \} = \{p_1, \ldots, p_m\}$. Moreover, suppose that $\mathsf{pre}(a)[p_1] = d_1, \ldots, \mathsf{pre}(a)[p_m] = d_m$. Then we denote the precondition $\mathsf{pre}(a)$ by $\mathsf{pre}(a) = \{p_1 \mapsto d_1, \ldots, p_m \mapsto d_m\}$. In particular, if $\mathsf{pre}(a)$ is the partial state such that $\mathsf{pre}(a)[v] = \mathbf{u}$ for each $v \in V$, we denote $\mathsf{pre}(a)$ by $\emptyset$. We use a similar notation for effects. Let a be the action with $\mathsf{pre}(a) = \{p_1 \mapsto d_1, \ldots, p_m \mapsto d_m\}$ and $\mathsf{eff}(a) = \{e_1 \mapsto d'_1, \ldots, e_\ell \mapsto d'_\ell\}$. We then use the notation $a : \{p_1 \mapsto d_1, \ldots, p_m \mapsto d_m\} \to \{e_1 \mapsto d'_1, \ldots, e_{m'} \mapsto d'_{m'}\}$ to describe the action a.

Let $a \in A$ be an action and $s \in D^n$ be a state. Then a is *valid in s* if for all $v \in V$, either $\mathsf{pre}(a)[v] = s[v]$ or $\mathsf{pre}(a)[v] = \mathbf{u}$. The *result of a in s* is the state $t \in D^n$ defined as follows. Tor all $v \in V$, $t[v] = \mathsf{eff}(a)[v]$ if $\mathsf{eff}(a)[v] \neq \mathbf{u}$ and $t[v] = s[v]$ otherwise. Let $s_0, s_\ell \in D^n$ and let $\omega = (a_1, \ldots, a_\ell)$ be a sequence of actions (of length ℓ). We say that ω is a *plan from s_0 to s_ℓ* if either (i) ω is the empty sequence (and $\ell = 0$, and thus $s_0 = s_\ell$), or (ii) there are states $s_1, \ldots, s_{\ell-1} \in D^n$ such that for each $i \in [\ell]$, a_i is valid in s_{i-1} and s_i is the result of a_i in s_{i-1}. A state $s \in D^n$ is a *goal state* if for all $v \in V$, either $G[v] = s[v]$ or $G[v] = \mathbf{u}$. An action sequence ω is a *plan for $\mathbb{P}$* if ω is a plan from I to a goal state.

In planning instances, often so-called *conditional effects* are permitted as effects [172]. A conditional effect is of the form $s \triangleright t$, where $s, t \in (D^+)^n$ are partial states. Intuitively, such a conditional effect ensures that the variable assignment t is only applied if the condition s is satisfied. When allowing conditional effects, the effect of an action is not a partial state $\mathsf{eff}(a) \in (D^+)^n$, but a set $\mathsf{eff}(a) = \{s_1 \triangleright t_1, \ldots, s_\ell \triangleright t_\ell\}$ of conditional effects. (For the sake of simplicity, we assume that the partial states $t_1, \ldots, t_\ell$ are non-conflicting—that is, there exist no $v \in V$ and no $i_1, i_2 \in [\ell]$ with $i_1 < i_2$ such that $\mathbf{u} \neq t_{i_1}[v] \neq t_{i_2}[v] \neq \mathbf{u}$.) The result of an action a with $\mathsf{eff}(a) = \{s_1 \triangleright t_1, \ldots, s_\ell \triangleright t_\ell\}$ in a state s (in which a is valid) is the state $t \in D^n$ that is defined as follows. For all $v \in V$, $t[v] = t_i[v]$ if there exists some $i \in [\ell]$ such that s_i is satisfied in s and $t_i[v] \neq \mathbf{u}$, and $t[v] = s[v]$ otherwise.

12.2 Planning with Uncertainty

In this section, we consider various parameterized problems that involve planning with uncertainty in the initial state. We start our analysis with the problem of deciding whether there exists a plan that works for all possible initial states, where the plan length is polynomially bounded in the input size. The unparameterized variant of this problem is known to be Σ_2^p-complete [12].

In this setting, in addition to the variables V, we consider a set V_u of n' variables. Intuitively, the value of the variables in V_u in the initial state is unknown. The question is whether there exists an action sequence ω such that for each state $I_0 \in D^{n+n'}$ that extends I—that is, $I_0[v] = I[v]$ for each $v \in V$—it holds that ω is a plan for the SAS$^+$ instance $(V \cup V_u, D, A, I_0, G)$.

Concretely, we consider the following parameterized problem:

POLYNOMIAL-PLANNING(uncertainty)
Instance: A planning instance $\mathbb{P} = (V, V_u, D, A, I, G)$ containing additional variables V_u that are *unknown* in the initial state, i.e., for all $v \in V_u$ it holds that $I(v) = \mathbf{u}$.
Parameter: $|V_u| + |D|$.
Question: Is there a plan of polynomial length for $\mathbb{P}$ that works for all complete initial states I_0, i.e., for each possible way of completing I with a combination of values for variables in V_u?

Intuitively, the problem POLYNOMIAL-PLANNING(uncertainty) involves finding a plan of polynomial length that works regardless of the values of the unknown variables in the initial state. We show that this problem is para-NP-complete.

Proposition 12.1. POLYNOMIAL-PLANNING(uncertainty) *is* para-NP-*complete.*

Proof. The basic idea behind the proof of membership in para-NP is that we can enumerate all possible (complete) initial states I_0 in fixed-parameter tractable time. Let n be the input size and $m = p(n)$ be a bound on the plan length where p is a polynomial that bounds the plan length. To show membership, recall that classical (SAS$^+$) planning is NP-complete if an explicit upper bound m on the plan length is given (in unary). Thus, for a classical planning instance $\mathbb{P}$ we can construct a propositional formula $\varphi[\mathbb{P}]$ in polynomial time that is satisfiable if and only if there is a plan of length at most m for $\mathbb{P}$. This formula $\varphi[\mathbb{P}]$ contains (among others) variables V_A that encode the choice of actions in the plan. Due to the uncertainty in the initial state in POLYNOMIAL-PLANNING(uncertainty), there is not a single initial state, but rather a set $\mathcal{I}$ of initial states of cardinality $|D|^{|V_u|}$. Now let $\mathbb{P}'$ be an instance of POLYNOMIAL-PLANNING(uncertainty) and $I_0 \in \mathcal{I}$. Then, let $\mathbb{P}'(I_0)$ denote the classical planning instance obtained by instantiating the unknown variables in $\mathbb{P}'$ with the values of the corresponding variables in I_0. To show membership, we construct the formula $\psi = \bigwedge_{I_0 \in \mathcal{I}} \varphi[\mathbb{P}'(I_0)]$ that consists of conjuncts that are variable-disjoint (with exception of the variables V_A, that are shared among all formulas $\varphi[\mathbb{P}'(I_0)]$). Then, ψ is satisfiable if and only if $\mathbb{P}'$ is a yes-instance.

Hardness for para-NP follows from a proof by Kronegger, Pfandler, and Pichler [140, Theorem 11], that shows NP-hardness for the problem even when $V_u = \emptyset$ and $|D| = 2$.

The next problem that we consider is also related to planning with uncertainty in the initial state. We consider the same setting, where we have additional variables V_u whose value in the initial state is unknown. The crucial difference is that we ask whether an action a_0 is *essential* for achieving the goal. That is, the question is whether the goal can be reached within the required number of steps (with a single plan that works for each possible complete initial state) when the action a_0 is available, and cannot be reached (with a single plan for all initial states) when a_0 is not available. Concretely, we consider the following parameterized problem.

POLYNOMIAL-PLANNING-ESSENTIAL-ACTION(uncertainty)
Instance: A planning instance $\mathbb{P} = (V, V_u, D, A, I, G)$ with unknown variables V_u.
Parameter: $|V_u| + |D|$.
Question: Is there a plan of polynomial length for $\mathbb{P}$ that uses a_0 and works for all complete initial states I_0, but there is no such plan for $\mathbb{P}$ without using a_0?

We show that this problem is para-DP-complete.

Proposition 12.2. *The parameterized problem* POLYNOMIAL-PLANNING-ESSENTIAL-ACTION(uncertainty) *is* para-DP-*complete.*

Proof. To establish hardness, we will show that we can encode an instance of the DP-complete problem SAT-UNSAT into an instance of POLYNOMIAL-PLANNING-ESSENTIAL-ACTION(uncertainty) with $V_u = \emptyset$ and $|D| = 2$. Let (φ, ψ) be a SAT-UNSAT instance. Recall that deciding whether a given planning instance has a plan of length at most m (where m is given in unary) is NP-complete. Moreover, NP-hardness holds even for the case where $D = D_{\text{bin}} = \mathbb{B}$. Therefore, there are two planning instances $\mathbb{P}_\varphi = (V_\varphi, V_u, D_{\text{bin}}, A_\varphi, I_\varphi, G_\varphi)$ and $\mathbb{P}_\psi = (V_\psi, V_u, D_{\text{bin}}, A_\psi, I_\psi, G_\psi)$ with $V_u = \emptyset$ and with disjoint variables and actions, such that $\mathbb{P}_\varphi$ (respectively, $\mathbb{P}_\psi$) has a plan of polynomial length if and only if φ (respectively, ψ) is satisfiable. We then use the action a_0 to verify that $\mathbb{P}_\psi$ has indeed no plan.

From $\mathbb{P}_\psi$ we construct an instance $\mathbb{P}'_\psi$ as follows: Let the set $A'_\psi = A_\psi \cup \{a_0 : \emptyset \to G_\psi\}$ of actions be defined by adding to A_ψ an additional action a_0 with empty precondition that immediately fulfills the goal G_ψ. Further, let $\mathbb{P}'_\psi = (V_\psi, V_u, D_{\text{bin}}, A'_\psi, I_\psi, G_\psi)$. We now combine $\mathbb{P}_\varphi$ and $\mathbb{P}'_\psi$ to a single planning instance $\mathbb{P}^*$. Notice that this can always be done as the instances are disjoint (they share only the domain D_{bin}). The instance of POLYNOMIAL-PLANNING-ESSENTIAL-ACTION(uncertainty) is then given by $(\mathbb{P}^*, a_0)$. It is now easy to verify that φ is satisfiable and ψ is unsatisfiable if and only if there is a plan for $\mathbb{P}^*$ that uses a_0 and there is no plan for $\mathbb{P}^*$ that does not use a_0.

We show membership in para-DP by giving a reduction to SAT-UNSAT. Let $(\mathbb{P}, a_0)$ be an instance of POLYNOMIAL-PLANNING-ESSENTIAL-ACTION(uncertainty), where $\mathbb{P} = (V, V_u, D, A, I, G)$, and let m be a bound on the plan length that is polynomial in the input size. Without loss of generality, we assume that $0, 1 \in D$. We have to check whether there is a plan that uses a_0 and that there is no plan without using a_0. Recall from the proof of Theorem 12.1 that we can construct in fpt-time for an arbitrary instance $\mathbb{P}$ of POLYNOMIAL-PLANNING(uncertainty) a propositional formula $\varphi[\mathbb{P}]$ that is satisfiable if and only if $\mathbb{P}$ is a yes-instance.

For the case where action a_0 is available, we simply consider the instance $\mathbb{P}$. Moreover, we consider the propositional formula $\varphi[\mathbb{P}]$, that is satisfiable if and only if $\mathbb{P}$ has a plan. Then, for the case where a_0 is not available, we consider the instance $\mathbb{P}'$ that is obtained from $\mathbb{P}$ by removing the action a_0 from A. Similarly, we consider the propositional formula $\varphi[\mathbb{P}']$. The instance of SAT-UNSAT is then given by $(\varphi[\mathbb{P}], \varphi[\mathbb{P}'])$. Then it holds that $(\varphi[\mathbb{P}'], \varphi[\mathbb{P}'']) \in$ SAT-UNSAT if and only if there is a plan (of length at most m) that uses action a_0, and that there is no plan (of length at most m) without using action a_0.

In the next parameterized problem that we investigate, we take the plan length as parameter. Concretely, we consider the following parameterized problem.

BOUNDED-UNCERTAIN-PLANNING
Instance: A planning instance $\mathbb{P} = (V, V_u, D, A, I, G)$ with unknown variables V_u, and an integer k.
Parameter: k.
Question: Is there a plan of length k for $\mathbb{P}$ that works for all complete initial states I_0?

We show that this problem is $\Sigma_2^p[k*]$-complete.

Proposition 12.3. BOUNDED-UNCERTAIN-PLANNING *is* $\Sigma_2^p[k*]$-*complete.*

Proof. For membership in $\Sigma_2^p[k*]$, we use the encoding of Rintanen [174] into QSAT$_2$ for planning instances with uncertainty in the initial state and a binary domain. In this encoding, a QBF of the form $\exists X.\forall Y.\psi$ (with ψ quantifier-free) is constructed, that is satisfiable if and only if there is a plan whose length is bounded by a given integer. Moreover, X only contains variables representing actions. Furthermore, for truth assignment $\alpha : X \to \mathbb{B}$ such that $\forall Y.\psi[\alpha]$ is true, the weight of α is equal to the plan length. The encoding of Rintanen assumes a binary domain. Here, we consider planning instances with an arbitrarily large domain. This reduction can straightforwardly be modified to work also for larger domains. Therefore, this allows us to reduce the problem to $\Sigma_2^p[k*]$-WSAT.

To show $\Sigma_2^p[k*]$-hardness, we give an fpt-reduction from $\Sigma_2^p[k*]$-WSAT(3DNF). In this reduction, we will make use of conditional effects. Let (φ, k) be an instance of $\Sigma_2^p[k*]$-WSAT(3DNF), where k is an integer, $\varphi = \exists X.\forall Y.\psi$, and $\psi = \bigvee_{i \in [m]} \bigwedge_{j \in [3]} l_{i,j}$ is a 3DNF formula over $X \cup Y$. We construct an instance $\mathbb{P} = (V, V_u, D, A, I, G)$ as follows. The variables of the planning instance $\mathbb{P}$ are $V = X \cup \{c_1, \ldots, c_k\} \cup \{e, g\}$. For every $v \in X$ and $i \in [k]$, we introduce an action $a_v^i : \{v \mapsto 0, c_i \mapsto 0, e \mapsto 0\} \to \{v \mapsto 1, c_i \mapsto 1\}$. Furthermore, we introduce two additional actions: $a_e : \{c_1 \mapsto 1, \ldots, c_k \mapsto 1\} \to \{e \mapsto 1\}$, and $a_g : \{e \mapsto 1\} \to \{\{l_{1,1} \mapsto 1, l_{1,2} \mapsto 1, l_{1,3} \mapsto 1\} \triangleright \{g \mapsto 1\}, \ldots, \{l_{m,1} \mapsto 1, l_{m,2} \mapsto 1, l_{m,3} \mapsto 1\} \triangleright \{g \mapsto 1\}\}$. The set of actions A is then given by $A = \{a_v^i : v \in X, i \in [k]\} \cup \{a_e, a_g\}$. Moreover, we let $V_u = Y$, $D = \mathbb{B}$, $I = 0^{|V|}$ and $G = \{g \mapsto 1\}$.

The intuition of this encoding is to first guess an assignment α of weight k using k distinct actions a_v^i, then to fix this assignment using action a_e and finally to evaluate α according to ϕ with the action a_g. It is straightforward to check that (ϕ, k) is a yes-instance of $\Sigma_2^p[k*]$-WSAT(3DNF) if and only if there is a plan of length $k' = k + 2$ for $\mathbb{P}$ that works for all possible (complete) initial states. $\square$

For the final parameterized problem that we consider, the parameter captures the extent to which the unknown variables can deviate from their default values. Let $\mathbb{P} = (V, V_u, D, A, I, G)$ be a planning instance with unknown variables V_u. Without loss of generality, we assume that $0 \in D$. We call 0 the *base value* for the variables in V_u. The parameter specifies an upper bound on the maximum number of unknown variables that deviate from this base value. Intuitively, in this setting, there can be many unknown variables, yet we only need to consider cases where few of them have unexpected values. That is, we consider initial states $I_0 \in D^{n+n'}$ that extend I and for which there are at most d variables in V_u such that $I_0[d] \neq 0$. The question is then whether there exists an action sequence ω such that for each such state I_0 it holds that ω is a plan for the SAS$^+$ instance $(V \cup V_u, D, A, I_0, G)$. Concretely, we consider the following parameterized problem.

POLYNOMIAL-PLANNING(bounded-deviation)
Instance: A planning instance $\mathbb{P} = (V, V_u, D, A, I, G)$ with unknown variables V_u, and an integer d.
Parameter: d.
Question: Is there a plan of polynomial length for $\mathbb{P}$ that works for each complete initial state I_0 where at most d unknown variables deviate from the base value?

We show that this problem is $\Sigma_2^p[*k, P]$-complete.

Proposition 12.4. *The parameterized problem* POLYNOMIAL-PLANNING(bounded-deviation) *is* $\Sigma_2^p[*k, P]$-*complete.*

Proof. To show membership in $\Sigma_2^p[*k, P]$, we describe an algorithm to solve the problem that can be implemented by an $\Sigma_2^p[*k, P]$-machine. The algorithm first guesses a sequence ω of m actions from A using $m \log |A|$ (binary) non-deterministic choices in the existential phase of the computation. Then, in the universal phase, it verifies whether this plan works for all cases where k of the unknown variables deviate from the base value in the initial state. Each such initial state can be specified using $k \log (|V_u| \cdot |D|)$ non-deterministic choices (in the universal phase), and for any such initial state I, checking whether the ω reaches a goal state from I can be done in polynomial time. This algorithm is correct and can be implemented by an $\Sigma_2^p[*k, P]$-machine. Therefore, by Proposition 6.29, membership in $\Sigma_2^p[*k, P]$ follows.

To show $\Sigma_2^p[*k, P]$-hardness, we give an fpt-reduction from $\Sigma_2^p[*k]$-WSAT(CIRC). Intuitively, in this reduction, we will emulate the evaluation of the circuit by a planning problem. Here, the goal is to set the output gate to true and the unknown variables are used to model the universally quantified variables of the circuit. With help of the actions, we sequentially evaluate the output of the gates in a fixed order to finally compute the value of the output gate of the circuit. Recall that an instance of $\Sigma_2^p[*k]$-WSAT(CIRC) consists of a Boolean circuit C over two sets of disjoint input variables X and Y, and an integer k. The question is whether there is an assignment α to the variables in X such that for any assignment β to the variables in Y of weight k, the circuit C is satisfied by $\alpha \cup \beta$.

Since a circuit can be seen as an acyclic directed graph, we may assume that the gates $g_1, \ldots, g_m$ are numbered in such a way that for each gate g_i, its inputs $g_{j_1}, \ldots, g_{j_\ell}$ are numbered in such a way that $j_1, \ldots, j_\ell < i$. This numbering gives a natural ordering of the gates, which ensures that all input values for a gate are already computed if the output value is to be determined. We create the set $L = \{l_1, \ldots, l_m\}$ of variables. Then, we introduce a new variable f and the action $a_f : \{f \mapsto 0\} \to \{f \mapsto 1\}$. Furthermore, for each variable $x \in X$, we create the action $a_x : \{f \mapsto 0\} \to \{x \mapsto 1\}$. For each gate g_i, we create an action a_{g_i}, where:

$$\mathsf{pre}(a_{g_i}) = \{f \mapsto 1, l_{i-1} \mapsto 1\}$$
$$\mathsf{eff}(a_{g_i}) = \{l_i \mapsto 1\} \cup \Gamma_{g_i}$$

Here Γ_{g_i} depends on the type of gate g_i and is obtained as follows. For each gate g_i we introduce the variable $o(g_i)$, representing the unnegated output of gate g_i. Let the unnegated input gates of gate g be $g_1', \ldots, g_p'$, and the negated ones be $g_1'', \ldots, g_l''$. To simplify the presentation, we assume that each variable in Y is also represented by a gate g, whose output is represented as $o(g)$. If g is an $\wedge$-gate we define:

$$\Gamma_{g_i} = \big\{\{o(g_1') \mapsto 1, \ldots, o(g_p') \mapsto 1, o(g_1'') \mapsto 0, \ldots, o(g_l'') \mapsto 0\} \triangleright \{o(g) \mapsto 1\}\big\}$$

If g is an $\vee$-gate, Γ_{g_i} is defined analogously. Notice that the effects in the set Γ_{g_i} are conditional. Intuitively, executing the actions $a_{g_1}, \ldots, a_{g_m}$ in order corresponds to evaluating the circuit using the given values for the variables in X and Y. Moreover, executing these actions is the only way to set $o(g_o)$ to 1, where g_o is the output gate of C.

To put things together, we set $V = X \cup L \cup \{f\} \cup \{o(g) : g \text{ is a gate in } C\}$, $V_u = Y$, $A = \{a_f\} \cup \{a_x : x \in X\} \cup \{a_g : g \text{ is a gate in } C\}$, $I(v) = 0$ for each $v \in V$, $G = \{o(g_o) \mapsto 1\}$, where g_o is the output gate of C, and $d = k$. Moreover, we let $m + |X|$ the upper bound on the plan length. Verifying the correctness of this reduction is straightforward.

12.3 Soft Goal Optimization

In this section, we investigate a parameterized problem that involves an optimization component. We consider planning instances that have two different types of goals: (1) a *hard goal* G_h that needs to be satisfied in each solution, and (2) a *soft goal* G_s, for which the number of satisfied variables needs to be maximized. We call a plan *optimal* with respect to some given bound m on the plan length if it satisfies G_h and there does not exist another plan with length at most m that satisfies G_h and that satisfies more variables according to the soft goal G_s. We analyze the problem of finding an optimal plan given a planning instance $\mathbb{P} = (V, D, A, I, G_h, G_s)$ and a bound m on the plan length, parameterized by $|G_s| = |\{v \in V : G_s(v) \neq \mathbf{u}\}|$.

In general, finding an optimal plan for $\mathbb{P}$ given a bound m on the plan length in polynomial time requires $O(\log |\mathbb{P}|)$ SAT queries. We can find such a plan by performing binary search on the number of fulfilled variables in the soft goal, asking whether at most ℓ variables in the soft goal can be fulfilled. We show that we can restrict the number of SAT queries to a function of the number $|G_s|$ of variables in the soft goal. In particular, we show that planning with soft goals can be solved in fixed-parameter tractable time using $\lceil \log |G_s| \rceil$ SAT queries.

Proposition 12.5. *Let* $\mathbb{P} = (V, D, A, I, G_h, G_s)$ *be a planning instance with a hard goal* G_h *and a soft goal* G_s, *and let m be an integer that is polynomial in the input size. Then finding a plan of length at most m that is optimal for* $\mathbb{P}$ *with respect to m (if it exists) can be done in fpt-time using* $\lceil \log |G_s| \rceil$ *SAT queries.*

Proof (sketch). The problem of deciding whether there is a plan of length at most k that reaches some state s' satisfying the hard goal G_h and that agrees with the soft goal G_s on at least u variables is in NP. Namely, one can guess such a plan, and verify whether it satisfies the requirements. Therefore, any instance of this problem can be encoded into an instance of SAT in polynomial time. Moreover, from a satisfying assignment (if one exists) for such a SAT instance, we can extract in polynomial time a plan that satisfies the requirements. Then we can find the maximum number u of variables contained in the soft goal that can be fulfilled using $\lceil \log |G_s| \rceil$ SAT queries, by using binary search. Moreover, using the satisfying assignments that are given by the SAT solver, we can find a plan that is optimal for $\mathbb{P}$ with respect to length m.

Secondly, we prove that the number of required SAT queries cannot be bounded by a constant. In order to show that we cannot find such an optimal plan in fixed-parameter tractable time using a constant number of SAT queries, we consider the following parameterized decision problem.

POLYNOMIAL-OPTIMIZATION-PLANNING(#soft.goals)
Instance: A planning instance $\mathbb{P} = (V, D, A, I, G_h, G_s)$ with a hard goal G_h and a soft goal G_s, and an integer m.
Parameter: $|G_s|$.
Question: Does there exist a plan of length at most m that is optimal for $\mathbb{P}$ (w.r.t. m), and that satisfies an odd number of variables of the soft goal?

We show that this problem is complete for the class $\text{FPT}^{\text{NP}}[\text{few}]$. It then follows from Proposition 7.12 that we cannot find an optimal plan in fpt-time using a constant number of SAT queries, unless the Polynomial Hierarchy collapses.

Proposition 12.6. *The parameterized problem* POLYNOMIAL-OPTIMIZATION-PLANNING (#soft.goals) *is* $\text{FPT}^{\text{NP}}[\text{few}]$*-complete.*

Proof (sketch). Membership in $\text{FPT}^{\text{NP}}[\text{few}]$ follows directly from Proposition 12.5. To show hardness, we give an fpt-reduction from the problem ODD-LOCAL-MAX-MODEL. Let (φ, X) specify an instance of ODD-LOCAL-MAX-MODEL. We may assume without loss of generality that φ is in CNF. Let u be the number of clauses of φ. We construct a planning instance $\mathbb{P} = (V, D, A, I, G_h, G_s)$, with a hard goal G_h and a soft goal G_s, and an integer m bounding the plan length as follows. We let $V = \text{Var}(\varphi) \cup \{ z_x : x \in \text{Var}(\varphi) \} \cup \{ y_\delta : \delta \text{ is a clause of } \varphi \}$, and $D = \mathbb{B}$. For each variable $x \in \text{Var}(\varphi)$, we introduce two actions: $a_0^x : \emptyset \to \{x \mapsto 0, z_x \mapsto 1\}$ and $a_1^x : \emptyset \to \{x \mapsto 1, z_x \mapsto 1\}$. Moreover, for each clause δ in φ and each literal l in δ, we introduce an action a_l^δ whose precondition requires all variables z_x to have value 1 and requires the variable of l to be set in such a way that δ

is satisfied, and whose effect ensures that y_δ is set to 1. We let $I(v) = 0$ for all $v \in V$, $G_h = \{ y_\delta \mapsto 1 : \delta$ is a clause of $\varphi \}$, and $G_s = \{ x \mapsto 1 : x \in X \}$. Finally, we let $m = |\text{Var}(\varphi)| + u$.

The intuition behind this reduction is that the actions a_i^x can be used to enforce a truth assignment over the variables in $\text{Var}(\varphi)$, and that the actions a_l^δ can be used to check that such a truth assignment satisfies all clauses δ of φ. It is straightforward to verify that those plans consisting of m actions—an action a_i^x (for some $i \in \mathbb{B}$) for each $x \in \text{Var}(\varphi)$, and an action a_l^δ (for some $l \in \delta$) for each $\delta \in \varphi$—correspond to truth assignments over $\text{Var}(\varphi)$ that satisfy φ. Moreover, only plans of this form can satisfy the hard goal. Then, finding such a plan that maximizes the number of fulfilled variables in the soft goal corresponds to a satisfying truth assignment that maximizes the number of satisfied variables in X. From this, the correctness of the reduction follows.

Notes

The results in this chapter appeared in a paper in the proceedings of IJCAI 2015 [109].

Chapter 13
Graph Problems

In this chapter, we conclude our parameterized complexity investigation of problems at higher levels of the Polynomial Hierarchy by looking at several parameterized variants of two Π_2^p-complete graph problems. We consider several parameterizations of the problem of deciding whether particular 3-colorings of the leaves of a graph can be extended to a proper 3-coloring of the entire graph. This problem is denoted by 3-COLORING-EXTENSION. Moreover, we consider a parameterized variant of the problem of deciding, given a graph $G = (V, E)$, whether each clique containing only vertices in a given subset V' of vertices of the graph can be extended to a larger clique (with a given lower bound) using vertices outside the set V'. This problem is denoted by CLIQUE-EXTENSION.

Outline of This Chapter
In Sect. 13.1, we analyze the complexity of several parameterized variants of 3-COLORING-EXTENSION. For two parameters, the problem is para-Π_2^p-complete. For another parameter, the problem is para-NP-complete, and for yet another parameter, the problem is $\Pi_2^p[k*]$-complete.

Then, in Sect. 13.2, we consider the parameterized variant of CLIQUE-EXTENSION, parameterized by the lower bound of the clique extension. We show that this parameterized problem is $\Pi_2^p[*k, 1]$-complete.

13.1 Extending Graph Colorings

In this section, we consider several parameterized variants of the well-known Π_2^p-complete problem 3-COLORING-EXTENSION, as considered by Ajtai, Fagin, and Stockmeyer [5].

Let $G = (V, E)$ be a graph. We will denote those vertices v that have degree 1 by *leaves*. We call a (partial) function $c : V \to \{1, 2, 3\}$ a 3-*coloring (of G)*. Moreover, we say that a 3-coloring c is *proper* if c assigns a color to every vertex $v \in V$, and if for each edge $e = \{v_1, v_2\} \in E$ holds that $c(v_1) \neq c(v_2)$.

© Springer-Verlag GmbH Germany, part of Springer Nature 2019
R. de Haan: Parameterized Complexity in the Polynomial Hierarchy, LNCS 11880,
https://doi.org/10.1007/978-3-662-60670-4_13

Consider the following decision problem.

3-COLORING-EXTENSION
Instance: A graph $G = (V, E)$ with n leaves, and an integer m.
Question: Can each 3-coloring that assigns a color to exactly m leaves of G (and to no other vertices) be extended to a proper 3-coloring of G?

We consider the following parameterizations for this problem.

- 3-COLORING-EXTENSION(degree), where the parameter is the degree of G, i.e., $k = \deg(G)$;
- 3-COLORING-EXTENSION(#leaves), where the parameter is the number of leaves of G, i.e., $k = n$;
- 3-COLORING-EXTENSION(#col.leaves), where the parameter is the number of leaves that are pre-colored, i.e., $k = m$; and
- 3-COLORING-EXTENSION(#uncol.leaves), where the parameter is the number of leaves that are not pre-colored, i.e., $k = n - m$.

We begin by showing para-Π_2^p-completeness for the problem parameterized by the degree of the graph.

Proposition 13.1. 3-COLORING-EXTENSION(degree) *is* para-Π_2^p-*complete.*

Proof. It is known that 3-COLORING-EXTENSION is already Π_2^p-hard when restricted to graphs of degree 4 [5]. From this, it follows immediately that 3-COLORING-EXTENSION is para-Π_2^p-hard [84]. Membership in para-Π_2^p follows directly from the para-Π_2^p-membership of 3-COLORING-EXTENSION.

Next, we show that the problem is para-NP-complete when parameterized by the number of leaves.

Proposition 13.2. 3-COLORING-EXTENSION(#leaves) *is* para-NP-*complete.*

Proof. To show membership in para-NP, we give an fpt-reduction to SAT. Let (G, m) be an instance of 3-COLORING-EXTENSION(#leaves), where k denotes the number of leaves of G. We construct a propositional formula that is satisfiable if and only if $(G, m) \in$ 3-COLORING-EXTENSION(#leaves). Let V' denote the set of leaves of G, and let C be the set of all 3-colorings that assigns m vertices in V' to any color. We know that $|C| \leq 3^k$. For each $c \in C$, we know that the problem of deciding whether c can be extended to a proper 3-coloring of G is in NP. Therefore, for each c, we can construct a propositional formula φ_c that is satisfiable if and only if c can be extended to a proper 3-coloring of G. We may assume without loss of generality that for any distinct $c, c' \in C$ it holds that φ_c and $\varphi_{c'}$ are variable-disjoint. We then let $\varphi = \bigwedge_{c \in C} \varphi_c$. Clearly, φ is satisfiable if and only $(G, m) \in$ 3-COLORING-EXTENSION. Moreover, φ is of size $O(3^k n^d)$, where n is the input size and d is some constant.

Hardness for para-NP follows directly from the NP-hardness of deciding whether a given graph has a proper 3-coloring, which corresponds to the restriction of 3-COLORING-EXTENSION to instances with $k = 0$, i.e., to graphs that have no leaves.

Then, we show that, when parameterized by the number of leaves that are not pre-colored, the problem is para-Π_2^p-complete.

Proposition 13.3. 3-COLORING-EXTENSION(#uncol.leaves) *is* para-Π_2^p-*complete.*

Proof. We know that 3-COLORING-EXTENSION is already Π_2^p-hard when restricted to instances where $n = m$ (and thus where $k = 0$) [5]. From this, it follows immediately that 3-COLORING-EXTENSION is para-Π_2^p-hard [84]. Membership in para-Π_2^p follows directly from the Π_2^p-membership of 3-COLORING-EXTENSION.

Finally, we show that the problem is $\Pi_2^p[k*]$-complete when parameterized by the number of leaves that are pre-colored.

Theorem 13.4. 3-COLORING-EXTENSION(#col.leaves) *is* $\Pi_2^p[k*]$-*complete.*

Proof. To show $\Pi_2^p[k*]$-hardness, it suffices to observe that the polynomial-time reduction from co-QSAT$_2$ to 3-COLORING-EXTENSION to show Π_2^p-hardness [5, Appendix] can be seen as an fpt-reduction from $\Pi_2^p[k*]$−WSAT to 3-COLORING-EXTENSION(#col.leaves).

To show membership, we give an fpt-reduction from 3-COLORING-EXTENSION (#col.leaves) to $\Pi_2^p[k*]$-MC. Let (G, m) be an instance of 3-COLORING-EXTENSION (#col.leaves), where V' denotes the set of leaves of G, and where $k = m$ is the number of edges that can be pre-colored. Moreover, let $V' = \{v_1, \ldots, v_n\}$ and let $V = V' \cup \{v_{n+1}, \ldots, v_u\}$. We construct an instance $(\mathcal{A}, \varphi)$ of $\Pi_2^p[k*]$-MC. We define the domain $A = \{a_{v,i} : v \in V', i \in [3]\} \cup \{1, 2, 3\}$. Next, we define $C^{\mathcal{A}} = \{1, 2, 3\}$, $S^{\mathcal{A}} = \{(a_{v,i}, a_{v,i'}) : v \in V', i, i' \in [3]\}$, and $F^{\mathcal{A}} = \{(j, j') : j, j' \in [3], j \neq j'\}$. Then, we can define the formula φ, by letting $\varphi = \forall x_1, \ldots, x_k. \exists y_1, \ldots, y_u.(\psi_1 \rightarrow (\psi_2 \wedge \psi_3 \wedge \psi_4))$, where $\psi_1 = \bigwedge_{j,j' \in [k], j < j'} \neg S(x_i, x_{i'})$, and $\psi_2 = \bigwedge_{j \in [u]} C(y_j)$, and $\psi_3 = \bigwedge_{v_j \in V', i \in [3]} \left(\left(\bigvee_{\ell \in [k]} (x_\ell = a_{v_j,i}) \right) \rightarrow (y_j = i) \right)$, and $\psi_4 = \bigwedge_{\{v_j, v_{j'}\} \in E} F(y_j, y_{j'})$. It is straightforward to verify that $(G, m) \in$ 3-COLORING-EXTENSION if and only if $\mathcal{A} \models \varphi$.

Intuitively, the assignments to the variables x_i correspond to the pre-colorings of the vertices in V'. This is done by means of elements $a_{v,i}$, which represent the coloring of vertex v with color i. The subformula ψ_1 is used to disregard any assignments where variables x_i are not assigned to the intended elements. Moreover, the assignments to the variables y_i correspond to a proper 3-coloring extending the pre-coloring. The subformula ψ_2 ensures that the variables y_i are assigned to a color in $\{1, 2, 3\}$, the subformula ψ_3 ensures that the coloring encoded by the assignment to the variables y_i extends the pre-coloring encoded by the assignment to the variables x_i, and the subformula ψ_4 ensures that the coloring is proper.

13.2 Extending Cliques

In this section, we analyze the complexity of a parameterized variant of the Π_2^p-complete problem CLIQUE-EXTENSION. We show that this parameterized problem is

$\Pi_2^p[*k, 1]$-complete. Moreover, we show how the $\Pi_2^p[*k, 1]$-hardness proof for this parameterized problem can be adapted to show that the problem CLIQUE-EXTENSION is Π_2^p-complete.

Let $G = (V, E)$ be a graph. A clique $C \subseteq V$ of G is a subset of vertices that induces a complete subgraph of G, i.e., $\{v, v'\} \in E$ for all $v, v' \in C$ such that $v \neq v'$. The W[1]-complete problem of determining whether a graph has a clique of size k is an important problem in the W-hierarchy, and is used in many W[1]-hardness proofs.

Consider the following parameterized variant of CLIQUE-EXTENSION.

SMALL-CLIQUE-EXTENSION
Instance: A graph $G = (V, E)$, a subset $V' \subseteq V$, and an integer k.
Parameter: k.
Question: Is it the case that for each clique $C \subseteq V'$, there is some k-clique D of G such that $C \cup D$ is a $(|C| + k)$-clique?

We show $\Pi_2^p[*k, 1]$-completeness for this problem SMALL-CLIQUE-EXTENSION.

Proposition 13.5. SMALL-CLIQUE-EXTENSION *is* $\Pi_2^p[*k, 1]$-*complete.*

Proof. To show $\Pi_2^p[*k, 1]$-hardness, we give an fpt-reduction from $\Pi_2^p[*k]$-WSAT (2CNF) to SMALL-CLIQUE-EXTENSION. Let (φ, k) be an instance of $\Pi_2^p[*k]$-WSAT (2CNF), where $\varphi = \forall Y.\exists X.\psi$. By step 2 in the proof of Theorem 6.21, we may assume without loss of generality that ψ is antimonotone in X, i.e., all literals of ψ that contain variables in X are negative.

We construct an instance (G, V', k) of SMALL-CLIQUE-EXTENSION as follows. We define:

$$G = (V, E);$$
$$V' = \{v_y, v_{\neg y} : y \in Y\};$$
$$V = V' \cup \{v_x : x \in X\};$$
$$E = E_Y \cup E_{XY};$$
$$E_Y = \{\{v_y, v'\} : y \in Y, v' \in V', v' \neq v_{\neg y}\} \cup$$
$$\qquad \{\{v_{\neg y}, v'\} : y \in Y, v' \in V', v' \neq v_y\}; \text{ and}$$
$$E_{XY} = \{\{v_x, v_{x'}\} : x \in X, x' \in X, \{\neg x, \neg x'\} \notin \psi\} \cup$$
$$\qquad \{\{v_x, v_y\} : x \in X, y \in Y, \{\neg x, \neg y\} \notin \psi\} \cup$$
$$\qquad \{\{v_x, v_{\neg y}\} : x \in X, y \in Y, \{\neg x, y\} \notin \psi\}.$$

We claim that $(\varphi, k) \in \Pi_2^p[*k]$-WSAT(2CNF) if and only if $(G, V', k) \in$ SMALL-CLIQUE-EXTENSION.

($\Rightarrow$) Assume that $(\varphi, k) \in \Pi_2^p[*k]$-WSAT(2CNF). Let $C \subseteq V'$ be an arbitrary clique of G. It suffices to consider maximal cliques C, i.e., assume there is no clique C' such that $C \subsetneq C' \subseteq V'$. If a maximal clique C can be extended with k elements in V to another clique, then clearly this holds for all its subsets as well.

We show that for all $y \in Y$, either $v_y \in C$ or $v_{\neg y} \in C$. Assume the contrary, i.e., assume that for some $y \in Y$ it holds that $v_y \notin C$ and $v_{\neg y} \notin C$. Then $C \cup \{v_y\} \supsetneq C$

is a clique. This contradicts our assumption that C is maximal. Now define the assignment $\alpha_C : Y \to \mathbb{B}$ by letting $\alpha_C(y) = 1$ if and only if $v_y \in C$. We then know that there exists an assignment β to the variables X of weight k such that $\alpha_C \cup \beta$ satisfies ψ. Consider the set $D_\beta = \{ v_x \in V : x \in X, \beta(x) = 1 \} \subseteq V$. Since β has weight k, we know that $|D_\beta| = k$. Also, $D_\beta \cap C = \emptyset$, therefore $|C \cup D_\beta| = |C| + k$. By the construction of E, and by the fact that $\alpha_C \cup \beta$ satisfies ψ, it follows that $C \cup D_\beta$ is a clique. To see this, assume the contrary, i.e., assume that there exist $v, v' \in C \cup D_\beta$ such that $\{v, v'\} \notin E$. We assume that $v = v_x$ and $v' = v_{x'}$ for $x, x' \in X$. The other cases are analogous. Then $\{\neg x, \neg x'\} \in \psi$. Since $v_x, v_{x'} \in D_\beta$, we know that $\beta(x) = \beta(x') = 1$. Then $\alpha_C \cup \beta$ does not satisfy ψ, which is a contradiction. From this we can conclude that $C \cup D_\beta$ is a $(|C| + k)$-clique of G, and thus that $(G, V', k) \in$ SMALL-CLIQUE-EXTENSION.

($\Leftarrow$) Assume $(G, V', k) \in$ SMALL-CLIQUE-EXTENSION. We show that for all assignments $\alpha : Y \to \mathbb{B}$ there exists an assignment $\beta : X \to \mathbb{B}$ of weight k such that $\psi[\alpha \cup \beta]$ evaluates to true. Let $\alpha : Y \to \mathbb{B}$ be an arbitrary assignment. Consider $C_\alpha = \{ v_y \in V : y \in Y, \alpha(y) = 1 \} \cup \{ v_{\neg y} : y \in Y, \alpha(y) = 0 \} \subseteq V'$. By construction of $E_Y \subseteq E$, it follows that C_α is a clique of G. We then know that there exists some set $D \subseteq V$ of size k such that $C_\alpha \cup D$ is a clique. We show that $D \subseteq V \backslash V'$. To show the contrary, assume that this is not the case, i.e., assume that there exists some $v \in D \cap V'$. By construction of C_α, we know that for each $y \in Y$, either $v_y \in C$ or $v_{\neg y} \in C_\alpha$. We also know that $v \in \{v_{y'}, v_{\neg y'}\}$ for some $y' \in Y$. Assume that $v = v_{y'}$; the other case is analogous. If $v_{y'} \in C_\alpha \cap D$, then $C_\alpha \cup D$ cannot be a clique of size $|C_\alpha| + k = |C_\alpha| + |D|$. Therefore, $v_{y'} \notin C_\alpha$, and thus $v_{\neg y'} \in C_\alpha$. We then know that $\{v_{y'}, v_{\neg y'}\} \subseteq C_\alpha \cup D$. However, $\{v_{y'}, v_{\neg y'}\} \notin E$, and therefore $C_\alpha \cup D$ is not a clique. This is a contradiction, and thus $D \subseteq V \backslash V'$.

We define $\beta_D : X \to \mathbb{B}$ as follows. We let $\beta_D(x) = 1$ if and only if $v_x \in D$. Clearly, β_D is of weight k. We show that $\psi[\alpha \cup \beta_D]$ evaluates to true. Consider an arbitrary clause c of ψ. Assume that $c = \{\neg x, y\}$ for some $x \in X$ and some $y \in Y$; the other cases are analogous. To show the contrary, assume that $\alpha \cup \beta_D$ does not satisfy c, i.e., $(\alpha \cup \beta_D)(x) = 1$ and $(\alpha \cup \beta_D)(y) = 0$. Then $v_x \in D$ and $v_{\neg y} \in C$. However $\{v_x, v_{\neg y}\} \notin E$, and thus $C_\alpha \cup D$ is not a clique. This is a contradiction, and therefore we can conclude that $\alpha \cup \beta_D$ satisfies c. Since c was arbitrary, we know that $\psi[\alpha \cup \beta_D]$ evaluates to true. Thus, $(\varphi, k) \in \Pi_2^p[*k]$-WSAT(2CNF).

To show membership in $\Pi_2^p[*k, 1]$, we give an fpt-reduction from SMALL-CLIQUE-EXTENSION to $\Pi_2^p[*k]$-WSAT($\Gamma_{1,3}$). Let (G, V', k) be an instance SMALL-CLIQUE-EXTENSION, with $G = (V, E)$. We construct an equivalent instance (C, k) of $\Pi_2^p[*k]$-WSAT($\Gamma_{1,3}$). We will define a circuit C, that has universally quantified variables Y and existentially quantified variables X. We present the circuit C as a propositional formula. We define:

$$Y = \{ y_{v'} : v' \in V' \};$$

$$X = \{ x_v : v \in V \};$$

$$C = C_1 \vee (C_2 \wedge C_3);$$

$$C_1 = \bigvee_{\{v_1, v_2\} \in (V' \times V') \setminus E} \left(y_{v_1} \wedge y_{v_2} \right);$$

$$C_2 = \bigwedge_{v' \in V'} (\neg y_{v'} \vee \neg x_{v'});$$

$$C_3 = \bigwedge_{e \in (V \times V) \setminus E} \chi_e;$$

$$\chi_e = \bigwedge_{\substack{z_1 \in \Theta(v_1) \\ z_2 \in \Theta(v_2)}} (\neg z_1 \vee \neg z_2), \qquad \text{for all } \{v_1, v_2\} = e \in (V \times V) \setminus E; \text{ and}$$

$$\Theta(v) = \begin{cases} \{x_v, y_v\} & \text{if } v \in V', \\ \{x_v\} & \text{otherwise,} \end{cases} \qquad \text{for all } v \in V.$$

We show that $(G, V', k) \in$ SMALL-CLIQUE-EXTENSION if and only if $(C, k) \in \Pi_2^p[*k]$-WSAT$(\Gamma_{1,3})$.

($\Rightarrow$) Assume that $(G, V', k) \in$ SMALL-CLIQUE-EXTENSION. This means that for each clique $F \subseteq (V' \times V')$ there exists some set $D \subseteq V$ of vertices such that $F \cup D$ is a $(|F| + k)$-clique. We show that for each assignment $\alpha : Y \to \mathbb{B}$ there exists an assignment $\beta : X \to \mathbb{B}$ of weight k such that $C[\alpha \cup \beta]$ evaluates to true. Let $\alpha : Y \to \mathbb{B}$ be an arbitrary assignment. We define $F_\alpha = \{ v \in V' : \alpha(y_v) = 1 \}$. We distinguish two cases: either (i) F_α is not a clique in G, or (ii) F_α is a clique in G. In case (i), we know that there exist $v_1, v_2 \in F_\alpha$ such that $\{v_1, v_2\} \notin E$. Then α satisfies $(y_{v_1} \wedge y_{v_2})$, and therefore α satisfies C_1 and C as well. Thus for every assignment $\beta : X \to \mathbb{B}$ of weight k, $\alpha \cup \beta$ satisfies C.

Consider case (ii). Let $m = |F_\alpha|$. We know that there exists a subset $D \subseteq V$ of vertices such that $F_\alpha \cup D$ is an $(m + k)$-clique. Define the assignment $\beta : X \to \mathbb{B}$ by letting $\beta(x_v) = 1$ if and only if $v \in D$. Clearly, β has weight k. We show that $\alpha \cup \beta$ satisfies C. We know there is no $v \in V$, such that $v \in F_\alpha \cap D$, since otherwise $F_\alpha \cup D$ could not be a clique of size $m + k$. Therefore, $\alpha \cup \beta$ satisfies C_2. Since $F_\alpha \cup D$ is a clique, we know that $\alpha \cup \beta$ satisfies C_3 as well. Thus $\alpha \cup \beta$ satisfies C. This concludes our proof that $(C, k) \in \Pi_2^p[*k]$-WSAT$(\Gamma_{1,3})$.

($\Leftarrow$) Assume that $(C, k) \in \Pi_2^p[*k]$-WSAT$(\Gamma_{1,3})$. This means that for each assignment $\alpha : Y \to \mathbb{B}$ there exists an assignment $\beta : X \to \mathbb{B}$ of weight k such that $C[\alpha \cup \beta]$ evaluates to true. Let $F \subseteq (V' \times V')$ be an arbitrary clique, and let $m = |F|$. We show that there exists a set $D \subseteq V$ of vertices such that $F \cup D$ is an $(m + k)$-clique. Let $\alpha_F : Y \to \mathbb{B}$ be the assignment defined by letting $\alpha_F(y_v) = 1$ if and only if $v \in F$. We know that there must exist an assignment $\beta : X \to \mathbb{B}$ of weight k such that $C[\alpha_F \cup \beta]$ evaluates to true. Since F is a clique, it is straightforward to verify that $\alpha_F \cup \beta$ does not satisfy C_1. Therefore, $\alpha_F \cup \beta$ must satisfy C_2 and C_3. Consider $D_\beta = \{ v \in V : \beta(x_v) = 1 \}$. Since β has weight k, $|D| = k$. Because $\alpha_F \cup \beta$ satisfies C_2, we know that $F \cap D_\beta = \emptyset$, and thus that $|F \cup D_\beta| = m + k$. Because $\alpha_F \cup \beta$ satisfies C_3, we know that $F \cup D_\beta$ is a clique in G. Since F was arbitrary, we can conclude that $(G, V', k) \in$ SMALL-CLIQUE-EXTENSION.

13.2.1 Π_2^p-Completeness for CLIQUE-EXTENSION

We show that the $\Pi_2^p[*k, 1]$-hardness proof for SMALL-CLIQUE-EXTENSION, that we gave above, can be adapted to show that CLIQUE-EXTENSION is Π_2^p-complete. In order to do so, we consider the following intermediate problem.

WEIGHTED-Π_2^p-SAT(2CNF)

Instance: A quantified Boolean formula $\varphi = \forall X.\exists Y.\psi$, where ψ is in 2CNF, and a positive integer u
Question: Is it the case for all truth assignments $\alpha : X \to \mathbb{B}$ there exists a truth assignment $\beta : Y \to \mathbb{B}$ of weight u such that $\psi[\alpha \cup \beta]$ is true?

We firstly show that WEIGHTED-Π_2^p-SAT(2CNF) is Π_2^p-complete.

Proposition 13.6. WEIGHTED-Π_2^p-SAT(*2CNF*) is Π_2^p-complete. Moreover, hardness already holds when restricted to instances (φ, u), where $\varphi = \forall X.\exists Y.\psi$ and where all variables in Y appear only negatively in ψ.

Proof. Membership in Π_2^p can be shown routinely. To show Π_2^p-hardness, we give a polynomial-time reduction from co-QSAT$_2$(3CNF). Let $\varphi = \forall X.\exists Y.\psi$ be an instance of co-QSAT$_2$ where $\psi = c_1 \wedge \cdots \wedge c_m$ is in 3CNF. Let $c_j = l_{j,1} \vee l_{j,2} \vee l_{j,3}$, for each $j \in [m]$. We construct an instance (φ', u) of WEIGHTED-Π_2^p-SAT(2CNF) with the required properties as follows.

We introduce the variables $Y' = \{ y' : y \in Y \}$, which intuitively represent the negation of the variables in Y. Moreover, we introduce the set $Z = \{ z_{j,i} : j \in [m], i \in [3] \}$ of variables, which we will use to encode whether a truth assignment satisfies the clauses $c_1, \ldots, c_m$. We define $u = m + n$. Moreover, we let $\varphi' = \forall X.\exists Y.\exists Y'.\exists Z.\chi$, where we define the 2CNF formula χ below. We let χ consist of the following clauses.

Firstly, we add the claus $(\neg y \vee \neg y')$, for each $y \in Y$. These clauses ensure that for each $y \in Y$, at most one of y and y' is true.

Then, we add the clauses $(\neg z_{j,1} \vee \neg z_{j,2})$, $(\neg z_{j,1} \vee \neg z_{j,3})$, and $(\neg z_{j,2} \vee \neg z_{j,3})$, for each $j \in [m]$. These clauses ensure that for each clause c_j, there is at most one variable $z_{j,i}$ that is true, which represents that the i-th literal in clause j is satisfied.

Moreover, since $u = m + n$, we know that for each $y \in Y$, exactly one of y and y' is true, and for each $j \in [m]$, exactly one of the variables $z_{j,1}, z_{j,2}$, and $z_{j,3}$ is true.

Finally, we add the clauses $(\neg z_{j,i} \vee \sigma(l_{j,i}))$, for each $j \in [m]$ and each $i \in [3]$, where σ is defined as follows.

$$\sigma(l) = \begin{cases} x & \text{if } l = x \text{ for } x \in X, \\ \neg x & \text{if } l = \neg x \text{ for } x \in X, \\ \neg y & \text{if } l = \neg y \text{ for } y \in Y, \text{ and} \\ \neg y' & \text{if } l = y \text{ for } y \in Y. \end{cases}$$

This concludes our construction of the instance (φ', u). We show that $\varphi \in$ co-QSAT$_2$ if and only if $(\varphi', u) \in$ WEIGHTED-Π_2^p-SAT(2CNF).

($\Rightarrow$) Suppose that $\varphi \in$ co-QSAT$_2$. We show that $(\varphi', u) \in$ WEIGHTED-Π_2^p-SAT (2CNF). Let $\alpha : X \to \mathbb{B}$ be an arbitrary truth assignment. We show that there exists a truth assignment $\beta : Y \cup Y' \cup Z \to \mathbb{B}$ of weight u such that $\chi[\alpha \cup \beta]$ is true. Since $\varphi \in$ co-QSAT$_2$, we know that there exists a truth assignment $\beta' : Y \to \mathbb{B}$ such that $\psi[\alpha \cup \beta']$ is true. We define the truth assignment β as follows. For each $y \in Y$, we let $\beta(y) = \beta'(y)$ and $\beta(y') = 1 - \beta'(y)$. Since $\psi[\alpha \cup \beta']$ is true, we know that there exists for each $j \in [m]$ some $i_j \in [3]$ such that β satisfies the literal l_{j,i_j}. For each $j \in [m]$ and each $i \in [3]$, we let $\beta(z_{j,i}) = 1$ if and only if $i = i_j$. It is straightforward to verify that $\chi[\alpha \cup \beta]$ is true. Therefore, we know that $(\varphi', u) \in$ WEIGHTED-Π_2^p-SAT(2CNF).

($\Leftarrow$) Conversely, suppose that $(\varphi', u) \in$ WEIGHTED-Π_2^p-SAT(2CNF). We show that $\varphi \in$ co-QSAT$_2$. Let $\alpha : X \to \mathbb{B}$ be an arbitrary truth assignment. We show that there exists a truth assignment $\beta : Y \to \mathbb{B}$ such that $\psi[\alpha \cup \beta]$ is true. Because $(\varphi', u) \in$ WEIGHTED-Π_2^p-SAT(2CNF), we know that there exists a truth assignment $\beta' : Y \cup Y' \cup Z \to \mathbb{B}$ of weight u such that $\chi[\alpha \cup \beta']$ is true. Moreover, as argued above, we know that for each $y \in Y$, the assignment β' sets exactly one of y and y' to 1, and for each $j \in [m]$, the assignment β' sets exactly one of $z_{j,1}, z_{j,2}$, and $z_{j,3}$ to 1. We define the truth assignment β as follows. For each $y \in Y$, we let $\beta(y) = \beta'(y)$. We show that $\psi[\alpha \cup \beta]$ is true. Let c_j be an arbitrary clause of ψ. Then, we know that there is some $i \in [3]$ such that $\beta'(z_{j,i}) = 1$. Because χ contains the clause $(\neg z_{j,i} \vee \sigma(l_{j,i}))$, we know that $\alpha \cup \beta'$ must satisfy $\sigma(l_{j,i})$. It is then readily verified that $\alpha \cup \beta$ satisfies the literal $l_{j,i}$. Since j was arbitrary, we can conclude that $\alpha \cup \beta$ satisfies all clauses c_j, and thus satisfies ψ. Therefore, $\varphi \in$ co-QSAT$_2$.

We are now ready to show Π_2^p-completeness for CLIQUE-EXTENSION.

Corollary 13.7. *The problem* CLIQUE-EXTENSION *is* Π_2^p-*complete.*

Proof. Membership in Π_2^p can be shown routinely. To show Π_2^p-hardness, it suffices to see that the reduction given in the proof of Proposition 13.5 can also be used as a polynomial-time reduction from the Π_2^p-hard problem WEIGHTED-Π_2^p-SAT(2CNF) (restricted to instances where existentially quantified variables only appear negatively) to the problem CLIQUE-EXTENSION.

Notes

The results in Sect. 13.1 appeared in a paper in the proceedings of SOFSEM 2015 [113]. The results in Sect. 13.2 appeared in a technical report [117].

Relation to Other Topics
in Complexity Theory

Chapter 14
Subexponential-Time Reductions

In Chaps. 4–7, we investigated several parameterized complexity classes for parameterized problems that admit (various types of) fpt-reductions to SAT. Moreover, we considered parameterized complexity classes to characterize the complexity of intractable problems, that seemingly do not admit fpt-reductions to SAT. We started a structural investigation of the relation between these classes. For instance, we showed that the classes $\Sigma_2^p[k*]$ and $\Pi_2^p[*k, t]$ are incomparable to para-NP, unless NP = co-NP (Propositions 6.32, 6.33, 6.36, and 6.37 in Sect. 6.4). However, a number of relevant questions remained unanswered in this structural complexity investigation. In this chapter, we answer these questions by separating the classes para-NP, para-co-NP and FPT^{NP}[few], on the one hand, and the classes A[2], $\Sigma_2^p[k*]$ and $\Sigma_2^p[*k, t]$, on the other hand. These separation results are contingent on some complexity-theoretic assumptions. In order to state these assumptions, we will use some notions and terminology from the area of *subexponential-time complexity*.

Generally speaking, subexponential-time complexity is the area of complexity theory that investigates what difficult search problems can be solved asymptotically faster than the brute force exponential-time algorithm. For instance, the problem 3SAT can be solved by a brute force algorithm in time $O(2^n)$, where n is the number of variables in the instance. A subexponential-time algorithm for 3SAT would run in time $2^{o(n)}$. Remember that a function $f(n)$ is $o(n)$ if $f(n) \leq n/s(n)$ for sufficiently large n and for some unbounded, nondecreasing, computable function s. It is commonly believed that there is no subexponential-time algorithm for 3SAT. This statement is known as the *Exponential Time Hypothesis (ETH)*, and is often used as a complexity-theoretic assumption. This assumption is stronger than the assumption that P $\neq$ NP; it could be the case, for instance, that 3SAT admits a $2^{\sqrt{n}}$ time algorithm, which is not polynomial, but is subexponential.

The area of subexponential-time complexity is related to parameterized complexity theory in several ways. Firstly, the subexponential-time complexity of problems can be studied within the framework of parameterized complexity. In order to set a baseline for the brute force exponential-time algorithm, one considers a *complexity* (or search space size) *parameter n*, in addition to the input size m. For the example of 3SAT, this parameter n would be the number of variables in the instance, whereas the

© Springer-Verlag GmbH Germany, part of Springer Nature 2019
R. de Haan: Parameterized Complexity in the Polynomial Hierarchy, LNCS 11880,
https://doi.org/10.1007/978-3-662-60670-4_14

input size m is the bitsize of the 3CNF formula. Then the exponential-time algorithm to which we compare possible subexponential-time algorithms runs in time $2^n m^{O(1)}$. In other words, subexponential-time complexity can be seen as a particular type of bounded fixed-parameter tractability (see also [85, Chapters 15 and 16]).

Another connection between the fields of subexponential-time complexity and parameterized complexity is the use of assumptions from subexponential-time complexity to separate parameterized complexity classes. One of the most prominent examples of this is the result that, assuming the ETH, it holds that FPT $\neq$ W[1] [37, 45–47, 150]. In this chapter, we will use a similar approach to separate some of the classes discussed in Chaps. 4–7. We show that problems that are hard for A[2], $\Sigma_2^p[k*]$ or $\Sigma_2^p[*k, t]$ do not admit fpt-reductions to SAT, assuming that there exist no subexponential-time reductions from various hard problems at the second level of the PH to SAT or UNSAT. (For more discussion on the use of such atypical complexity-theoretic assumptions, we refer to Sect. 1.5.3.)

To avoid confusion, we point out that the subexponential-time reductions that we refer to throughout this chapter are slightly more admissive than the type of reductions usually used in the area of subexponential-time complexity (the latter are called *serf-reductions*, or *subexponential reduction families*). These serf-reductions require that the complexity parameter of the resulting instance is linear in the parameter of the original instance. For the reductions that we refer to in this chapter, there are no constraints on the size of the complexity parameter of the resulting instance.

Outline of This Chapter

Firstly, in Sect. 14.1, we briefly review a result from the literature that implies that FPT $\neq$ W[1], unless the ETH fails. We describe the general idea behind its proof, because the proofs of the results in the remainder of the chapter are based on the same idea.

Then, in Sect. 14.2, we work out the separation results for the classes A[2], $\Sigma_2^p[k*]$ and $\Sigma_2^p[*k, t]$.

Finally, in Sect. 14.3, we show how some of the separation results can be used to argue that the classes $\Sigma_2^p[k*]$, $\Pi_2^p[k*]$, $\Sigma_2^p[*k, t]$ and $\Pi_2^p[*k, t]$ are different from each other.

14.1   A Known Separation Result

We begin with reviewing a result from the literature. This result implies that FPT $\neq$ W[1], unless the ETH fails. More specifically, the result states that if the W[1]-complete parameterized problem WSAT(3CNF) can be solved in time $f(k)m^{o(k)}$, where k is the parameter value and m is the input size, then the ETH fails [45–47]. Because the proofs of the results in Sect. 14.2 are based on the same idea that is behind the proof of this result, we will explain the general lines of this idea.

Suppose that there is an algorithm A that solves the problem WSAT(3CNF) in time $f(k)m^{o(k)}$. We will use this algorithm to construct a $2^{o(n)}$ time algorithm for 3SAT. Let φ be an arbitrary instance of 3SAT with n variables.

Firstly, we consider the inverse f^{-1} of the computable function f, that is defined by letting $f^{-1}(h) = \max\{q : f(q) \leq h\}$. By assuming without loss of generality that f is nondecreasing and unbounded, we get that f^{-1} is a nondecreasing, unbounded, computable function.

We then divide the set $\mathrm{Var}(\varphi)$ of variables into $k = f^{-1}(n)$ partitions $X_1, \ldots, X_k$, each of roughly the same size. Each such partition then contains approximately $r = \frac{n}{f^{-1}(n)}$ variables.

We construct an instance ψ of WSAT(3CNF) as follows. For each set X_i of variables, we consider all possible truth assignments to these variables. There are roughly 2^r of these assignments. For each such truth assignment α, we introduce a propositional variable y_α to the instance of WSAT(3CNF). We add clauses to ψ that ensure that no two variables y_{α_1} and y_{α_2}, for different assignments α_1, α_2 to the same set X_i of variables can both be set to true. Moreover, for each clause of φ we add a number of clauses to ψ. Take a clause c of φ, and suppose that the 3 variables occurring in c appear in different sets $X_{i_1}, X_{i_2}, X_{i_3}$. Then, for each triple $(\alpha_1, \alpha_2, \alpha_3)$ of truth assignments to the sets $X_{i_1}, X_{i_2}, X_{i_3}$ (respectively) that together do not satisfy c, we add a clause to ψ that ensures that the variables $y_{\alpha_1}, y_{\alpha_2}, y_{\alpha_3}$ cannot all be set to true. We then get that there is a satisfying assignment for ψ of weight k if and only if φ is satisfiable.

We can now decide if φ is satisfiable by using the algorithm A to decide if $(\psi, k) \in$ WSAT(3CNF). The instance ψ has at most $n' = k2^r$ variables, and at most $m' = (n')^3$ clauses. Moreover, the parameter value is $k = f^{-1}(n)$. By assumption, A runs in time $f(k)(m')^{o(k)}$. Due to our choice of k, we get that $f(k) \leq n$. Moreover, we can assume without loss of generality that $f(h) \geq 2^h$ for all h, and thus that $k = f^{-1}(n) \leq \log n$. Then, by some careful analysis, we get that the running time $f(k)(m')^{o(k)}$ of A is $2^{o(n)}$, witnessing that 3SAT can be solved in subexponential time, and thus that the ETH fails.

In the remainder of this chapter, we will show several results that are based on the same idea as the one described above. That is, assuming the existence of a parameterized algorithm for a logic problem with bounds on its running time similar to $f(k)m^{o(k)}$ (e.g., an algorithm to reduce a parameterized problem to SAT), we show that a subexponential-time algorithm exists that solves a particular computational task (e.g., an algorithm to reduce various problems at the second level of the PH to SAT) by partitioning the set of variables in the input problem into roughly $k = f^{-1}(n)$ sets $X_1, \ldots, X_k$, and constructing an equivalent instance of the appropriate problem by iterating over all truth assignments to the variables in each of these sets X_i.

14.2 More Separation Results

In this section, we will use the idea described in Sect. 14.1 to show various separation results for the classes A[2], $\Sigma_2^p[k*]$ and $\Sigma_2^p[*k, t]$, on the one hand, and para-NP, para-co-NP and FPT^{NP}[few], on the other hand.

14.2.1 Separation Results for A[2]

We begin with separating the parameterized complexity class A[2] from the classes para-NP, para-co-NP and $FPT^{NP}[few]$. Firstly, we show that if A[2] $\subseteq$ para-NP, then there exists a subexponential-time reduction from $QSAT_2(3DNF)$ to SAT. We prove the following slightly stronger result.

Theorem 14.1. *If there exists an* $f(k)m^{o(k^{1/3})}$ *time reduction from* $MC(\Sigma_2)$ *to SAT, where k denotes the parameter value, m denotes the instance size, and f is some computable function, then there exists a subexponential-time reduction from* $QSAT_2(3DNF)$ *to SAT, i.e., a reduction that runs in time* $2^{o(n)}$, *where n denotes the number of variables.*

Proof. Suppose that there is a reduction R from $MC(\Sigma_2)$ to SAT that runs in time $f(k)m^{o(k^{1/3})}$, that is, it runs in time $f(k)m^{k^{1/3}/\lambda(k)}$ for some computable function f and some computable, unbounded and nondecreasing function λ. We construct a reduction from $QSAT_2(3DNF)$ to SAT that, for sufficiently large values of n, runs in time $2^{o(n)}$, where n is the number of variables. Let (φ, k) be an instance of $QSAT_2$, where $\varphi = \exists X_1.\forall X_2.\psi$ and where ψ is in 3DNF. We will construct an equivalent instance $(\mathcal{A}, \varphi')$ of $MC(\Sigma_2)$ from the instance (φ, k) of $QSAT_2(3DNF)$, and we will then use the reduction from $MC(\Sigma_2)$ to SAT to transform $(\mathcal{A}, \varphi')$ to an equivalent instance of SAT in subexponential time.

We may assume without loss of generality that $|X_1| = |X_2|$; we can add dummy variables to X_1 and X_2 to ensure this. Then, let $n = |X_1| = |X_2|$. We denote the size of ψ by m. Let $X = X_1 \cup X_2$. We may assume without loss of generality that $f(\ell) \geq 2^\ell$ for all $\ell \geq 1$, and that f is nondecreasing and unbounded.

In order to construct $(\mathcal{A}, \varphi')$, we will use several auxiliary definitions. We define the function g as follows:

$$g(\ell) = f(c \cdot \ell^3);$$

we will define the value of the constant c below. Then, define the function g^{-1} as follows:

$$g^{-1}(h) = \max\{q : g(2q + 1) \leq h\}.$$

Since the function f is nondecreasing and unbounded, the functions g and g^{-1} are also nondecreasing and unbounded. Moreover, we get that $g(2g^{-1}(h) + 1) \leq h$. Also, since $f(\ell) \geq 2^\ell$ for all $\ell \geq 1$, we get that $g(\ell) \geq 2^\ell$ for all $\ell \geq 1$, and therefore also that $g^{-1}(h) \leq \log h$, for all $h \geq 1$.

We then choose the integers r and k as follows.

$$r = \lfloor n/g^{-1}(n) \rfloor \quad \text{and} \quad k = \lceil n/r \rceil.$$

Due to this choice for k and r, we get the following inequalities:

$$r \leq \frac{n}{g^{-1}(n)}, \quad k \geq g^{-1}(n), \quad r \geq \frac{n}{2g^{-1}(n)}, \quad \text{and} \quad k \leq 2g^{-1}(n) + 1.$$

Next, we construct an instance $(\mathcal{A}, \varphi')$ of $\mathrm{MC}(\Sigma_2)$ such that $\mathcal{A} \models \varphi'$ if and only if φ is a yes-instance of QSAT_2. In order to do so, for each $i \in [2]$, we split X_i into k disjoint sets $X_{i,1}, \ldots, X_{i,k}$. We do this in such a way that each set $X_{i,j}$ has at most n/k elements, i.e., $|X_{i,j}| \leq 2r$ for each $i \in [2]$ and each $j \in [k]$. To construct φ', we will introduce a first-order variable $y_{i,j}$ for each $i \in [2]$ and each $j \in [k]$. Intuitively, these variables will be used to quantify over truth assignments to the variables in the sets $X_{i,j}$. For each $i \in [2]$, we let $Y_i = \{ y_{i,j} : j \in [k] \}$. Moreover, for the sake of convenience, we introduce alternative names for these variables. Let $Z = Y_1 \cup Y_2 = \{z_1, \ldots, z_{2k}\}$. Also, for each $i \in [2]$ and each $j \in [k]$, we introduce a binary predicate symbol $S_{i,j}$. In addition, we introduce a ternary predicate symbol R. We then define the first-order formula φ' as follows:

$$
\begin{aligned}
\varphi' &= \exists Y_1. \forall Y_2. (\psi_1' \wedge (\psi_2' \to \psi_3')); \\
\psi_i' &= \bigwedge_{j \in [k]} S_{i,j}(y_{i,j}) && \text{for each } i \in [2]; \text{ and} \\
\psi_3' &= \bigvee_{j_1, j_2, j_3 \in [2k], j_1 < j_2 < j_3} R(z_{j_1}, z_{j_2}, z_{j_3}).
\end{aligned}
$$

We define the relational structure $\mathcal{A}$ as follows. The universe A of $\mathcal{A}$ is defined as follows:

$$
A = \{ \alpha : i \in [2], j \in [k], (\alpha : X_{i,j} \to \mathbb{B}) \}.
$$

Then, for each $i \in [2]$ and each $j \in [k]$, the interpretation of the relation $S_{i,j}$ is defined as follows:

$$
S_{i,j}^{\mathcal{A}} = \{ \alpha : (\alpha : X_{i,j} \to \mathbb{B}) \}.
$$

Finally, the interpretation of the relation R is defined as follows:

$$
R^{\mathcal{A}} = \{ (\alpha_1, \alpha_2, \alpha_3) : \text{the assignment } \alpha_1 \cup \alpha_2 \cup \alpha_3 \text{ satisfies some term in } \psi \}.
$$

Let $m' = |\mathcal{A}| + |\varphi'|$, and $k' = |\varphi'|$. Observe that by the construction of $\mathcal{A}$, we know that $|\mathcal{A}| \leq 2k2^{2r} + 8k^3 2^{6r}$. Moreover, we have that $k' = |\varphi'| = O(k^3)$. In fact, we can straightforwardly construct φ' in such a way that $k' = c \cdot k^3$, for some constant c. We let this constant c be the constant used for the definition of the function g above.

We verify that $\varphi \in \mathrm{QSAT}_2$ if and only if $\mathcal{A} \models \varphi'$.

($\Rightarrow$) Suppose that $\varphi \in \mathrm{QSAT}_2$. Then there is a truth assignment $\beta_1 : X_1 \to \mathbb{B}$ such that for all truth assignment $\beta_2 : X_2 \to \mathbb{B}$ it holds that $\psi[\beta_1 \cup \beta_2]$ is true. We show that $\mathcal{A} \models \varphi'$. We define the assignment $\gamma_1 : Y_1 \to A$ as follows. For each $j \in [k]$, we let $\gamma_1(y_{1,j}) = \alpha_{1,j}$, where $\alpha_{1,j}$ is the restriction of β_1 to the variables in $X_{1,j}$. We show that for each assignment $\gamma_2 : Y_2 \to A$ it holds that $\mathcal{A}, \gamma_1 \cup \gamma_2 \models \psi'$. Take an arbitrary assignment $\gamma_2 : Y_2 \to A$. Let $\gamma = \gamma_1 \cup \gamma_2$. Clearly, γ satisfies ψ_1'. Suppose that γ satisfies ψ_2'. We need to show that then γ also satisfies ψ_3'. For each $i \in [2]$ and each $j \in [k]$, we have that $\gamma(y_{i,j})$ is a truth assignment to the propositional variables $X_{i,j}$. Now consider the truth assignment $\beta : X_1 \cup X_2 \to \mathbb{B}$ that is defined as follows. For each $x \in X_{i,j}$, we let $\beta(x) = \alpha_{i,j}(x)$, where $\alpha_{i,j} = \gamma(y_{i,j})$. By construction of γ_1 and β, we know that β agrees with β_1 on the variables in X_1. Therefore, we know that β must satisfy ψ, that is, β must satisfy some term in ψ. Since each term contains at most three literals, we know that there are some $j_1, j_2, j_3 \in [2k]$ with $j_1 < j_2 < j_3$ such that $R(z_{j_1}, z_{j_2}, z_{j_3})$ is satisfied by γ. Therefore, γ satisfies ψ_3'. Since γ_2 was arbitrary, we can conclude that $\mathcal{A} \models \varphi'$.

($\Leftarrow$) Conversely, suppose that $\mathcal{A} \models \varphi'$. That is, there is some assignment $\gamma_1 : Y_1 \to A$ such that for all assignments $\gamma_2 : Y_2 \to A$ it holds that $\mathcal{A}, \gamma_1 \cup \gamma_2 \models \psi_1' \wedge (\psi_2' \to \psi_3')$. We show that $\varphi \in \text{QSAT}_2$. Since ψ_1' contains only variables in Y_1, we know that γ_1 satisfies ψ_1'. Consider the truth assignment $\beta_1 : X_1 \to \mathbb{B}$ that is defined as follows. For each $x \in X_{1,j}$, we let $\beta_1(x) = \alpha_{1,j}(x)$, where $\alpha_{1,j} = \gamma_1(y_{1,j})$. We know that β_1 is well defined, because γ_1 satisfies ψ_1'. We show that for all truth assignments $\beta_2 : X_2 \to \mathbb{B}$ it holds that $\beta_1 \cup \beta_2$ satisfies ψ. Take an arbitrary truth assignment $\beta_2 : X_2 \to \mathbb{B}$. Then, we define $\gamma_2 : Y_2 \to A$ as follows. For each $j \in [k]$, we let $\gamma_2(y_{2,j}) = \alpha_{2,j}$, where $\alpha_{2,j}$ is the restriction of β_2 to the variables in $X_{2,j}$. Let $\gamma = \gamma_1 \cup \gamma_2$. Clearly, γ satisfies ψ_2'. Therefore, we know that γ also satisfies ψ_3'. By construction of ψ_3' and $\mathcal{A}$, there must be some $j_1, j_2, j_3 \in [2k]$ with $j_1 < j_2 < j_3$ such that $R(z_{j_1}, z_{j_2}, z_{j_3})$ is satisfied by γ. From this, we can conclude that $\beta_1 \cup \beta_2$ satisfies some term in ψ. Since β_2 was arbitrary, we can conclude that $\varphi \in \text{QSAT}_2$.

Since $(\mathcal{A}, \varphi')$ is an instance of $\text{MC}(\Sigma_2)$, we can apply the reduction R to obtain an equivalent instance φ'' of SAT. By first constructing $(\mathcal{A}, \varphi')$ from φ, and then constructing φ'' from $(\mathcal{A}, \varphi')$, we get a reduction R' from QSAT_2 to SAT. We analyze the running time of this reduction R' in terms of the values n.

Firstly, constructing $(\mathcal{A}, \varphi')$ can be done in time:

$$O(2k2^{2r} + 8k^3 2^{6r} \cdot |\psi|) = 2^{o(n)}.$$

Then, applying the reduction R to obtain φ'' from $(\mathcal{A}, \varphi')$ takes time $f(k')$ $(m')^{(k')^{1/3}/\lambda(k')}$. We analyze the different factors of this expression in terms of n. Firstly:

$$f(k') = f(ck^3) = g(k) \le g(2g^{-1}(n) + 1) \le n.$$

In our analysis of the running time of R, we will use an auxiliary function λ'. In particular, it will turn out that in this analysis we need the following inequality to hold:

$$\lambda'(n) \le \frac{\lambda(ck^3)g^{-1}(n)}{6(ck^3)^{1/3}}.$$

We do so by defining λ' as follows:

$$\lambda'(h) = \frac{\lambda(cg^{-1}(h)^3)g^{-1}(h)}{6c^{1/3}(2g^{-1}(h) + 1)}.$$

We will also need λ' to be unbounded. In order to see that λ' is unbounded, we observe the following inequality:

$$\lambda'(h) \geq \frac{\lambda(cg^{-1}(h)^3)g^{-1}(h)}{6c^{1/3}3g^{-1}(h)} = \frac{\lambda(cg^{-1}(h)^3)}{18c^{1/3}}.$$

Then, in order to analyze the second factor $(m')^{(k')^{1/3}/\lambda(k')}$ in terms of n, we firstly consider the following inequality:

$$(2^{6r})^{(k')^{1/3}/\lambda(k')} \leq 2^{6n(k')^{1/3}/(\lambda(k')g^{-1}(n))} = 2^{n6(ck^3)^{1/3}/(\lambda(ck^3)g^{-1}(n))}$$
$$\leq 2^{n/\lambda'(n)} = 2^{o(n)}.$$

We then get:

$$(m')^{(k')^{1/3}/\lambda(k')} \leq (k' + 2k2^{2r} + 8k^3 2^{6r})^{(k')^{1/3}/\lambda(k')}$$
$$\leq (k')^{k'}(2k)^{k'}2^{2r(k')^{1/3}/\lambda(k')}(8k^3)^{k'}2^{6r(k')^{1/3}/\lambda(k')}.$$

Then, since $(k')^{k'} \leq O(\log^3 n)^{O(\log^3 n)} = 2^{o(n)}$, we know that the factors $(k')^{k'}$ and $(2k)^{k'}$ are $2^{o(n)}$. By a similar argument, we know that the factor $(8k^3)^{k'}$ is $2^{o(n)}$. Moreover, because we know that $(2^{6r})^{(k')^{1/3}/\lambda(k')} \leq 2^{o(n)}$ we know that the factors $2^{2r(k')^{1/3}/\lambda(k')}$ and $2^{6r(k')^{1/3}/\lambda(k')}$ are also $2^{o(n)}$. Therefore, we know that $(m')^{(k')^{1/3}/\lambda(k')}$ is $2^{o(n)}$. Concluding, the reduction R' from $QSAT_2(3DNF)$ to SAT runs in time $2^{o(n)}$.

Theorem 14.1 directly gives us the following corollary, separating A[2] from para-NP.

Corollary 14.2. *If* A[2] $\subseteq$ para-NP, *then there exists a subexponential-time reduction from* $QSAT_2(3DNF)$ *to* SAT.

The proof of Theorem 14.1 can straightforwardly be modified to separate A[2] also from para-co-NP.

Corollary 14.3. *If* A[2] $\subseteq$ para-co-NP, *then there exists a subexponential-time reduction from* $QSAT_2(3DNF)$ *to* UNSAT.

We can extend these results to the more general case of A[t] and para-Σ_i^p (or para-Π_i^p), for arbitrary $t \geq 2$ and $i \geq 1$.

Proposition 14.4. *Let* $t \geq 2$ *and* $i \geq 1$. *If* A[t] $\subseteq$ para-Σ_i^p, *then there exists a subexponential-time reduction from* $QSAT_t(3DNF)$ *to* $QSAT_i$ *for even* t, *and from* $QSAT_t$ (3CNF) *to* $QSAT_i$ *for odd* t. *If* A[t] $\subseteq$ para-Π_i^p, *then there exists a subexponential-time reduction from* $QSAT_t(3DNF)$ *to* co-$QSAT_i$ *for even* t, *and from* $QSAT_t(3CNF)$ *to* co-$QSAT_i$ *for odd* t.

Proof (sketch). The proof of Theorem 14.1 straightforwardly generalizes to arbitrary $t \geq 2$ and arbitrary $i \geq 1$, both for para-Σ_i^p and para-Π_i^p.

We can also use a modification of the proof of Theorem 14.1 to separate A[2] from FPT.

Proposition 14.5. *Let $t \geq 2$. If* A[t] = FPT, *then* QSAT$_t$ *restricted to instances whose matrix is in* 3DNF *if t is even, and in* 3CNF *if t is odd, can be solved in time* $2^{o(n)}$, *where n denotes the number of variables.*

Proof (sketch). The proof of Theorem 14.1 can straightforwardly be modified to show this result.

Next, we separate A[2] from FPT^{NP}[few]. We do so by showing that if A[2] $\subseteq$ FPT^{NP}[few], then there exists a subexponential-time Turing reduction from QSAT$_2$ (3DNF) to SAT. We prove the following slightly stronger result.

Theorem 14.6. *If there exists an $f(k)m^{o(k^{1/3})}$ time Turing reduction from* MC(Σ_2) *to* SAT, *where k denotes the parameter value, m denotes the instance size, and f is some computable function, then there exists a subexponential-time Turing reduction from* QSAT$_2$(3DNF) *to* SAT, *i.e., a Turing reduction that runs in time $2^{o(n)}$, where n denotes the number of variables.*

Proof. Assuming the existence of an $f(k)m^{o(k^{1/3})}$ time Turing reduction from MC(Σ_2) to SAT, we can construct a subexponential-time Turing reduction from QSAT$_2$(3DNF) to SAT analogously to the construction in the proof of Theorem 14.1. Firstly, we construct an instance $(\mathcal{A}, \varphi')$ as described in the proof of Theorem 14.1 in time $2^{o(n)}$. Then, we apply the reduction R to $(\mathcal{A}, \varphi')$ to get a Turing reduction R' from QSAT$_2$(3DNF) to SAT. Since the reduction R runs in time $f(k')(m')^{o(k^{1/3})} = 2^{o(n)}$, the reduction R' also runs in time $2^{o(n)}$.

Theorem 14.6 directly gives us the following corollary, separating A[2] from FPT^{NP}[few].

Corollary 14.7. *If* A[2] $\subseteq$ FPT^{NP}[few], *then there exists a subexponential-time Turing reduction from* QSAT$_2$(3DNF) *to* SAT.

The above results also allow us to separate the classes $\Sigma_2^p[k*]$ and $\Sigma_2^p[*k, t]$ from the parameterized complexity classes XNP and Xco-NP.

Corollary 14.8. *If (i)* $\Sigma_2^p[k*]$ = XNP, *(ii)* $\Sigma_2^p[k*]$ = Xco-NP, *(iii)* $\Sigma_2^p[*k, P]$ = XNP, *or (iv)* $\Sigma_2^p[*k, P]$ = Xco-NP, *then for each $t \geq 3$ there is a subexponential-time reduction from* QSAT$_t$(3CNF) *to* QSAT$_2$ *and from* QSAT$_t$(3DNF) *to* QSAT$_2$.

Proof. Suppose that one of the statements (i)–(iv) holds. Then for each $t \geq 3$ it holds that A[t] $\subseteq$ XP $\subseteq$ XNP $\cap$ Xco-NP $\subseteq$ para-Σ_2^p. Therefore, the required consequence follows from Proposition 14.4.

14.2.2 Separation Results for $\Sigma_2^p[k*]$ and $\Sigma_2^p[*k, t]$

Next, we show that the above separation results for A[2] can be strengthened for the cases of $\Sigma_2^p[k*]$, $\Sigma_2^p[*k, 2]$ and $\Sigma_2^p[*k, P]$. We begin with the case for $\Sigma_2^p[*k, 2]$, and show the following slightly stronger result.

Theorem 14.9. *If there exists an $f(k)n^{o(k)}m^{O(1)}$ time reduction from $\Sigma_2^p[kk]$-WSAT $(\Gamma_{2,4})$ to SAT, where k denotes the parameter value, n denotes the number of variables and m denotes the instance size, then there exists a subexponential-time reduction from $QSAT_2(DNF)$ to SAT, i.e., a reduction that runs in time $2^{o(n)}m^{O(1)}$, where n denotes the number of variables and m denotes the instance size.*

Proof. Assume that there exists a reduction R from $\Sigma_2^p[kk]$-WSAT$(\Gamma_{2,4})$ to SAT that runs in time $f(k)n^{k/\lambda(k)}m^{O(1)}$, for some computable function f and some nondecreasing and unbounded computable function λ.

We now construct a reduction from $QSAT_2(DNF)$ to SAT that runs in time $2^{o(n)}$ $m^{O(1)}$, where n is the number of variables, and m is the instance size. Let $\varphi = \exists X_1.\forall X_2.\psi$ be an instance of $QSAT_2(DNF)$, where ψ is in DNF. We will construct an equivalent instance (φ', k) of $\Sigma_2^p[kk]$-WSAT$(\Gamma_{2,4})$ from the instance (φ, k) of $QSAT_2(DNF)$, and we will then use the reduction from $\Sigma_2^p[kk]$-WSAT$(\Gamma_{2,4})$ to SAT to transform (φ', k) to an equivalent instance of SAT in subexponential time.

We may assume without loss of generality that $|X_1| = |X_2| = n$; we can add dummy variables to X_1 and X_2 to ensure this. We denote the size of ψ by m. Let $X = X_1 \cup X_2$. We may assume without loss of generality that $f(k) \geq 2^k$ and that f is nondecreasing and unbounded.

In order to construct (φ', k), we will use several auxiliary definitions. Define f^{-1} as follows:

$$f^{-1}(h) = \max\{q : f(2q+1) \leq h\}.$$

Since the function f is nondecreasing and unbounded, the function f^{-1} is also nondecreasing and unbounded. Also, we know that $f(2f^{-1}(h)+1) \leq h$, and since $f(k) \geq 2^k$, we know that $f^{-1}(h) \leq \log h$. We then choose the integers r and k as follows.

$$r = \lfloor n/f^{-1}(n) \rfloor \quad \text{and} \quad k = \lceil n/r \rceil.$$

Due to this choice for k and r, we get the following inequalities:

$$r \leq \frac{n}{f^{-1}(n)}, \quad k \geq f^{-1}(n), \quad r \geq \frac{n}{2f^{-1}(n)}, \quad \text{and} \quad k \leq 2f^{-1}(n)+1.$$

We firstly construct an instance (φ', k) of $\Sigma_2^p[kk]$-WSAT$(\Gamma_{2,4})$ that is a yes-instance if and only if φ is a yes-instance of $QSAT_2(DNF)$. We will describe φ' as a quantified Boolean formula whose matrix corresponds to a circuit of depth 4 and weft 2. In order to do so, for each $i \in [2]$, we split X_i into k disjoint sets $X_{i,1}, \ldots, X_{i,k}$. We do this in such a way that each set $X_{i,j}$ has at most n/k elements, i.e., $|X_{i,j}| \leq 2r$ for all $i \in [2]$ and all $j \in [k]$. Now, for each truth assignment $\alpha : X_{i,j} \to \mathbb{B}$ we introduce a new variable $y_{i,j}^\alpha$. Formally, we define a set of variables $Y_{i,j}$ for each $i \in [2]$ and each $j \in [k]$:

$$Y_{i,j} = \{ y_{i,j}^\alpha : (\alpha : X_{i,j} \to \mathbb{B}) \}.$$

We have that $|Y_{i,j}| \leq 2^{2^r}$, for each $i \in [2]$ and each $j \in [k]$. We let $Y_i = \bigcup_{j \in [k]} Y_{i,j}$, and we let $Y = Y_1 \cup Y_2$.

We continue the construction of the formula φ'. For each $i \in [2]$, we define the formula ψ_{Y_i} as follows:

$$\psi_{Y_i} = \bigwedge_{j \in [k]} \bigwedge_{\substack{\alpha, \alpha' : X_{i,j} \to \mathbb{B} \\ \alpha \neq \alpha'}} \left(\neg y_{i,j}^{\alpha} \vee \neg y_{i,j}^{\alpha'} \right).$$

Then we define the auxiliary functions $\sigma_0, \sigma_1 : X \to 2^Y$, that map variables in X to sets of variables in Y. For each $x \in X_{i,j}$, we let:

$$\sigma_0(x) = \{ y_{i,j}^{\alpha} : (\alpha : X_{i,j} \to \mathbb{B}), \alpha(x) = 0 \}, \text{ and}$$
$$\sigma_1(x) = \{ y_{i,j}^{\alpha} : (\alpha : X_{i,j} \to \mathbb{B}), \alpha(x) = 1 \}.$$

Intuitively, for $b \in \mathbb{B}$, $\sigma_b(x)$ corresponds to those variables $y_{i,j}^{\alpha}$ where α is an assignment that sets x to b.

Now, we construct a formula ψ'', by transforming the formula ψ in the following way. We replace each occurrence of a positive literal $x \in X_{i,j}$ in ψ by the formula χ_x, that is defined as follows:

$$\chi_x = \bigwedge_{y_{i,j}^{\alpha} \in \sigma_0(x)} \neg y_{i,j}^{\alpha}.$$

Moreover, we replace each occurrence of a negative literal $\neg x$ in ψ (for $x \in X_{i,j}$) by the formula $\chi_{\neg x}$, that is defined as follows:

$$\chi_{\neg x} = \bigwedge_{y_{i,j}^{\alpha} \in \sigma_1(x)} \neg y_{i,j}^{\alpha}.$$

We can now define the quantified Boolean formula φ'. We let $\varphi' = \exists Y_1. \forall Y_2. \psi'$, where ψ' is defined as follows:

$$\psi' = \psi_{Y_1} \wedge (\psi_{Y_2} \to \psi'').$$

The formula ψ' can be seen as a circuit of depth 4 and weft 2. In the remainder, we will refer to ψ' as a circuit.

We verify that $\varphi \in \mathrm{QSAT}_2(\mathrm{DNF})$ if and only if $(\varphi', k) \in \Sigma_2^p[kk]\text{-WSAT}(\Gamma_{2,4})$.

($\Rightarrow$) Assume that $\varphi \in \mathrm{QSAT}_2(\mathrm{DNF})$, i.e., that there exists a truth assignment $\beta_1 : X_1 \to \mathbb{B}$ such that for all truth assignments $\beta_2 : X_2 \to \mathbb{B}$ it holds that $\psi[\beta_1 \cup \beta_2]$ is true. We show that $(\varphi', k) \in \Sigma_2^p[kk]\text{-WSAT}$. We define the truth assignment $\gamma_1 : Y_1 \to \mathbb{B}$ by letting $\gamma_1(y_{1,j}^{\alpha}) = 1$ if and only if β_1 coincides with α on the variables $X_{1,j}$, for each $j \in [k]$ and each $\alpha : X_{1,j} \to \mathbb{B}$. Clearly, γ_1 has weight k. Moreover, γ_1 satisfies ψ_{Y_1}. We show that for each truth assignment $\gamma_2 : Y_2 \to \mathbb{B}$ of weight k it holds that $\psi'[\gamma_1 \cup \gamma_2]$ is true. Let γ_2 be an arbitrary truth assignment of weight k.

We distinguish two cases: either (i) γ_2 does not satisfy ψ_{Y_2}, or (ii) γ_2 does satisfy ψ_{Y_2}. In case (i), clearly, $\psi'[\gamma_1 \cup \gamma_2]$ is true. In case (ii), we know that for each $j \in [k]$, there is exactly one $\alpha_j : X_{2,j} \to \mathbb{B}$ such that $\gamma_2(y_{2,j}^\alpha) = 1$. Now let $\beta_2 : X_2 \to \mathbb{B}$ be the assignment that coincides with α_j on the variables $Y_{2,j}$, for each $j \in [k]$. We know that $\psi[\beta_1 \cup \beta_2]$ is true. Then, by definition of ψ'', it follows that $\psi''[\gamma_1 \cup \gamma_2]$ is true as well. Since γ_2 was arbitrary, we can conclude that $(\varphi', k) \in \Sigma_2^p[kk]$-WSAT.

($\Leftarrow$) Conversely, assume that $(\varphi', k) \in \Sigma_2^p[kk]$-WSAT, i.e., that there exists a truth assignment $\gamma_1 : Y_1 \to \mathbb{B}$ of weight k such that for all truth assignments $\gamma_2 : Y_2 \to \mathbb{B}$ of weight k it holds that $\psi'[\gamma_1 \cup \gamma_2]$ is true. We show that $\varphi \in \text{QSAT}_2$. Since ψ_{Y_1} contains only variables in Y_1, we know that γ_1 satisfies ψ_{Y_1}, i.e., that for each $j \in [k]$ there is a unique $\alpha_j : X_{1,j} \to \mathbb{B}$ such that $\gamma_1(y_{1,j}^{\alpha_j}) = 1$. We define the truth assignment $\beta_1 : X_1 \to \mathbb{B}$ to be the unique truth assignment that coincides with α_j for each $j \in [k]$. We show that for all truth assignments $\beta_2 : X_2 \to \mathbb{B}$ it holds that $\psi[\beta_1 \cup \beta_2]$ is true. Let β_2 be an arbitrary truth assignment. We construct the truth assignment $\gamma_2 : Y_2 \to \mathbb{B}$ by letting $\gamma_2(y_{2,j}^\alpha) = 1$ if and only if β_2 coincides with α on the variables in $Y_{2,j}$, for each $j \in [k]$ and each $\alpha : X_{2,j} \to \mathbb{B}$. Clearly, γ_2 has weight k. Moreover, γ_2 satisfies ψ_{Y_2}. Therefore, since we know that $\psi'[\gamma_1 \cup \gamma_2]$ is true, we know that $\psi''[\gamma_1 \cup \gamma_2]$ is true. Then, by definition of ψ'', it follows that $\psi[\beta_1 \cup \beta_2]$ is true as well. Since β_2 was arbitrary, we can conclude that $\varphi \in \text{QSAT}_2(\text{DNF})$.

We observe some properties of the quantified Boolean formula $\varphi' = \exists Y_1.\forall Y_2.\psi'$. Each Y_i, for $i \in [2]$, contains at most $n' = k2^{2r}$ variables. Furthermore, the circuit ψ' has size $m' \leq O(k2^{4r} + 2^{2r}m) \leq O(k2^{4r}m)$. Finally, it is straightforward to verify that the circuit ψ' can be constructed in time $O((m')^2)$.

Since (φ', k) is an instance of $\Sigma_2^p[kk]$-WSAT$(\Gamma_{2,4})$, we can apply the reduction R to obtain an equivalent instance (φ'', k'') of SAT. This reduction runs in time $f(k)(n')^{k/\lambda(k)}(m')^{O(1)}$. By first constructing (φ', k) from φ, and then constructing φ'' from (φ', k), we get a reduction R' from QSAT$_2$(DNF) to SAT, that runs in time $f(k)(n')^{k/\lambda(k)}(m')^{O(1)} + O((m')^2)$. We analyze the running time of this reduction R' in terms of the values n and m. Firstly:

$$f(k) \leq f(2f^{-1}(n) + 1) \leq n.$$

In our analysis of the running time of R', we will use an auxiliary function λ'. In particular, it will turn out that in this analysis we need the following inequality to hold:

$$\lambda'(n) \leq \frac{\lambda(k)}{6}.$$

We do so by defining λ' as follows:

$$\lambda'(h) = \frac{\lambda(f^{\text{rev}}(h))}{6}.$$

We will also need λ' to be unbounded. Since both λ and f^{rev} are nondecreasing and unbounded, λ' is a nondecreasing and unbounded function.

We have,

$$(n')^{k/\lambda(k)} = (k2^{2r})^{k/\lambda(k)} \le k^k 2^{2kr/\lambda(k)} \le k^k 2^{2kn/(\lambda(k)f^{-1}(n))} \le k^k 2^{6n/\lambda(k)}$$
$$\le k^k 2^{n/\lambda'(n)} = O(\log n)^{O(\log n)} 2^{n/\lambda'(n)} = 2^{o(n)}.$$

Finally, consider the factor m'. Since f^{-1} is nondecreasing and unbounded,

$$m' \le O(k2^{4r}m) = O(\log n 2^{4n/f^{-1}(n)}m) = 2^{o(n)}m.$$

Therefore, both terms $(m')^{O(1)}$ and $O((m')^2)$ in the running time of R' are bounded by $2^{o(n)}m^{O(1)}$. Combining all these, we conclude that the running time $f(k)(n')^{k/\lambda(k)}$ $(m')^{O(1)} + O((m')^2)$ of R' is bounded by $2^{o(n)}m^{O(1)}$. Therefore, R' is a subexponential-time reduction from $\mathrm{QSAT}_2(\mathrm{DNF})$ to SAT. This completes our proof.

Theorem 14.9 gives us the following corollary, separating $\Sigma_2^p[*k, 2]$ from para-NP.

Corollary 14.10. *If* $\Sigma_2^p[*k, 2] \subseteq$ *para-NP, then there exists a subexponential-time reduction from* $\mathrm{QSAT}_2(\mathrm{DNF})$ *to* SAT.

Proof. The parameterized problem $\Sigma_2^p[kk]\text{-WSAT}(\Gamma_{2,4})$ is in $\Sigma_2^p[*k, 2]$. Therefore, the result for $\Sigma_2^p[*k, 2]$ follows directly from Theorem 14.9.

Finally, the proof of Theorem 14.9 extends to even stronger results for the cases of $\Sigma_2^p[*k, \mathrm{SAT}]$ and $\Sigma_2^p[*k, \mathrm{P}]$.

Corollary 14.11. *If* $\Sigma_2^p[*k, \mathrm{SAT}] \subseteq$ *para-NP, then there exists a subexponential-time reduction from* QSAT_2 *to* SAT. *Moreover, if* $\Sigma_2^p[*k, \mathrm{P}] \subseteq$ *para-NP, then there exists a subexponential-time reduction from the extension of* QSAT_2 *to quantified Boolean circuits to* SAT.

Proof (sketch). The proof of Theorem 14.9 can straightforwardly be modified to show this result.

A similar result as for the case of $\Sigma_2^p[*k, \mathrm{P}]$ holds for the case of $\Sigma_2^p[k*]$.

Corollary 14.12. *If* $\Sigma_2^p[k*] \subseteq$ *para-co-NP, then there exists a subexponential-time reduction from the extension of* QSAT_2 *to quantified Boolean circuits to* UNSAT.

Proof (sketch). The proof of Theorem 14.9 can straightforwardly be modified to show this result.

14.3   Relating $\Sigma_2^p[k*]$ and $\Sigma_2^p[*k, t]$ to Each Other

In this section, we use some of the separation results that we developed in Sect. 14.2 to argue that the classes $\Sigma_2^p[k*]$, $\Pi_2^p[k*]$, $\Sigma_2^p[*k, t]$ and $\Pi_2^p[*k, t]$ are different from each other.

We begin with some results that are based on the assumption that NP $\ne$ co-NP.

Proposition 14.13. *Assuming that* NP $\neq$ co-NP, *the following statements hold:*

(i) $\Sigma_2^p[k*] \not\subseteq \Pi_2^p[k*]$;
(ii) $\Sigma_2^p[*k, 1] \not\subseteq \Pi_2^p[*k, P]$;
(iii) $\Sigma_2^p[k*] \not\subseteq \Sigma_2^p[*k, P]$; *and*
(iv) $\Sigma_2^p[*k, 1] \not\subseteq \Sigma_2^p[k*]$.

Proof. These results all follow from the facts that para-co-NP $\subseteq$ XNP implies NP = co-NP, and that para-NP $\subseteq$ Xco-NP implies NP = co-NP. Take for instance result (i). Suppose that $\Sigma_2^p[k*] \subseteq \Pi_2^p[k*]$. Since para-co-NP $\subseteq \Sigma_2^p[k*]$ and $\Pi_2^p[k*] \subseteq$ XNP, we get that para-co-NP $\subseteq$ XNP, and thus that NP = co-NP. The other results can be proven analogously.

All that remains to show now is that $\Sigma_2^p[k*] \not\subseteq \Pi_2^p[*k, P]$ and $\Sigma_2^p[*k, 1] \not\subseteq \Pi_2^p[k*]$. Proposition 14.4 allows us to show these results, under the assumption that there is no subexponential-time reduction from QSAT$_2$(3DNF) to co-QSAT$_2$.

Proposition 14.14. *Assuming that there is no subexponential-time reduction from* QSAT$_2$(3DNF) *to* co-QSAT$_2$, *the following statements hold:*

(v) $\Sigma_2^p[k*] \not\subseteq \Pi_2^p[*k, P]$; *and*
(vi) $\Sigma_2^p[*k, 1] \not\subseteq \Pi_2^p[k*]$.

Proof. We give a proof for result (v). The other result can be proven analogously.

Suppose that $\Sigma_2^p[k*] \subseteq \Pi_2^p[*k, P]$. Then, since A[2] $\subseteq \Sigma_2^p[k*]$ and $\Pi_2^p[*k, P] \subseteq$ para-Π_2^p, we get that A[2] $\subseteq$ para-Π_2^p. Then, by Proposition 14.4, we get that there exists a subexponential-time reduction from QSAT$_2$(3DNF) to co-QSAT$_2$.

Notes

Theorem 14.9 appeared in a technical report [117]. Many of the results in this section appeared in an article appearing in the Journal of Computer and System Sciences [114].

Chapter 15
Non-uniform Parameterized Complexity

Traditional computational complexity research has shown that there is a close relation between the Polynomial Hierarchy and the concept of non-uniformity. There are various ways to define non-uniform algorithms (and the non-uniform complexity classes that are based on them), one of which is by means of advice strings. For each input size n, the algorithm gets an advice string α_n—whose length is bounded by a particular function $z(n)$—that can be used to solve the problem for inputs of length n. This advice string is given for free, and it is possibly hard to compute (or even uncomputable). The close relation between advice and the PH is illustrated by the seminal result that if SAT is solvable in polynomial time using advice strings of polynomial length, then the PH collapses. This result is known as the Karp-Lipton Theorem [127, 128].

In this chapter, we investigate non-uniform variants of several parameterized complexity classes, and their relation to some of the parameterized variants of the complexity classes at the second level of the PH that we introduced in Chap. 6. This investigation culminates in several parameterized analogues of the Karp-Lipton Theorem.

We give an overview of the different possible ways to define non-uniform variants of parameterized complexity classes. Because parameterized complexity is a two-dimensional complexity framework, there are more ways to obtain natural non-uniform complexity classes than in the classical one-dimensional complexity framework. Moreover, we relate various non-uniform parameterized complexity classes that have been considered in the literature to each other (see, e.g., [44, 67, 85]). To our knowledge, this is the first structured overview of non-uniform parameterized complexity in the literature.

Moreover, to further motivate the investigation of non-uniform parameterized complexity, we show how some of the non-uniform complexity classes that we consider can be used in the setting of parameterized compilability. In particular, we show a number of parameterized incompilability results that are based on the assumption that some non-uniform parameterized complexity classes are different.

© Springer-Verlag GmbH Germany, part of Springer Nature 2019
R. de Haan: Parameterized Complexity in the Polynomial Hierarchy, LNCS 11880,
https://doi.org/10.1007/978-3-662-60670-4_15

Outline of This Chapter

We begin in Sect. 15.1 with defining the most prominent non-uniform parameterized complexity classes. These include the non-uniform parameterized complexity classes that have been considered in the literature, as well as most of the parameterized complexity classes that play a role in the parameterized analogues of the Karp-Lipton Theorem. Most of these non-uniform classes can be defined using the concept of advice.

In Sect. 15.2, we prove some basic results about the non-uniform parameterized complexity classes that we consider. We provide several alternative characterizations of some of the non-uniform classes. Moreover, we relate several of these classes to each other, by showing both inclusion and separation results.

Then, in Sect. 15.3, we show how non-uniform parameterized complexity can be used in the setting of parameterized compilability. We briefly review a parameterized compilability framework, that can be used to investigate questions about knowledge compilation from a parameterized complexity perspective. This parameterized framework has been proposed by Chen [44], and builds forth on the compilability framework by Cadoli, Donini, Liberatore and Schaerf [34]. Moreover, we introduce a novel non-uniform parameterized complexity class that can be used to investigate parameterized compilability questions that could not be addressed adequately before.

Finally, in Sect. 15.4, we develop the parameterized analogues of the Karp-Lipton Theorem, relating several parameterized complexity classes introduced in Chap. 6 to non-uniform parameterized complexity classes.

15.1 Non-uniform Parameterized Complexity Classes

In this section, we give a definition of the most important non-uniform parameterized complexity classes that we discuss in this chapter. Most classes can be defined naturally in a homogeneous way, namely by using advice. Only for the classes FPT_{nu} and XP_{nu}—natural non-uniform variants of the parameterized complexity classes FPT and XP—the most natural definition is of a different form.

An overview of all (non-uniform) parameterized complexity classes that we consider in this chapter, including the classes defined in this section, can be found in Fig. 15.1 on page 298.

Whenever we speak of an arbitrary parameterized complexity class K in this chapter, we assume that this class is closed under fpt-reductions. We will repeat this assumption for some results, to emphasize that this assumption is needed to establish these results.

15.1.1 Fpt-Size and Xp-Size Advice

We begin our exposition of non-uniform parameterized complexity classes with several classes that are based on advice depending on both the input size n and the parameter value k. Such classes, where the advice string α is of *fpt-size*, have been considered in the context of parameterized knowledge compilation [44]. We define these classes, as well as a natural variant where *xp-size advice* is allowed. We begin with the non-uniform parameterized complexity classes based on fpt-size advice.

Definition 15.1 (fpt-size advice). *Let* K *be a parameterized complexity class. We define* K/fpt *to be the class of all parameterized problems* Q *for which there exists a parameterized problem* $Q' \in$ K, *a computable function* f *and a constant* c *such that for each* $(n, k) \in \mathbb{N} \times \mathbb{N}$ *there exists some* $\alpha(n, k) \in \Sigma^*$ *of size* $f(k)n^c$ *with the property that for all instances* (x, k) *it holds that* $(x, k) \in Q$ *if and only if* $(x, \alpha(|x|, k), k) \in Q'$.

The classes K/fpt have been defined by Chen as K/ppoly [44] (for *parameterized polynomial-size*). Next, we turn to the non-uniform parameterized complexity classes based on xp-size advice.

Definition 15.2 (xp-size advice). *Let* K *be a parameterized complexity class. We define* K/xp *to be the class of all parameterized problems* Q *for which there exists a parameterized problem* $Q' \in$ K *and a computable function* f *such that for each* $(n, k) \in \mathbb{N} \times \mathbb{N}$ *there exists some* $\alpha(n, k) \in \Sigma^*$ *of size* $n^{f(k)}$ *with the property that for all instances* (x, k) *it holds that* $(x, k) \in Q$ *if and only if* $(x, \alpha(|x|, k), k) \in Q'$.

The above definition has the following direct consequence.

Observation 15.3. FPT/xp = XP/xp.

We will use the notation XP/xp for the class FPT/xp = XP/xp to emphasize that an $n^{f(k)}$ running time is allowed for algorithms that witness membership in this class.

15.1.2 Slice-Wise Advice

We continue our exposition with non-uniform variants of parameterized complexity classes based on (computable-size) advice for each slice. We say that the resulting non-uniform parameterized complexity classes are based on *slice-wise advice*.

Definition 15.4 (slice-wise advice). *Let* K *be a parameterized complexity class. We define* K/slice *to be the class of all parameterized problems* Q *for which there exists a parameterized problem* $Q' \in$ K *and a computable function* f *such that for each* $k \in \mathbb{N}$ *there exists some advice string* $\alpha(k) \in \Sigma^*$ *of size at most* $f(k)$ *with the property that for all instances* (x, k) *it holds that* $(x, k) \in Q$ *if and only if* $(x, \alpha(k), k) \in Q'$.

Intuitively, the difference between the definitions of the classes K/slice and K/xp can be explained as follows by taking as example K = XP. For problems in the class XP/slice, for each parameter value k there must be a single (uniformly defined) polynomial-time algorithm (whose descriptions must be of size computable from k). For problems in the class XP/xp, for each parameter value k there can be a (non-uniformly defined) polynomial-time algorithm (with similar size bounds).

We observe the following properties.

Observation 15.5. *The class* FPT/slice *can straightforwardly be shown to be equivalent to* P/slice. *Here we consider* P *as the parameterized complexity class consisting of all parameterized problems that can be solved in polynomial time.*

Observation 15.6. *For each parameterized complexity class* K, *it holds that* K $\subseteq$ K/slice $\subseteq$ K/fpt $\subseteq$ K/xp.

15.1.3 Poly-size and Kernel-Size Advice

Next, we present some additional natural non-uniform variants of parameterized complexity classes that are also based on advice. These variants are natural notions of non-uniform complexity that come up in analogy to the classes defined in Sect. 15.1.1, and are based on *polynomial-size* (or *poly-size*) and *kernel-size advice*. We begin with defining the classes based on poly-size advice.

Definition 15.7 (poly-size advice). *Let* K *be a parameterized complexity class. We define* K/poly *to be the class of all parameterized problems* Q *for which there exists a parameterized problem* $Q' \in K$ *and a constant c such that for each* $(n, k) \in \mathbb{N} \times \mathbb{N}$ *there exists some* $\alpha(n, k) \in \Sigma^*$ *of size* n^c *with the property that for all instances* (x, k) *it holds that* $(x, k) \in Q$ *if and only if* $(x, \alpha(|x|, k), k) \in Q'$.

This definition directly gives us the following relation to the classes K/fpt.

Observation 15.8. *For each parameterized complexity class* K, *we have that* K $\subseteq$ K/poly $\subseteq$ K/fpt.

Next, we define the classes based on kernel-size advice.

Definition 15.9 (kernel-size advice). *Let* K *be a parameterized complexity class. We define* K/kernel *to be the class of all parameterized problems* Q *for which there exists a parameterized problem* $Q' \in K$ *and a computable function f such that for each* $(n, k) \in \mathbb{N} \times \mathbb{N}$ *there exists some* $\alpha(n, k) \in \Sigma^*$ *of size* $f(k)$ *with the property that for all instances* (x, k) *it holds that* $(x, k) \in Q$ *if and only if* $(x, \alpha(|x|, k), k) \in Q'$.

The classes K/kernel bear a similar relation to the classes K/fpt as the classes K/poly do.

Observation 15.10. *For each parameterized complexity class* K, *we have that* K $\subseteq$ K/kernel $\subseteq$ K/fpt.

15.1.4 Slice-Wise Non-uniformity

The last variant of non-uniformity that we consider can be called *slice-wise non-uniformity*. Firstly, we consider the class FPT_{nu}, where problems are required to be solvable in fpt-time, but the algorithm for each slice can be different.

Definition 15.11 ([67]). *The parameterized complexity class* FPT_{nu} *is defined as the class of all parameterized problems Q for which there exists a (possibly uncomputable) function f and a constant c such that for every $k \in \mathbb{N}$, the k-th slice Q_k of Q is decidable in time $f(k)n^c$, where n is the size of the instance.*

The class FPT_{nu} is the class of parameterized problems that are *non-uniformly fixed-parameter tractable* in the sense that is discussed in textbooks [67, 85]. Moreover, it coincides with a variant of the class FPT/slice, where the function f bounding the size of the advice $\alpha(k)$ is not required to be computable (see [36, Theorem 1.3]).

Finally, we give a definition of the most prominent non-uniform variant of XP, that is also presented in textbooks on parameterized complexity [67, 85].

Definition 15.12 ([67]). *The parameterized complexity class* XP_{nu} *is defined as the class of all parameterized problems Q for which for each $k \in \mathbb{N}$, the slice $Q_k = \{x : (x, k) \in Q\}$ is polynomial-time solvable.*

We point out that in the definition of the class XP_{nu}, the order of the polynomial that is a bound on the running time of the algorithm that solves slice Q_k of a problem $Q \in \text{XP}_{\text{nu}}$ is allowed to vary for different values of k, and does not have to be bounded by a computable function of k. In contrast, for problems Q in FPT_{nu}, each slice Q_k must be solvable in time $O(n^c)$ for some fixed constant c. (Here the factor $f(k)$ hidden by the big-oh notation depends on k, and must be computable from k.)

15.2 Basic Results

In this section, we provide some basic results about the different non-uniform parameterized complexity classes that we defined in the previous section. We begin with giving several alternative characterizations of some of the classes. Then we relate several of the classes to each other, by giving some separation (non-inclusion) results. A graphical overview of the relations between the classes can be found in Fig. 15.1.

15.2.1 Alternative Characterizations

We provide alternative characterizations of the non-uniform complexity classes defined in Sect. 15.1, in terms of fpt-reductions with advice, in terms of Boolean circuits, by using slice-wise solvability, and in terms of potentially uncomputable-size advice.

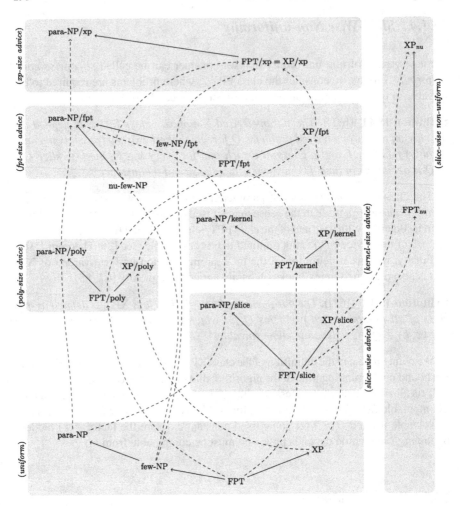

Fig. 15.1 Non-uniform parameterized complexity classes. Most classes are defined in Sect. 15.1. For a definition of few- NP, nu-few-NP and few- NP/fpt, we refer to Sect. 15.3.3.

Fpt-Reductions with Advice

We show that the classes K/fpt, K/kernel and K/poly can be defined by means of fpt-reductions with advice of appropriate size. We start with defining fpt-reductions with fpt-size advice.

Definition 15.13 (fpt-reductions with fpt-size advice). *Let Q, Q' be parameterized problems. We say that an fpt-algorithm R is an fpt-reduction with fpt-size advice from Q to Q' if there exist computable functions f, h and a constant c such that for each $(n, k) \in \mathbb{N} \times \mathbb{N}$ there is an advice string $\alpha(n, k)$ of length $f(k)n^c$ such that for* *each instance* $(x, k) \in \Sigma^* \times \mathbb{N}$ *it holds that (1)* $(x', k') \in Q'$ *if and only if* $(x, k) \in Q$, *and (2)* $k' \le h(k)$, *where* $(x', k') = R(x, \alpha(|x|, k), k)$.

When there exists an fpt-reduction with fpt-advice from Q to Q', we say that Q is fpt-reducible to Q' with fpt-size advice.

We show that K/fpt can be characterized using reductions with fpt-size advice.

Proposition 15.14. *Let* K *be a parameterized complexity class that is closed under fpt-reductions. Then* K/fpt *coincides with the class of parameterized problems that are fpt-reducible with fpt-size advice to some problem in* K.

Proof. ($\Rightarrow$) Take an arbitrary problem $Q \in$ K/fpt. Then, by definition, there exists some $Q' \in$ K such that for each $(n, k) \in \mathbb{N} \times \mathbb{N}$ there exists some advice string $\alpha(n, k)$ of fpt-size with the property that for each $(x, k) \in \Sigma^* \times \mathbb{N}$ it holds that $(x, k) \in Q$ if and only if $(x, \alpha(|x|, k), k) \in Q'$. We construct an fpt-reduction R with fpt-size advice from Q to Q'. For each (n, k), we let the advice string for R be the string $\alpha(n, k)$. Moreover, we let $R(x, k) = (x, \alpha(n, k), k)$. This proves that Q is fpt-reducible to some problem in K with fpt-size advice.

($\Leftarrow$) Conversely, take an arbitrary parameterized problem Q that is fpt-reducible to some problem $Q' \in$ K, by an fpt-reduction R with fpt-size advice. By definition, then, for each $(n, k) \in \mathbb{N} \times \mathbb{N}$ there exists some fpt-size advice string $\alpha(n, k)$ with the property that for each $(x, k) \in \Sigma^* \times \mathbb{N}$ it holds that $(x', k') \in Q'$ if and only if $(x, k) \in Q$, where $(x', k') = R(x, \alpha(|x|, k), k)$. We show that $Q \in$ K/fpt by specifying fpt-size advice for each $(n, k) \in \mathbb{N} \times \mathbb{N}$ and giving a problem $Q'' \in$ K such that for each $(x, k) \in \Sigma^* \times \mathbb{N}$ it holds that $(x, k) \in Q$ if and only if $(x, \alpha(|x|, k), k) \in Q''$. For each $(n, k) \in \mathbb{N} \times \mathbb{N}$ we let the advice string be $\alpha(n, k)$. Moreover, we let $Q'' = \{ (x, y, k) \in \Sigma^* \times \Sigma^* \times \mathbb{N} : R(x, y, k) \in Q' \}$. Since K is closed under fpt-reductions, we know that $Q'' \in$ K. Therefore, $Q \in$ K/fpt. $\square$

Next, we extend this characterization to the cases of K/kernel and K/poly by introducting fpt-reductions with kernel-size and polynomial-size advice, and showing that these can be used to characterize K/kernel and K/poly, respectively. We begin with the case of kernel-size advice.

Definition 15.15 (fpt-reductions with kernel-size advice). *Let* Q, Q' *be parameterized problems. We say that an fpt-algorithm* R *is* an fpt-reduction with kernel-size advice from Q to Q' *if there exist computable functions* f, h *such that for each* $(n, k) \in \mathbb{N} \times \mathbb{N}$ *there is an advice string* $\alpha(n, k)$ *of length* $f(k)$ *such that for each instance* $(x, k) \in \Sigma^* \times \mathbb{N}$ *it holds that (1)* $(x', k') \in Q'$ *if and only if* $(x, k) \in Q$, *and (2)* $k' \le h(k)$, *where* $(x', k') = R(x, \alpha(|x|, k), k)$.

Proposition 15.16. *Let* K *be a parameterized complexity class that is closed under fpt-reductions. Then* K/kernel *coincides with the class of parameterized problems that are fpt-reducible with kernel-size advice to some problem in* K.

Proof. The proof of Proposition 15.14 can straightforwardly be modified to show this result. $\square$

Next, we turn to the case of poly-size advice.

Definition 15.17 (fpt-reductions with poly-size advice). *Let Q, Q' be parameterized problems. We say that an fpt-algorithm R is an fpt-reduction with poly-size advice from Q to Q' if there exist a computable function h and a constant c such that for each $(n, k) \in \mathbb{N} \times \mathbb{N}$ there is an advice string $\alpha(n, k)$ of length n^c such that for each instance $(x, k) \in \Sigma^* \times \mathbb{N}$ it holds that (1) $(x', k') \in Q'$ if and only if $(x, k) \in Q$, and (2) $k' \le h(k)$, where $(x', k') = R(x, \alpha(|x|, k), k)$.*

Proposition 15.18. *Let* K *be a parameterized complexity class that is closed under fpt-reductions. Then* K/poly *coincides with the class of parameterized problems that are fpt-reducible with poly-size advice to some problem in* K.

Proof. The proof of Proposition 15.14 can straightforwardly be modified to show this result.

Circuits

We provide the alternative characterizations of the classes FPT/fpt and para-NP/fpt, in terms of (families of) circuits. Let C be a circuit with m input nodes, and let $x \in \mathbb{B}^*$ be a bitstring of length $n \le m$. Then, by $C[x]$ we denote the circuit obtained from C by instantiating the first n input nodes according to the n values in x.

We begin with the case of FPT/fpt.

Proposition 15.19. *The class* FPT/fpt *coincides with the set of all parameterized problems Q for which there exists a computable function f and a constant c such that for each $(n, k) \in \mathbb{N} \times \mathbb{N}$ there is some circuit $C_{n,k}$ of size $f(k)n^c$ that decides, for each input $x \in \mathbb{B}^n$ of length n, whether $(x, k) \in Q$.*

Proof (sketch). Let Q be a problem that is solvable in fpt-time using an fpt-size advice string $\alpha(n, k)$ that only depends on the input size n and the parameter value k. Then for any (n, k), one can "hard-code" this advice and its use by the algorithm in an fpt-size circuit $C_{n,k}$ that decides for inputs x of length n whether $(x, k) \in Q$.

Conversely, if for a problem Q, for each (n, k) there exists an fpt-size circuit $C_{n,k}$ that decides for inputs x of length n whether $(x, k) \in Q$, then one can decide Q in fpt-time by taking a description of this circuit as the advice string $\alpha(n, k)$, and simulating the circuit on any input x of length n.

Next, we address the case of para-NP/fpt.

Proposition 15.20. *The class* para-NP/fpt *coincides with the set of all parameterized problems Q for which there exists a computable function f and a constant c such that for each $(n, k) \in \mathbb{N} \times \mathbb{N}$ there is some circuit $C_{n,k}$ of size $f(k)n^c$ such that, for each input $x \in \mathbb{B}^n$ of length n, it holds that $x \in Q$ if and only if $C_{n,k}[x]$ is satisfiable.*

Proof (idea). An argument similar to the one in the proof of Proposition 15.19 can be used to show this result. In addition, one needs to use the well-known correspondence between non-deterministic algorithms and satisfiability of circuits, as used in the Cook-Levin Theorem [54, 144].

Slice-Wise Polynomial-Time Solvability

We show that the class XP/slice can be characterized using polynomial-time algorithms for the different slices of the problem. Here, the order of the polynomials in the running time must be bounded by a computable function.

Proposition 15.21. *Let the class* $\text{XP}_{\text{nu}}^{\text{comp}}$ *consist of all parameterized problems* Q *for which there exists some computable function* f *such that for each* $k \in \mathbb{N}$, *the slice* Q_k *is solvable in time* $n^{f(k)}$, *where* $n = |x|$. *It holds that* $\text{XP}_{\text{nu}}^{\text{comp}} = \text{XP/slice}$.

Proof. ($\Rightarrow$) Let $Q \in \text{XP}_{\text{nu}}^{\text{comp}}$. Then there exists some computable function f such that for each k there exists an algorithm A_k that solves Q_k in time $n^{f(k)}$. Then for each k, we define the advice string $\alpha(k)$ to be a description of A_k. On input $(x, \alpha(k), k) = (x, A_k, k)$, we can then simulate the algorithm A_k on input (x, k), which runs in time $|x|^{f(k)}$. Therefore, the problem of deciding whether $(x, k) \in Q$, given $(x, \alpha(k), k)$, is in XP. In other words, $Q \in \text{XP/slice}$.

($\Leftarrow$) Let $Q \in \text{XP/slice}$. Then for each $k \in \mathbb{N}$ there is some advice string $\alpha(k)$ such that deciding if $(x, k) \in Q$, given $(x, \alpha(k), k)$, is in XP. In other words, there is an algorithm A that decides whether $(x, k) \in Q$, given $(x, \alpha(k), k)$, in time $|x|^{f(k)}$, for some computable function f. Take an arbitrary $k \in \mathbb{N}$. Let the algorithm A_k be the algorithm A that has "hard-coded" access to the advice string $\alpha(k)$. Then A_k decides the slice Q_k of Q in time $n^{f(k)}$, where n denotes the input size. Thus $Q \in \text{XP}_{\text{nu}}^{\text{comp}}$.

Uncomputable-Size Advice

Finally, we show that the class XP_{nu} can be characterized using slice-wise advice, where the size of the advice strings is not required to be bounded by a computable function.

Proposition 15.22. *Let the class* XP/u-slice *consist of all parameterized problems* Q *for which there exists a parameterized problem* $Q' \in \text{XP}$ *such that for each* $k \in \mathbb{N}$ *there exists some advice string* $\alpha(k) \in \Sigma^*$ *with the property that for all instances* (x, k) *it holds that* $(x, k) \in Q$ *if and only if* $(x, \alpha(k), k) \in Q'$. *Then* $\text{XP}_{\text{nu}} = \text{XP/u-slice}$.

Proof (sketch). This result can be shown with an argument that is entirely similar to the proof of Proposition 15.21.

15.2.2 Separations

Next, we give some (conditional and unconditional) non-inclusion results for some of the non-uniform complexity classes that we defined in Sect. 15.1. Several of these results can be shown rather straightforwardly, whereas others need a bit more technical machinery.

We begin with relating the classes XP_{nu} and XP/slice. The class XP_{nu} clearly contains the class XP/slice, and its subclasses. We show that this containment is strict.

Proposition 15.23. *There exists a parameterized problem $Q \in (\mathrm{XP}_{nu}) \backslash (\mathrm{XP}/\mathrm{slice})$.*

Proof. Let g be an increasing function that grows faster asymptotically than every computable function. For instance, we can let g be the busy-beaver function, where for each $n \in \mathbb{N}$ the value $g(n)$ is the maximum number of steps performed by any Turing machine with n states that halts when given the empty string as input. By the Time Hierarchy Theorem [120], we know that for each $m \in \mathbb{N}$ there exists some problem P_m that is solvable in time $O(n^m)$ but not in time $O(n^{m-1})$, where n denotes the input size. We then let Q be the following parameterized problem:

$$Q = \{ (x, k) : k \in \mathbb{N}, x \in P_{g(k)} \}.$$

For each value k, we know by assumption that $P_{g(k)}$ is solvable in polynomial-time, namely in time $O(n^{g(k)})$. Therefore, $Q \in \mathrm{XP}_{nu}$.

We claim that $Q \notin \mathrm{XP}/\mathrm{slice}$. We proceed indirectly, and suppose that $Q \in \mathrm{XP}/\mathrm{slice}$. Then there exist a computable function f such that for each $k \in \mathbb{N}$, there exists some advice string $\alpha(k)$ such that the problem of deciding if $(x, k) \in Q$, given $(x, \alpha(k), k)$, is solvable in time $n^{f(k)}$, where $n = |x|$ denotes the input size. We know that g grows faster asymptotically than f. Therefore, there exists some ℓ such that $g(\ell) > f(\ell)$. Consider the slice Q_ℓ of Q. By "hard-coding" the advice string $\alpha(\ell)$ in an algorithm A_ℓ, we can then solve Q_ℓ in time $n^{f(\ell)}$. However, by construction of Q, the slice Q_ℓ is not solvable in time $O(n^{g(\ell)-1})$, and thus also not solvable in time $O(n^{f(\ell)})$, which is a contradiction. Therefore, we can conclude that $Q \notin \mathrm{XP}/\mathrm{slice}$.

Corollary 15.24. $\mathrm{XP}/\mathrm{slice} \subsetneq \mathrm{XP}_{nu}$.

Next, we will show that the different notions of non-uniformity that are used for classes like FPT/fpt, on the one hand, and for the class XP_{nu}, on the other hand, are incomparable. In order to show this, we begin by exhibiting a parameterized problem that is in FPT/fpt (and that is in fact also contained in FPT/kernel and FPT/poly) but that is not in XP_{nu}.

Proposition 15.25. *There exists a parameterized problem $Q \in (\mathrm{FPT}/\mathrm{fpt}) \backslash (\mathrm{XP}_{nu})$.*

Proof. Let S be an undecidable unary set, e.g., the set of all strings 1^m such that the m-th Turing machine (in some enumeration of all Turing machines) halts when given the empty string as input. We define the following parameterized problem Q:

$$Q = \{ (s, 1) : s \in S \}.$$

Clearly, $Q \in \mathrm{FPT}/\mathrm{fpt}$, since for input size n there is at most one string s of length n such that $(s, 1) \in Q$. We claim that $Q \notin \mathrm{XP}_{nu}$. We proceed indirectly, and we suppose that $Q \in \mathrm{XP}_{nu}$. Then there is an algorithm A that decides the slice $Q_1 = \{ x : (x, 1) \in Q \}$ in polynomial-time. However, then the algorithm A can straightforwardly be modified to decide the undecidable set S, which is a contradiction. Therefore, $Q \notin \mathrm{XP}_{nu}$.

The problem that we constructed that is not in XP_{nu}, is also contained in FPT/poly and FPT/kernel.

Corollary 15.26. *There exists a parameterized problem* $Q \in (FPT/poly) \setminus (XP_{nu})$ *and there exists a parameterized problem* $Q \in (FPT/kernel) \setminus (XP_{nu})$.

Proof (sketch). The problem Q constructed in the proof of Proposition 15.25 is in fact contained both in (FPT/poly) and in (FPT/kernel), and not in XP_{nu}.

In order to complete our incomparability result, we show that there exists also a problem that is contained in XP_{nu}, but that is not contained in XP/xp. This result can also be seen as a strenghtened version of Proposition 15.23.

Proposition 15.27. *There exists a parameterized problem* $Q \in (XP_{nu}) \setminus (XP/xp)$.

Proof. In order to show this, we will use a few theoretical tools. Firstly, we will consider the busy-beaver function $g : \mathbb{N} \to \mathbb{N}$, where $g(n)$ is defined to be the maximum number of steps performed by any Turing machine with n states that halts on the empty string. It is well-known and it can be shown straightforwardly that g grows faster asymptotically than any computable function f.

Next, we will make use of the following facts. For each $n \in \mathbb{N}$, there are 2^{2^n} possible Boolean functions $F : \mathbb{B}^n \to \mathbb{B}$. Each such function is computed by a circuit of size at most $c2^n$, for some fixed constant c. (By the size of a circuit, we denote the number of bits required in a binary representation of the circuit.) Moreover, there are at most $2^{f(n)}$ circuits with n input nodes of size $f(n)$. Additionally, we can enumerate all circuits $C_1^n, C_2^n, \ldots$ with n input nodes (possibly allowing repetitions) in such a way that:

- for each $m \in \mathbb{N}$, computing the m-th circuit C_m^n in this enumeration can be done in time $O(m \cdot |C_m^n|^2)$; and
- for each $\ell, \ell' \in \mathbb{N}$ such that $\ell < \ell'$ it holds that all circuits of size ℓ come before all circuits of size ℓ' in this enumeration.

With these theoretical tools in place, we can begin constructing the problem Q. Since we want $Q \in XP_{nu}$, we need to ensure that all slices Q_k of Q are solvable in polynomial-time. However, the order of the polynomials that bounds the running time of the algorithms solving the slices Q_k does not need to be bounded by a computable function. This is exactly the property that we will exploit. We will construct Q in such a way that the running time needed to solve the slices Q_k grows so fast that any XP/xp algorithm will make a mistake on some slice Q_k.

In addition to g, we will consider two other functions h_1, h_2, that are defined as follows. Consider the following (infinite) sequences:

$$h_1 = (\overbrace{g(1)^2, \ldots, g(1)^2}^{g(1) \text{ times}}, \overbrace{g(2)^2, \ldots, g(2)^2}^{g(2) \text{ times}}, \overbrace{g(3)^2, \ldots, g(3)^2}^{g(3) \text{ times}}, \ldots), \text{ and}$$

$$h_2 = (1, 2, \ldots, g(1), 1, 2, \ldots, g(2), 1, 2, \ldots, g(3), 1, 2, \ldots).$$

We define $h_1(x)$ to be the x-th element in the sequence h_1 and $h_2(x)$ to be the x-th element in the sequence h_2.

We then define $Q \subseteq \mathbb{B}^* \times \mathbb{N}$ to be the parameterized problem that is recognized by the following algorithm A. We make sure that A runs in polynomial time on each slice Q_k. On input $(x, k) \in \mathbb{B}^* \times \mathbb{N}$, where $|x| = n$, the algorithm A uses the enumeration $C_1^n, C_2^n, \ldots$ of circuits on n input nodes, and tries to compute the $h_2(k)$-th circuit $C_{h_2(k)}^n$. If this takes more than $h_1(k)$ steps, the algorithm accepts (x, k); otherwise, the algorithm simulates $C_{h_2(k)}^n$ on the input $x \in \mathbb{B}^n$, and accepts if and only if x satisfies $C_{h_2(k)}^n$. The algorithm A runs in time $O(h_1(k)^2 \cdot n)$, so for each $k \in \mathbb{N}$, it decides Q_k in linear time.

We now show that $Q \notin \mathrm{XP/xp}$. We proceed indirectly, and suppose that $Q \in \mathrm{XP/xp}$. Then there exists some computable function $f : \mathbb{N} \to \mathbb{N}$ such that for each $(n, k) \in \mathbb{N} \times \mathbb{N}$ there exists a circuit $C_{n,k}$ of size $n^{f(k)}$ that decides for inputs $x \in \mathbb{B}^n$ whether $(x, k) \in Q$. We now identify a pair $(n_0, k_0) \in \mathbb{N} \times \mathbb{N}$ of values such that $(n_0)^{f(k_0)} < 2^{2^{n_0}}$ and $g(k_0) \geq c2^{n_0}$. In other words, we want (n_0, k_0) to satisfy the property that $f(k_0) < 2^{n_0}/\log n_0$ and $c2^{n_0} \leq g(k_0)$. We can do this as follows. Since $g(k)$ grows faster asymptotically than any computable function, we know that there exists some k_0 such that $g(k_0) > c2^{f(k_0)}$. Moreover, we let $n_0 = f(k_0)$. We then get that $g(k_0) > c2^{n_0}$ and $f(k_0) = n_0 < 2^{n_0}/\log n_0$. Now, let $k_1 \in \mathbb{N}$ be the least integer such that $h_1(k_1) = g(k_0)$. Then, by construction of h_1, for each $\ell \in [0, g(k_0) - 1]$, it holds that $h_1(k_1 + \ell) = g(k_0)$.

Now, since we know that $(n_0)^{f(k_0)} < 2^{2^{n_0}}$, we know that there exists some Boolean function $F_0 : \mathbb{B}^{n_0} \to \mathbb{B}$ on n_0 variables that is not computed by any circuit of size $(n_0)^{f(k_0)}$. However, this function F_0 can be computed by some circuit of size $c2^{n_0}$. Therefore, because $g(k_0) > c2^{n_0}$, there is some $\ell \in [0, g(k_0) - 1]$ such that the circuit $C_{h_2(k_1+\ell)}^{n_0}$ computes F_0. Moreover, we can choose ℓ in such a way that $C_{h_2(k_1+\ell)}^{n_0}$ is of size at most $c2^{n_0} < g(k_0)$. Therefore, computing $C_{h_2(k_1+\ell)}^{n_0}$, using the enumeration $C_1^{n_0}, C_2^{n_0}, \ldots$, can be done in time $h_2(k_1 + \ell)^2 \leq h_1(k_1 + \ell) = g(k_0)$.

Now consider the pair $(n_0, k_0) \in \mathbb{N} \times \mathbb{N}$ of values. We assumed that there is some circuit C_{n_0,k_0} of size $(n_0)^{f(k_0)}$ that decides for inputs $x \in \mathbb{B}^{n_0}$ whether $(x, k_0) \in Q$. By construction of Q and by the choice of (n_0, k_0), we know that for each $x \in \mathbb{B}^{n_0}$ it must hold that $(x, k_0) \in Q$ if and only if $F_0(x) = 1$. This means that the circuit C_{n_0,k_0} computes F_0. However, since C_{n_0,k_0} is of size $(n_0)^{f(k_0)}$, this is a contradiction with the fact that F_0 is not computable by a circuit of size $(n_0)^{f(k_0)}$. Therefore, we can conclude that there exists no family of circuits $C_{n,k}$ of size $n^{f(k)}$ for $(n, k) \in \mathbb{N} \times \mathbb{N}$. In other words, $Q \notin \mathrm{XP/xp}$.

The above results (Propositions 15.25 and 15.27 and Corollary 15.26) give us the incomparability results that we were after, which can be summarized as follows.

Corollary 15.28. *Let* K *be a parameterized complexity class such that* $\mathrm{FPT/poly} \subseteq$ K $\subseteq \mathrm{XP/xp}$. *Then* K *and* $\mathrm{XP_{nu}}$ *are incomparable w.r.t. set-inclusion, i.e.,* K $\not\subseteq \mathrm{XP_{nu}}$ *and* $\mathrm{XP_{nu}} \not\subseteq$ K.

Corollary 15.29. *Let* K *be a parameterized complexity class such that* FPT/kernel $\subseteq$ K $\subseteq$ XP/xp. *Then* K *and* XP_{nu} *are incomparable w.r.t. set-inclusion, i.e.,* K $\nsubseteq$ XP_{nu} *and* XP_{nu} $\nsubseteq$ K.

In fact, the proof of Proposition 15.27 shows that there is a problem Q that is in FPT_{nu} but not in XP/xp.

Corollary 15.30. *There exists a problem* $Q \in (FPT_{nu}) \backslash (XP/xp)$.

Proof. The problem Q from the proof of Proposition 15.27 is in $(FPT_{nu}) \backslash (XP/xp)$.

This observation gives us the following incomparability result.

Corollary 15.31. *Let* K *be a parameterized complexity class such that either* FPT/poly $\subseteq$ K $\subseteq$ XP/xp *or* FPT/kernel $\subseteq$ K $\subseteq$ XP/xp. *Then* K *and* FPT_{nu} *are incomparable w.r.t. set-inclusion, i.e.,* K $\nsubseteq$ FPT_{nu} *and* FPT_{nu} $\nsubseteq$ K.

Next, we explore the power of the non-uniformity resulting from kernel-size and polynomial-size advice. Firstly, we give an easy proof that both forms of non-uniformity are more powerful than the uniform setting.

Proposition 15.32. *For every (non-trivial) decidable parameterized complexity class* K, *there is a parameterized problem* $Q \in (K/poly) \backslash K$ *and a parameterized problem* $Q' \in (K/kernel) \backslash K$.

Proof. The problem Q as constructed in the proof of Proposition 15.25 is also contained in K/poly and K/kernel, for each non-trivial K. However, since Q is undecidable, we know that $Q \notin K$.

Secondly, we show that the non-uniform parameterized complexity classes FPT/kernel and FPT/poly are incomparable. We start by identifying a problem that is in FPT/poly but not in FPT/kernel.

Proposition 15.33. *There is a parameterized problem* $Q \in (FPT/poly) \backslash (FPT/kernel)$.

Proof. To show this result, it suffices to give a (classical) problem Q that is in P/poly but that is not solvable in polynomial time using constant-size advice. If we have such a problem Q, then the parameterized problem $\{(x, 1) : x \in Q\}$ is in $(FPT/poly) \backslash (FPT/kernel)$.

We use a diagonalization argument to construct the problem Q. Let (A_1, c_1), $(A_2, c_2), \ldots$ be an enumeration of all pairs (A_i, c_i) consisting of a polynomial-time algorithm A_i and a constant c_i. We will construct Q in such a way that for each algorithm A_i and each advice string y of length c_i, the algorithm A_i, when using the advice string y, makes a mistake on some input. We construct Q in stages: one stage for each (A_i, c_i). In stage i, we make sure that A_i makes a mistake for each advice string y of length c_i.

We now describe how to construct Q in a given stage i. Let (A_i, c_i) be the pair consisting of a polynomial-time algorithm A_i and a constant c_i. Take an input size n that has not yet been considered in previous stages, and such that $n \geq 2^{c_i}$. Since we have infinitely many input sizes at our disposal, we can always find such an n. Next, consider $2^{c_i} = u$ (arbitrary, but different) input strings $x_1, \ldots, x_u$ of length n. Moreover, consider all possible advice strings $y_1, \ldots, y_u$ of length c_i. For each $j \in [u]$, we let $x_j \in Q$ if and only if the algorithm A_i, using the advice string y_j, rejects the input string x_j. For all other input strings x of length n, we let $x \notin Q$. Note that Q contains at most $2^{c_i} \leq n$ strings of length n. This completes stage i. For all input lengths n that are not considered in any stage of our construction, we let Q contain no strings of length n.

We show that $Q \in P/poly$. We define the advice for input size n to be a table consisting of all strings in Q of length n. Since for each input length n, Q contains at most n strings of this length, this table is of polynomial size. Deciding whether a string of length n is in Q can clearly be done in polynomial time, when given such a table. Therefore, $Q \in P/poly$.

On the other hand, we claim that Q is not solvable in polynomial time using a constant number of bits as advice. We proceed indirectly, and suppose that there is a polynomial-time algorithm A that decides Q, when given c bits of advice for each input size, for some fixed constant c. We know that (A, c) appears in the enumeration of all pairs of polynomial-time algorithms and constants. Let $(A_i, c_i) = (A, c)$, for some $i \geq 1$. Then consider the input size n that is used in stage i of the construction of Q. Moreover, let y be the advice string (of length c) that the algorithm A uses to solve inputs of size n. We know that y appeared as some string y_j of length c_i in stage i of the construction of Q. Now consider input x_j. By definition of Q, we know that the algorithm A gives the wrong output for x_j, when using y_j as advice. Thus, A does not solve Q using c bits of advice. Since A and c were arbitrary, we can conclude that Q is not solvable in polynomial time using a constant number of bits as advice.

Next, we identify a problem that is in FPT/kernel but not in FPT/poly.

Proposition 15.34. *There is a parameterized problem $Q \in (FPT/kernel) \setminus (FPT/poly)$.*

Proof. We use a diagonalization argument to construct the problem Q. Let (A_1, p_1), $(A_2, p_2), \ldots$ be an enumeration of all pairs (A_i, p_i) consisting of an fpt-algorithm A_i and a polynomial p_i. We will construct Q in such a way that for each algorithm A_i and each polynomial p_i, there is some input size n, such that for each advice string y of size $p_i(n)$, the algorithm A_i makes a mistake on some input x of length n when using y as advice. We construct Q in stages: one stage for each (A_i, p_i). In stage i, we make sure that A_i makes such a mistake for the polynomial p_i.

We now describe how to construct Q in a given stage i. Let (A_i, p_i) be the pair consisting of an fpt-algorithm A_i and a polynomial p_i. We may assume without loss of generality that for each such polynomial p_i it holds for all n that $p_i(n) \geq n$. Take an input size n that has not yet been considered in previous stages. Since we have infinitely many input sizes at our disposal, we can always find such an n. Moreover, fix the parameter value $k = p_i(n)$. Then, consider all possible advice strings $y_1, \ldots, y_u$ of length $p_i(n)$, where $u = 2^k$. Also, consider u (arbitrary, but different) input strings $x_1, \ldots, x_u$ of length n. For each $j \in [u]$, we let $(x_j, k) \in Q$ if and only if the algorithm A_i, using the advice string y_j, rejects the input string x_j. For all other input strings x of length n, we let $(x, k) \notin Q$. Note that Q contains at most $2^k \leq n$ pairs (x, k), where x is a string of length n. This completes stage i. For all pairs (n, k) consisting of an input length n and a parameter value k that are not considered in any stage of our construction, we let Q contain no pairs (x, k), where x is a string of length n.

We show that $Q \in \text{FPT/kernel}$. We define the advice for the pair (n, k), consisting of input size n and parameter value k, to be a table consisting of all strings x of length n such that $(x, k) \in Q$. Since for each input length n, Q contains at most 2^k pairs (x, k), where x is a string of length $n \leq k$, this table is of size $f(k) = O(k2^k)$. Given a pair (x, k) with x of length n, and given such a table for (n, k), deciding whether $(x, k) \in Q$ can clearly be done in fpt-time. Therefore, $Q \in \text{FPT/kernel}$.

On the other hand, we claim that $Q \notin \text{FPT/poly}$. We proceed indirectly, and suppose that there is an fpt-algorithm A that decides Q, when given $p(n)$ bits of advice for each input size n, for some fixed polynomial p. We know that (A, p) appears in the enumeration of all pairs of fpt-algorithms and polynomials. Let $(A_i, p_i) = (A, p)$, for some $i \geq 1$. Then consider the input size n and the parameter value $k = p(n)$ that are used in stage i of the construction of Q. Moreover, let y be the advice string (of length $p(n)$) that the algorithm A uses to solve inputs of size n. We know that y appeared as some string y_j of length $p(n)$ in stage i of the construction of Q. Now consider input x_j. By definition of Q, we know that the algorithm A gives the wrong output for (x_j, k), when using y_j as advice. Thus, A does not solve Q using $p(n)$ bits of advice. Since A and p were arbitrary, we can conclude that $Q \notin \text{FPT/poly}$.

As a consequence of the above results, the following relations hold between non-uniform variants of the class FPT.

Corollary 15.35. *It holds that:*

- FPT $\subsetneq$ FPT/poly $\subsetneq$ FPT/fpt;
- FPT $\subsetneq$ FPT/kernel $\subsetneq$ FPT/fpt;
- FPT/poly $\not\subseteq$ FPT/kernel; *and*
- FPT/kernel $\not\subseteq$ FPT/poly.

In fact, this picture can be generalized to arbitrary (decidable) classes that contain FPT. We begin by generalizing Propositions 15.33 and 15.34.

Proposition 15.36. *For every decidable parameterized complexity class K, there is a parameterized problem $Q \in (\text{FPT/poly}) \setminus (\text{K/kernel})$ and a parameterized problem $Q' \in (\text{FPT/kernel}) \setminus (\text{K/poly})$.*

Proof. The proofs of Propositions 15.33 and 15.34 can straightforwardly be modified to show this result, by using an enumeration of all algorithms (rather than enumerating all polynomial-time or fpt-time algorithms).

This gives us the following result, relating non-uniform variants of arbitrary classes $K \supseteq FPT$.

Corollary 15.37. *For every decidable parameterized complexity class* K *such that* $FPT \subseteq K$, *it holds that:*

- $K \subsetneq K/poly \subsetneq K/fpt;$
- $K \subsetneq K/kernel \subsetneq K/fpt;$
- $K/poly \nsubseteq K/kernel;$ *and*
- $K/kernel \nsubseteq K/poly.$

Finally, we show that (under various complexity-theoretic assumptions), the class para-NP is not contained in any of the classes $K \subseteq XP/xp$. The first case that we consider is that of $XP/kernel$.

Proposition 15.38. *If* para-NP $\subseteq XP/kernel$, *then* $P = NP$.

Proof. Consider the para-NP-complete language $SAT_1 = \{ (\varphi, 1) : \varphi \in SAT \}$. By our assumption that para-NP $\subseteq XP/kernel$, we get that $SAT_1 \in XP/kernel$. That is, there is some computable function f such that for each $n \in \mathbb{N}$ there exists some $\alpha(n) \in \Sigma^*$ of size $f(k)$ with the property that for each instance $(x, 1)$ of SAT_1 with $|x| = n$, we can decide in fpt-time, given $(x, \alpha(n, 1), 1)$, whether $(x, 1)$ is a yes-instance of SAT_1. In other words, SAT is solvable in polynomial time using a constant number of bits of advice for each input size n.

Then, by self-reducibility of SAT, in polynomial time, using constant-size advice, we can also construct an algorithm A that computes a satisfying assignment of a propositional formula, if it exists, and fails otherwise. We now show how to use this to construct an algorithm B that solves SAT in polynomial time. The idea is the following. Since we only need a constant number of bits of advice for algorithm A, we can simply try out all (constantly many) possible advice strings in a brute force fashion. Given a propositional formula φ, algorithm B iterates over all (constantly many) possible advice strings. For each such string α, algorithm B simulates algorithm A on φ using α. If A outputs a truth assignment for φ, algorithm B verifies whether this assignment satisfies φ. If it does, then clearly φ is satisfiable, and so B accepts φ. If it does not, B continues with the next possible advice string. If for no advice string, the simulation of A outputs a truth assignment, B rejects φ. In this case, we can safely conclude that φ is unsatisfiable. Since at least one advice string leads to correct behavior of A, we know that if φ were satisfiable, a satisfying assignment would have been constructed in some simulation of A. Thus, we can conclude that $P = NP$.

The second case that we consider is that of XP/xp.

Proposition 15.39. *If* para-NP $\subseteq XP/xp$, *then the PH collapses to the second level.*

Proof. Assume that para-NP $\subseteq XP/xp$. Consider the NP-complete language $SAT_1 = \{ (\varphi, 1) : \varphi \in SAT \}$. This is then in $P/poly$. Thus, $NP \subseteq P/poly$. By the Karp-Lipton Theorem [127, 128], it follows that $PH = \Sigma_2^p$.

This result gives us the following corollary.

Corollary 15.40. *It holds that* para-NP $\not\subseteq$ FPT/fpt, para-NP $\not\subseteq$ XP/fpt, para-NP $\not\subseteq$ FPT/poly *and* para-NP $\not\subseteq$ XP/poly, *unless the PH collapses to the second level.*

15.3 Relation to Parameterized Knowledge Compilation

In this section, we show how several non-uniform parameterized complexity classes can be used to establish negative results within the framework of parameterized knowledge compilation.

Knowledge compilation is a technique of dealing with intractable problems where part of the input stays stable for an extended amount of time (see, e.g., [34, 58, 160, 180]). Positive knowledge compilation results amount to the (computationally expensive) preprocessing of the stable part of the input, so that for each possible remainder of the input (the varying part), the problem can be solved efficiently (using the preprocessed information). A natural scenario where knowledge compilation can be used, for instance, is that of a (stable) database D containing some body of knowledge, and (varying) queries q that are posed against the database. Unfortunately, it turns out that many important problems do not lead to positive compilation results. Parameterized knowledge compilation refers to the use of the parameterized complexity framework to knowledge compilation, with the aim of increasing the possibility of positive compilation results [44].

However, unsurprisingly, there are also problems that cannot be compiled even in the parameterized setting. Most of these negative results are in fact conditional on some non-uniform parameterized complexity classes being different. We give an overview of the relation between parameterized incompilability results and non-uniform parameterized complexity classes. In addition, we define some new non-uniform parameterized complexity classes that can be used to get additional (conditional) parameterized incompilability results.

The (parameterized) compilability framework that we use is based on the work by Cadoli et al. [34] and Chen [44]. Relating non-uniform parameterized complexity to the more recent framework of compilability by Chen [43] remains a topic for future research.

15.3.1 Parameterized Knowledge Compilation

We begin with reviewing the basic notions from (parameterized) knowledge compilability theory.

Compilability

We provide some basic notions from the theory of compilability. For more details, we refer to the work of Cadoli et al. [34]. In the following, we fix an alphabet Σ. We begin with the basic definitions of compilation problems (formalized as *knowledge representation formalisms*) and the class of polynomial-size compilable problems (comp-P).

Definition 15.41. *A* knowledge representation formalism (KRF) *is a subset of* $\Sigma^* \times \Sigma^*$.

Definition 15.42. *We say that a function* $f : \Sigma^* \to \Sigma^*$ *is* poly-size *if there exists a constant c such that for each* $(x, k) \in \Sigma^*$ *it holds that* $|f(x)| \leq n^c$, *where* $n = |x|$.

Definition 15.43. *Let K be a complexity class. A KRF F belongs to* comp-K *if there exist (1) a computable poly-size function* $f : \Sigma^* \to \Sigma^*$ *and a KRF F' in K such that for all pairs* $(x, y) \in \Sigma^* \times \Sigma^*$ *it holds that* $(x, y) \in F$ *if and only if* $(f(x), y) \in F'$.

Note that, unlike the original definition by Cadoli et al. [34], we require the compilation function f to be computable. This is a reasonable requirement for practically useful compilability results. To the best of our knowledge, there are no natural problems where this distinction (between computable and possibly uncomputable compilation functions) makes a difference.

Next, we turn to the definitions that are needed to establish incompilability results (par-nucomp-K-hardness under nucomp-reductions).

Definition 15.44. *Let K be a complexity class. A KRF F belongs to* nucomp-K *if there exist (1) a poly-size function* $f : \Sigma^* \times 1^* \to \Sigma^*$ *and a KRF F' in K such that for all pairs* $(x, y) \in \Sigma^* \times \Sigma^*$ *it holds that* $(x, y) \in F$ *if and only if* $(f(x, 1^{|y|}), y) \in F'$.

Definition 15.45. *A KRF F is* nucomp-reducible *to a KRF F' if there exist poly-size functions* $f_1, f_2 : \Sigma^* \times 1^* \to \Sigma^*$, *an poly-time function* $g : \Sigma^* \times \Sigma^* \to \Sigma^*$, *such that for all pairs* $(x, y) \in \Sigma^* \times \Sigma^*$ *it holds that* $(x, y) \in F$ *if and only if* $(f_1(x, 1^{|y|}), g(f_2(x, 1^{|y|}), y)) \in F'$.

We do not require the compilation function f witnessing membership in nucomp-K, and the compilation functions f_1 and f_2, nucomp-reducibility, to be computable. The reason for this is that the class nucomp-K and the corresponding nucomp-reductions are intended to show incompilability results, which are even stronger when possibly uncomputable compilation functions are considered.

We will use the following notation of *compilation problems*. Consider the compilation problem F specified as follows.

F
Offline instance: $x \in \Sigma^*$ with property P_1.
Online instance: $y \in \Sigma^*$ with property P_2.
Question: does (x, y) have property P_3?

We use this notation to denote the KRF $F = \{ (x, y) \in \Sigma^* \times \Sigma^* : (x, y)$ satisfies properties $P_1, P_2, P_3 \}$.

Finally, we consider compilation problems that are complete for the classes nucomp-K.

Definition 15.46. *Let L be a decision problem. Then the KRF ϵL is defined as follows:*

$$\epsilon L = \{\epsilon\} \times L = \{ (\epsilon, x) : x \in L \},$$

where ϵ is the empty string.

Proposition 15.47 ([34, **Theorem 2.9**]). *Let K be a complexity class that is closed under polynomial-time reductions. Let S be a problem that is complete for K (under polynomial-time reductions). Then ϵS is complete for nucomp-K under (polynomial-time) nucomp-reductions.*

Parameterized Compilability

Next, we revisit the basic notions of parameterized compilability. For more details, we refer to the work of Chen [44]. We begin with the basic definitions of parameterized compilation problems (formalized as *parameterized knowledge representation formalisms*) and the class of fpt-size compilable problems (par-comp-FPT).

Definition 15.48. *A* parameterized knowledge representation formalism (PKRF) *is a subset of $\Sigma^* \times \Sigma^* \times \mathbb{N}$. A PKRF can also be seen as a parameterized problem by pairing together the first two strings of each triple.*

Definition 15.49. *We say that a function $f : \Sigma^* \times \mathbb{N} \to \Sigma^*$ is fpt-size if there exists a constant c and a computable function $h : \mathbb{N} \to \mathbb{N}$ such that for each $(x, k) \in \Sigma^* \times \mathbb{N}$ it holds that $|f(x, k)| \leq h(k)n^c$, where $n = |x|$.*

Definition 15.50. *Let C be a parameterized complexity class. A PKRF F belongs to* par-comp-C *if there exist an fpt-size computable function $f : \Sigma^* \times \mathbb{N} \to \Sigma^*$ and a PKRF F' in C such that for all triples $(x, y, k) \in \Sigma^* \times \Sigma^* \times \mathbb{N}$ and all natural numbers $m \geq |y|$ it holds that $(x, y, k) \in F$ if and only if $(f(x, k), y, k) \in F'$.*

Note that we require the compilation function f witnessing membership in par-comp-C to be computable. This is a reasonable requirement for practically useful compilability results.

Next, we turn to the definitions that are needed to establish incompilability results (par-nucomp-K-hardness under fpt-nucomp-reductions).

Definition 15.51. *Let C be a parameterized complexity class. A PKRF F belongs to* par-nucomp-C *if there exist an fpt-size function $f : \Sigma^* \times 1^* \times \mathbb{N} \to \Sigma^*$ and a PKRF F' in C such that for all triples $(x, y, k) \in \Sigma^* \times \Sigma^* \times \mathbb{N}$ and all natural numbers $m \geq |y|$ it holds that $(x, y, k) \in F$ if and only if $(f(x, 1^m, k), y, k) \in F'$.*

Definition 15.52. *A PKRF F is fpt-nucomp-reducible to a PKRF F' (denoted by $F \leq_{\text{nucomp}}^{\text{fpt}} F'$) if there exist fpt-size functions $f_1, f_2 : \Sigma^* \times 1^* \times \mathbb{N} \to \Sigma^*$, an fpt-time function $g : \Sigma^* \times \Sigma^* \times \mathbb{N} \to \Sigma^*$, and a computable function $h : \mathbb{N} \to \mathbb{N}$ such that for all triples $(x, y, k) \in \Sigma^* \times \Sigma^* \times \mathbb{N}$ and all natural numbers $m \geq |y|$ it holds that $(x, y, k) \in F$ if and only if $(f_1(x, 1^m, k), g(f_2(x, 1^m, k), y, k), h(k)) \in F'$.*

Intuitively, the function f_1 transforms the offline instance x of F into an fpt-size offline instance of F', and the function f_2 transforms the offline instance x of F into an auxiliary offline instance x'. The function g transforms the online instance y of F (together with the auxiliary offline instance x') into an online instance of F'. Finally, the function h ensures that the parameter value for F' is not unbounded.

Note that, like the original definition by Chen [44], we do not require the compilation function f witnessing membership in par-nucomp-K, and the compilation functions f_1 and f_2, fpt-nucomp-reducibility, to be computable. The reason for this is that the class par-nucomp-K and the corresponding fpt-nucomp-reductions are intended to show incompilability results, which are even stronger when possibly uncomputable compilation functions are considered. Moreover, allowing possibly uncomputable compilation functions allows us to establish closer connections to several non-uniform parameterized complexity classes.

We will use the following notation of *parameterized compilation problems.* Consider the parameterized compilation problem F specified as follows.

F
Offline instance: $x \in \Sigma^*$ with property P_1.
Online instance: $y \in \Sigma^*$ with property P_2.
Parameter: $f(x, y) = k$.
Question: does (x, y) have property P_3?

We use this notation to denote the PKRF $F = \{ (x, y, k) \in \Sigma^* \times \Sigma^* \times \mathbb{N} : k = f(x, y), (x, y, k) \text{ satisfies properties } P_1, P_2, P_3 \}$.

Finally, we consider compilation problems that are complete for the classes par-nucomp-K.

Definition 15.53. *Let L be a parameterized decision problem. Then the PKRF ϵL is defined as follows:*

$$\epsilon L = \{ (\epsilon, x, k) : (x, k) \in L \},$$

where ϵ is the empty string.

The following result is given without proof by Chen [44]. We provide a proof. The main lines of the proof follow that of the proof of Proposition 15.47 (cf. [34, Theorem 2.9]).

Proposition 15.54 ([44, **Theorem 17**]). *Let K be a parameterized complexity class. Let Q be a parameterized problem that is complete for K (under fpt-reductions). Then ϵQ is complete for* par-nucomp-K *(under fpt-nucomp-reductions).*

Proof. Let Q be a parameterized problem that is K-complete under fpt-reductions. Moreover, let A be an arbitrary PKRF in par-nucomp-K. Then there exist an fpt-size function $f : \Sigma^* \times 1^* \times \mathbb{N} \to \Sigma^*$ and a PKRF Q' in K such that for all triples $(x, y, k) \in \Sigma^* \times \Sigma^* \times \mathbb{N}$ and all natural numbers $m \geq |y|$ it holds that:

$$(x, y, k) \in A \quad \text{if and only if} \quad (f(x, 1^m, k), y, k) \in Q'.$$

But $Q' \in$ K, so it can be reduced to Q by means of an fpt-reduction R:

$$(x, y, k) \in Q' \quad \text{if and only if} \quad R(x, y, k) \in Q.$$

We now define an fpt-nucomp-reduction from A to ϵQ. We specify fpt-size functions f_1, f_2, an fpt-time function g, and a computable function h, as follows:

$$
\begin{aligned}
f_1(x, 1^m, k) &= \epsilon, \\
f_2(x, 1^m, k) &= f(x, 1^m, k), \\
g(x', y, k) &= \pi_1(R(x', y, k)), \text{ and} \\
h(k) &= \pi_2(R(x', y, k)).
\end{aligned}
$$

Here π_i represents projection to the i-th element of a tuple. Since R is an fpt-reduction, we know that $\pi_2(R(x', y, k))$ only depends on k, and can thus be expressed as a function h that only depends on k.

We then get that for each $m \geq |y|$:

$(x, y, k) \in A$ if and only if $(f(x, 1^m, k), y, k) \in Q'$
  if and only if $R(f(x, 1^m, k), y, k) \in Q$
  if and only if $(\epsilon, R(f(x, 1^m, k), y, k)) \in \epsilon Q$
  if and only if $(f_1(x, 1^m, k), g(f_2(x, 1^m, k), y, k), h(k)) \in \epsilon Q.$

Thus, this is a correct fpt-nucomp-reduction from A to ϵQ. Since A was an arbitrary PKRF in par-nucomp-K-complete, we can conclude that ϵQ is par-nucomp-K-complete. ∎

15.3.2 (Conditional) Incompilability Results

To exemplify what role non-uniform parameterized complexity can play in parameterized incompilability results, we give two examples of parameterized problems that do not allow an fpt-size compilation, unless $W[1] \subseteq$ FPT/fpt.

Constrained Clique
As first example, we will take the following problem CONSTRAINED-CLIQUE. The definition of this problem is analogous to the definition of the Constrained Vertex Cover problem in the seminal paper by Cadoli et al. [34].

CONSTRAINED- CLIQUE
Offline instance: A graph $G = (V, E)$.
Online instance: Two subsets $V_1, V_2 \subseteq V$ of vertices, and a positive integer $u \geq 1$.
Question: Is there a clique $C \subseteq V$ in G of size u such that $C \cap V_1 = \emptyset$ and $V_2 \subseteq C$?

We consider the following parameterization of this problem, where the parameter is the number of vertices that have to be added to V_2 to create a clique of size u.

CONSTRAINED- CLIQUE(sol.-size)
Offline instance: A graph $G = (V, E)$.
Online instance: Two subsets $V_1, V_2 \subseteq V$ of vertices, and a positive integer $u \geq 1$.
Parameter: $k = u - |V_2|$.
Question: Is there a clique $C \subseteq V$ in G of size u such that $C \cap V_1 = \emptyset$ and $V_2 \subseteq C$?

We begin by showing that CONSTRAINED-CLIQUE(sol.-size) is par-nucomp-W[1]-complete. In order to do so, we will use the W[1]-compete problem MULTI-COLORED CLIQUE (for short: MCC) [80]. Instances of MCC are tuples (V, E, k), where k is a positive integer, V is a finite set of vertices partitioned into k subsets $V_1, \ldots, V_k$, and (V, E) is a simple graph. The parameter is k. The question is whether there exists a k-clique in (V, E) that contains a vertex in each set V_i.

Proposition 15.55. CONSTRAINED-CLIQUE(sol.-size) *is* par-nucomp-W[1]-*complete. Moreover, hardness holds already for the restricted case where* $V_2 = \emptyset$ *(and so* $u = k$).

Proof. To show membership in par-nucomp-W[1], it suffices to show membership in W[1]. We give an fpt-reduction to CLIQUE, which is contained in W[1]. Let $(G, (V_1, V_2, u), k)$ be an instance of CONSTRAINED-CLIQUE(sol.-size). The reduction outputs the following instance (G', k) of CLIQUE. The graph G' is obtained from G by removing all vertices in V_1, removing all vertices in V_2, and all vertices that are not connected to each vertex $v \in V_2$.

To show hardness for par-nucomp-W[1], we give a fpt-nucomp-reduction from ϵMCC to CONSTRAINED-CLIQUE(sol.-size). Since MCC is W[1]-hard, we know that the problem ϵMCC is par-nucomp-W[1]-hard (under fpt-nucomp-reductions). We specify fpt-size functions f_1, f_2, a computable function h and an fpt-time function g such that for each instance (δ, G, k) of ϵMCC it holds that for each $m \geq \|G\|$: $(\delta, G, k) \in \epsilon$MCC if and only if $(f_1(\delta, 1^m, k), g(f_2(\delta, 1^m, k), G, k), h(k)) \in$ CONSTRAINED-CLIQUE(sol.-size).

Let (δ, G, k) be an instance of ϵMCC, where $G = (V, E)$, where the set V of vertices is partitioned into subsets $V_1, \ldots, V_k$, and where $V_i = \{v_1^i, \ldots, v_n^i\}$, for each $i \in [k]$. We define $f_1(\delta, 1^m, k)$ to be the graph $G' = (V', E')$. Here we let $V' = \bigcup_{i \in [k]} V^i \cup \bigcup_{i, j \in [k], i < j} W^{i,j}$, where for each $i \in [k]$ we let $V^i = \{v_\ell^i : \ell \in [m]\}$, and for each $i, j \in [k]$ with $i < j$ we let $W^{i,j} = \{w_{i, \ell_1, j, \ell_2} : \ell_1, \ell_2 \in [m]\}$. We let E' consist of the following edges. First, we connect all vertices between different sets V^i, i.e., for each $i, j \in [k]$ with $i < j$, we connect each vertex $v_\ell^i \in V^i$ with each vertex $v_{\ell'}^j$. Next, we connect all vertices between different sets $W^{i,j}$, i.e., for each $i, j \in [k]$ with $i < j$ and each $i', j' \in [k]$ with $i' < j'$, we connect each vertex w_{i,ℓ_1,j,ℓ_2} with each vertex $w_{i',\ell_1',j',\ell_2'}$. Then, we connect vertices in $W^{i,j}$ and in V^b for $b \neq i$ and $b \neq j$, i.e., for each $i, j \in [k]$ with $i < j$, and each $b \in [k]$ such that $b \notin \{i, j\}$, we connect each vertex w_{i,ℓ_1,j,ℓ_2} with each vertex v_ℓ^b. Finally, we describe the edges between vertices in $W^{i,j}$ and vertices in $V^i \cup V^j$. Let $i, j \in [k]$ with $i < j$, and let $\ell_1, \ell_2 \in [m]$. Then we connect w_{i,ℓ_1,j,ℓ_2} with $v_{\ell_1}^i$ and with $v_{\ell_2}^j$. Then, we define $h(k) = k' = k + \binom{k}{2}$. We define $f_2(\delta, 1^m, k) = 1^m$. Finally, we let $g(1^m, G, k) = (U_1, U_2, u)$. Here we let $U_1 = \{ w_{i,\ell_1,j,\ell_2} : i, j \in [k], i < j, (\ell_1 > n \vee \ell_2 > n) \} \cup \{ w_{i,\ell_1,j,\ell_2} : i, j \in [k], i < j, \{v_{\ell_1}^i, v_{\ell_2}^j\} \notin E \}$. Moreover, we let $U_2 = \emptyset$, and we let $u = |U_2| + k' = k'$. Clearly f_1 and f_2 are fpt-size functions, g is an fpt-time computable function, and h is a computable function.

All that remains to show is that G has a (multi-colored) clique of size k if and only if G' has a clique of size u that contains no vertex in V_1. One direction is easy. Let $C = \{v_{\ell_1}^1, \ldots, v_{\ell_k}^k\}$ be a multicolored clique in G. Then the set $C \cup \{ w_{i,\ell_i,j,\ell_j} : i, j \in [k], i < j \}$ is a clique in G' of size k' that contains no vertices in V_1. Conversely, assume that G' has a clique C' of size k' that contains no vertices in V_1. We know that the set V' of vertices of G' consists of k' subsets for which holds that there are no edges between any two vertices of the same subset. Namely, these subsets are $V^1, \ldots, V^k, W^{1,2}, \ldots, W^{k-1,k}$. Therefore C' must contain one vertex from each of these subsets. Let $C = C' \cap V$. We know that C contains no vertex of the form v_ℓ^i for $\ell \in [n+1, m]$, because for each $\ell \in [n+1, m]$, the vertex v_ℓ^i is not connected to any vertex $W = \bigcup_{i,j \in [k], i < j} W^{i,j}$ that is not contained in V_1. Therefore, we know that $C \subseteq V$ and $|C| = k$. Let $C = \{v_{\ell_1}^1, \ldots, v_{\ell_k}^k\}$, where for each $i \in [k]$ we have $C \cap V_i = \{v_{\ell_i}^i\}$. By definition of G', we know that C' must contain for each $i, j \in [k]$ with $i < j$ the vertex w_{i,ℓ_i,j,ℓ_j}. Then, because we know that w_{i,ℓ_i,j,ℓ_j} is not in V_1, by construction of V_1 we know that $\{v_{\ell_i}^i, v_{\ell_j}^j\} \in E$. Therefore, we can conclude that C' is a (multicolored) clique in V. This concludes the correctness proof of our fpt-nucomp-reduction from ϵMCC to CONSTRAINED-CLIQUE(sol.-size).

We can then use this par-nucomp-W[1]-completeness result to relate the fpt-size (in)compilability of this problem to the inclusion between W[1] and the non-uniform parameterized complexity class FPT/fpt. In order to do so, we will first need the following results. These results were already stated without proof by Chen [44]. We provide a proof here.

Proposition 15.56 ([44, **Proposition 16**]). *Let* K *be a parameterized complexity class. If a parameterized problem* S *is in* K/fpt, *then* $\epsilon S \in$ par-nucomp-K.

Proof. Let S be a parameterized problem in K/fpt. Then, by definition, there exists an fpt-size function f and a parameterized problem $S' \in$ K such that:

$$(y, k) \in S \quad \text{if and only if} \quad (f(1^{|y|}, k), y, k) \in S'.$$

We show that $\epsilon S \in$ par-nucomp-K. We specify an fpt-size function f' such that:

$$(\epsilon, y, k) \in \epsilon S \quad \text{if and only if} \quad (f'(\epsilon, 1^{|y|}, k), y, k).$$

We let $f'(x, y, k) = f(1^{|y|}, k)$. It is straightforward to modify f' in such a way that it works for each $m \geq |y|$, rather than just for $|y|$. This witnesses that $\epsilon S \in$ par-nucomp-K.

The proof of the next proposition follows the main lines of the proof of its counterpart in classical compilability (cf. [34, Theorem 2.12]).

Proposition 15.57 ([44, **Theorem 18**]). *Let* K *and* K$'$ *be parameterized complexity classes that are closed under fpt-reductions and that have complete problems. The inclusion* par-nucomp-K $\subseteq$ par-nucomp-K$'$ *holds if and only if the inclusion* K/fpt $\subseteq$ K$'$/fpt *holds.*

Proof. First, we show that K/fpt $\subseteq$ K$'$/fpt implies that par-nucomp-K $\subseteq$ par-nucomp -K$'$. Suppose that K/fpt $\subseteq$ K$'$/fpt. Moreover, let S be a problem that is K-complete under fpt-reductions. Then, by Proposition 15.54, ϵS is par-nucomp-K-complete. Also, since $S \in$ K/fpt and K/fpt $\subseteq$ K$'$/fpt, we know that $S \in$ K$'$/fpt. Therefore, by Proposition 15.56, $\epsilon S \in$ par-nucomp-K$'$. Then, since there is a par-nucomp-K-complete problem that is in par-nucomp-K$'$, we get the inclusion par-nucomp-K $\subseteq$ par-nucomp-K$'$.

Conversely, we show that par-nucomp-K $\subseteq$ par-nucomp-K$'$ implies that K/fpt $\subseteq$ K$'$/fpt. Suppose that par-nucomp-K $\subseteq$ par-nucomp-K$'$. Let S be an arbitrary problem in K/fpt. We show that $S \in$ K$'$/fpt. By definition, there exists an fpt-size function f and a PRKF $S' \in$ K such that:

$$(y, k) \in S \quad \text{if and only if} \quad (f(1^{|y|}, k), k) \in S'.$$

From this it follows that the PKRF ϵS is in par-nucomp-K. Therefore, it is also in par-nucomp-K$'$. This means that there is an fpt-size function f' and a PKRF $S'' \in$ K$'$ such that for all instances (x, y, k):

$$(x, y, k) \in \epsilon S \quad \text{if and only if} \quad (f'(x, 1^{|y|}, k), y, k) \in S''.$$

Then, we get that:

$$(y, k) \in S \quad \begin{array}{ll} \text{if and only if} & (\epsilon, y, k) \in \epsilon S \\ \text{if and only if} & (f'(\epsilon, 1^{|y|}, k), y, k) \in S''. \end{array}$$

Since $f'(\epsilon, 1^{|y|}, k)$ is an fpt-size function that depends on $1^{|y|}$ and k alone, and since $S'' \in$ K$'$, we conclude that $S \in$ K$'$/fpt. Since S was an arbitrary problem in K/fpt, we can conclude that K/fpt $\subseteq$ K$'$/fpt.

Additionally, we show that W[1]/fpt $\subseteq$ FPT/fpt if and only if W[1] $\subseteq$ FPT/fpt.

Proposition 15.58. *Let* K *and* K$'$ *be parameterized complexity classes. Then it holds that* K/fpt $\subseteq$ K$'$/fpt *if and only if* K $\subseteq$ K$'$/fpt.

Proof. Since K ⊆ K/fpt, one inclusion follows immediately. We prove the converse inclusion, namely that K ⊆ K'/fpt implies that K/fpt ⊆ K'/fpt.

Let Q be an arbitrary problem in K/fpt. This means that there exists a problem $Q' \in$ K and fpt-size advice strings $\alpha(n, k)$ for each $(n, k) \in \mathbb{N} \times \mathbb{N}$, such that:

$$(x, k) \in Q \text{ if and only if } (x, \alpha(|x|, k), k) \in Q'.$$

Moreover, since $Q' \in$ K, by assumption, we know that $Q' \in$ K'/fpt. Thus, there exists a problem $Q'' \in$ K' and fpt-size advice strings $\beta(n, k)$ for each $(n, k) \in \mathbb{N} \times \mathbb{N}$, such that:

$$(y, k) \in Q' \text{ if and only if } (y, \beta(|y|, k), k) \in Q'',$$

where $y = (x, \alpha(|x|, k))$. From this, we get that $(x, k) \in Q$ if and only if $(x, \alpha(|x|, k), \beta(|y|, k), k) \in Q''$. Therefore, $Q \in$ K'/fpt. ∎

This now gives us the following corollary.

Corollary 15.59. *It holds that* par-nucomp-W[1] ⊆ par-nucomp-FPT *if and only if* W[1] ⊆ FPT/fpt.

From this, we get the following result that CONSTRAINED-CLIQUE(sol.-size) is fpt-size incompilable, under the assumption that W[1] ⊄ FPT/fpt.

Proposition 15.60. CONSTRAINED-CLIQUE(sol.-size) *is not in* par-nucomp-FPT, *and so it is not in* par-comp-FPT, *unless* W[1] ⊆ FPT/fpt.

Weighted Clause Entailment

As second example, we consider a parameterized variant of the clause entailment problem. The clause entailment problem plays a central role in the literature on knowledge compilation [34, 58, 160], and is defined as follows.

CLAUSE-ENTAILMENT
Offline instance: A CNF formula φ.
Online instance: A clause c, i.e., a disjunction of literals.
Question: $\varphi \models c$?

We consider a weighted variant of the problem CLAUSE-ENTAILMENT. In this variant, we consider logical entailment with respect to truth assignments of a certain weight. In particular, we say that a CNF formula φ entails a clause c with respect to truth assignments of weight w, written $\varphi \models_w c$, if for all truth assignments α to φ of weight w it holds that if α satisfies φ, then α satisfies c as well. We consider the following parameterized compilation problem.

CLAUSE-ENTAILMENT(weight)
Offline instance: A CNF formula φ, and a positive integer w.
Online instance: A clause c.
Parameter: w.
Question: $\varphi \models_w c$?

We show that CLAUSE-ENTAILMENT(weight) is not fpt-size compilable unless $W[1] \subseteq FPT/fpt$. Similarly to the case of CONSTRAINED-CLIQUE(sol.-size), we show that CLAUSE-ENTAILMENT(weight) is par-nucomp-co-$W[1]$-hard to derive this result.

Proposition 15.61. CLAUSE-ENTAILMENT(weight) *is not in* par-comp-FPT, *unless* $W[1] \subseteq FPT/fpt$.

Proof. We show that the problem is par-nucomp-co-$W[1]$-hard. This suffices, since any parameterized compilation problem is fpt-size compilable if and only if its co-problem (the problem consisting of all the no-instances) is fpt-size compilable. We do so by giving an fpt-nucomp-reduction from the problem ϵco-MCC $= \{ (\epsilon, G, k) : (G, k) \notin MCC \}$. Let (ϵ, G, k) be an instance of ϵco-MCC, where G is a graph whose vertex set is partitioned into $V_1, \ldots, V_k$, and let $m \geq |G|$. Moreover, without loss of generality we may assume that all sets V_i have the same cardinality; for each $i \in [k]$, let $V_i = \{v_{i,1}, \ldots, v_{i,n}\}$. To describe the fpt-nucomp-reduction, we specify suitable functions f_1, f_2, g, h. We let $f_1(\epsilon, 1^m, k)$ be the CNF formula φ that we will define below. Moreover, we let $f_2(\epsilon, 1^m, k) = 1^m$ and we let $g(1^m, G, k)$ be the clause δ that we will define below. Finally, we let $h(k) = k' = k + \binom{k}{2}$.

We let $\text{Var}(\varphi) = \bigcup_{i \in [k]} X_i \cup \bigcup_{i, j \in [k], i < j} Y_{i,j}$. Here, for each i we let $X_i = \{ x_{i,\ell} : \ell \in [m] \}$. Also, for each $i, j \in [k]$ with $i < j$, we let $Y_{i,j} = \{ y_{i,\ell_1,j,\ell_2} : \ell_1 \in [k], \ell_2 \in [k] \}$. Intuitively, the variables $x_{i,\ell}$ encode the choice of vertices in a clique, and the variables y_{i,ℓ_1,j,ℓ_2} encode the choice of edges. Then, for each set $Z \in \{ X_i, Y_{i,j} : i \in [k], j \in [i+1, k] \}$ of variables, and for each two variables $z_1, z_2 \in Z$, we add the clause $(\neg z_1 \vee \neg z_2)$ to φ. This enforces that each satisfying truth assignment of φ of weight k' must satisfy exactly one variable in each set Z. Moreover, for each $i, j \in [k]$ with $i < j$, each $\ell_1 \in [m]$ and each $\ell_2 \in [m]$, we add the clauses $(\neg y_{i,\ell_1,j,\ell_2} \vee x_{i,\ell_1})$ and $(\neg y_{i,\ell_1,j,\ell_2} \vee x_{j,\ell_2})$ to φ. Intuitively, these clauses enforce that the choice of edges is compatible with the choice of vertices.

We define the clause $\delta = g(1^m, G, k)$ as follows. Let n be the number of vertices in G. For each $i, j \in [k]$ with $i < j$, each $\ell_1 \in [m]$, and each $\ell_2 \in [m]$, we add the literal y_{i,ℓ_1,j,ℓ_2} to δ if one of the following cases holds: (i) either $\ell_1 > n$ or $\ell_2 > n$, or (ii) there is no edge in G between v_{i,ℓ_1} and v_{j,ℓ_2}.

We claim that satisfying truth assignments of $\psi \wedge \neg \delta$ of weight k' are in one-to-one correspondence with cliques in G of size k containing exactly one vertex in each V_i. For each such clique $V' = \{v_{1,\ell_1}, \ldots, v_{k,\ell_k}\}$ in G (with $v_{i,\ell_i} \in V_i$), one can obtain the satisfying assignment that sets exactly those variables in the set $X' = \{ x_{i,\ell_i} : i \in [k] \} \cup \{ y_{i,\ell_i,j,\ell_j} : i, j \in [k], i < j \}$ to true. Vice versa, from each satisfying assignment, one can construct a suitable clique in G. With this correspondence, one can verify straightforwardly that $\varphi \models_{k'} \delta$ if and only if $(G, k) \notin MCC$. This shows the correctness of our reduction.

The result now follows from Corollary 15.59, which itself follows from Propositions 15.57 and 15.58.

Additionally, we give a direct proof that par-nucomp-co-$W[1]$-hardness entails incompilability (under the assumption that $W[1] \not\subseteq FPT/fpt$). Suppose that weighted clause entailment is fpt-size compilable. We show that $W[1] \subseteq FPT/fpt$, by showing that MCC can be solved in fpt-time using fpt-size advice. For an instance (G, k) of MCC we firstly construct φ, δ and k' according to the construction discussed above. We showed that $\varphi \not\models_{k'} \delta$ if and only if $(G, k) \in$ MCC. Since the offline instance (φ, k) of weighted clause entailment depends only on the pair (n, k), where $n = |G|$, we can use the fpt-size compilation $c(\varphi, k)$ as the advice string to solve the problem MCC in fpt-time.

15.3.3 Restricting the Instance Space

In order to connect the parameterized compilability of another natural parameterized variant of CONSTRAINED-CLIQUE (that we will define in Sect. 15.3.4) to non-uniform parameterized complexity, we will need some additional non-uniform parameterized complexity classes. In this section, we develop these classes.

Concretely, we will define two new parameterized complexity classes: few-NP and nu-few-NP. In order to define these classes, we will consider a particular type of functions, and a parameterized decision problem based on such functions.

We point out that these classes are not directly connected to the class FEWP consisting of all NP problems with a polynomially-bounded number of solutions [6]. The classes few-NP and nu-few-NP consist of problems that have few NP instances (each of which can have many solutions).

We begin with defining the notion of (SAT instance) generators.

Definition 15.62 (generators). *We say that a function* $\gamma : \mathbb{N}^3 \to \Sigma^*$ *is a (SAT instance) generator if for each* $(n, \ell, k) \in \mathbb{N}^3$ *it holds that:*

- *if* $\ell \in [n^k]$, *then* $\gamma(n, \ell, k)$ *is a SAT instance, i.e., a propositional formula;*
- *otherwise* $\gamma(n, \ell, k)$ *is the trivial SAT instance* $\emptyset$.

We say that a generator γ *is* nice *if for each* (n, ℓ, k) *the formula* $\gamma(n, \ell, k)$ *has exactly* n *variables. We say that a generator is a 3CNF generator if all the (non-trivial) instances that it generates are propositional formulas in 3CNF.*

Moreover, we say that a generator is (uniformly) fpt-time computable *if there exists an algorithm A, a computable function* f *and a constant* c *such that for each* $(n, \ell, k) \in \mathbb{N}^3$ *the algorithm A computes* $\gamma(n, \ell, k)$ *in time* $f(k)n^c$.

We say that a generator is non-uniformly fpt-time computable *if there exist a computable function* f *and a constant* c *such that for each* $(n, k) \in \mathbb{N}^2$, *there exists an algorithm* $A_{(n,k)}$ *that for each* $\ell \in \mathbb{N}$ *computes* $\gamma(n, \ell, k)$ *in time* $f(k)n^c$ *(this corresponds to the non-uniformity notion of fpt-size advice).*

Using the notion of generators, we can now define the following two (schemes of) parameterized problems.

Definition 15.63. *Let γ be a generator. We define the parameterized decision problem* FEWSAT$_\gamma$ *as follows.*

FEWSAT$_\gamma$
Instance: $(n, \ell, k) \in \mathbb{N}^3$, where n is given in unary and ℓ, k are given in binary.
Parameter: k.
Question: is $\gamma(n, \ell, k)$ satisfiable?

Definition 15.64. *Let γ be a generator. We define the parameterized decision problem* FEWUNSAT$_\gamma$ *as follows.*

FEWUNSAT$_\gamma$
Instance: $(n, \ell, k) \in \mathbb{N}^3$, where n is given in unary and ℓ, k are given in binary.
Parameter: k.
Question: is $\gamma(n, \ell, k)$ unsatisfiable?

Definition 15.65. *We define the following parameterized complexity class* few-NP*:*

$$\text{few-NP} = [\{\, \text{FEWSAT}_\gamma : \gamma \text{ is a uniformly fpt-time computable} \\ \text{nice 3CNF generator} \,\}]_{\text{fpt}}.$$

By using Definition 15.1, we could obtain the non-uniform variant few-NP/fpt of this parameterized complexity class few-NP. However, the following definition of nu-few-NP captures a more natural non-uniform version of the class few-NP, that turns out to be more useful in the setting of parameterized knowledge compilation.

Definition 15.66. *We define the following parameterized complexity class* nu-few-NP*:*

$$\text{nu-few-NP} = [\{\, \text{FEWSAT}_\gamma : \gamma \text{ is a non-uniformly fpt-time computable} \\ \text{nice 3CNF generator} \,\}]_{\text{fpt}}.$$

The intuition behind the classes few-NP and nu-few-NP is the following. Both the classes $W[t]$ of the Weft hierarchy and the classes few-NP and nu-few-NP contain problems that are restrictions of problems in NP. For the classes $W[t]$ the set of possible witnesses is restricted in number from $2^{O(n)}$ to $n^{O(k)}$. The classes few-NP and nu-few-NP are based on a dual restriction. For these classes, not the set of possible witnesses is restricted, but the set of instances (for each input size n and parameter value k) is restricted in number from $2^{O(n)}$ to $n^{O(k)}$.

We illustrate the notion of (fpt-time computable) 3CNF generators using the following example.

Example 15.67. For each $n \in \mathbb{N}$, there are 2^{8n^3} possible different 3CNF formulas over the variables $x_1, \ldots, x_n$, each of which can thus be described by $8n^3$ bits. On the other hand, for each $(n, k) \in \mathbb{N}^2$, each integer $\ell \in [n^k]$ can be described by $k \log n$ bits. We consider the following example of an fpt-time computable 3CNF generator. Using an *pseudorandom generator*, one can construct a function γ that for each $(n, k) \in \mathbb{N}^2$ and each bitstring of length $k \log n$ (representing the number ℓ) produces a seemingly random bitstring of length $8n^3$ that is then interpreted as a 3CNF formula over the variables $x_1, \ldots, x_n$.   ⊣

Before we put these newly introduced non-uniform parameterized complexity classes to use by employing them to characterize the (in)compilability of certain parameterized compilation problems, we make a few digressions and provide some alternative characterizations of few-NP and nu-few-NP and relate them to other non-uniform parameterized complexity classes.

Normalization Results

It will be useful to develop some normalization results for the classes few-NP and nu-few-NP. In this section, we will do so. We begin by showing that any uniformly fpt-time computable 3CNF generator can be transformed into a nice 3CNF generator.

Proposition 15.68. *Let γ be a uniformly fpt-time computable* 3CNF *generator that is not necessarily nice. Then* FEWSAT$_\gamma \in$ few-NP.

Proof. We construct a nice 3CNF generator γ' and we give an fpt-reduction from FEWSAT$_\gamma$ to FEWSAT$_{\gamma'}$. Let f be a computable function and let c be a constant such that for each (n, ℓ, k), $\gamma(n, \ell, k)$ can be computed in time $f(k)n^c$. We consider the function π that is defined by $\pi(n', k) = \lfloor \sqrt[c]{n'/f(k)} \rfloor$, i.e., for each $(n', k) \in \mathbb{N}^2$ it holds that $n' = f(k)n^c$ where $n = \pi(n', k)$. This function π is fpt-time computable. Also, for each $(n', k) \in \mathbb{N}^2$ it holds that $\pi(n', k) \leq n'$.

Next, we construct the uniformly fpt-time computable 3CNF generator γ' as follows. We describe the algorithm that computes γ'. On input (n', ℓ, k), the algorithm first computes $n = \pi(n', k)$. Then, it computes $\varphi = \gamma(n, k, \ell)$. The formula φ is of size at most $f(k)n^c$, and thus contains at most $f(k)n^c$ variables. The algorithm then transform φ into an equivalent formula φ' that has exactly $f(k)n^c$ variables by adding dummy variables. Finally, the algorithm returns φ'.

All that remains is to specify an fpt-reduction from FEWSAT$_\gamma$ to FEWSAT$_{\gamma'}$. Given an instance (n, ℓ, k) of FEWSAT$_\gamma$, the reduction returns the instance $(f(k)n^c, \ell, k)$ of FEWSAT$_{\gamma'}$. Clearly, this is computable in fpt-time. Since we know that $\gamma'(f(k)n^c, \ell, k) \equiv \gamma(n, \ell, k)$, this reduction is correct.

This result can straightforwardly be extended to the case of non-uniformly fpt-time computable 3CNF generators.

Proposition 15.69. *Let γ be a non-uniformly fpt-time computable* 3CNF *generator that is not necessarily nice. Then* FEWSAT$_\gamma \in$ nu-few-NP.

Proof. Completely analogous to the proof of Proposition 15.68

Next, we show that any fpt-time computable generator can be transformed into a 3CNF generator.

Proposition 15.70. *Let γ be a uniformly fpt-time computable generator (that is not necessarily a* 3CNF *generator). Then* FEWSAT$_\gamma \in$ few-NP.

Proof. We construct a uniformly fpt-time computable 3CNF generator γ' as follows. We describe the algorithm that computes γ'. On input (n, ℓ, k), the algorithm first computes $\varphi = \gamma(n, k, \ell)$. Using the well-known Tseitin transformation, the algorithm constructs a formula φ' in 3CNF that is equivalent to φ. The algorithm then returns $\gamma'(n, \ell, k) = \varphi'$.

The generator γ' is a uniformly fpt-time computable 3CNF generator that is not necessarily nice. The identity mapping is then an fpt-reduction from FEWSAT$_\gamma$ to FEWSAT$_{\gamma'}$. We know by Proposition 15.68 that FEWSAT$_{\gamma'} \in$ few- NP. From this we can conclude that FEWSAT$_\gamma \in$ few- NP.

This result can also straightforwardly be extended to the case of non-uniformly fpt-time computable generators.

Proposition 15.71. *Let γ be a non-uniformly fpt-time computable generator (that is not necessarily a 3CNF generator). Then* FEWSAT$_\gamma \in$ nu-few-NP.

Proof. Completely analogous to the proof of Proposition 15.70 □

The proofs of Propositions 15.68 and 15.70 (and of Propositions 15.69 and 15.71) can then be straightforwardly combined to get the following alternative characterization of few- NP and nu-few-NP.

Corollary 15.72. *The class* few-NP *consists of all parameterized problems that are fpt-reducible to* FEWSAT$_\gamma$, *for some uniformly fpt-time computable generator. Similarly, the class* nu-few-NP *consists of all parameterized problems that are fpt-reducible to* FEWSAT$_\gamma$, *for some non-uniformly fpt-time computable generator.*

Relating few- NP **and** nu-few-NP **to other classes**
We situate the classes few- NP and nu-few-NP in the landscape of (non-uniform) parameterized complexity classes that we considered in Sect. 15.1. We do so by giving inclusion as well as separation results.

We begin with the following observation.

Observation 15.73. *If* P $=$ NP, *then* few-NP $\subseteq$ FPT *and* nu-few-NP $\subseteq$ FPT/fpt.

Next, we establish some basic results relating few- NP and nu-few-NP to the classes FPT, para-NP, para-NP/fpt and XP/xp.

Proposition 15.74. *We have the following inclusions:*

1. FPT $\subseteq$ few-NP;
2. few-NP $\subseteq$ nu-few-NP;
3. few-NP $\subseteq$ para-NP;
4. nu-few-NP $\subseteq$ para-NP/fpt;
5. few-NP $\subseteq$ XP/xp;
6. nu-few-NP $\subseteq$ XP/xp;

Proof. Inclusions 1 and 2 are trivial. Inclusion 3 can be shown as follows. For each uniformly fpt-time computable generator γ, the problem FEWSAT$_\gamma$ can be decided in non-deterministic fpt-time as follows. Firstly, compute $\varphi = \gamma(n, \ell, k)$ in deterministic fpt-time. Then, using non-determinism, decide if φ is satisfiable.

Next, Inclusion 4 can be shown as follows. For each non-uniformly fpt-time computable generator γ, the problem FEWSAT_γ can be decided in non-deterministic fpt-time using fpt-size advice as follows. Firstly, compute $\varphi = \gamma(n, \ell, k)$ in deterministic fpt-time using fpt-size advice. Then, using non-determinism, decide if φ is satisfiable.

Inclusion 6 can be shown as follows. For each non-uniformly fpt-time computable generator γ, the problem FEWSAT_γ can be decided in xp-time using xp-size advice as follows. For each (n, k), the advice $\alpha(n, k)$ is a lookup-table T of size $O(n^k)$ that contains for each $\ell \in [n^k]$ a bit $T_\ell \in \mathbb{B}$ that represents whether $\gamma(n, \ell, k)$ is satisfiable or not.

Then, Inclusion 5 follows directly from Inclusions 2 and 6.

The inclusion few-$\text{NP} \subseteq$ para-NP is likely to be strict.

Observation 15.75. *By Proposition 15.39, we then have that* few-$\text{NP} \subsetneq$ para-NP, *unless the PH collapses to the second level.*

In fact, we can get an even stronger result.

Proposition 15.76. *If* para-$\text{NP} =$ few-NP, *then* $P = NP$.

Proof. Consider the NP-complete language $\text{SAT}_1 = \{ (\varphi, 1) : \varphi \in \text{SAT} \}$. We know that SAT_1 is para-NP-complete. By our assumption that para-$\text{NP} =$ few-NP, we get that there exists some uniformly fpt-time computable generator γ such that SAT_1 is fpt-reducible to FEWSAT_γ. Call the fpt-reduction that witnesses this R. We then know that there exists a computable function h such that for each instance $(\varphi, 1)$ of SAT_1, the resulting instance $R(\varphi, 1)$ of FEWSAT_γ is of the form (n, ℓ, k'), for some $k' \in [h(1)]$. Let Q be the restriction of FEWSAT_γ to instances (n, ℓ, k') where $k' \in [h(1)]$. Because $h(1)$ is a constant, we then have that Q, when considered as a classical (nonparameterized) decision problem, is sparse, i.e., there is a constant c such that for each input size n there are at most n^c yes-instances of Q of length n. The fpt-reduction R then also witnesses that SAT_1, when considered as a classical (nonparameterized) decision problem, is polynomial-time reducible to Q. Since SAT_1 is NP-complete and Q is sparse, by Mahaney's Theorem [152], it then follows that $P = NP$.

Corollary 15.77. *It holds that* few-$\text{NP} =$ para-NP *if and only if* $P = NP$.

Interestingly, one can prove—under the assumption that $E \neq NE$—that the class few-NP lies strictly between FPT and para-NP [118, 162]. Here E denotes the class of all problems that can be solved in deterministic time $2^{O(n)}$, and NE denotes the class of all problems that can be solved in non-deterministic time $2^{O(n)}$, where n is the input size.

Finally, we show that nu-few-NP is incomparable to the class XP_{nu}. We do so by identifying a problem that is in nu-few-NP, but not in XP_{nu}. By the fact that nu-few-$\text{NP} \subseteq \text{XP/xp}$ and by Proposition 15.27, this suffices to show that nu-few-$\text{NP} \not\subseteq \text{XP/xp}$ and $\text{XP/xp} \not\subseteq$ nu-few-NP.

Proposition 15.78. *There exists a parameterized problem* $Q \in$ (nu-few-NP)\ (XP_{nu}).

Proof. (sketch). One can straightforwardly show that the problem constructed in the proof of Proposition 15.25 is contained in nu-few-NP, but not in XP_{nu}.

Corollary 15.79. *It holds that* nu-few-NP $\not\subseteq$ XP/xp *and* XP/xp $\not\subseteq$ nu-few-NP.

Differences Between Non-Uniform Variants of few-NP

Above, we (implicitly) provided two different possible definitions of a non-uniformly defined variant of few-NP: few-NP/fpt and nu-few-NP. We briefly discuss how these relate to each other.

The difference between few-NP/fpt and nu-few-NP can be described as follows. The class few-NP/fpt consists of all parameterized problems that are *non-uniformly* fpt-reducible to the problem FEWSAT$_\gamma$, for some *uniformly* fpt-time computable generator γ. The class nu-few-NP consists of all parameterized problems that are *uniformly* fpt-reducible to the problem FEWSAT$_\gamma$, for some *non-uniformly* fpt-time computable generator γ. For the usual parameterized complexity classes (such as FPT), adding non-uniformity (in the form of fpt-size advice) either (i) to reductions or (ii) to the algorithms witnessing membership in a complexity class, yields the same outcome. This is because one can simply "save" the advice in the problem input. However, since for any problem in few-NP, the problem input can only contain $O(k \log n)$ bits, for any fixed (n, k), you cannot "save" the (fpt-size) advice string in the problem input.

The class nu-few-NP is the non-uniform variant of few-NP that results from the "circuit non-uniformity" point of view (for each n and k, there is a different algorithm). The class few-NP/fpt is the non-uniform variant of few-NP that results from the "advice non-uniformity" point of view. Interestingly, for the parameterized complexity class few-NP these two different notions of non-uniformity seem to differ. For all other parameterized complexity classes that we have considered, these two notions of non-uniformity give rise to the same class.

Characterizing few-NP in Terms of 3-Colorability

To illustrate that for the classes few-NP and nu-few-NP, it does not matter what NP-complete base problem one chooses, we give a different characterization of these classes in terms of a problem based on 3-colorability of graphs.

Definition 15.80 (graph generators). *We say that a function* $\gamma : \mathbb{N}^3 \to \Sigma^*$ *is a graph generator if for each* $(n, \ell, k) \in \mathbb{N}^3$ *it holds that:*

- *if* $\ell \in [n^k]$, *then* $\gamma(n, \ell, k)$ *is a graph;*
- *otherwise* $\gamma(n, \ell, k)$ *is the trivial empty graph* $(\emptyset, \emptyset)$.

We say that a graph generator γ *is nice if for each* (n, ℓ, k) *the graph* $\gamma(n, \ell, k)$ *has exactly n vertices.*

Moreover, we say that a generator is (uniformly) *fpt-time computable if there exists an algorithm* A, *a computable function* f *and a constant* c *such that for each* $(n, \ell, k) \in \mathbb{N}^3$ *the algorithm* A *computes* $\gamma(n, \ell, k)$ *in time* $f(k)n^c$.

We say that a generator is non-uniformly fpt-time computable *if there exist a computable function f and a constant c such that for each $(n, k) \in \mathbb{N}^2$, there exists an algorithm $A_{(n,k)}$ that for each $\ell \in \mathbb{N}$ computes $\gamma(n, \ell, k)$ in time $f(k)n^c$.*

Using the concept of graph generators, we can now define the following parameterized problem.

Definition 15.81. *Let γ be a graph generator. We define the parameterized decision problem* FEW3COL$_\gamma$ *as follows.*

FEW3COL$_\gamma$
Instance: $(n, \ell, k) \in \mathbb{N}^3$, where n is given in unary and ℓ, k are given in binary.
Parameter: k.
Question: is $\gamma(n, \ell, k)$ 3-colorable?

We can then derive the following characterization of few-NP.

Proposition 15.82. *It holds that* few-NP $= [\{$ FEW3COL$_\gamma$: γ *is a uniformly fpt-time computable graph generator* $\}]_{\text{fpt}}$.

Proof. Firstly, we show that for each uniformly fpt-time computable graph generator γ it holds that FEW3COL$_\gamma \in$ few-NP. Let γ be such a graph generator. We construct the following uniformly fpt-time computable SAT instance generator γ'. We describe the algorithm that computes γ'. On input (n, ℓ, k), the algorithm firstly computes the graph $\gamma(n, \ell, k)$. By using the standard polynomial-time reduction from 3-colorability to propositional satisfiability, we can construct (in polynomial time) a propositional formula φ that is satisfiable if and only if $\gamma(n, \ell, k)$ is 3-colorable. The algorithm then returns $\gamma'(n, \ell, k) = \varphi$. It is then straightforward to verify that the identity mapping is an fpt-reduction from FEW3COL$_\gamma$ to FEWSAT$_{\gamma'}$. Therefore, we can conclude that FEW3COL$_\gamma \in$ few-NP.

Next, we show that for each problem $Q \in$ few-NP, it holds that Q is fpt-reducible to FEW3COL$_\gamma$ for some uniformly fpt-time computable graph generator γ. For this, it suffices to show that for each uniformly fpt-time computable nice 3CNF generator γ there exists some uniformly fpt-time computable graph generator γ' such that FEWSAT$_\gamma$ is fpt-reducible to FEW3COL$_{\gamma'}$.

Let γ be a uniformly fpt-time computable nice 3CNF generator. We construct the following uniformly fpt-time computable graph generator γ'. We describe the algorithm that computes γ'. On input (n, ℓ, k), the algorithm firstly computes the 3CNF formula $\varphi = \gamma(n, \ell, k)$. By using the standard polynomial-time reduction from satisfiability of 3CNF formulas to 3-colorability, we can construct (in polynomial time) a graph G that is 3-colorable if and only if φ is satisfiable. The algorithm then returns $\gamma'(n, \ell, k) = G$. It is then straightforward to verify that the identity mapping is an fpt-reduction from FEWSAT$_\gamma$ to FEW3COL$_{\gamma'}$. This concludes our proof.

We get a similar characterization of nu-few-NP.

Proposition 15.83. nu-few-NP $= [\{$ FEW3COL$_\gamma$: γ *is a non-uniformly fpt-time computable graph generator* $\}]_{\text{fpt}}$.

Proof. The proof is analogous to the proof of Proposition 15.82.

Characterizing few-NP by Filtering para-NP Problems
To further illustrate the robustness of the classes few-NP and nu-few-NP, we give a
different characterization of these classes, in terms of applying what can be called
"well-behaved efficiently computable filters" to para-NP problems.

We begin with the definition of xp-numberings. These functions will be the foundation of our notion of "filters."

Definition 15.84. *Let Q be a parameterized problem. We say that an xp-numbering
of Q is a function $\rho : \Sigma^* \times \mathbb{N} \to \mathbb{N}$ for which there exists a computable function $f :
\mathbb{N} \to \mathbb{N}$ such that:*

- *for each $(n, k) \in \mathbb{N} \times \mathbb{N}$, ρ maps each instance (x, k) of Q with $|x| = n$ to a
  number $\ell \in [0, n^{f(k)}]$;*
- *for any instance (x, k) of Q it holds that if $\rho(x, k) = 0$, then $(x, k) \notin Q$; and*
- *for each $(n, k) \in \mathbb{N} \times \mathbb{N}$ and any two instances $(x, k), (x', k) \in \Sigma^* \times \mathbb{N}$ with $|x| =
  |x'| = n$ it holds that $\rho(x, k) = \rho(x', k)$ implies that $(x, k) \in Q$ if and only
  if $(x', k) \in Q$.*

Moreover, we say that an xp-numbering is an fpt-time xp-numbering *if it is computable in fixed-parameter tractable time.*

In the remainder of this section, we will consider the following class of parameterized problems:

$$\{ A \cap B : A \text{ has an fpt-time xp-numbering, } B \in \text{para-NP} \}.$$

Intuitively, in this definition, one can consider fpt-time xp-numberings as efficiently computable well-behaved filters. They are efficiently computable, because
they run in fpt-time. They are filters, because they reduce the para-NP set to a set where
for each (n, k), all instances can be decided by solving the problem for only $n^{f(k)}$
instances. Finally, they are well-behaved, because one can identify for each instance
if it boils down to one of these $n^{f(k)}$ "crucial" instances, and if so, to which one of
these instances (indicated by a number $\ell \in [n^{f(k)}]$).

In fact, we will show that this class coincides with few-NP.

Proposition 15.85. $\{ A \cap B : A \text{ has an fpt-time xp-numbering, } B \in \text{para-NP} \} \subseteq$
few-NP.

Proof. Let $Q = A \cap B$ for some problem A that has an fpt-time xp-numbering ρ and
some $B \in$ para-NP. We show that $Q \in$ few-NP by showing that Q is fpt-reducible
to the problem FEWSAT_γ for some uniformly fpt-time computable generator γ. The
main idea of this proof is to encode the concatenation of the inverse computation of
the xp-numbering ρ and the non-deterministic check of membership in B into the
SAT instance generated for some number $\ell \in [n^k]$.

Let f be the computable function such that for each input (x, k) with $|x| = n$
it holds that $\rho(x, k) \leq n^{f(k)}$. We will construct the uniformly fpt-time generator γ below. Firstly, we will specify the fpt-reduction R from Q to FEWSAT_γ.

Let (x, k) be an instance of Q, with $|x| = n$. We let $R(x, k) = (n, \ell, f(k))$, where $\ell = \rho(x, k) \leq n^{f(k)}$. Next, we will specify the generator γ. Since it is more informative to describe the non-deterministic computation encoded in the SAT instance that γ outputs, we do so without spelling out the resulting SAT instance. On input (n, ℓ, k), the computation $\gamma(n, \ell, k)$ firstly computes $k' = f^{-1}(k)$, i.e., the k' such that $f(k') = k$. This can be done in deterministic fpt-time. Next, the computation $\gamma(n, \ell, k)$ non-deterministically computes an instance x of length n for which it holds that $\rho(x, k) = \ell$. Since for all instances x, x' with $\rho(x, k) = \rho(x', k)$ it holds that $(x, k) \in A$ if and only if $(x', k) \in A$, we can allow the computation to find any such instance x. Finally, the computation $\gamma(n, \ell, k)$ non-deterministically verifies that $(x, k) \in B$. Since $B \in$ para-NP, this can be done in non-deterministic fpt-time. As explained above, technically, γ encodes this non-deterministic fpt-time computation $\gamma(n, \ell, k)$ in a SAT instance of fpt-size. We get for each instance (x, k) with $|x| = n$ that $(x, k) \in Q$ if and only if $R(x, k) = (n, \ell, f(k)) \in \text{FEWSAT}_\gamma$. Therefore, $Q \in$ few-NP.

Next, to establish an inclusion in the converse direction, we prove the following technical lemma.

Lemma 15.86. *Let* $Q = A \cap B$ *be a parameterized problem, where A is a problem that has an fpt-time xp-numbering, and $B \in$ para-NP. Moreover, let Q' be a parameterized problem that is fpt-reducible to Q. Then $Q' = A' \cap B'$ for some problem A' that has an fpt-time xp-numbering and some $B' \in$ para-NP.*

Proof. We construct A' and B'. Let R be the fpt-reduction from Q' to Q. We let A' be the set consisting of the following instances. We let $(x, k) \in A'$ if and only if $R(x, k) \in A$. We show that A' has an fpt-time xp-numbering, by constructing such an xp-numbering ρ'. We know that A has an fpt-time xp-numbering ρ. On input (x, k), the xp-numbering ρ' returns the value $\ell = \rho(R(x, k))$. Since R is an fpt-reduction, is straightforward to construct a computable function f' such that $\ell \leq n^{f'(k)}$ for each instance (x, k) with $|x| = n$. Moreover, since both R and ρ are fpt-time computable, the xp-numbering ρ' is also fpt-time computable.

Next, we construct the set B', in a similar way. For each instance (x, k), we let $(x, k) \in B'$ if and only if $R(x, k) \in B$. Since $B \in$ para-NP and since para-NP is closed under fpt-reductions, we get that $B' \in$ para-NP.

These definitions then have the consequence that for each instance (x, k) it holds that $(x, k) \in Q'$ if and only if $(x, k) \in A' \cap B'$. Namely, we get that $(x, k) \in Q'$ if and only if $R(x, k) \in Q$, if and only if both $R(x, k) \in A$ and $R(x, k) \in B$, if and only if both $(x, k) \in A'$ and $(x, k) \in B'$, if and only if $(x, k) \in A' \cap B'$.

Using the above lemma, we can now prove that each problem in few-NP can be described as the intersection of a problem in para-NP and a problem that has an fpt-time xp-numbering.

Proposition 15.87. few-NP $\subseteq \{ A \cap B : A$ *has an fpt-time xp-numbering, $B \in$ para-NP $\}$.*

Proof. Let $Q \in$ few-NP. We know that there is an fpt-reduction R from Q to FEWSAT$_\gamma$, for some uniformly fpt-time computable generator γ. We show that Q is fpt-reducible to a problem $Q' = A \cap B$, for some problem A that has an fpt-time xp-numbering and some $B \in$ para-NP. By Lemma 15.86, this suffices to show that Q itself is of the required form.

We specify the fpt-reduction R', the problem A with its fpt-time xp-numbering ρ, and the problem $B \in$ para-NP. We begin with the fpt-reduction R'. On input (x, k) with $|x| = n$, the reduction first computes $R(x, k) = (n', \ell', k')$. Here we know that $k' \leq f(k)$ for some computable function f. Next, it computes $\gamma(n', \ell', k') = \varphi$. The reduction then outputs $R(x, k) = (x, \varphi, \ell', k)$. We let the set A consist of all instances (x, φ, ℓ', k) such that $\gamma(n', \ell', k') = \varphi$, where $R(x, k) = (n', \ell', k')$, i.e., A checks whether φ and ℓ' are computed correctly from (x, k), according to R and γ. Next, we specify the xp-numbering ρ of A. On input (x, φ, ℓ', k), the xp-numbering ρ checks whether $(x, \varphi, \ell', k) \in A$; if so, it outputs $\rho(x, \varphi, \ell', k') = \ell'$; otherwise, it outputs $\rho(x, \varphi, \ell', k') = 0$. Since $\ell' \leq |x|^{f(k)}$, we know that ρ is in fact an xp-numbering. Moreover, ρ is fpt-time computable. Finally, we specify the set $B \in$ para-NP. For each input (x, φ, ℓ', k), we let $(x, \varphi, \ell', k) \in B$ if and only if φ is satisfiable. Clearly, then, $B \in$ para-NP. Now, since A checks whether φ is computed correctly from (x, k) according to R and γ, and since B checks whether φ is satisfiable, we get that $(x, k) \in Q$ if and only if $R'(x, k) \in A \cap B$. This concludes our proof.

We now get the following characterization of few-NP.

Theorem 15.88. *The class* few-NP *can be characterized in the following way:*

$$\text{few-NP} = \{ A \cap B : A \text{ has an fpt-time xp-numbering}, B \in \text{para-NP} \}.$$

Proof. The result follows directly from Propositions 15.85 and 15.87.

In an entirely analogous way, we can now derive the following characterization of nu-few-NP, using xp-numberings ρ that can be computed by FPT/fpt algorithms.

Proposition 15.89. *The class* nu-few-NP *can be characterized in the following way:*

$$\text{nu-few-NP} = \{ A \cap B : A \text{ has an xp-numbering } \rho \text{ that is computable}$$
$$\text{by a FPT/fpt algorithm}, B \in \text{para-NP} \}.$$

Proof. The proof is entirely analogous to the proofs of Lemma 15.86, Propositions 15.85 and 15.87, and Theorem 15.88.

We illustrate this characterization of few-NP in terms of filtering by using it to show that the following example problem is in few-NP. We consider the binary expansion of irrational numbers. For instance, take the square root of two. We can compute the first n bits of the binary expansion of $\sqrt{2}$ in time polynomial in n. Let bits$(n, \sqrt{2})$ denote the first n bits in the binary expansion of $\sqrt{2}$.

Example 15.90. The following parameterized problem $3\text{SAT}\lfloor\text{dist-}\sqrt{2}\rfloor$ is in few-NP:

$$3\text{SAT}\lfloor\text{dist-}\sqrt{2}\rfloor = \{ \ (x, k) \in \mathbb{B}^* \times \mathbb{N} : x \text{ encodes a satisfiable}$$
$$\text{propositional formula, } |x| = n, \text{ the Hamming}$$
$$\text{distance between } x \text{ and bits}(n, \sqrt{2}) \text{ is at most } k \ \}.$$

To see this, we express $3\text{SAT}\lfloor\text{dist-}\sqrt{2}\rfloor$ as a problem of the form $A \cap B$, where A has an fpt-time computable xp-numbering and where $B \in$ para-NP. We let $B = 3\text{SAT} \times \mathbb{N}$, which is clearly in para-NP. Then we describe A and its fpt-time xp-numbering ρ. On input (x, k), with $|x| = n$, we first compute $y = \text{bits}(n, \sqrt{2})$. Then we compute $z = x \oplus y$ (where $\oplus$ denotes the bitwise exclusive-or operator). This can be done in polynomial time. Then, if z has more than k ones, we let $(x, k) \notin A$, and $\rho(x, k) = 0$. Otherwise, if z has at most k ones, we let $(x, k) \in A$ and we compute $\rho(x, k)$ as follows. Let $i_1, \ldots, i_u$, with $u \leq k$, be the indices of the bits in $z = z_1 \ldots z_n$ with ones, i.e., $z_{i_j} = 1$ for all $j \in [u]$. We can extend this sequence of indices to a sequence $i'_1, \ldots, i'_k$ by adding $k - u$ zeroes to the beginning. We then concatenate $i'_1, \ldots, i'_k$ (each described using $\log n$ bits) into a single bitstring i, and interpret this as a number ℓ. We then let $\rho(x, k) = \ell + 1$. Since each i'_j is described by $\log n$ bits, we have that i is of length $k \log n$, and thus $\ell \in [n^k]$. Thus ρ is an xp-numbering. ⊣

In fact, there is nothing special about the number $\sqrt{2}$ (well, there is, but not for this example), except for the fact that the first n bits of its binary expansion are computable in polynomial time. Consequently, for any other real number r for which we can compute the first n bits of its binary expansion in polynomial time (in n), the similarly defined problem $\text{SAT}\lfloor\text{dist-}r\rfloor$ is in few-NP.

Circuit Characterization of nu-few-NP

Finally, we observe that the non-uniform parameterized complexity class nu-few-NP can be characterized using Boolean circuits.

Proposition 15.91. *The class* nu-few-NP *coincides with the set of all parameterized problems Q for which there exists a computable function f and a constant c such that for each $(n, k) \in \mathbb{N} \times \Sigma^*$ there is some circuit $C_{n,k}$ of size $f(k)n^c$ such that, for each $\ell \in \mathbb{B}^{k \log n}$, it holds that $(n, \ell, k) \in Q$ if and only if $C_{n,k}[\ell]$ is satisfiable.*

Proof (idea). An argument similar to the one in the proof of Proposition 15.20 can be used to show this result.

15.3.4 The Parameterized Compilability of Finding Small Cliques

Using the non-uniform parameterized complexity class nu-few-NP defined in the previous section, we can analyze the (in)compilability of several additional parameterized compilation problems. We use another natural parameterized variant of CONSTRAINED-CLIQUE as an example to illustrate this.

We consider the following parameterized compilation problem.

CONSTRAINED-CLIQUE(constr.-size)
Offline instance: a graph $G = (V, E)$.
Online instance: two subsets $V_1, V_2 \subseteq V$ of vertices, and a positive integer $u \geq 1$.
Parameter: $k = |V_1| + |V_2|$.
Question: is there a clique $C \subseteq V$ in G of size u such that $C \cap V_1 = \emptyset$ and $V_2 \subseteq C$?

We characterize the compilability of this parameterized compilation problem using the parameterized complexity classes nu-few-NP and FPT/fpt. In order to do so, we consider the following auxiliary parameterized compilation problem.

SATQUERY(query-size)
Offline instance: a propositional formula φ.
Online instance: a set L of literals.
Parameter: $k = |L|$.
Question: is $\varphi \wedge \bigwedge_{l \in L} l$ satisfiable?

We firstly show that we can focus on the restriction of the problem SATQUERY(query-size) to 3CNF formulas.

Proposition 15.92. *The problem* SATQUERY(query-size) *is equivalent to the restriction of* SATQUERY(query-size) *to offline instances in* 3CNF *(under fpt-nucomp-reductions).*

Proof. We give an fpt-nucomp-reduction from SATQUERY(query-size) to itself, where the resulting offline instance is in 3CNF. In order to do so, we need to specify fpt-size functions f_1, f_2, a computable function h and an fpt-time function g such that for each instance (φ, L, k) of SATQUERY(query-size) it holds that for each $m \geq |L|$, $(\varphi, L, k) \in$ SATQUERY(query-size) if and only if $(f_1(\varphi, 1^m, k), g(f_2(\varphi, 1^m, k), L, k), h(k)) \in$ SATQUERY(query-size). By using the standard Tseitin transformations [188], we can in polynomial time transform the formula φ into a 3CNF formula φ' with the property that $\text{Var}(\varphi) \subseteq \text{Var}(\varphi')$ and that for each truth assignment $\alpha : \text{Var}(\varphi) \to \mathbb{B}$ it holds that $\varphi[\alpha]$ is true if and only if $\varphi'[\alpha]$ is satisfiable. We then let $f_1(\varphi, 1^m, k) = \varphi'$. Moreover, we let $f_2(\varphi, 1^m, k) = \emptyset$, we let $g(\emptyset, L, k) = L$, and we let $h(k) = k$. It is straightforward to verify the correctness of this fpt-nucomp-reduction.

Next, we relate the problem SATQUERY(query-size) to the parameterized problem FEWSAT$_\gamma$, for any non-uniformly fpt-time computable generator γ. Here we consider FEWSAT$_\gamma$ as a parameterized compilation problem.

Proposition 15.93. *Let γ be a non-uniformly fpt-time computable generator. Then* FEWSAT$_\gamma$ *is fpt-nucomp-reducible to* SATQUERY(query-size).

Proof (sketch). We need to specify fpt-size functions f_1, f_2, a computable function h and an fpt-time function g such that for each instance (n, ℓ, k) of FEWSAT$_\gamma$ it holds that for each $m \geq |\ell|$:

$$(n, \ell, k) \in \text{FEWSAT}_\gamma \quad \text{if and only if}$$
$$(f_1(n, 1^m, k), g(f_2(n, 1^m, k), \ell, k), h(k)) \in \text{SATQUERY(query-size)}.$$

We let $f_2(n, 1^m, k) = n$. In addition, we let $g(n, \ell, k)$ be an encoding L_ℓ of ℓ in terms of a set of literals of size k over the variables $x_{i,j}$, for $i \in [n]$, $j \in [k]$. Then, we let $f_1(n, 1^m, k) = \varphi$ be a 3CNF formula that is satisfiable in conjunction with L_ℓ if and only if $\gamma(n, \ell, k)$ is satisfiable, for each $\ell \in [n^k]$. This can be done as follows.

By Proposition 15.71, we may assume without loss of generality that γ is a 3CNF generator. Firstly, the formula φ contains clauses to ensure that at most k variables $x_{i,j}$ are true. Then, we add $8n^3$ variables y_i, corresponding to the $8n^3$ possible clauses $c_1, \ldots, c_{8n^3}$ of size 3 over the variables $x_1, \ldots, x_n$. Then, since for each $(n, k) \in \mathbb{N}^2$, the function $\gamma(n, \cdot, k)$ is computable in fpt-time, we can construct fpt-many clauses that ensure that, whenever L_ℓ is satisfied, the variables y_i must be set to true for the clauses $c_i \in \gamma(n, \ell, k)$. Finally, for each such possible clause c_i, we add clauses to ensure that whenever y_i is set to true, then the clause c_i must be satisfied.

Next, we show that for the problems FEWSAT$_\gamma$ membership in par-nucomp-FPT is equivalent to membership in FPT/fpt.

Proposition 15.94. *Let γ be a generator. Then* FEWSAT$_\gamma \in$ par-nucomp-FPT *if and only if* FEWSAT$_\gamma \in$ FPT/fpt.

Proof. ($\Rightarrow$) Assume that there exists an fpt-size function f_1, an fpt-time function g a computable function h and a parameterized problem $Q \in$ FPT such that for each instance (n, ℓ, k) of FEWSAT$_\gamma$ and each $m \geq |\ell|$ it holds that:

$$(n, \ell, k) \in \text{FEWSAT}_\gamma \quad \text{if and only if} \quad (f_1(n, 1^m, k), g(\ell, k), h(k)) \in Q.$$

We show that FEWSAT$_\gamma \in$ FPT/fpt. Take an arbitrary $(n, k) \in \mathbb{N}^2$. Consider the string $\alpha = f_1(n, 1^m, k)$ as advice, for some $m \geq |\ell| = k \log n$. Then deciding for some $\ell \in [n^k]$ whether $(n, \ell, k) \in$ FEWSAT$_\gamma$ can be done in fpt-time using the advice string α, by checking whether $(\alpha, g(\ell, k), h(k)) \in Q$.

($\Leftarrow$) Conversely, assume that FEWSAT$_\gamma \in$ FPT/fpt, i.e., that there exist a computable function f and a constant c such that for each $(n, k) \in \mathbb{N}^2$ there exists some advice string $\alpha(n, k)$ of length $f(k)n^c$ such that, given $(\alpha(n, k), n, \ell, k)$, deciding whether $(n, \ell, k) \in$ FEWSAT$_\gamma$ is fixed-parameter tractable. We then construct a fpt-size function f_1, an fpt-time function g and a computable function h such that there exists a parameterized problem $Q \in$ FPT such that for each instance (n, ℓ, k) of FEWSAT$_\gamma$ and each $m \geq |\ell|$ it holds that:

$$(n, \ell, k) \in \text{FEWSAT}_\gamma \quad \text{if and only if} \quad (f_1(n, 1^m, k), g(\ell, k), h(k)) \in Q.$$

We let $f_1(n, 1^m, k) = (\alpha(n, k), n)$, we let $g(\ell, k) = \ell$ and we let $h(k) = k$. Then, by assumption, given $(f_1(n, 1^m, k), g(\ell, k), h(k))$, deciding whether $(n, \ell, k) \in \text{FEWSAT}_\gamma$ is fixed-parameter tractable.

Finally, we show that nu-few-NP $\subseteq$ FPT/fpt implies that SATQUERY(query-size) $\in$ par-nucomp-FPT.

Proposition 15.95. *If* FEWSAT$_\gamma \in$ FPT/fpt *for each non-uniformly fpt-time computable generator* γ, *then* SATQUERY(query-size) $\in$ par-nucomp-FPT.

Proof. In order to show that SATQUERY(query-size) $\in$ par-nucomp-FPTpar-nucomp -FPT, we specify an fpt-size function f_1, an fpt-time function g, a computable function h and a parameterized problem $Q \in$ FPT such that for all instances (φ, L, k) of SATQUERY(query-size) and each $m \geq |L| = k$ it holds that:

$$(\varphi, L, k) \in \text{SATQUERY(query-size)} \quad \text{if and only if}$$
$$(f_1(\varphi, 1^m, k), g(L, k), h(k)) \in Q.$$

Assume without loss of generality that φ contains the variables $x_1, \ldots, x_n$. There are (at most) $2n^k \leq n^{2k}$ sets of k literals over the variables $x_1, \ldots, x_n$. Let $L_1, \ldots, L_u$ be an enumeration of these clauses. We consider the following non-uniformly fpt-time computable generator γ_φ. Given a triple $(n', \ell, k') \in \mathbb{N}^3$, we define $\gamma_\varphi(n', \ell, k') = \emptyset$ if $n' \neq n$, or $k' \neq 2k$, or $\ell > u$; otherwise, we define $\gamma_\varphi(n', \ell, k') = \gamma_\varphi(n, \ell, 2k)$ to be a propositional formula φ' that is satisfiable if and only if $\varphi \wedge \bigwedge L_\ell$ is satisfiable. Then, $(n', \ell, k') \in \text{FEWSAT}_{\gamma_\varphi}$ if and only if $(\varphi, L_\ell, k) \in$ SATQUERY(query-size). By assumption, FEWSAT$_{\gamma_\varphi} \in$ FPT/fpt, and therefore there exists some algorithm $A_{(n',k')}$ that decides whether $(n', \ell, k') \in$ FEWSAT$_{\gamma_\varphi}$ in fpt-time. We now let $f_1(\varphi, 1^m, k) = \alpha_{(n',k')}$ be a description of the algorithm $A_{(n',k')}$. Since $A_{(n',k')}$ runs in fixed-parameter tractable time, we know that its description $\alpha_{(n',k')}$ is of fpt-size. Moreover, we let $g(L, k) = \ell$, for $L = L_\ell$; and we let $h(k) = k' = 2k$. The problem Q consists of simulating the algorithm $A_{(n',k')}$ on input (n', ℓ, k'), which is fixed-parameter tractable. This completes our fpt-nucomp-reduction.

The above results together give us the following theorem.

Theorem 15.96. SATQUERY(query-size) $\in$ par-nucomp-FPT *if and only if* nu-few-NP $\subseteq$ FPT/fpt.

Proof. ($\Rightarrow$) Assume that SATQUERY(query-size) $\in$ par-nucomp-FPT. Then by Proposition 15.93, we know that FEWSAT$_\gamma \in$ par-nucomp-FPT, for each non-uniformly fpt-time computable generator γ. Then, by Proposition 15.94, we know that FEWSAT$_\gamma \in$ FPT/fpt, for each non-uniformly fpt-time computable generator γ. In other words, nu-few-NP $\subseteq$ FPT/fpt.

($\Leftarrow$) Conversely, assume that nu-few-NP $\subseteq$ FPT/fpt. Then for each non-uniformly fpt-time computable generator γ, it holds that FEWSAT$_\gamma \in$ FPT/fpt. Then, by Proposition 15.95, we know that SATQUERY(query-size) $\in$ par-nucomp-FPT.

With this characterization of the parameterized compilability of SATQUERY (query-size), we can return to the problem CONSTRAINED-CLIQUE(constr.-size).

Proposition 15.97. *The problem* CONSTRAINED-CLIQUE(constr.-size) *is equivalent to* SATQUERY(size) *under fpt-nucomp-reductions.*

Proof. Firstly, we give an fpt-nucomp-reduction from SATQUERY(query-size) to CONSTRAINED-CLIQUE(constr.-size). Before we specify this reduction, we will look at a well-known polynomial-time reduction from 3SAT to CLIQUE. Let $\varphi = c_1 \wedge \cdots \wedge c_b$, and let $\mathrm{Var}(\varphi) = \{x_1, \ldots, x_n\}$. We construct a graph $G = (V, E)$ as follows. We let $V = \{(c_j, l) : j \in [b], l \in c_j\} \cup \{x_i, \overline{x_i} : i \in [n]\}$. We describe the edges of G as three sets: (1) the edges between two vertices of the form (c_j, l), (2) the edges between two vertices of the form $x_i, \overline{x_i}$, and (3) the edges between a vertex of the form (c_j, l) and a vertex of the form $x_i, \overline{x_i}$. We describe set (1). Let (c_j, l) and $(c_{j'}, l')$ be two vertices. These vertices are connected by an edge if and only if both $j \neq j'$ and $l \neq \overline{l'}$. Next, we describe set (2). Two vertices $v_1, v_2 \in \{x_i, \overline{x_i} : i \in [n]\}$ are connected by an edge if and only if v_1 and v_2 are not complementary literals x_i and $\overline{x_i}$, for some $i \in [n]$. Finally, we describe set (3). Let (c_j, l) and $v \in \{x_i, \overline{x_i} : i \in [n]\}$ be two vertices. Then (c_j, l) and v are connected by an edge if and only if l and v are not complementary literals x_i and $\overline{x_i}$, for some $i \in [n]$. We have that G has a clique of size $n + b$ if and only if φ is satisfiable. Moreover, each satisfying assignment $\alpha : \{x_1, \ldots, x_n\} \to \mathbb{B}$ of φ corresponds to a clique $C = \{l \in \{x_i, \overline{x_i} : i \in [n]\} : \alpha(l) = 1\} \cup D$ of size $n + b$, for some $D \subseteq \{(c_j, l) \in V : \alpha(l) = 1\}$, and vice versa. This direct correspondence between cliques and satisfying truth assignments has the following consequence. Let L be a set of literals. There exists a clique of size $n + b$ in G that does not contain the vertices $l \in L$ if and only if there exists a satisfying truth assignment of φ that does not satisfy any literal $l \in L$, which is the case if and only if there exists a satisfying truth assignment of φ that satisfies the complement $\overline{l}$ of each $l \in L$.

We are now ready to specify our fpt-nucomp-reduction. In order to do so, we need to specify fpt-size functions f_1, f_2, a computable function h and an fpt-time function g. Let (φ, L, k) be an instance of SATQUERY(query-size). We may assume without loss of generality that φ is in 3CNF. We let $f_1(\varphi, 1^m, k) = G$, where G is the graph constructed from φ as in the polynomial-time reduction from 3SAT to CLIQUE that is described above. We let $f_2(\varphi, 1^m, k) = u = n + b$; we let $g(u, L, k) = (u, V_1, \emptyset)$, where $V_1 = \{\overline{l} : l \in L\}$ is the set of vertices $\overline{l}$ in V corresponding to the complements of the literals in L. Finally, we let $h(k) = k$. As argued above, we have that $\varphi \wedge \bigwedge_{l \in L} l$ is satisfiable if and only if G has a clique of size u that does not contain any vertex in V_1. Therefore, this reduction is correct.

Next, we give an fpt-nucomp-reduction from CONSTRAINED-CLIQUE(constr.-size) to SATQUERY(query-size). In order to do so, we consider a polynomial-time reduction from CLIQUE to 3SAT. It is straightforward to construct, given a graph $G = (V, E)$ and an integer $u \in [|V|]$, a 3CNF formula φ_u in polynomial time, such that φ_u is satisfiable if and only if G contains a clique of size u. Moreover, we can do this in such a way that the formulas $\varphi_1, \ldots, \varphi_{|V|}$ share a set of variables $\{x_v : v \in V\}$ with the property that for each $u \in [|V|]$ and each two subset $V_1, V_2 \subseteq V$ it holds that G has a clique $C \subseteq V$ of size u that contains no vertex in V_1 and that contains all vertices in V_2 if and only if the formula $\varphi_u \wedge \bigwedge l \in L$ is satisfiable, where $L = \{ \overline{x_v} : v \in V_1 \} \cup \{ x_v : v \in V_2 \}$.

Now, we are ready to specify our fpt-nucomp-reduction. In order to do so, we need to specify fpt-size functions f_1, f_2, a computable function h and an fpt-time function g. Let $(G, (V_1, V_2, u), k)$ be an instance of CONSTRAINED-CLIQUE(constr.-size). We let $f_1(G, 1^m, k) = \psi$, where we define $\psi = \bigwedge_{j \in [|V|]}(y_j \rightarrow \varphi_j)$, where the formulas φ_j are constructed as in the polynomial-time reduction from CLIQUE to 3SAT that is described above. We let $f_2(G, 1^m, k) = \emptyset$, we let $g(G, (V_1, V_2, u), k) = L \cup \{y_u\}$, where $L = \{ \overline{x_v} : v \in V_1 \} \cup \{ x_v : v \in V_2 \}$, and we let $h(k) = k + 1$. As argued above, we have that G has a clique of size u that does not contain any vertex in V_1 and that contains all vertices in V_2 if and only if $\varphi \wedge \bigwedge_{l \in L} l \wedge y_u$ is satisfiable. Therefore, this reduction is correct.

Then, by Theorem 15.96 and Proposition 15.97, we know that CONSTRAINED-CLIQUE(constr.-size) is not fpt-size compilable, unless nu-few-NP $\subseteq$ FPT/fpt.

Corollary 15.98. *It holds that* CONSTRAINED-CLIQUE(constr.-size) $\in$ par-comp-FPT *implies that* nu-few-NP $\subseteq$ FPT/fpt.

Clique Size as Part of the Offline Instance
In fact, requiring that the size of the cliques is part of the offline instance (rather than part of the online instance) does not make a difference for the compilability of the problem. Consider the following variant of CONSTRAINED-CLIQUE(constr.-size), where the size of the cliques is part of the offline instance.

CONSTRAINED-CLIQUE$^{\text{offline-size}}$(constr.-size)
Offline instance: a graph $G = (V, E)$, and a positive integer $u \geq 1$.
Online instance: two subset V_1, $V_2 \subseteq V$ of vertices.
Parameter: $k = |V_1| + |V_2|$.
Question: is there a clique $C \subseteq V$ in G of size u such that $C \cap V_1 = \emptyset$ and $V_2 \subseteq C$?

This problem is also not fpt-size compilable, unless nu-few-NP $\subseteq$ FPT/fpt.

Proposition 15.99. CONSTRAINED-CLIQUE$^{\text{offline-size}}$(constr.-size) *is equivalent to* SATQUERY(query-size) *under fpt-nucomp-reductions.*

Proof. The proof of Proposition 15.97 can be easily modified to show this result.

Corollary 15.100. *It holds that if* CONSTRAINED-CLIQUE$^{\text{offline-size}}$(constr.-size) $\in$ par-comp-FPT *then* nu-few-NP $\subseteq$ FPT/fpt.

15.3.5   Other Parameterized Compilation Problems

In this section, we identify several other parameterized compilation problems to which the problem SATQUERY(query-size) (or its co-problem) can be fpt-nucomp-reduced. As a result of Theorem 15.96, these parameterized compilation problems are not fpt-size compilable, unless nu-few-NP $\subseteq$ FPT/fpt.

Small Clause Entailment

Firstly, we consider another parameterized variant of the compilation problem
CLAUSE-ENTAILMENT.

CLAUSE-ENTAILMENT(clause-size)
Offline instance: A CNF formula φ, and a positive integer s.
Online instance: A clause c.
Parameter: s.
Question: $|c| \leq s$ and $\varphi \models c$?

We observe that co-SATQUERY(query-size) can straightforwardly be fpt-nucomp-reduced to this problem.

Proposition 15.101. *The parameterized compilation problems co-SATQUERY
(query-size)* and CLAUSE-ENTAILMENT(clause-size) *are equivalent under fpt-nucomp-reductions.*

Proof (sketch). An fpt-nucomp-reduction can be constructed straightforwardly by
using the fact that for any propositional formula φ and any set L of literals it holds
that $\varphi \wedge \bigwedge_{l \in L} l$ is satisfiable if and only if $\varphi \not\models c$, where $c = \bigvee_{l \in L} \bar{l}$. Using a similar
argument, an fpt-nucomp-reduction in the other direction can also be constructed
straightforwardly.

Hamiltonian Paths and the Travelling Salesperson Problem

Using ideas similar to the ones used in Sect. 15.3.4, we can show for some parameterized compilation problems related to finding Hamiltonian paths and related to the
Travelling Salesperson Problem that they are equivalent to SATQUERY(query-size)
under fpt-nucomp-reductions. The main idea that we use to show these equivalences
is that the standard polynomial-time reductions from 3SAT to finding Hamiltonian
paths in (undirected or directed) graphs or to the TSP result in instances where specific edges in the solutions correspond to the assignment of specific literals to specific
truth values in satisfying assignments of the original 3CNF formula. We consider the
following parameterized compilation problems.

CONSTRAINED-HP- UNDIRECTED(constr.-size)
Offline instance: an undirected graph $G = (V, E)$.
Online instance: two subsets $E_1, E_2 \subseteq E$ of edges.
Parameter: $k = |E_1| + |E_2|$.
Question: is there a Hamiltonian path π in G such that π includes no edge in E_1 and includes
each edge in E_2?

CONSTRAINED-HP- DIRECTED(constr.-size)
Offline instance: a directed graph $G = (V, E)$.
Online instance: two subsets $E_1, E_2 \subseteq E$ of edges.
Parameter: $k = |E_1| + |E_2|$.
Question: is there a Hamiltonian path π in G such that π includes no edge in E_1 and includes
each edge in E_2?

CONSTRAINED-TSP(constr.-size)
Offline instance: a directed graph $G = (V, E)$, and a cost $c(e) \in \mathbb{N}$ (given in binary) for each edge $e \in E$.
Online instance: two subsets $E_1, E_2 \subseteq E$ of edges, and a positive integer $u \geq 1$ (given in binary).
Parameter: $k = |E_1| + |E_2|$.
Question: is there a Hamiltonian cycle π in G of total cost $\leq u$ such that π includes no edge in E_1 and includes each edge in E_2?

Proposition 15.102. *The following three parameterized compilation problems are equivalent to* SATQUERY(query-size) *(under fpt-nucomp-reductions):*

- CONSTRAINED-HP- UNDIRECTED(constr.-size),
- CONSTRAINED-HP- DIRECTED(constr.-size), *and*
- CONSTRAINED-TSP(constr.-size).

Proof. (sketch). The problems of finding a Hamiltonian path (in an undirected or in a directed graph), finding a Hamiltonian cycle (in an undirected or in a directed graph), and finding a tour of length $\leq u$ for an instance of the TSP are classic NP-complete problems. The well-known polynomial-time reductions from these problems to 3SAT and from 3SAT to these problems have a direct correspondence between solutions of these problems and satisfying assignments of the 3SAT instances, similar to the correspondence used in the proof of Proposition 15.97. Using arguments that are completely analogous to the arguments used in these proofs, we can show that the problems CONSTRAINED-HP- UNDIRECTED(constr.-size), CONSTRAINED-HP- DIRECTED(constr.-size), and CONSTRAINED-TSP(constr.-size) are equivalent under fpt-nucomp-reductions to SATQUERY(query-size).

We observe that the standard reductions can be modified straightforwardly to work also for the restrictions of CONSTRAINED-HP- UNDIRECTED(constr.-size), CONSTRAINED-HP- DIRECTED(constr.-size), and CONSTRAINED-TSP(constr.-size), where $E_1 = \emptyset$, as well as for the restriction where $E_2 = \emptyset$.

Graph Colorability

Next, we look at parameterized compilation problems that are based on graph colorability problems. Again, by exploiting the well-known polynomial-time reductions to and from 3SAT, and the direct correspondence between solution to the coloring problems and the satisfiability problem, we can establish equivalence between the following problems and SATQUERY(query-size). Consider the following parameterized compilation problem, where $r \geq 1$ is an arbitrary positive integer.

CONSTRAINED-r- COLORING(constr.-size)
Offline instance: a graph $G = (V, E)$.
Online instance: a partial r-coloring ρ, that assigns a color in $\{1, 2, 3\}$ to a subset $V' \subseteq V$ of vertices.
Parameter: $k = |V'|$.
Question: is there a proper r-coloring ρ' of G that extends ρ?

We then get the following result (for which we omit the straightforward proof).

Proposition 15.103. *Let* $r \geq 1$ *be a positive integer. Then the parameterized compilation problem* CONSTRAINED-r- COLORING(constr.-size) *is equivalent to* SATQUERY(query-size) *under fpt-nucomp-reductions.*

Other Problems

Again, using similar ideas, exploiting well-known polynomial time reductions to and from 3SAT, we observe that the following parameterized compilation problems are equivalent to SATQUERY(query-size) under fpt-nucomp-reductions. Consider the following parameterized compilation problems.

CONSTRAINED-IS(constr.-size)
Offline instance: a graph $G = (V, E)$.
Online instance: two subsets $V_1, V_2 \subseteq V$ of vertices, and a positive integer $u \geq 1$.
Parameter: $k = |V_1| + |V_2|$.
Question: is there an independent set $C \subseteq V$ in G of size $\geq u$ such that $C \cap V_1 = \emptyset$ and $V_2 \subseteq C$?

CONSTRAINED-VC(constr.-size)
Offline instance: a graph $G = (V, E)$.
Online instance: two subsets $V_1, V_2 \subseteq V$ of vertices, and a positive integer $u \geq 1$.
Parameter: $k = |V_1| + |V_2|$.
Question: is there a vertex cover $C \subseteq V$ of G of size $\leq u$ such that $C \cap V_1 = \emptyset$ and $V_2 \subseteq C$?

CONSTRAINED-DS(constr.-size)
Offline instance: a graph $G = (V, E)$.
Online instance: two subsets $V_1, V_2 \subseteq V$ of vertices, and a positive integer $u \geq 1$.
Parameter: $k = |V_1| + |V_2|$.
Question: is there a dominating set $C \subseteq V$ of G of size $\leq u$ such that $C \cap V_1 = \emptyset$ and $V_2 \subseteq C$?

CONSTRAINED-HS(constr.-size)
Offline instance: a finite set T, and a collection $\mathcal{S} = \{S_1, \ldots, S_b\}$ of subsets of T.
Online instance: two subsets $T_1, T_2 \subseteq T$ of elements, and a positive integer $u \geq 1$.
Parameter: $k = |T_1| + |T_2|$.
Question: is there a hitting set $H \subseteq T$ of $\mathcal{S}$ of size $\leq u$ such that $H \cap T_1 = \emptyset$ and $T_2 \subseteq H$?

A *kernel* in a directed graph $G = (V, E)$ is a set $K \subseteq V$ of vertices such that (1) no two vertices in K are adjacent, and (2) for every vertex $u \in V \setminus K$ there is a vertex $v \in K$ such that $(u, v) \in E$.

CONSTRAINED-KERNEL(constr.-size)
Offline instance: a directed graph $G = (V, E)$.
Online instance: two subsets $V_1, V_2 \subseteq V$ of vertices.
Parameter: $k = |V_1| + |V_2|$.
Question: is there a kernel $K \subseteq V$ of G such that $K \cap V_1 = \emptyset$ and $V_2 \subseteq C$?

CONSTRAINED-EC(constr.-size)
Offline instance: a finite set T, and a collection $\mathcal{S} = \{S_1, \ldots, S_b\}$ of subsets of T.
Online instance: two subsets $\mathcal{S}_1, \mathcal{S}_2 \subseteq \mathcal{S}$.
Parameter: $k = |\mathcal{S}_1| + |\mathcal{S}_2|$.
Question: is there an exact cover $C \subseteq \mathcal{S}$ of T, i.e., a set C such that $\bigcup C = T$ and for each S, $S' \in C$ with $S \neq S'$ it holds that $S \cap S' = \emptyset$, such that $C \cap \mathcal{S}_1 = \emptyset$ and $\mathcal{S}_2 \subseteq C$?

CONSTRAINED-EC3S(constr.-size)
Offline instance: a finite set T, and a collection $\mathcal{S} = \{S_1, \ldots, S_b\}$ of subsets of T, each of size 3.
Online instance: two subsets $\mathcal{S}_1, \mathcal{S}_2 \subseteq \mathcal{S}$.
Parameter: $k = |\mathcal{S}_1| + |\mathcal{S}_2|$.
Question: is there an exact cover $C \subseteq \mathcal{S}$ of T, i.e., a set C such that $\bigcup C = T$ and for each S, $S' \in C$ with $S \neq S'$ it holds that $S \cap S' = \emptyset$, such that $C \cap \mathcal{S}_1 = \emptyset$ and $\mathcal{S}_2 \subseteq C$?

CONSTRAINED-KNAPSACK(constr.-size)

Offline instance: a finite set T of elements, and for each $t \in T$ a cost $c(t) \in \mathbb{N}$ and a value $v(t)$ $\in \mathbb{N}$, both given in binary.

Online instance: two subsets $T_1, T_2 \subseteq T$, and two positive integers $u_1, u_2 \geq 1$, both given in binary.

Parameter: $k = |T_1| + |T_2|$.

Question: is there an subset $S \subseteq T$ such that $\sum_{s \in S} c(s) \leq u_1$, such that $\sum_{s \in S} v(s) \geq u_2$, and such that $S \cap T_1 = \emptyset$ and $T_2 \subseteq S$?

We get the following results (for which we omit the straightforward proofs).

Proposition 15.104. *The following parameterized compilation problems are equivalent to* SATQUERY(query-size) *under fpt-nucomp-reductions:*

- CONSTRAINED-IS(constr.-size),
- CONSTRAINED-VC(constr.-size),
- CONSTRAINED-DS(constr.-size),
- CONSTRAINED-HS(constr.-size),
- CONSTRAINED-KERNEL(constr.-size),
- CONSTRAINED-EC(constr.-size),
- CONSTRAINED-EC3S(constr.-size), *and*
- CONSTRAINED-KNAPSACK(constr.-size).

15.4 Parameterized Variants of the Karp-Lipton Theorem

Finally, we turn our attention to developing parameterized analogues of the Karp-Lipton Theorem, which relate inclusions between certain non-uniform parameterized complexity classes to inclusions between some of the parameterized variants of the PH that we developed in Chap. 6. In order to develop these results, we consider another non-uniform variant of parameterized complexity classes, based on advice of size $f(k) \log n$ (we call this *log-kernel-size advice*).

Definition 15.105 (*log-kernel-size advice*). *Let* K *be a parameterized complexity class. We define* K/log-kernel *to be the class of all parameterized problems* Q *for which there exists a parameterized problem* $Q' \in K$ *and a computable function* f *such that for each* $(n, k) \in \mathbb{N} \times \mathbb{N}$ *there exists some* $\alpha(n, k) \in \Sigma^*$ *of size* $f(k) \log n$ *with the property that for all instances* (x, k) *with* $|x| = n$, *it holds that* $(x, k) \in Q$ *if and only if* $(x, \alpha(|x|, k), k) \in Q'$.

Our first analogue of the Karp-Lipton Theorem involves an inclusion between W[P] and FPT/log-kernel.

Proposition 15.106. *If* W[P] $\subseteq$ FPT/log-kernel, *then* $\Pi_2^p[*k, P] \subseteq \Sigma_2^p[k*]$.

Proof. Suppose that W[P] $\subseteq$ FPT/log-kernel. Then there exists a computable function f such that for each $(n, k) \in \mathbb{N} \times \mathbb{N}$ there is some advice string $\alpha(n, k)$ of length $f(k) \log n$ such that for any instance (x, k) of WSAT(CIRC) with $|x| = n$, deciding whether $(x, k) \in \text{WSAT}(\text{CIRC})$, given $(x, \alpha(n, k), k)$ is fixed-parameter tractable. Then, by self-reducibility of $\text{WSAT}(\text{CIRC})$, we can straightforwardly transform this fpt-algorithm A_1 that decides whether $(x, k) \in \text{WSAT}(\text{CIRC})$ into an fpt-algorithm A_2 that, given $(x, \alpha(n, k), k)$, computes a satisfying assignment for x of weight k, if it exists, and fails otherwise. Here we assume without loss of generality that instances of $\text{WSAT}(\text{CIRC})$ are encoded in such a way that we can instantiate variables to truth values without changing the size of the instance.

Our assumption that $\text{W[P]} \subseteq \text{FPT}/\text{log-kernel}$ only implies that for each (n, k), a suitable advice string $\alpha(n, k)$ exists. The idea of this proof is that we can guess this advice string using (bounded weight) existential quantification. For any guessed string $\alpha(n, k)$, we can use it to execute the fpt-algorithm A_2 to compute a satisfying assignment (of weight k) for an instance of $\Pi_2^p[*k]$-$\text{WSAT}(\text{CIRC})$ and to verify that this is in fact a correct satisfying assignment. Moreover, we know that (at least) one of the advice strings $\alpha(n, k)$ is correct, so if no guess leads to a satisfying assignment of weight k, we can conclude that no such assignment exists.

We show that the $\Pi_2^p[*k, \text{P}]$-complete problem $\Pi_2^p[*k]$-$\text{WSAT}(\text{CIRC})$ is contained in $\Sigma_2^p[k*]$. This suffices to show that $\Pi_2^p[*k, \text{P}] \subseteq \Sigma_2^p[k*]$. We do so by giving an algorithm that solves $\Pi_2^p[*k]$-$\text{WSAT}(\text{CIRC})$ and that can be implemented by a $\Sigma_2^p[k*]$-machine (see Sect. 6.2.3), i.e., an algorithm that can use $f(k) \log n$ existential non-deterministic steps, followed by $f(k)n^c$ universal non-deterministic steps, where n is the input size, f is some computable function and c is some constant.

Take an arbitrary instance of $\Pi_2^p[*k]$-$\text{WSAT}(\text{CIRC})$ consisting of the quantified Boolean instance $\forall X.\exists Y.C(X, Y)$ and the positive integer k. Let n denote the size of this instance. We construct the following algorithm B. Firstly, B guesses an advice string $\alpha(n, k) \in \mathbb{B}^{f(k) \log n}$. Since we have an upper bound of $f(k) \log n$ on the length of this string, we can guess this string in the $f(k) \log n$ existential non-deterministic steps of the algorithm. Next, the algorithm needs to verify that for all truth assignments γ (of any weight) to the variables in X, the instance $C(X, Y)[\gamma]$ of $\text{WSAT}(\text{CIRC})$ is a yes-instance. Using universal non-deterministic steps, the algorithm B guesses a truth assignment γ to the variables in X. We assume without loss of generality that for any γ the instance $C(X, Y)[\gamma]$ is of size n; if this instance is smaller, we can simply add some padding to ensure that it is of the right size. Then, using the guessed string $\alpha(n, k)$, the algorithm simulates the (deterministic) fpt-algorithm A_2 to decide whether $(C(X, Y)[\gamma], k) \in \text{WSAT}(\text{CIRC})$. Moreover, it verifies whether the truth assignment of weight k that is constructed by A_2 (if any) in fact satisfies $C(X, Y)[\gamma]$. Clearly, this can be done using universal non-deterministic steps, since deterministic steps are a special case of universal non-deterministic steps. Finally, the algorithm accepts if and only if $C(X, Y)[\gamma]$ has a satisfying assignment of weight k.

To see that the algorithm B correctly decides $\Pi_2^p[*k]$-$\text{WSAT}(\text{CIRC})$, we observe the following. If the algorithm accepts an instance (C, k), then it must be the case that for all truth assignments γ to the universal variables of C the algorithm verified that there exists a satisfying assignment of weight k for $C[\gamma]$. Therefore, $(C, k) \in \Pi_2^p[*k]$-$\text{WSAT}(\text{CIRC})$. Conversely, if B rejects an instance (C, k), it means that for all advice strings $\alpha(n, k)$ of length $f(k) \log n$, there is some truth assignment γ to the universal variables of C such that simulating A_2 with $\alpha(n, k)$

does not yield a satisfying assignment. Then, since by assumption, at least one such advice string $\alpha(n, k)$ leads to correct behavior of A_2, we get that there exists some truth assignment γ to the universal variables of C such that $C[\gamma]$ does not have a satisfying assignment of weight k. In other words, $(C, k) \notin \Pi_2^p[*k]$-WSAT(CIRC).

Since FPT/slice $\subseteq$ FPT/kernel $\subseteq$ FPT/log-kernel, this result directly gives us the following corollary.

Corollary 15.107. W[P] $\not\subseteq$ FPT/kernel *and* W[P] $\not\subseteq$ FPT/slice, *unless* $\Pi_2^p[*k, P] \subseteq \Sigma_2^p[k*]$.

The result of Proposition 15.106 can straightforwardly be extended to the other levels of the Weft hierarchy, resulting in additional analogues of the Karp-Lipton Theorem.

Proposition 15.108. *For each* $t \in \mathbb{N}^+ \cup \{SAT\}$, *it holds that* W[$t$] $\subseteq$ FPT/log-kernel *implies* $\Pi_2^p[*k, t] \subseteq \Sigma_2^p[k*]$.

Proof. The proof of Proposition 15.106 can straightforwardly be modified to show this result.

Corollary 15.109. *For each* $t \in \mathbb{N}^+ \cup \{SAT\}$, *it holds that* W[$t$] $\not\subseteq$ FPT/kernel *and* W[t] $\not\subseteq$ FPT/slice, *unless* $\Pi_2^p[*k, t] \subseteq \Sigma_2^p[k*]$.

In addition, the proof of Proposition 15.106 can straightforwardly be extended to the case of FPT/fpt, resulting in even more parameterized analogues of the Karp-Lipton Theorem.

Proposition 15.110. *For each* $t \in \mathbb{N}^+ \cup \{SAT, P\}$ *it holds that* W[t] $\not\subseteq$ FPT/fpt, *unless* $\Pi_2^p[*k, t] \subseteq$ para-Σ_2^p.

Proof. The proof of Proposition 15.106 can straightforwardly be modified to show this result.

To illustrate the connections that these parameterized analogues of the Karp-Lipton Theorem make between different areas of parameterized complexity theory (that at first sight might seem distant), we give an example. Consider the result of Proposition 15.60 (in Sect. 15.3.2 on page 318). This result states that the parameterized compilation problem CONSTRAINED-CLIQUE(sol.-size) is not fpt-size compilable, unless W[1] $\subseteq$ FPT/fpt. Using the result of Proposition 15.110, we can now connect the (in)compilability of this parameterized compilation problem to the relation between the parameterized complexity classes $\Pi_2^p[*k, 1]$ and para-Σ_2^p.

Corollary 15.111. CONSTRAINED-CLIQUE(sol.-size) *is not fpt-size compilable, unless* $\Pi_2^p[*k, 1] \subseteq$ para-Σ_2^p.

Notes

The results in this chapter appeared in a technical report [105].

Conclusions

Chapter 16
Open Problems and Future Research Directions

The theoretical investigation that we performed in this thesis, leaves a number of open questions and brings forth plenty of topics for future research. This includes theoretical issues, as well as questions about the possibility of applying the theoretical results in this thesis in an applied zhmic setting. In this chapter, we point out several of the most important topics for future research.

Outline of This Chapter

We begin in Sect. 16.1 by discussing several topics that are to be investigated in future research in order to study to what extent the theoretical results in this thesis can benefit practical solving methodologies for problems at the second level of the PH or higher.

Then, in Sect. 16.2, we discuss various theoretical (and technical) questions that the research in this thesis has brought forward.

In Sect. 16.3, we sketch a possible direction for future research that investigates various refined notions of fpt-reducibility to SAT that are based on bounds on the number of variables in the produced SAT instances (or to put it differently, based on bounds on the amount of non-determinism used).

In Sect. 16.4, we discuss topics for future research that are related to fpt-algorithms that have access to witness-producing SAT oracles. We briefly discussed this oracle model in Sect. 7.5. We also consider several questions that arise from our work and that are related to identifying (lower and upper) bounds on the number of calls to a SAT oracle that any fpt-algorithm needs to make to solve certain problems.

Finally, in Sect. 16.5, we propose several ideas that could be used to develop a theoretical framework that is able to provide a complexity distinction between problems in para-NP. These ideas are based on a non-deterministic generalization of the concept of generalized kernelization [28].

© Springer-Verlag GmbH Germany, part of Springer Nature 2019
R. de Haan: Parameterized Complexity in the Polynomial Hierarchy, LNCS 11880,
https://doi.org/10.1007/978-3-662-60670-4_16

16.1   Improving Solving Methods

The theoretical research in this thesis is unmistakably motivated by the prospect of using the concept of fpt-reductions to problems such as SAT—for which optimized, well-performing algorithms are available off-the-shelf—to develop solving methods for many problems at higher levels of the PH that come up in real-world settings. Therefore, one of the most important directions for future research is to employ the theoretical tools and results in this thesis to identify such cases. Our hope is that the algorithmic approaches of (1) using the viewpoint of parameterized complexity to pinpoint structure in the input that can be used to speed up algorithms and (2) putting to use the enormous engineering advances in SAT solvers (and similar algorithms for other problems) by encoding problem instances into SAT—that have both separately been tremendously successful—will lead to even further improvements for the efficiency of solving algorithms for problems in practically relevant settings when combined. Even if algorithms based on this combined methodology outperform existing algorithms only in particular settings, this could lead to tangible benefits.

At this point, there is only a single example of such a case where the combination of employing a fixed-parameter tractable algorithm to encode a problem into SAT, and subsequently solving the problem by calling a SAT solver, led to concrete, implemented algorithms that can compete with other state-of-the-art algorithms. This example is the problem of model checking for a fragment of the temporal logic LTL on very large Kripke structures (that are encoded succinctly). We discussed this problem and the potential of fpt-reductions to SAT to solve this problem in Sect. 4.2.4 and Chapter 9. Even though this is only a single example of a case where the possibility of fpt-reductions to SAT coincides with the development of practical algorithmic techniques based on SAT solvers, we are optimistic that this example indicates a more general relation between positive theoretical results and the possibility of solving methods that work well in real-world settings.

There are several lines of research that could be fruitful in the search for more cases where the combination of fixed-parameter tractable algorithms and SAT solvers can be used to design algorithms that improve over existing algorithms for problems that are relevant in practical settings. The first line of research that we like to point out consists of taking a theoretical result that establishes the possibility of an fpt-reduction to SAT and empirically investigating how well algorithms based on this result perform in practice. This involves implementing an algorithm that performs the reduction witnessed by the theoretical result, and that subsequently solves the problem by calling a SAT solver. Moreover, the implemented algorithm needs to be optimized its efficiency needs to be empirically tested on inputs that are representative for one or more particular problem settings.

A second promising research direction that has the potential of pointing out further settings where algorithms based on fpt-reductions to SAT can lead to practical improvements is to establish more theoretical results that establish the possibility of fpt-reductions to SAT. As a natural part of this pursuit, the structured parameterized complexity investigation of practically motivated problems at higher levels of the PH—that we initiated in this thesis—can be extended to additional parameterizations and to more problems. Moreover, to further facilitate this parameterized complexity investigation, it would be beneficial to develop additional general techniques that can be used to establish positive results in many different settings. A good example of such a general technique could be a 'meta-theorem' that states that every parameterized problem that can be expressed in a particular logic language is fpt-reducible to SAT (when parameterized by a particular parameter).

16.2 Open Theoretical Problems

We presented many technical results in this thesis. We introduced a lot of technical machinery, and obtaining a thorough understanding of this machinery involved overcoming many non-trivial obstacles. We tried to be as meticulous as is possible within the scope of a doctoral thesis. We aimed to give at least a basic answer to all of the most fundamental questions that presented themselves. Of course, we had to leave some technical questions to be answered in future research. Moreover, our results directly give rise to a range of interesting theoretical questions that are to be addressed in future research.

In this section, we describe a number of topics that came up in the various chapters of this thesis and that would be interesting to investigate in the future. In the remaining sections of this chapter, we turn to several interesting research questions that emerge from our research and that we want to address in more detail than the topics that we mention in this section.

16.2.1 Further Study of the $\Sigma_2^p[*k, t]$ Hierarchy

As we showed in Sect. 6.2, the k-$*$ hierarchy collapses into a single classe $\Sigma_2^p[k*]$. This collapse allowed us to focus on the single case of $\Sigma_2^p[k*]$, rather than spreading our attention over a whole hierarchy of subtly different classes, as in the case of the $\Sigma_2^p[*k, t]$ hierarchy. Moreover, the class $\Sigma_2^p[k*]$ enjoys some properties that made it easier to establish several technical results. For instance, the fact that the second quantifier block for the canonical problem $\Sigma_2^p[k*]$-WSAT is unweighted allowed us to show that $\Sigma_2^p[k*]$-hardness for this problem already holds when the input is restricted to 3DNF formulas. For the classes $\Sigma_2^p[*k, t]$, on the contrary, several striking questions remain open. We discuss several of them.

For the weighted satisfiability problems $\Sigma_2^p[*k]$-WSAT($\mathcal{C}$) that are canonical for the different classes $\Sigma_2^p[*k, t]$, we showed in several settings that the circuits in $\mathcal{C}$ can be transformed into one of various types of normal forms. For example, the problem $\Sigma_2^p[*k]$-WSAT(2DNF) is already $\Sigma_2^p[*k, 1]$-hard. However, for the classes

$\Sigma_2^p[*k, t]$, for $2 \le t \in \mathbb{N}^+$, we do not have such normalization results. It would be interesting to investigate whether normalization results hold that are similar to the ones for the classes W[t], for $2 \le t \in \mathbb{N}^+$, as is the case for $t = 1$. The techniques that we used to show the normalization result for $\Sigma_2^p[*k, 1]$, in Sect. 6.3.1, would need to be augmented significantly to obtain such results.

Another obvious question that arises is whether the $\Sigma_2^p[*k, t]$ hierarchy is proper, or whether any two distinct levels in fact coincide. It remains unclear even whether the equality $\Sigma_2^p[*k, t] = \Sigma_2^p[*k, t + 1]$, for any $t \in \mathbb{N}^+$, implies a collapse at higher levels than t. It would also be fascinating to determine what consequences a collapse of the $\Sigma_2^p[*k, t]$ would have. For instance, does the collapse of this hierarchy imply that other parameterized complexity classes coincide with each other or with classes of the $\Sigma_2^p[*k, t]$ hierarchy? Finding answers to these questions is likely to be very difficult, because similar questions remain open also for the Weft hierarchy. Since the $\Sigma_2^p[*k, t]$ hierarchy generalizes the Weft hierarchy, it is conceivable that answering any of these questions will involve more tedious technical work than answering the corresponding question for the Weft hierarchy.

A third aspect of the $*$-k hierarchy that would be interesting to study in future work is the way in which the classes in the hierarchy can be characterized using notions and problems other than the ones used to define the classes—that is, the weighted quantified satisfiability problems $\Sigma_2^p[*k]$-WSAT($\mathcal{C}$). For instance, can we characterize the classes $\Sigma_2^p[*k, t]$ using alternating Turing machines. For the class $\Sigma_2^p[*k, \text{P}]$, we provided an alternating Turing machine characterization in Sect. 6.3.4, but for the remaining classes of the $*$-k hierarchy, such a characterization is absent. Characterizations of the classes $\Sigma_2^p[*k, 1]$ and $\Sigma_2^p[*k, 2]$ in terms of alternating Turing machines could be similar in nature to the machine characterizations of the parameterized complexity classes W[1] and W[2] (see, e.g., [85]).

16.2.2 Further Study of the Relation Between Classes

We investigated how the new parameterized classes $\Sigma_2^p[k*]$ and $\Sigma_2^p[*k, t]$ relate to each other and to other parameterized complexity classes. In fact, we provided results that the classes $\Sigma_2^p[k*]$ and $\Sigma_2^p[*k, t]$ are indeed distinct from the classes depicted above and below the dashed gray lines in Fig. 5.1. For these separation results, we used different complexity-theoretic assumptions. Some of these assumptions are very commonly used, such as the assumption that NP $\ne$ co-NP (that we used to show that para-NP $\not\subseteq \Sigma_2^p[k*]$ and $\Sigma_2^p[k*] \not\subseteq$ para-NP in Sect. 6.4.1). For other results, we used assumptions that might not sound unreasonable, but that are stronger and less standard. An example of such an unusual complexity-theoretic assumption is the assumption that there exists no $2^{o(n)}$ time reduction from QSAT$_2$ to UNSAT, where n denotes the number of variables in the QSAT$_2$ instance. We used this assumption to show that $\Sigma_2^p[k*] \not\subseteq$ para-co-NP, in Sect. 14.2.2.

It would considerably strengthen our conviction that the parameterized complexity classes that we introduced are indeed different from classes that were known from the literature, if we could establish the separation results mentioned above using weaker and more common complexity-theoretic assumptions. On the other hand, it could also be the case that some of our separation results actually do not hold, and thus that some of the less usual complexity-theoretic assumptions that we used to demonstrate these results are also false. A proof that confirms the falsity of any of the complexity-theoretic assumptions that we used would certainly be interesting in and of itself, as we discussed in Sect. 1.5.3.

Another interesting topic for future research is the relation between the parameterized complexity classes $\Sigma_2^p[kk, t]$ and $A[2,t]$, for $t \in \mathbb{N}^+$. As we pointed out in Sect. 6.1.2, it holds that $A[2,t] \subseteq \Sigma_2^p[kk, t]$, for each $t \in \mathbb{N}^+$. Whether the converse inclusion also holds remains open.

16.2.3 Other Topics

In Sects. 7.3 and 14.2, we developed techniques to establish lower bounds on the number of NP oracle queries needed by fpt-algorithms to solve particular problems. For example, we showed that hardness for the class $\mathrm{FPT}^{NP}[few]$ can be used to show that a problem cannot be solved by an fpt-algorithm that uses $O(1)$ queries to an NP oracle. Moreover, this bound can be improved by inspecting the fpt-reductions used to show $\mathrm{FPT}^{NP}[few]$-hardness. For instance, for the problem MAJORITY-SAFETY-(agenda-size), we showed in Sect. 11.2.2 that any fpt-algorithm needs $\Omega(\log k)$ NP oracle queries to solve the problem. However, for all the problems that we considered in this thesis, there is a notable gap between the lower bounds and the upper bounds that we obtained for the number of oracle queries needed. As an illustration of this, for the problem MAJORITY-SAFETY(agenda-size), the lower bound is $\Omega(\log k)$, and the upper bound is $2^{O(k)}$. It would be interesting to investigate how to obtain tighter (lower and upper) bounds on the number of oracle queries needed by fpt-algorithms to solve various problems.

Another fascinating topic for future research is the relation between the non-uniform parameterized complexity classes—that we considered in Chap. 15—and other notions in the field of parameterized complexity theory. We investigated various connections that these classes have with the concept of parameterized compilability (in Sect. 15.3) and the classes $\Sigma_2^p[*k]$ and $\Sigma_2^p[*k, t]$ (in Sect. 15.4). An example of another notion whose relation to non-uniform parameterized complexity would be interesting to study is that of randomized fpt-algorithms.

Yet another idea for future research is to study the power of fpt-algorithms that can query an oracle that is weaker than the NP oracles that we considered. For instance, instead of equipping fpt-algorithms with an NP oracle, we could provide access to a W[1] oracle (and put various bounds on the number of oracle queries that the algorithm can make). This would give rise to the class $\mathrm{FPT}^{W[1]}[few]$ (and variants of it). Clearly, it holds that $\mathrm{FPT}^{W[1]}[few] \subseteq \mathrm{FPT}^{NP}[few]$, and this inclusing is strict (unless $P = NP$). However, the exact relation between $\mathrm{FPT}^{W[1]}[few]$ and other parameterized complexity classes seems to be more intricate to establish. Also, it would be interesting to identify natural parameterized problems that are complete for $\text{FPT}^{W[1]}[\text{few}]$ (and its variants).

16.2.4   Generalizing to Higher Levels of the PH

As we explained in Sect. 1.5.2, the focus in this thesis lies on problems at the second level of the PH (rather than at higher levels of the PH). However, the concept of using fpt-reductions to a problem in NP can be used for problems at any level of complexity. Moreover, one could also investigate whether problems at higher levels of the PH can be fpt-reduced to a lower level of the PH. For example, for a Σ_3^p-complete problem, an fpt-reduction to QSAT_2 could be useful for developing improved algorithms.

The parameterized complexity classes $\Sigma_2^p[k*]$ and $\Sigma_2^p[*k, t]$ can straightforwardly be generalized to higher levels of the PH. In Appendix B, we begin an investigation of variants of these parameterized complexity classes at arbitrary levels of the PH, and their relation to each other and to other parameterized complexity classes. It would be interesting to extend the exploratory study in Appendix B to a more extensive analysis. Additionally, it would be interesting to find natural problems that are complete for the variants of $\Sigma_2^p[k*]$ and $\Sigma_2^p[*k, t]$ at higher levels of the PH. This is likely to be challenging, as natural problems that are complete for the third level of the PH are already in short supply.

16.3   Limited Non-determinism

For the reductions to SAT that we studied in this thesis, we only considered bounds on the running time of the reductions (namely, we required them to run in fixed-parameter tractable time), but we did not consider bounds on the size of the SAT instances produced by the reductions. This means that the SAT instances produced by these reductions can contain $f(k)n^c$ propositional variables, where k is the parameter value, n is the input size, f is a computable function and c is a constant. In many cases, modern SAT solvers have no trouble dealing with instances that contain many variables. However, the larger the number of variables, the larger the search space that the SAT solvers have to navigate. For this reason, it might well be that a more refined notion of fpt-reducibility to SAT more adequately captures the cases where the combination of fixed-parameter tractable algorithms and SAT solving algorithms leads to practically efficient solving methods. Such a refined notion of fpt-reducibility to SAT could place restrictions on the number of variables in the produced SAT instances. For instance, one could consider reductions that produce SAT instances containing only $O(n)$ variables.

Such restrictions on the number of variables in the resulting satisfiability instances can also be characterized differently, when considering the satisfiability problem for Boolean circuits (CIRCUIT-SAT) as target of the fpt-reductions. Fixed-parameter tractable reductions to CIRCUIT-SAT correspond to non-deterministic fpt-algorithms that solve the problem directly—this is a known characterization of the class para-NP [85, Chapter 2]. Moreover, the number of variables in the instance of CIRCUIT-SAT that is produced by the reduction is exactly the same as the number of non-deterministic bits used by the fpt-algorithm that solves the problem. For instance, an fpt-reduction from a parameterized problem Q to CIRCUIT-SAT that produces instances with only $O(n)$ variables corresponds to a non-deterministic fpt-algorithm that solves Q using $O(n)$ non-deterministic bits.

Various bounds on the number of variables in the produced SAT instances present themselves as reasonable choices for refined notions of fpt-reducibility to SAT. For instance, we could require the fpt-reductions to produce instances with only $O(n)$ variables, or with only $f(k)n$ variables, where n is the input size, k is the parameter value, and f is a computable function. Other reasonable options are to require the number of variables to be bounded by $O(n^2)$ or $f(k)n^2$.

One might be tempted to also consider the restriction that for each reduction there must be some constant c such that the propositional formulas produced by this reduction contain at most $O(n^c)$ variables. This, however, is not a restriction from the setting where instances are allowed to contain $f(k)n^c$ variables, for the following reason. From the inequality $a \cdot b \leq a^2 + b^2$ (for all $a, b \in \mathbb{N}$), we get that $f(k)n^c \leq f(k)^2 + n^{2c}$. Suppose that we have an fpt-reduction that produces a formula φ with $f(k)n^c \leq f(k)^2 + n^{2c}$ variables. We can then straightforwardly modify this reduction as follows. After producing φ, it (arbitrarily) chooses $f(k)^2$ variables, and iterates over all possible truth assignments to these variables (there are $2^{f(k)^2}$ such assignments). Then, for each such truth assignment α, it computes the formula $\varphi[\alpha]$ that results from applying α to φ and simplifying the formula. Then, the resulting formula φ' is the disjunction of all these formulas $\varphi[\alpha]$. The formula φ' can be constructed in fpt-time, and contains at most n^{2c} variables. By a similar argument, we know that each fpt-reduction that produces propositional formulas with $f(k)n$ variables can be simulated by a reduction that produces formulas with only n^2 variables.

It would be riveting and useful to investigate what problems we can solve using these various restricted notions of fpt-reducibility to SAT. In order to satisfactorily study this, a structured investigation would need to be set up. This investigation would include finding natural parameterized problems that can be solved using the different notions of fpt-reducibility to SAT. Moreover, a robust framework would have to be developed with parameterized complexity classes (within para-NP) that capture the problems that are fpt-reducible to SAT, with various restrictions on the number of variables in the formulas produced by the fpt-reductions. The relation between these parameterized complexity classes and their relation to other parameterized complexity classes could then be analyzed, in order to establish the possibilities and limits of the different notions of fpt-reducibility to SAT in more detail. Finally, an important aspect of the different refined notions of fpt-reducibility to SAT that would have to be investigated experimentally is the potential that these notions have for practical applications.

16.4   Witness Oracles

In Sect. 7.5, we already briefly considered fixed-parameter tractable algorithms that have access to a SAT oracle that is more powerful than the typical oracles that only answer whether a propositional formula is satisfiable or unsatisfiable. These more powerful oracles return a satisfying assignment for propositional formulas that are satisfiable. (For a more formal definition of such *witness oracles*, we refer to Sect. 7.5.) Arguably, this oracle model is more reasonable in a theoretical framework that is motivated by the great performance of SAT solving algorithms in practice. This is because SAT solvers actually do return a satisfying assignment when given a satisfiable formula as input. An important direction for future research is to investigate the power that this more powerful theoretical oracle model yields.

As we showed in Sect. 7.5, when considering decision problems, witness oracles do not give additional power over yes-no oracles—at least not for the upper bounds on the number of oracle queries that we considered: $f(k)$, $O(\log n)$, or $O(n^c)$, where n denotes the input size, k is the parameter value, f is a computable function and c is a constant. Therefore, in order to adequately study the power offered by witness oracles, one needs to consider search problems (and consequently, consider reductions that transform solutions of one problem into solutions of another problem).

It is well known that even a single call to a witness oracle is more powerful than $O(\log n)$ calls to a yes-no oracle, unless $P \neq NP$ [99, Theorem 5.4]. From this, it immediately follows that fixed-parameter tractable algorithms that can query a witness oracle $f(k)$ times can solve strictly more problems than fixed-parameter tractable algorithms that can query a yes-no oracle $f(k)$ times, for instance. However, to obtain a more detailed understanding of the solving power of fixed-parameter tractable algorithms with access to a witness NP oracle, a structured investigation is needed.

The main product of such an investigation would be a fine-grained framework with parameterized complexity classes that capture what parameterized (search) problems can be solved in fixed-parameter tractable time using different amounts of queries to a witness NP oracle. Moreover, such a framework of parameterized complexity classes could be supported by identifying natural parameterized search problems that populate the various parameterized complexity classes. In addition, an essential part of such a framework would be easy-to-use lower bound tools, that can be used to determine whether a particular parameterized problem cannot be solved in fixed-parameter tractable time using a particular number of queries to a witness NP oracle. Possibly, such lower bound results could take the form of hardness for parameterized complexity classes that are developed for this purpose.

In addition to returning satisfying assignments for satisfiable formulas, essentially all modern SAT solvers are also able to return a proof of unsatisfiability for unsatisfiable formulas. In fact, from the execution trace of a CDCL SAT solver when given an unsatisfiable CNF formula as input, one can directly extract a resolution refutation of the formula. Therefore, it might also be interesting to investigate an even stronger oracle model, where the oracle returns a satisfying assignment for satisfiable formulas and returns a proof of unsatisfiability for unsatisfiable formulas. However, in the worst case, such proofs of unsatisfiability are of super-polynomial size, unless NP = co-NP. Therefore, it might be reasonable to place restrictions on the answers given by the oracle. For instance, one could consider oracles that return a polynomial-size proof of unsatisfiability when given an unsatisfiable formula, if such a proof exists, and otherwise simply returns "no".

In classical complexity theory, typically only search problems are considered where the size of solutions grows linearly (or polynomially) in the input size. For instance, the size of solutions for SAT (which consist of satisfying assignments for a propositional formula) are linear in the size of the input. However, in the setting of parameterized problems, it often makes sense to consider search problems where solutions are of size $f(k) \log n$, where n denotes the input size, k is the parameter value, and f is some computable function. A good example of such a problem is the problem of finding a repair set of size at most k, given an unsatisfiable set Φ of formulas, if such a repair set exists—we considered this problem in Sect. 10.2. Solutions for this problem consist of a list of k pointers to formulas in Φ, which can be described using $k \log n$ bits. It would be interesting to see if the additional power yielded by witness oracles differs for search problems where the size of solutions is, say, $f(k) \log n$, rather than $O(n)$ or $O(n^c)$.

Another interesting theoretical question for future research is what the power is of parallel queries to a witness NP oracle. For instance, what is the relative power of fixed-parameter tractable algorithms that can make $f(k)$ (adaptive) queries to a witness NP oracle and fixed-parameter tractable algorithms that can make $f(k)$ queries to a witness NP oracle in parallel?

16.5 Non-deterministic Kernelization

In Sect. 16.3, we contemplated the idea that a more refined notion of fpt-reducibility to SAT more adequately captures the cases where the combination of using fixed-parameter tractable algorithms and employing SAT solvers could lead to practically efficient solving methods. There, we considered restricted variants of fpt-reductions to SAT where the number of variables in the produced SAT instances is bounded. In this section, we look at another possible way of defining a refined notion of fpt-reducibility to SAT that is based on a non-deterministic variant of the concept of generalized kernelization [28]. We hope that the ideas that we describe in this section can be used to develop a framework to make more fine-grained complexity distinctions between problems in para-NP, in a similar way that the notion of kernelizations (with different size bounds) can be used to distinguish between problems in FPT.

Moreover, such a framework could potentially be used to help establish bounds on the function f in the running time of fpt-reductions to SAT for different parameterized problems. An fpt-reduction to SAT that runs in time $f(k)n^c$, where f is a function that is computable but that grows extremely quickly—for instance, the Ackermann function—is likely to be inefficient even for small parameter values, even though it qualifies as an fpt-reduction to SAT. Therefore, it would be useful to develop techniques that can be used to provide bounds on the functions f needed in the running time of fpt-reductions to SAT for different problems.

We introduce the following notion of non-deterministic kernelization. Essentially, this is a non-deterministic version of the known concept of generalized kernelization, which allows the target problem to be different from the original problem [28]. Intuitively, a non-deterministic kernelization for a parameterized problem Q takes non-deterministic time polynomial in the size of the original instance x to compute an equivalent instance x' of the problem Q' that is of size at most $f(k)$.

Definition 16.1 (Non-deterministic kernelization). *Let $Q \subseteq \Sigma^* \times \mathbb{N}$ be a parameterized problem. Let C be a classical complexity class. We call an algorithm A a non-deterministic kernelization for L into C (or ndK into C, for short) if there exists a problem $Q' \in C$, polynomials p, q, and a computable function f such that for each instance $(x, k) \in \Sigma^* \times \mathbb{N}$, it holds that:*

1. *for each $\beta \in \mathbb{B}^{p(|x|)}$, $A(x, k, \beta)$ is computable in time $q(|x|)$;*
2. *$(x, k) \in Q$ if and only if for some $\beta \in \mathbb{B}^{p(|x|)}$ it holds that $A(x, k, \beta) \in Q'$; and*
3. *for each $\beta \in \mathbb{B}^{p(|x|)}$, $|A(x, k, \beta)| \leq f(k)$.*

The notion of non-deterministic kernelization allows us to give an alternative characterization of the class para-NP.

Proposition 16.2. *Let DEC be the class of all decidable decision problems. Then para-NP coincides with the class of all parameterized problems that have an ndK into DEC.*

Proof. Firstly, we show that any parameterized problem that has an ndK into DEC is in para-NP. Take a parameterized problem Q that has an ndK into DEC. That is, there are an algorithm A, polynomials p, q, a computable function f, and a problem $Q' \in$ DEC as specified in the definition of non-deterministic kernelizations. We describe a non-deterministic fpt-time algorithm to solve Q. Take an instance $(x, k) \in \Sigma^* \times \mathbb{N}$ of Q. Firstly, the algorithm non-deterministically guesses a string $\beta \in \mathbb{B}^{p(|x|)}$. Then, the algorithm computes $x' = A(x, k, \beta)$ in polynomial time. We know that $|x'| \leq f(k)$. Moreover, we know that $x' \in Q'$ if and only if $(x, k) \in Q$. Finally, the algorithm decides whether $x' \in Q'$. This is possible, since $Q' \in$ DEC. Moreover, since $|x'| \leq f(k)$, the time it takes to decide whether $x' \in Q'$ depends only on k. Therefore, this algorithm runs in non-deterministic fixed-parameter tractable time. From this, we can conclude that $Q \in$ para-NP.

Conversely, suppose that $Q \in$ para-NP. We show that Q has an ndK into DEC. We know that there exists a non-deterministic fpt-time algorithm B that solves Q. That is, for each instance $(x, k) \in \Sigma^* \times \mathbb{N}$, the algorithm B runs in time $g(k)r(n)$, where n denotes the size of x, g is a computable function, and r is a polynomial. Without loss of generality, we may assume that $r(n) \geq n$ for all $n \in \mathbb{N}$. We describe an algorithm A, polynomials p, q, a computable function f, and a problem $Q' \in$ DEC as specified in the definition of non-deterministic kernelizations.

Without loss of generality, we may assume that Q is non-trivial, i.e., that $Q_1 \neq \emptyset$ and $Q_1 \neq \Sigma^*$, where $Q_1 = \{x \in \Sigma^* : (x, k) \in Q, k \in \mathbb{N}\}$. We fix arbitrary $x_0 \in$

$\Sigma^* \backslash Q_1$ and $x_1 \in Q_1$. We set $p(n) = r(n)^2$, we let $Q' = Q_1$. Moreover, we let $f(k) = g(k) + |x_0| + |x_1|$. The algorithm A uses the bits in $\beta \in \mathbb{B}^{p(n)}$ to simulate the first $p(n)$ steps of the execution of B on (x, k). If B halts within $p(n)$ steps and rejects the input, then A returns x_0. Similarly, if B halts within $p(n)$ steps and accepts the input, then A returns x_1. Otherwise, if B does not halt within $p(n)$ steps, A returns x. Clearly, q can be chosen in such a way that A runs in time $q(n)$.

We show that for each instance $(x, k) \in \Sigma^* \times \mathbb{N}$, it holds that $(x, k) \in Q$ if and only if for some $\beta \in \mathbb{B}^{p(n)}$ it holds that $A(x, k, \beta) \in Q'$, and that for each $\beta \in \mathbb{B}^{p(n)}$, $|A(x, k, \beta)| \leq f(k)$. Suppose that $(x, k) \in Q$. Moreover, suppose that there is some sequence of non-deterministic choices that lead B to accept (x, k) in time $p(n)$. Let β be a string that corresponds to this sequence of non-deterministic choices. Then, A returns $x_1 \in Q'$ when given the string β. Clearly, $|x_1| \leq f(k)$. Alternatively, suppose that for every sequence of non-deterministic choices, B needs time more than $p(n) = r(n)^2$ to accept (x, k). Then, $n \leq r(n) \leq g(k)$. Moreover, then there is some β such that $A(x, k, \beta) = x$. Since $(x, k) \in Q$, it holds that $x \in Q'$. Also, since $n \leq g(k)$, we have that $|x| \leq f(k)$. Conversely, suppose that $(x, k) \notin Q$. Take an arbitrary $\beta \in \mathbb{B}^{p(n)}$. Suppose that A halts within $p(n)$ steps when given (x, k) and β. Then A returns $x_0 \notin Q'$. Clearly, $|x_0| \leq f(k)$. Alternatively, suppose that A needs more than $p(n) = r(n)^2$ steps to halt when given (x, k) and β. Then, $n \leq r(n) \leq g(k)$. Moreover, then A returns x. Since $(x, k) \notin Q$, we know that $x \notin Q'$. Also, since $n \leq g(k)$, we have that $|x| \leq f(k)$. This concludes our description of the ndK for Q into DEC.

In order to define different levels of complexity within the class para-NP, we consider various restricted variants of non-deterministic kernelizations. There are various natural ways to define such restricted notions of non-deterministic kernelizations. Firstly, one can restrict the computational complexity of the target problems. That is, one can consider non-deterministic kernelizations into different complexity classes $\mathcal{C} \subseteq$ DEC. For instance, one can consider the subclass of para-NP consisting of all parameterized problems that have an ndK into PSPACE.

Secondly, one can consider restricted variants of non-deterministic kernelizations by putting restrictions on the size of the product of the kernelization function A. That is, we can restrict the function f in Definition 16.1. At this point, we consider two such size bounds on the kernelizations. Let A be a non-deterministic kernelization into some complexity class $\mathcal{C}$, as specified in Definition 16.1. Moreover, let f be the function that bounds the size of the result of A, i.e., $|A(x, k, \beta)| \leq f(k)$ for all x, k, β. If f is a polynomial, we call A a *non-deterministic polynomial kernelization into* $\mathcal{C}$ (or *ndPK into* $\mathcal{C}$). If f is a linear function, we call A a *non-deterministic linear kernelization into* $\mathcal{C}$ (or *ndLK into* $\mathcal{C}$). We can also combine these two different ways of restricting non-deterministic kernelizations to obtain additional restricted variants of ndKs. For instance, one could consider ndPKs into PSPACE.

In many cases, the existence of such restricted variants of non-deterministic kernelizations for some parameterized problem directly yields upper bounds on the running time needed for fpt-reductions to SAT for this problem. Examples of this are the following two results.

Proposition 16.3. *Let* EXP *be the class of all decision problems that can be decided in time* $2^{n^{O(1)}}$. *Moreover, let* Q *be a parameterized problem that has an ndPK into* EXP. *Then there exists a* $2^{k^{O(1)}} n^{O(1)}$ *time encoding into SAT for* Q.

Proof (sketch). By an argument similar to the one used in the proof of Proposition 16.2, it can be shown that any parameterized problem Q that has an ndPK into EXP can be solved by a non-deterministic algorithm that runs in time $2^{k^{O(1)}} n^{O(1)}$. We know that the class of parameterized problems that can be solved by a non-deterministic algorithm that runs in time $2^{k^{O(1)}} n^{O(1)}$ coincides with the class of parameterized problems for which there exists a $2^{k^{O(1)}} n^{O(1)}$ time encoding into SAT. Therefore, we know that there exists a $2^{k^{O(1)}} n^{O(1)}$ time encoding into SAT for Q.

Proposition 16.4. *Let* E *be the class of all decision problems that can be decided in time* $2^{O(n)}$. *Moreover, let* Q *be a parameterized problem that has an ndLK into* E. *Then there exists a* $2^{O(k)} n^{O(1)}$ *time encoding into SAT for* Q.

Proof (sketch). By an argument similar to the one used in the proof of Proposition 16.2, it can be shown that any parameterized problem Q that has an ndLK into E can be solved by a non-deterministic algorithm that runs in time $2^{O(k)} n^{O(1)}$. We know that the class of parameterized problems that can be solved by a non-deterministic algorithm that runs in time $2^{O(k)} n^{O(1)}$ coincides with the class of parameterized problems for which there exists a $2^{O(k)} n^{O(1)}$ time encoding into SAT. Therefore, we know that there exists a $2^{O(k)} n^{O(1)}$ time encoding into SAT for Q.

As these restricted variants of non-deterministic kernelizations directly give rise to a more fine-grained distinction between parameterized problems within para-NP, it would be an interesting topic for future research to develop a robust theory to classify parameterized problems in para-NP according to what kind of non-deterministic kernelizations they admit. For this, one needs to investigate the various different types of restricted variants of non-deterministic kernelizations in more detail.

Moreover, one would need to explore appropriate notions of reductions—that is, reductions that satisfy the property that if a parameterized problem Q has a certain kind of non-deterministic kernelization and another parameterized problem Q' is reducible to Q, then Q' also has this kind of non-deterministic kernelization. One example of such an appropriate concept of reductions is the notion of polynomial time and parameter transformations, where the reduction runs in polynomial (deterministic) time and the parameter value increases only by a polynomial amount [29]. Whenever a parameterized problem Q has an ndPK into some class C and there is a polynomial time and parameter transformation from another parameterized problem Q' to Q, then Q' also has an ndPK into C. To obtain a similar preservation property for ndLKs, one could consider a variant of such transformations where the running time and increase in parameter value are linear, rather than polynomial. Additionally, one could consider non-deterministic generalizations of these polynomial (or linear) time and parameter transformations.

A central research question in the development of a structured theory of the possibilities and limits of the different types of non-deterministic kernelizations would be to identify natural parameterized problems that inhabit the different levels of complexity within para-NP characterized by the different types of non-deterministic kernelizations. Also, it would be useful to develop theoretical tools that can be used to show that particular parameterized problems do not admit certain types of non-deterministic kernelizations. Finally, in order to establish how well the theoretical levels of complexity within para-NP—that are induced by the various kinds of non-deterministic kernelizations—correspond to how well solving methods based on fpt-reductions to SAT perform in practical settings, extensive experimental research is needed.

Chapter 17
Conclusion

In this chapter, we finish the main part of the thesis. We do so by summarizing our results, in Sect. 17.1. Moreover, in Sect. 17.2, we consider the impact of our research (and of future research based on the work in this thesis) for the area of computer science and engineering. To allow a wider audience to understand and appreciate our results, we describe the research impact in layperson's terms.

17.1 Summary

In this thesis, we took on the problem of enabling and starting a structured complexity-theoretic investigation of the possibilities and limits of using fpt-reductions to SAT to solve intractable problems at higher levels of the Polynomial Hierarchy. Concretely, we addressed four shortcomings of previous parameterized complexity research related to fpt-reductions to SAT.

1. Of the numerous possible incarnations of the general scheme of fpt-reductions to SAT, only the simplest had been considered—and only in very few cases.
2. Theoretical tools to rule out fpt-reductions to SAT were underdeveloped. That is, only in obvious cases previously known parameterized complexity tools could be used to show that an fpt-reduction to SAT is not possible.
3. A fine-grained parameterized complexity toolbox to adequately characterize the complexity of parameterized variants of problems at higher levels of the Polynomial Hierarchy was lacking.
4. Only for a very small number of problems it had been studied whether they admit an fpt-reduction to SAT.

We confronted these shortcomings with the following contributions. We established a better understanding of the different possible forms of fpt-reductions to SAT that can be employed, and the power that they provide.

- We provide the first structured parameterized complexity investigation that concentrates on fpt-reducibility to SAT as a positive result.

© Springer-Verlag GmbH Germany, part of Springer Nature 2019
R. de Haan: Parameterized Complexity in the Polynomial Hierarchy, LNCS 11880,
https://doi.org/10.1007/978-3-662-60670-4_17

- We considered parameterized complexity classes that capture problems that can be solved using various types of fpt-reductions to SAT: para-NP, para-co-NP, para-BH, $FPT^{NP}[few]$, para-Θ_2^p, and para-Δ_2^p.
- We developed theoretical tools that can be used to provide lower bounds on the number of oracle queries that is needed for any fpt-algorithm to solve certain problems.

We enriched the theoretical machinery that can be used to rule out fixed-parameter tractable reductions to SAT. The additional tools that we develop for this purpose can also be used in subtle cases where previously known tools were not applicable.

- We show that parameterized problems that are hard for the parameterized complexity class A[2] do not admit fpt-reductions to SAT—under the assumption that there exists no subexponential-time reduction from the problem $QSAT_2(3DNF)$ to SAT.

We developed new parameterized complexity classes that can be used to accurately characterize the complexity of parameterized variants of problems at higher levels of the Polynomial Hierarchy. Using these new classes, subtly different levels of complexity can be distinguished, which could not be done using previously known parameterized complexity classes.

- We firstly argued that new parameterized complexity classes were needed, by pointing out natural parameterized variants of a problem known from the literature, and showing that these parameterized problems are not complete for any of the known classes (under various complexity-theoretic assumptions).
- We develop the parameterized complexity classes $\Sigma_2^p[k*]$ and $\Sigma_2^p[*k, t]$—which are generalizations of the well-known classes of the Weft hierarchy—that map out the parameterized complexity landscape between the first and the second level of the Polynomial Hierarchy.
- We provided a solid understanding of the phenomena underlying these classes by giving alternative characterizations of the classes in terms of (i) first-order logic model checking, and (ii) alternating Turing machines—the latter characterization yields an analogue of the Cook-Levin Theorem.
- We showed that the new parameterized complexity classes can be used to characterize the complexity of the parameterized problems that we showed earlier not to be complete for any previously known parameterized complexity classes.
- We further substantiated the completeness theory yielded by the newly developed parameterized complexity classes by showing that there are many natural parameterized variants of known problems at higher levels of the Polynomial Hierarchy that are complete for the new classes.
- We demonstrated that the new classes add to the parameterized complexity toolbox for providing evidence that fpt-reductions to SAT are not possible in certain cases. We did so by showing that hardness for these classes rules out fpt-reductions to SAT under weaker assumptions than for the case of hardness for A[2].

- We related the newly developed parameterized complexity classes to other areas of (parameterized) complexity theory. In particular, we drew connections to the topics of subexponential-time algorithms and non-uniform parameterized complexity—the latter connection resulted in parameterized analogues of the Karp-Lipton Theorem.
- We generalized the new classes to arbitrary levels of the Polynomial Hierarchy—resulting in a parameterized complexity landscape that can provide a very fine-grained analysis of parameterized variants of problems at any level of the Polynomial Hierarchy.
- We additionally developed another new parameterized complexity class that is a natural parameterized variant of the classical complexity class PSPACE.

Using the parameterized complexity tools that are based on previously known parameterized complexity classes as well as on the newly developed parameterized complexity classes, we initiated a structured analysis of the potential of fpt-reductions to SAT to solve problems at higher levels of the Polynomial Hierarchy that originate in various domains of computer science and artificial intelligence.

- We demonstrated that several non-trivial results from the literature can in fact be considered as fpt-reductions to SAT. In the setting of symbolic temporal logic model checking, one of the most competitive algorithms to solve the problem can in fact be seen as an fpt-reduction to SAT.
- We showed that productive techniques from parameterized complexity—that is, the concepts of treewidth and backdoors—can be used to develop fpt-reductions to SAT.
- We investigated for what parameters fpt-reductions to SAT are possible for many natural parameterizations of a wide range of problems from various areas of computer science and artificial intelligence—the problems that we investigated include problems related to answer set programming, temporal logic model checking, propositional planning, and minimization of propositional formulas.

In short, in this thesis we paved the way for future parameterized complexity research that further investigates whether the method of fpt-reductions to SAT can be used to develop useful algorithms for problems from computer science, artificial intelligence, and other domains. We developed a parameterized complexity toolbox that contains tools both for (1) identifying cases where various possible implementations of the general scheme of fpt-reductions to SAT can be used, and for (2) providing a detailed characterization of the complexity of parameterized problems for which fpt-reductions to SAT are unlikely to exist. We demonstrated how this toolbox can be used for many relevant parameterized problems to determine whether an fpt-reduction to SAT is possible, and if so, what kind of fpt-reduction to SAT is needed to solve the problem. Using the new toolbox, we initiated a structured investigation of the parameterized complexity of problems at higher levels of the Polynomial Hierarchy by considering many natural parameterized variants of such problems and analyzing their complexity.

17.2 Research Impact

We describe the impact of our research (and of future research that builds upon our results) for the area of computer science and engineering. To allow readers without a background in theoretical computer science and computational complexity to better understand and value our results, we avoid specialized technical language as much as possible in this description. Moreover, to allow this section to be read by a general audience, independently from the rest of the thesis, we also briefly repeat our main contributions.

Intractable Problems
Computers are amazingly efficient for finding solutions to many important problems, but there are also extremely many important problems for which computers often need a paralyzing amount of time to find a solution (centuries or longer). Such *intractable problems* do not only show up in many places in the sciences—physics, biology, and economics, to name a few—but can also have life or death consequences—for example, for the problem of finding the optimal match of organ donors to recipients. There is no universal way of quickly finding solutions for these problems, according to our current belief and understanding of computation. So it is extremely important to develop methods that can solve these problems as fast as possible in as many cases as possible.

Combining Methods
Computer science research has resulted in the development of many solving methods that perform well in different cases. One of these methods is the use of so-called *fixed-parameter tractable algorithms*, that are designed to work well whenever the problem input shows one of various kinds of structure. Another of these methods involves *encoding problems into* one of several *key problems*, for which concentrated efforts have produced highly optimized algorithms. An example of such a key problem is the problem of propositional satisfiability (or SAT, for short). Both of these methods have been studied intensively over the last two decades, and produce solutions in a practical amount of time in many settings.

It has been proposed recently to combine these two approaches in a potentially more powerful solving method. This combined method could be useful in many more cases where the separate approaches both have trouble getting solutions. In order to study the possibilities and limits of this new combined approach, new mathematical theory is needed. Current theoretical tools are aimed at the analysis of both of the methods separately, but do not work well when studying the combination of the two methods.

A New Theoretical Framework
We extended current theoretical tools from the paradigm of *parameterized complexity theory* to a new theoretical framework that makes it possible to investigate the potential and the limits of the approach of solving problems using *fixed-parameter tractable encodings into SAT* (or any of the other key problems). Using this frame-

work, it is possible to better identify the cases where this approach might work in practice, and the cases where other methods are needed to solve the problem.

We also started to use this framework to analyze under what conditions the new combined approach could work well for many intractable problems from a wide range of settings in computer science and artificial intelligence.

Impact for Applications

With the many possible solving methods that are available, computer scientists and engineers that work on finding solutions for concrete intractable problems benefit greatly from guidance in what methods will work well for their problems. The theoretical research in this thesis can be used to advance the knowledge that developers use to guide their choices for designing algorithms. For example, when there is little structure in the problem inputs faced by a developer that could be used to come up with a fixed-parameter tractable encoding into SAT, the developer is probably better off directing their efforts towards trying other methods to solve the problem.

It can be argued that the guidance that is offered directly by mathematical computational complexity research (which includes the results in this thesis) is limited for practical algorithm design. One reason for this is the fact that this research involves abstracting away from many of the real-world details that play a role in the practice of solving computational problems. Still, computational complexity analysis is needed to provide the foundations for a better understanding of what methods work well in practice for solving what problems and under what conditions. The use of theoretical knowledge for computer science applications is neither to be overestimated, nor to be underestimated.

The theoretical analysis of concrete problems from many settings in computer science and artificial intelligence—using the new theoretical framework—that we began in this thesis, could serve as a starting point for future research on the practical knowledge that algorithm developers need to make better design choices.

Impact for Fundamental Computer Science Research

Our present understanding of the practical limit of what computers can and cannot do has been greatly influenced by theoretical work that has been done over the last handful of decades. The work in this thesis is a continuation of this foundational research, and forms a step forward in our collective effort of understanding the power of computation.

More concretely, the tools that we constructed and the results that we established in this thesis contribute to a more detailed understanding of the subtle levels of difficulty that are inherent to intractable problems. The parameterized complexity classes that we developed map out a finer landscape of the inherent difficulty in finding solutions for a range of intractable problems than was previously known. Also, we illustrated how these classes can be used to carry out a detailed complexity investigation for computational problems from diverse domains.

This thesis is the first work that systematically investigates the combination of several research topics in the area of theoretical computer science. As such, it naturally brings to light many concrete, relevant, and interesting questions for future research. We hope that this work sparks a rich body of future research.

Appendix A
Compendium of Parameterized Problems

In Chaps. 8–13, we used the new parameterized complexity classes that we introduced in Chaps. 6 and 7 (and in Sect. 9.3) to analyze natural problems from numerous domains in artificial intelligence and knowledge representation and reasoning, and computer science at large. In this appendix, we give an overview of parameterized complexity results for parameterized variants of problems in the PH, in the form of a compendium. This compendium is similar in concept to the compendia by Schaefer and Umans [179] and Cesati [38], that also list problems along with their computational complexity. The problems in this list are categorized by area.

A.1 Propositional Logic

A.1.1 Weighted Quantified Boolean Satisfiability

The following problems are parameterized variants of the problem $QSAT_2$ where the weight of assignments for the first quantifier block is restricted.

$\Sigma_2^p[k*]$-WSAT(CIRC)

Instance: A Boolean circuit C over two disjoint sets X and Y of variables, and a positive integer k.
Parameter: k.
Question: Does there exist a truth assignment α to X of weight k such that for all truth assignments β to Y the assignment $\alpha \cup \beta$ satisfies C?

Complexity: $\Sigma_2^p[k*]$-complete (by definition).

© Springer-Verlag GmbH Germany, part of Springer Nature 2019
R. de Haan: Parameterized Complexity in the Polynomial Hierarchy, LNCS 11880,
https://doi.org/10.1007/978-3-662-60670-4

$\Sigma_2^p[k*]$-WSAT(FORM), or simply $\Sigma_2^p[k*]$-WSAT

Instance: A quantified Boolean formula $\phi = \exists X.\forall Y.\psi$, where ψ is quantifier-free, and an integer k.
Parameter: k.
Question: Does there exist a truth assignment α to X of weight k such that for all truth assignments β to Y the assignment $\alpha \cup \beta$ satisfies ψ?

Complexity: $\Sigma_2^p[k*]$-complete (Theorem 6.2).

$\Sigma_2^p[k*]$-WSAT(3DNF)

Instance: A quantified Boolean formula $\phi = \exists X.\forall Y.\psi$, where ψ is a quantifier-free formula in 3DNF, and an integer k.
Parameter: k.
Question: Does there exist a truth assignment α to X of weight k such that for all truth assignments β to Y the assignment $\alpha \cup \beta$ satisfies ψ?

Complexity: $\Sigma_2^p[k*]$-complete (Theorem 6.2).

$\Sigma_2^p[k*]$-WSAT$^{\leq k}$

Instance: A quantified Boolean formula $\phi = \exists X.\forall Y.\psi$, where ψ is quantifier-free, and an integer k.
Parameter: k.
Question: Does there exist an assignment α to X with weight at most k, such that for all truth assignments β to Y the assignment $\alpha \cup \beta$ satisfies ψ?

Complexity: $\Sigma_2^p[k*]$-complete (Proposition 6.9).

$\Sigma_2^p[k*]$-WSAT$^{\geq k}$

Instance: A quantified Boolean formula $\phi = \exists X.\forall Y.\psi$, where ψ is quantifier-free, and an integer k.
Parameter: k.
Question: Does there exist an assignment α to X with weight at least k, such that for all truth assignments β to Y the assignment $\alpha \cup \beta$ satisfies ψ?

Complexity: para-Σ_2^p-complete (Proposition 6.10).

$\Sigma_2^P[k*]$-WSAT^{n-k}

Instance: A quantified Boolean formula $\phi = \exists X.\forall Y.\psi$, where ψ is quantifier-free and where $|X| = n$, and an integer k.
Parameter: k.
Question: Does there exist an assignment α to X of weight $n - k$, such that for all truth assignments β to Y the assignment $\alpha \cup \beta$ satisfies ψ?

Complexity: $\Sigma_2^P[k*]$-complete (Claim 1 in the proof of Proposition 10.4).

The following problems are parameterized variants of the problem QSAT$_2$ where the weight of assignments for the second quantifier block is restricted.

$\Sigma_2^P[*k]$-WSAT(CIRC)

Instance: A Boolean circuit C over two disjoint sets X and Y of variables, and a positive integer k.
Parameter: k.
Question: Does there exist a truth assignment α to X such that for all truth assignments β to Y of weight k the assignment $\alpha \cup \beta$ satisfies C?

Complexity: $\Sigma_2^P[*k, \text{P}]$-complete (by definition).

$\Sigma_2^P[*k]$-WSAT(CIRC$_{t,u}$)

Instance: A Boolean circuit C over two disjoint sets X and Y of variables that has depth u and weft t, and a positive integer k.
Parameter: k.
Question: Does there exist a truth assignment α to X such that for all truth assignments β to Y of weight k the assignment $\alpha \cup \beta$ satisfies C?

Complexity: $\Sigma_2^P[*k, t]$-complete (by definition).

$\Sigma_2^P[*k]$-WSAT(FORM)

Instance: A quantified Boolean formula $\phi = \exists X.\forall Y.\psi$, where ψ is quantifier-free, and an integer k.
Parameter: k.
Question: Does there exist a truth assignment α to X such that for all truth assignments β to Y of weight k the assignment $\alpha \cup \beta$ satisfies C?

Complexity: $\Sigma_2^P[*k, \text{SAT}]$-complete (by definition).

$\Sigma_2^p[*k]$-WSAT(2DNF)

Instance: A quantified Boolean formula $\phi = \exists X.\forall Y.\psi$, where ψ is a quantifier-free formula in 2DNF, and an integer k.
Parameter: k.
Question: Does there exist a truth assignment α to X such that for all truth assignments β to Y of weight k the assignment $\alpha \cup \beta$ satisfies C?

Complexity: $\Sigma_2^p[*k, 1]$-complete (Theorem 6.21).

A.1.2 Quantified Boolean Satisfiability

The following problems are parameterized variants of the problem QSAT (and its restriction QSAT$_2$). For the definition of universal and existential incidence treewidth of quantified Boolean formulas of the form $\exists X.\forall Y.\psi$, where ψ is a quantifier-free formula in DNF, we refer to Sect. 4.2.1.

QSAT(#∀-vars)

Instance: A quantified Boolean formula φ.
Parameter: The number of universally quantified variables of φ.
Question: Is φ true?

Complexity: para-NP-complete (Proposition 4.3).

QSAT$_2$(∀-itw)

Instance: A quantified Boolean formula $\varphi = \exists X.\forall Y.\psi$, where ψ is a quantifier-free formula in DNF, with universal incidence treewidth k.
Parameter: k.
Question: Is φ true?

Complexity: para-NP-complete (Proposition 4.4).

QSAT$_2$(∃-itw)

Instance: A quantified Boolean formula $\varphi = \exists X.\forall Y.\psi$, where ψ is a quantifier-free formula in DNF, with existential incidence treewidth k.
Parameter: k.
Question: Is φ true?

Complexity: para-Σ_2^p (Proposition 4.8).

QSAT(level)

Instance: A quantified Boolean formula $\varphi = \exists X_1 \forall X_2 \exists X_3 \ldots Q_k X_k \psi$, where Q_k is a universal quantifier if k is even and an existential quantifier if k is odd, and where ψ is quantifier-free.
Parameter: k.
Question: Is φ true?

Complexity: PH(level)-complete (by definition).

A.1.3 Minimization for DNF Formulas

The following parameterized problems are related to minimizing DNF formulas and minimizing implicants of DNF formulas. For a definition of term-wise subformulas of DNF formulas and implicants, we refer to Sect. 10.1.

DNF-MINIMIZATION(reduction size)

Instance: A DNF formula φ of size n, and an integer m.
Parameter: $n - m$.
Question: Does there exist a term-wise subformula φ' of φ of size m such that $\varphi \equiv \varphi'$?

Complexity: $\Sigma_2^p[k*]$-complete (Proposition 10.5).

DNF-MINIMIZATION(core size)

Instance: A DNF formula φ of size n, and an integer m.
Parameter: m.
Question: Does there exist a term-wise subformula φ' of φ of size m such that $\varphi \equiv \varphi'$?

Complexity: in $\Sigma_2^p[k*]$ (Proposition 10.7), in $\mathrm{FPT}^{\mathrm{NP}}[\mathrm{few}]$ (Proposition 10.11), para-co-NP-hard (Proposition 10.6).

SHORTEST-IMPLICANT-CORE(implicant size)

Instance: A DNF formula φ, an implicant C of φ of size n, and an integer m.
Parameter: n.
Question: Does there exist an implicant $C' \subseteq C$ of φ of size m?

Complexity: para-co-NP-complete (Proposition 10.1).

SHORTEST-IMPLICANT-CORE(core size)

Instance: A DNF formula φ, an implicant C of φ of size n, and an integer m.
Parameter: m.
Question: Does there exist an implicant $C' \subseteq C$ of φ of size m?

Complexity: $\Sigma_2^p[k*]$-complete (Proposition 10.3).

SHORTEST-IMPLICANT-CORE(reduction size)

Instance: A DNF formula φ, an implicant C of φ of size n, and an integer m.
Parameter: $n - m$.
Question: Does there exist an implicant $C' \subseteq C$ of φ of size m?

Complexity: $\Sigma_2^p[k*]$-complete (Proposition 10.4).

A.1.4 Other

The following problems are various other parameterized problems that are related to propositional satisfiability. For a definition of the problems BH_i-SAT, we refer to Sect. 7.2.1. For a definition of $\max^1(\varphi, X)$, we refer to Sect. 7.4. For a definition of (lexicographically) X-maximal models, we refer to Sect. 7.2.2. Finally, for a definition of the minimum repair size of a set of formulas, we refer to Sect. 10.2.

BH(level)-SAT

Instance: A positive integer k and a sequence $(\varphi_1, \ldots, \varphi_k)$ of propositional formulas.
Parameter: k.
Question: $(\varphi_1, \ldots, \varphi_k) \in BH_k$-SAT?

Complexity: $FPT^{NP}[few]$-complete (Theorem 7.5).

BOUNDED-SAT-UNSAT-DISJUNCTION

Instance: A family $(\varphi_i, \varphi_i')_{i \in [k]}$ of pairs of propositional formulas.
Parameter: k.
Question: Is there some $\ell \in [k]$ such that $(\varphi_\ell, \varphi_\ell') \in$ SAT-UNSAT?

Complexity: $FPT^{NP}[few]$-complete (Proposition 7.8).

LOCAL-MAX-MODEL

Instance: A satisfiable propositional formula φ, a subset $X \subseteq \text{Var}(\varphi)$ of variables, and a variable $w \in X$.
Parameter: $|X|$.
Question: Is there a model of φ that sets a maximal number of variables in X to true (among all models of φ) and that sets w to true?

Complexity: $\text{FPT}^{\text{NP}}[\text{few}]$-complete (Proposition 7.11).

LOCAL-MAX-MODEL-COMPARISON

Instance: Two satisfiable propositional formulas φ_1 and φ_2, a positive integer k, and a subset $X \subseteq \text{Var}(\varphi_1) \cap \text{Var}(\varphi_2)$ of k variables.
Parameter: k.
Question: $\max^1(\varphi_1, X) = \max^1(\varphi_2, X)$?

Complexity: $\text{FPT}^{\text{NP}}[\text{few}]$-complete (Proposition 7.14).

ODD-LOCAL-MAX-MODEL

Instance: A propositional formula φ, and a subset $X \subseteq \text{Var}(\varphi)$ of variables.
Parameter: $|X|$.
Question: Do the X-maximal models of φ set an odd number of variables in X to true?

Complexity: $\text{FPT}^{\text{NP}}[\text{few}]$-complete (Proposition 7.9).

ODD-LOCAL-LEX-MAX-MODEL

Instance: A propositional formula φ, and a subset $X \subseteq \text{Var}(\varphi)$ of variables.
Parameter: $|X|$.
Question: Does the lexicographically X-maximal model of φ set an odd number of variables in X to true?

Complexity: $\text{FPT}^{\text{NP}}[\text{few}]$-complete (Proposition 7.10).

ODD-BOUNDED-REPAIR-SET

Instance: A set Φ of propositional formulas, and a positive integer k.
Parameter: k.
Question: Is the minimum repair size of Φ both odd and at most k?

Complexity: $\text{FPT}^{\text{NP}}[\text{few}]$-complete (Proposition 10.15).

ODD-BOUNDED-INCONSISTENT-SET

Instance: An inconsistent set Φ of propositional formulas, and a
positive integer k.
Parameter: k.
Question: Is the minimum size of an inconsistent subset Φ' of Φ both odd
and at most k?

Complexity: $\Sigma_2^p[k*]$-hard (Proposition 10.14).

A.2   Knowledge Representation and Reasoning and Artificial Intelligence

A.2.1   Disjunctive Answer Set Programming

The following problems are parameterized variants of the consistency problem for
disjunctive answer set programming. For a detailed definition of disjunctive answer
set programming and of these parameterized problems, we refer to Sect. 5.1.

ASP-CONSISTENCY(norm.bd-size)

Instance: A disjunctive logic program P.
Parameter: The size of the smallest normality-backdoor for P.
Question: Does P have an answer set?

Complexity: para-NP-complete [82].

ASP-CONSISTENCY(#cont.atoms)

Instance: A disjunctive logic program P.
Parameter: The number of contingent atoms of P.
Question: Does P have an answer set?

Complexity: para-co-NP-complete (Proposition 5.3).

ASP-CONSISTENCY(#cont.rules)

Instance: A disjunctive logic program P.
Parameter: The number of contingent rules of P.
Question: Does P have an answer set?

Complexity: $\Sigma_2^p[k*]$-complete (Theorem 6.3).

ASP-CONSISTENCY(#disj.rules)

Instance: A disjunctive logic program P.
Parameter: The number of disjunctive rules of P.
Question: Does P have an answer set?

Complexity: $\Sigma_2^p[*k, \text{P}]$-complete (Theorem 6.24).

ASP-CONSISTENCY(#non-dual-normal.rules)

Instance: A disjunctive logic program P.
Parameter: The number of non-dual-normal rules of P.
Question: Does P have an answer set?

Complexity: $\Sigma_2^p[*k, \text{P}]$-complete (Theorem 6.26).

ASP-CONSISTENCY(max.atom.occ.)

Instance: A disjunctive logic program P.
Parameter: The maximum number of times that any atom occurs in P.
Question: Does P have an answer set?

Complexity: para-Σ_2^p-complete (Proposition 5.4).

A.2.2 Abductive Reasoning

The following problems are related to abductive reasoning. For more details about abduction and the parameters that are used, we refer to Sect. 8.2.

ABDUCTION(Horn-bd-size)

Input: An abduction instance $\mathcal{P} = (V, H, M, T)$, and a positive integer m.
Parameter: The size of the smallest Horn-backdoor of T.
Question: Does there exist a solution S of $\mathcal{P}$ of size at most m?

Complexity: para-NP-complete [173].

ABDUCTION(Krom-bd-size)

Input: An abduction instance $\mathcal{P} = (V, H, M, T)$, and a positive integer m.
Parameter: The size of the smallest Krom-backdoor of T.
Question: Does there exist a solution S of $\mathcal{P}$ of size at most m?

Complexity: para-NP-complete [173].

ABDUCTION(#non-Horn-clauses)

Input: An abduction instance $\mathcal{P} = (V, H, M, T)$, and a positive integer m.
Parameter: The number of clauses of T that are not Horn clauses.
Question: Does there exist a solution S of $\mathcal{P}$ of size at most m?

Complexity: $\Sigma_2^p[*k, \text{P}]$-complete (Proposition 8.3).

ABDUCTION(#non-Krom-clauses)

Input: An abduction instance $\mathcal{P} = (V, H, M, T)$, and a positive integer m.
Parameter: The number of clauses of T that are not Krom clauses.
Question: Does there exist a solution S of $\mathcal{P}$ of size at most m?

Complexity: $\Sigma_2^p[*k, 1]$-complete (Proposition 8.6).

A.2.3 Constraint Satisfaction

The following problems are related to constraint satisfaction. For a definition of k-robust satisfiability, we refer to Sect. 8.3.

SMALL-CSP-UNSAT-SUBSET

Instance: A CSP instance $\mathcal{I}$, and a positive integer k.
Parameter: k.
Question: Is there an unsatisfiable subset $\mathcal{I}' \subseteq \mathcal{I}$ of size k?

Complexity: A[2]-complete (Proposition 4.11).

ROBUST-CSP-SAT

Instance: A CSP instance $\mathcal{I}$, and a positive integer k.
Parameter: k.
Question: Is $\mathcal{I}$ k-robustly satisfiable?

Complexity: $\Pi_2^p[k*]$-complete (Proposition 8.8).

A.2.4 Planning

The following problems are related to planning with uncertainty in the initial state and to a variant of planning where the number of achieved goals is to be maximized. For more details, we refer to Chap. 12.

POLYNOMIAL-PLANNING(uncertainty)

Instance: A planning instance $\mathbb{P} = (V, V_u, D, A, I, G)$ containing additional variables V_u that are *unknown* in the initial state, i.e., for all $v \in V_u$ it holds that $I(v) = \mathbf{u}$.
Parameter: $|V_u| + |D|$.
Question: Is there a plan of polynomial length for $\mathbb{P}$ that works for all complete initial states I_0, i.e., for each possible way of completing I with a combination of values for variables in V_u?

Complexity: para-NP-complete (Proposition 12.1).

POLYNOMIAL-PLANNING-ESSENTIAL-ACTION(uncertainty)

Instance: A planning instance $\mathbb{P} = (V, V_u, D, A, I, G)$ with unknown variables V_u.
Parameter: $|V_u| + |D|$.
Question: Is there a plan of polynomial length for $\mathbb{P}$ that uses a_0 and works for all complete initial states I_0, but there is no such plan for $\mathbb{P}$ without using a_0?

Complexity: para-DP-complete (Proposition 12.2).

BOUNDED-UNCERTAIN-PLANNING

Instance: A planning instance $\mathbb{P} = (V, V_u, D, A, I, G)$ with unknown variables V_u, and an integer k.
Parameter: k.
Question: Is there a plan of length k for $\mathbb{P}$ that works for all complete initial states I_0?

Complexity: $\Sigma_2^p[k*]$-complete (Proposition 12.3).

POLYNOMIAL-PLANNING(bounded-deviation)

Instance: A planning instance $\mathbb{P} = (V, V_u, D, A, I, G)$ with unknown variables V_u, and an integer d.
Parameter: d.
Question: Is there a plan of polynomial length for $\mathbb{P}$ that works for each complete initial state I_0 where at most d unknown variables deviate from the base value?

Complexity: $\Sigma_2^p[*k, \mathrm{P}]$-complete (Proposition 12.4).

POLYNOMIAL-OPTIMIZATION-PLANNING(#soft.goals)

Instance: A planning instance $\mathbb{P} = (V, D, A, I, G_h, G_s)$ with a hard goal G_h and a soft goal G_s, and an integer m.
Parameter: $|G_s|$.
Question: Does there exist a plan of length at most m that is optimal for $\mathbb{P}$ (w.r.t. m), and that satisfies an odd number of variables of the soft goal?

Complexity: $\text{FPT}^{\text{NP}}[\text{few}]$-complete (Proposition 12.6).

A.3 Model Checking

A.3.1 First-Order Logic

The following problems are parameterized variants of the model checking problem for first-order logic.

$\Sigma_2^{\text{p}}[k*]$-MC

Instance: A first-order logic sentence $\varphi = \exists x_1, \ldots, x_k.\forall y_1, \ldots, y_n.\psi$ over a vocabulary τ, where ψ is quantifier-free, and a finite τ-structure $\mathcal{A}$.
Parameter: k.
Question: Is it the case that $\mathcal{A} \models \varphi$?

Complexity: $\Sigma_2^{\text{p}}[k*]$-complete (Theorem 6.6).

MC[FO](quant.alt.)

Instance: A first-order logic sentence $\varphi = Q_1 x_1.Q_2 x_2 \ldots Q_n x_n.\psi$ over a vocabulary τ, where ψ is quantifier-free, and a finite τ-structure $\mathcal{A}$.
Parameter: The number of quantifier alternations of φ, i.e., $|\{ i \in [n-1] : Q_i \neq Q_{i+1} \}|$.
Question: Is it the case that $\mathcal{A} \models \varphi$?

Complexity: PH(level)-complete (Proposition 9.5).

A.3.2 Temporal Logics

The following problems are parameterized variants of the model checking problem for (various fragments) of the temporal logics LTL, CTL and CTL*. For more details, we refer to Chap. 9.

SYMBOLIC-MC*[$\mathcal{L}$]

Input: A symbolically represented Kripke structure $\mathcal{M}$, $rd(\mathcal{M})$ in unary, and an $\mathcal{L}$ formula φ.
Parameter: $|\varphi|$.
Question: $\mathcal{M} \models \varphi$?

Complexity: para-NP-complete for $\mathcal{L} = \text{LTL}\backslash\text{U},\text{X}$ (Proposition 9.3), PH(level)-complete for each $\mathcal{L} \in \{\text{CTL}, \text{CTL}\backslash\text{U}, \text{CTL}\backslash\text{X}, \text{CTL}\backslash\text{U},\text{X}, \text{CTL}^\star\backslash\text{U},\text{X}\}$ (Theorems 9.9 and 9.10), and para-PSPACE-complete for each $\mathcal{L} \in \{\mathcal{L}_0, \mathcal{L}_0\backslash\text{U}, \mathcal{L}_0\backslash\text{X} : \mathcal{L}_0 \in \{\text{LTL}, \text{CTL}^\star\}\}$ (Theorems 9.6 and 9.7).

SYMBOLIC-MC[$\mathcal{L}$]

Input: A symbolically represented Kripke structure $\mathcal{M}$, and an $\mathcal{L}$ formula φ.
Parameter: $|\varphi|$.
Question: $\mathcal{M} \models \varphi$?

Complexity: para-PSPACE-complete, for each $\mathcal{L} \in \{\mathcal{L}_0, \mathcal{L}_0\backslash\text{U}, \mathcal{L}_0\backslash\text{X}, \mathcal{L}_0\backslash\text{U},\text{X} : \mathcal{L}_0 \in \{\text{LTL}, \text{CTL}, \text{CTL}^\star\}\}$ (Proposition 9.2).

A.4 Computational Social Choice

A.4.1 Agenda Safety in Judgment Aggregation

The following problems are related to the problem of agenda safety for the majority rule in judgment aggregation. For more details, we refer to Sect. 11.2.

MAJORITY-SAFETY(formula-size)

Instance: An agenda Φ.
Parameter: $\ell = \max\{|\varphi| : \varphi \in \Phi\}$.
Question: Is Φ safe for the majority rule?

Complexity: para-Π_2^p-complete (Proposition 11.5).

MAJORITY-SAFETY(degree)

Instance: An agenda Φ.
Parameter: The degree d of Φ.
Question: Is Φ safe for the majority rule?

Complexity: para-Π_2^p-complete (Proposition 11.5).

MAJORITY-SAFETY(degree + formula size)

Instance: An agenda Φ.
Parameter: $d + \ell$.
Question: Is Φ safe for the majority rule?

Complexity: para-Π_2^p-complete (Proposition 11.5).

MAJORITY-SAFETY(agenda-size)

Instance: An agenda Φ.
Parameter: $|\Phi|$.
Question: Is Φ safe for the majority rule?

Complexity: FPT^{NP}[few]-complete (Propositions 11.6 and 11.8).

MAJORITY-SAFETY(f-tw)

Instance: An agenda Φ, where each $\varphi \in \Phi$ is a CNF formula.
Parameter: The formula primal treewidth of Φ.
Question: Is Φ safe for the majority rule?

Complexity: in FPT (Proposition 11.10).

MAJORITY-SAFETY(c-tw)

Instance: An agenda Φ, where each $\varphi \in \Phi$ is a CNF formula.
Parameter: The clausal primal treewidth of Φ.
Question: Is Φ safe for the majority rule?

Complexity: para-Π_2^p-complete (Proposition 11.11).

MAJORITY-SAFETY(f-tw*)

Instance: An agenda Φ, where each $\varphi \in \Phi$ is a CNF formula.
Parameter: The formula incidence treewidth of Φ.
Question: Is Φ safe for the majority rule?

Complexity: para-Π_2^p-complete (Proposition 11.12).

MAJORITY-SAFETY(c-tw*)

Instance: An agenda Φ, where each $\varphi \in \Phi$ is a CNF formula.
Parameter: The clausal incidence treewidth of Φ.
Question: Is Φ safe for the majority rule?

Complexity: para-Π_2^p-complete (Proposition 11.13).

MAJORITY-SAFETY(c.e.-size)

Instance: An agenda Φ, and a positive integer k.
Parameter: k.
Question: Does every inconsistent subset Φ' of Φ of size k itself have an inconsistent subset of size at most 2?

Complexity: $\Pi_2^p[k*]$-hard (Proposition 11.15).

A.4.2 Computing Outcomes in Judgment Aggregation

The following problems are related to the problem of computing outcomes for the Kemeny rule in judgment aggregation. In particular, these are parameterized variants of the problems FB-OUTCOME-KEMENY and CB-OUTCOME-KEMENY. For more details, we refer to Sect. 11.3.

The parameters that we consider for the problem FB-OUTCOME-KEMENY are defined as follows. For an instance $(\Phi, \Gamma, J, L, L_1, \ldots, L_u)$ of FB-OUTCOME-KEMENY with $J = (J_1, \ldots, J_p)$, we let $n = |[\Phi]|$, $m = \max\{ |\varphi| : \varphi \in [\Phi] \}$, $c = |\Gamma|$, $p = |J|$, and $h = \max\{ d(J_i, J_{i'}) : i, i' \in [p] \}$.

FB-OUTCOME-KEMENY(c, n, m)

Instance: An agenda Φ with an integrity constraint Γ, a profile $J \in \mathcal{J}(\Phi, \Gamma)^+$ and subsets $L, L_1, \ldots, L_u \subseteq \Phi$ of the agenda, with $u \geq 0$.
Parameter: $c + n + m$.
Question: Is there a judgment set $J^* \in \text{Kemeny}(J)$ such that $L \subseteq J^*$ and $L_i \not\subseteq J^*$ for each $i \in [u]$?

Complexity: in FPT (Proposition 11.18).

FB-OUTCOME-KEMENY(h, p)

Instance: An agenda Φ with an integrity constraint Γ, a profile $J \in \mathcal{J}(\Phi, \Gamma)^+$ and subsets $L, L_1, \ldots, L_u \subseteq \Phi$ of the agenda, with $u \geq 0$.
Parameter: $h + p$.
Question: Is there a judgment set $J^* \in \text{Kemeny}(J)$ such that $L \subseteq J^*$ and $L_i \not\subseteq J^*$ for each $i \in [u]$?

Complexity: FPT^{NP}[few]-complete (Propositions 11.16 and 11.23–11.25).

FB-OUTCOME-KEMENY(n)

Instance: An agenda Φ with an integrity constraint Γ, a profile $\boldsymbol{J} \in \mathcal{J}(\Phi, \Gamma)^+$ and subsets $L, L_1, \ldots, L_u \subseteq \Phi$ of the agenda, with $u \geq 0$.
Parameter: n.
Question: Is there a judgment set $J^* \in \text{Kemeny}(\boldsymbol{J})$ such that $L \subseteq J^*$ and $L_i \nsubseteq J^*$ for each $i \in [u]$?

Complexity: $\text{FPT}^{\text{NP}}[\text{few}]$-complete (Propositions 11.17 and 11.23–11.25).

FB-OUTCOME-KEMENY(h, n, m, p)

Instance: An agenda Φ with an integrity constraint Γ, a profile $\boldsymbol{J} \in \mathcal{J}(\Phi, \Gamma)^+$ and subsets $L, L_1, \ldots, L_u \subseteq \Phi$ of the agenda, with $u \geq 0$.
Parameter: $h + n + m + p$.
Question: Is there a judgment set $J^* \in \text{Kemeny}(\boldsymbol{J})$ such that $L \subseteq J^*$ and $L_i \nsubseteq J^*$ for each $i \in [u]$?

Complexity: $\text{FPT}^{\text{NP}}[\text{few}]$-complete (Propositions 11.16, 11.17 and 11.23).

FB-OUTCOME-KEMENY(c, h, n, p)

Instance: An agenda Φ with an integrity constraint Γ, a profile $\boldsymbol{J} \in \mathcal{J}(\Phi, \Gamma)^+$ and subsets $L, L_1, \ldots, L_u \subseteq \Phi$ of the agenda, with $u \geq 0$.
Parameter: $c + h + n + p$.
Question: Is there a judgment set $J^* \in \text{Kemeny}(\boldsymbol{J})$ such that $L \subseteq J^*$ and $L_i \nsubseteq J^*$ for each $i \in [u]$?

Complexity: $\text{FPT}^{\text{NP}}[\text{few}]$-complete (Propositions 11.16, 11.17 and 11.24).

FB-OUTCOME-KEMENY(c, h, m, p)

Instance: An agenda Φ with an integrity constraint Γ, a profile $\boldsymbol{J} \in \mathcal{J}(\Phi, \Gamma)^+$ and subsets $L, L_1, \ldots, L_u \subseteq \Phi$ of the agenda, with $u \geq 0$.
Parameter: $c + h + m + p$.
Question: Is there a judgment set $J^* \in \text{Kemeny}(\boldsymbol{J})$ such that $L \subseteq J^*$ and $L_i \nsubseteq J^*$ for each $i \in [u]$?

Complexity: $\text{FPT}^{\text{NP}}[\text{few}]$-complete (Propositions 11.16, 11.17 and 11.25).

FB-OUTCOME-KEMENY(c, h, m)

Instance: An agenda Φ with an integrity constraint Γ, a profile $J \in \mathcal{J}(\Phi, \Gamma)^+$ and subsets $L, L_1, \ldots, L_u \subseteq \Phi$ of the agenda, with $u \geq 0$.
Parameter: $c + h + m$.
Question: Is there a judgment set $J^* \in$ Kemeny(J) such that $L \subseteq J^*$ and $L_i \not\subseteq J^*$ for each $i \in [u]$?

Complexity: para-Θ_2^p-complete (Corollary 11.21).

FB-OUTCOME-KEMENY(c, m, p)

Instance: An agenda Φ with an integrity constraint Γ, a profile $J \in \mathcal{J}(\Phi, \Gamma)^+$ and subsets $L, L_1, \ldots, L_u \subseteq \Phi$ of the agenda, with $u \geq 0$.
Parameter: $c + m + p$.
Question: Is there a judgment set $J^* \in$ Kemeny(J) such that $L \subseteq J^*$ and $L_i \not\subseteq J^*$ for each $i \in [u]$?

Complexity: para-Θ_2^p-complete (Proposition 11.22).

The parameters that we consider for the problem CB-OUTCOME-KEMENY are defined as follows. For an instance $(\mathcal{I}, \Gamma, r, l, l_1, \ldots, l_u)$ of CB-OUTCOME-KEMENY with $r = (r_1, \ldots, r_p)$, we let $n = |\mathcal{I}|$, $c = |\Gamma|$, $p = |r|$, and $h = \max\{d(r_i, r_{i'}) : i, i' \in [p]\}$.

CB-OUTCOME-KEMENY(c)

Instance: A set $\mathcal{I}$ of issues with an integrity constraint Γ, a profile $r \in \mathcal{R}(\mathcal{I}, \Gamma)^+$ and partial ballots $l, l_1, \ldots, l_u$ (for $\mathcal{I}$), with $u \geq 0$.
Parameter: c.
Question: Is there a ballot $r^* \in$ Kemeny(r) such that l agrees with r^* and each l_i does not agree with r^*?

Complexity: in FPT (Proposition 11.27).

CB-OUTCOME-KEMENY(n)

Instance: A set $\mathcal{I}$ of issues with an integrity constraint Γ, a profile $r \in \mathcal{R}(\mathcal{I}, \Gamma)^+$ and partial ballots $l, l_1, \ldots, l_u$ (for $\mathcal{I}$), with $u \geq 0$.
Parameter: n.
Question: Is there a ballot $r^* \in$ Kemeny(r) such that l agrees with r^* and each l_i does not agree with r^*?

Complexity: in FPT (Proposition 11.26).

CB-OUTCOME-KEMENY(h)

Instance: A set $\mathcal{I}$ of issues with an integrity constraint Γ, a profile
$r \in \mathcal{R}(\mathcal{I}, \Gamma)^+$ and partial ballots $l, l_1, \ldots, l_u$ (for $\mathcal{I}$), with $u \geq 0$.
Parameter: h.
Question: Is there a ballot $r^* \in \text{Kemeny}(r)$ such that l agrees with r^* and
each l_i does not agree with r^*?

Complexity: in XP (Proposition 11.28).

CB-OUTCOME-KEMENY(h, p)

Instance: A set $\mathcal{I}$ of issues with an integrity constraint Γ, a profile
$r \in \mathcal{R}(\mathcal{I}, \Gamma)^+$ and partial ballots $l, l_1, \ldots, l_u$ (for $\mathcal{I}$), with $u \geq 0$.
Parameter: $h + p$.
Question: Is there a ballot $r^* \in \text{Kemeny}(r)$ such that l agrees with r^* and
each l_i does not agree with r^*?

Complexity: W[SAT]-hard (Proposition 11.29).

CB-OUTCOME-KEMENY(p)

Instance: A set $\mathcal{I}$ of issues with an integrity constraint Γ, a profile
$r \in \mathcal{R}(\mathcal{I}, \Gamma)^+$ and partial ballots $l, l_1, \ldots, l_u$ (for $\mathcal{I}$), with $u \geq 0$.
Parameter: p.
Question: Is there a ballot $r^* \in \text{Kemeny}(r)$ such that l agrees with r^* and
each l_i does not agree with r^*?

Complexity: para-Θ_2^p-complete (Proposition 11.30).

A.5   Graph Theory

A.5.1   *Extending 3-Colorings*

The following parameterized problems are related to extending 3-colorings of the
leaves of a graph to proper 3-colorings of the entire graph.

3-COLORING-EXTENSION(degree)

Instance: A graph $G = (V, E)$ with n leaves, and an integer m.
Parameter: The degree of G.
Question: Can each 3-coloring that assigns a color to exactly m leaves of G (and to no other vertices) be extended to a proper 3-coloring of G?

Complexity: para-Π_2^p-complete (Proposition 13.1).

3-COLORING-EXTENSION(#leaves)

Instance: A graph $G = (V, E)$ with n leaves, and an integer m.
Parameter: n.
Question: Can each 3-coloring that assigns a color to exactly m leaves of G (and to no other vertices) be extended to a proper 3-coloring of G?

Complexity: para-NP-complete (Proposition 13.2).

3-COLORING-EXTENSION(#col.leaves)

Instance: A graph $G = (V, E)$ with n leaves, and an integer m.
Parameter: m.
Question: Can each 3-coloring that assigns a color to exactly m leaves of G (and to no other vertices) be extended to a proper 3-coloring of G?

Complexity: $\Pi_2^p[k*]$-complete (Theorem 13.4).

3-COLORING-EXTENSION(#uncol.leaves)

Instance: A graph $G = (V, E)$ with n leaves, and an integer m.
Parameter: $n - m$.
Question: Can each 3-coloring that assigns a color to exactly m leaves of G (and to no other vertices) be extended to a proper 3-coloring of G?

Complexity: para-Π_2^p-complete (Proposition 13.3).

A.5.2 Extending Cliques

The following parameterized problem is related to extending cliques of a subgraph of a graph to cliques of the entire graph.

SMALL-CLIQUE-EXTENSION
Instance: A graph $G = (V, E)$, a subset $V' \subseteq V$, and an integer k. *Parameter:* k. *Question:* Is it the case that for each clique $C \subseteq V'$, there is some k-clique D of G such that $C \cup D$ is a $(|C| + k)$-clique?
Complexity: $\Pi_2^p[*k, 1]$-complete (Proposition 13.5).

A.6 Alternating Turing Machines

The following parameterized problems are related to the halting problem for alternating Turing machines. For more details, we refer to Sect. 6.2.3.

$\Sigma_2^p[k*]$-TM-HALT^m
Instance: An $\exists\forall$-machine $\mathbb{M}$ with m tapes, and positive integers $k, t \geq 1$. *Parameter:* k. *Question:* Does $\mathbb{M}$ halt on the empty string with existential cost k and universal cost t?
Complexity: $\Sigma_2^p[k*]$-complete (Theorem 6.11).

$\Sigma_2^p[k*]$-TM-HALT*
Instance: An $\exists\forall$-machine $\mathbb{M}$ with m tapes, and positive integers $k, t \geq 1$. *Parameter:* k. *Question:* Does $\mathbb{M}$ halt on the empty string with existential cost k and universal cost t?
Complexity: $\Sigma_2^p[k*]$-complete (Theorem 6.11).

Appendix B
Generalization to Higher Levels of the Polynomial Hierarchy

Ow! My brains!

— Douglas Adams, *The Hitchhiker's Guide to the Galaxy [4]*

In Parts II and III of this thesis, we developed tools that enable a comprehensive parameterized complexity investigation of problems in the Polynomial Hierarchy, and we carried out such an investigation for many natural problems that arise in a variety of settings in computer science and artificial intelligence. In this investigation, we concentrated on problems at the second level of the PH, as the second level is most densely populated with natural problems (see, e.g., [179]). The tools that we developed are also mainly aimed at the second level of the PH. For instance, in Chap. 6, we introduced the parameterized complexity classes $\Sigma_2^p[k*]$ and $\Sigma_2^p[*k, t]$ that are based on weighted variants of the quantified Boolean satisfiability problem that is canonical for the second level of the Polynomial Hierarchy. However, there is no reason why this research should remain confined to the second level of the PH.

In this appendix, we generalize the parameterized complexity tools that we developed to higher levels of the PH. In particular, we consider variants of the classes $\Sigma_2^p[k*]$ and $\Sigma_2^p[*k, t]$ for higher levels of the PH.

Outline of This Chapter

In Sect. B.1, we define the complexity classes $\Sigma_i^p[w, t]$, for arbitrary $i \in \mathbb{N}$, $w \in \{*, k\}^i$, and $t \in \mathbb{N}^+ \cup \{P, SAT\}$. These are parameterized weighted variants of the classes Σ_i^p at arbitrary levels of the PH. The word w describes the weight restriction to the quantified Boolean satisfiability problems on which these parameterized complexity classes are based.

In Sect. B.2, we observe some basic properties of these classes and their relation to each other and to known classes. In particular, we show that for various choices of $i \in \mathbb{N}$ and $w \in \{*, k\}^i$, the classes $\Sigma_i^p[w, t]$ coincide with known parameterized complexity classes. Moreover, we identify some conditions for $i_1, i_2 \in \mathbb{N}$, $w_1 \in \{*, k\}^{i_1}$ and $w_2 \in \{*, k\}^{i_2}$ under which it holds that $\Sigma_{i_1}^p[w_1, t] \subseteq \Sigma_{i_2}^p[w_2, t]$.

Finally, in Sect. B.3, we pinpoint some conditions on i_1, i_2, w_1, w_2 for which we can rule out that $\Sigma_{i_1}^p[w_1, t] \subseteq \Sigma_{i_2}^p[w_2, t]$ (under various complexity-theoretic

© Springer-Verlag GmbH Germany, part of Springer Nature 2019
R. de Haan: Parameterized Complexity in the Polynomial Hierarchy, LNCS 11880,
https://doi.org/10.1007/978-3-662-60670-4

assumptions), and we describe the settings for which it remains open whether $\Sigma_{i_1}^P[w_1, t] \subseteq \Sigma_{i_2}^P[w_2, t]$.

B.1 The Parameterized Complexity Classes $\Sigma_i^P[w, t]$

We begin with defining the parameterized complexity classes $\Sigma_i^P[w, t]$, for $i \in \mathbb{N}$, $w \in \mathbb{B}^i$, and $t \in \mathbb{N}^+ \cup \{\text{SAT}, \text{P}\}$. These classes generalize the parameterized complexity classes $\Sigma_2^P[k*]$ and $\Sigma_2^P[*k, t]$—that we introduced in Chap. 6—to higher levels of the Polynomial Hierarchy. In order to do so, we define the parameterized problem $\Sigma_i^P[w]$-WSAT($\mathcal{C}$), for several classes $\mathcal{C}$ of Boolean circuits.

Intuitively, for any $i \in \mathbb{N}$ and any word $w \in \{*, k\}^i$ over the symbols $*$ and k—where $w = v_1 v_2 \ldots v_i$—the problem $\Sigma_i^P[w]$-WSAT is a parameterized, weighted variant of the problem QSAT$_i$, where the word w specifies the weight restriction for the different quantifier blocks—that is, truth assignments for the j-th quantifier block are restricted to weight k if and only if $v_j = k$, and truth assignments for the j-th quantifier block are unrestricted if and only if $v_j = *$. Below, we give a formal recursive definition of the problems $\Sigma_i^P[w]$-WSAT and their dual counterparts $\Pi_i^P[w]$-WSAT.

In order to formally define these parameterized problems, we introduce the concept of partially quantified Boolean circuits.

Definition B.1. *A partially quantified Boolean circuit $D = Q_1 X_1 \ldots Q_n X_n.C$ consists of a Boolean circuit C over a set X of variables, and a quantifier prefix $Q_1 X_1 \ldots Q_n X_n$, where for each $i \in [n]$, $Q_i \in \{\exists, \forall\}$, and where $(\bigcup_{i \in [n]} X_i) \subseteq X$. We let $\text{Free}(D) = X \backslash (\bigcup_{i \in [n]} X_i)$. If $\text{Free}(D) = \emptyset$, we say that D is fully quantified.*

For any truth assignment $\alpha : \text{Free}(D) \to \mathbb{B}$, truth of the fully quantified circuit $D[\alpha]$ is defined as usual.

We now turn to the recursive definition of the problems $\Sigma_i^P[w]$-WSAT and $\Pi_i^P[w]$-WSAT.

Definition B.2 ($\Sigma_i^P[w]$-WSAT **and** $\Pi_i^P[w]$-WSAT). *Let $i \in \mathbb{N}$, and let $w \in \{*, k\}^i$ be a finite word of length i over the symbols $*$ and k, and let $\mathcal{C}$ be a class of Boolean circuits. We define the parameterized decision problems $\Sigma_i^P[w]$-WSAT($\mathcal{C}$) and $\Pi_i^P[w]$-WSAT($\mathcal{C}$) by induction on i and on the structure of w as follows. Here we let ϵ denote the empty word. We begin with the base case, where $i = 0$ and $w = \epsilon$.*

$\Sigma_0^P[\epsilon]$-WSAT($\mathcal{C}$)
Instance: A Boolean circuit $C \in \mathcal{C}$, a truth assignment $\gamma : \text{Var}(C) \to \mathbb{B}$; and a positive integer k.
Parameter: k.
Question: Does $C[\gamma]$ evaluate to 1?

Next, we turn to the inductive cases. Let $i > 0$ and $w' \in \{*, k\}^{i-1}$.

$\Sigma_i^p[*w']$-WSAT($\mathcal{C}$)

Instance: A partially quantified Boolean circuit $D = \exists X_1.\forall X_2 \ldots Q_i X_i.C$, where $C \in \mathcal{C}$ is a Boolean circuit, and where $Q_i = \exists$ if i is odd and $Q_i = \forall$ if i is even, a truth assignment $\gamma : \text{Free}(D) \to \mathbb{B}$, and a positive integer k.

Parameter: k.

Question: Does there exist a truth assignment $\alpha : X_1 \to \mathbb{B}$ such that $(\forall X_2 \ldots Q_i X_i.C, \alpha \cup \gamma, k) \in \Pi_{i-1}^p[w']$-WSAT($\mathcal{C}$)?

$\Pi_i^p[*w']$-WSAT($\mathcal{C}$)

Instance: A partially quantified Boolean circuit $D = \forall X_1.\exists X_2 \ldots Q_i X_i.C$, where $C \in \mathcal{C}$ is a Boolean circuit, and where $Q_i = \forall$ if i is odd and $Q_i = \exists$ if i is even, a truth assignment $\gamma : \text{Free}(D) \to \mathbb{B}$, and a positive integer k.

Parameter: k.

Question: Is it the case that for all truth assignments $\alpha : X_1 \to \mathbb{B}$ it holds that $(\exists X_2 \ldots Q_i X_i.C, \alpha \cup \gamma, k) \in \Sigma_{i-1}^p[w']$-WSAT($\mathcal{C}$)?

$\Sigma_i^p[kw']$-WSAT($\mathcal{C}$)

Instance: A partially quantified Boolean circuit $D = \exists X_1.\forall X_2 \ldots Q_i X_i.C$, where $C \in \mathcal{C}$ is a Boolean circuit, and where $Q_i = \exists$ if i is odd and $Q_i = \forall$ if i is even, a truth assignment $\gamma : \text{Free}(D) \to \mathbb{B}$, and a positive integer k.

Parameter: k.

Question: Does there exist a truth assignment $\alpha : X_1 \to \mathbb{B}$ of weight k such that $(\forall X_2 \ldots Q_i X_i.C, \alpha \cup \gamma, k) \in \Pi_{i-1}^p[w']$-WSAT($\mathcal{C}$)?

$\Pi_i^p[kw']$-WSAT($\mathcal{C}$)

Instance: A partially quantified Boolean circuit $D = \forall X_1.\exists X_2 \ldots Q_i X_i.C$, where $C \in \mathcal{C}$ is a Boolean circuit, and where $Q_i = \forall$ if i is odd and $Q_i = \exists$ if i is even, a truth assignment $\gamma : \text{Free}(D) \to \mathbb{B}$, and a positive integer k.

Parameter: k.

Question: Is it the case that for all truth assignments $\alpha : X_1 \to \mathbb{B}$ of weight k it holds that $(\exists X_2 \ldots Q_i X_i.C, \alpha \cup \gamma, k) \in \Sigma_{i-1}^p[w']$-WSAT($\mathcal{C}$)?

This concludes our definition of the parameterized decision problems $\Sigma_i^p[w]$-WSAT$iw(\mathcal{C})$.

This definition results in a slightly different definition for the problems $\Sigma_2^p[*k]$-WSAT($\mathcal{C}$) and $\Sigma_2^p[k*]$-WSAT($\mathcal{C}$) than the ones defined in Chap. 6. (For instance, for the definition of $\Sigma_2^p[k*]$-WSAT that we gave in this section, the input consists of a partially quantified Boolean circuit D and a truth assignment $\alpha : \text{Free}(D) \to \mathbb{B}$, whereas for the definition that we gave in Sect. 6.1, the input can be seen as a fully quantified Boolean circuit.) However, the two problems are easily shown to be fpt-reducible to each other (for all classes $\mathcal{C}$ of Boolean circuits that are closed under partial instantiation). Adding a partial truth assignment γ to the input in the formulation of the problems $\Sigma_i^P[w]$-WSAT and $\Pi_i^P[w]$-WSAT allows for an easier recursive definition.

Definition B.3 ($\Sigma_i^P[w, t]$ **and** $\Pi_i^P[w, t]$). *In order to define the complexity classes $\Sigma_i^P[w, t]$ and $\Pi_i^P[w, t]$, we define the auxiliary function $\mathbb{C}(t)$ as follows:*

$$\mathbb{C}(t) = \begin{cases} \{\, \text{CIRC}_{t,d} : d \geq t \,\} & \text{if } t \in \mathbb{N}^+, \\ \{\text{FORM}\} & \text{if } t = \text{SAT}, \\ \{\text{CIRC}\} & \text{if } t = \text{P}. \end{cases}$$

Then, let $i \in \mathbb{N}$, let $w \in \{*, k\}^i$ be a finite word of length i over the symbols $*$ and k, and let $t \in \mathbb{N}^+ \cup \{\text{SAT}, \text{P}\}$. The parameterized complexity classes $\Sigma_i^P[w, t]$ and $\Pi_i^P[w, t]$ are defined as follows:

$$\Sigma_i^P[w, t] = [\, \{\, \Sigma_i^P[w]\text{-WSAT}(\mathcal{C}) : \mathcal{C} \in \mathbb{C}(t) \,\} \,]^{\text{fpt}}, \text{ and}$$
$$\Pi_i^P[w, t] = [\, \{\, \Pi_i^P[w]\text{-WSAT}(\mathcal{C}) : \mathcal{C} \in \mathbb{C}(t) \,\} \,]^{\text{fpt}}.$$

Here the notation $[\, \mathcal{Q} \,]^{\text{fpt}}$ denotes the class of all parameterized problems that are fpt-reducible to some problem $Q \in \mathcal{Q}$.

A visual overview of some of the classes $\Sigma_i^P[w, t]$ and $\Pi_i^P[w, t]$, for $i \leq 3$, can be found in Fig. B.1.

B.2 Inclusion Results

We continue with a number of results and observations that establish equality and inclusion between various classes $\Sigma_i^P[w, t]$ and $\Pi_i^P[w, t]$. We begin with a number of trivial equalities, that are straightforward to verify.

Observation B.4. *The following equalities hold, for $i > 0$, $w \in \{*, k\}^i$, and $t \in \mathbb{N}^+ \cup \{\text{SAT}, \text{P}\}$:*

$$\Pi_i^P[w, t] = \text{co-}\Sigma_i^P[w, t];$$

$$\Sigma_0^P[\epsilon, 1] = \Sigma_0^P[\epsilon, 2] = \cdots = \Sigma_0^P[\epsilon, \text{SAT}] = \Sigma_0^P[\epsilon, \text{P}] =$$
$$\Pi_0^P[\epsilon, 1] = \Pi_0^P[\epsilon, 2] = \cdots = \Pi_0^P[\epsilon, \text{SAT}] = \Pi_0^P[\epsilon, \text{P}] = \text{FPT};$$
$$\Sigma_1^P[k, t] = \text{W}[t];$$
$$\Sigma_1^P[*, \text{P}] = \text{para-NP};$$
$$\Pi_1^P[*, \text{P}] = \text{para-co-NP};$$
$$\Sigma_2^P[**, \text{P}] = \text{para-}\Sigma_2^P;$$
$$\Pi_2^P[**, \text{P}] = \text{para-}\Pi_2^P;$$
$$\Sigma_i^P[*^i, \text{P}] = \text{para-}\Sigma_i^P; \text{ and}$$
$$\Pi_i^P[*^i, \text{P}] = \text{para-}\Pi_i^P.$$

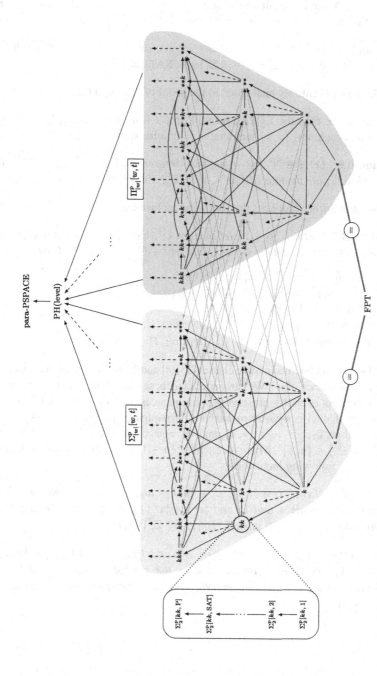

Fig. B.1 The classes $\Sigma_i^p[w, t]$ and $\Pi_i^p[w, t]$. Each word $w \in \{*, k\}^*$ represents the hierarchy of parameterized complexity classes $\Sigma_{|w|}^p[w, t]$ or $\Pi_{|w|}^p[w, t]$. Arrows indicate inclusion relations.

In Chap. 6, we already saw that $\Sigma_2^P[k*, 1] = \Sigma_2^P[k*, 2] = \cdots = \Sigma_2^P[k*, P]$ (Theorem 6.2). A similar result holds for every $i > 1$ and $w \in \{*, k\}^i$ where w ends with a $*$—that is, such that $w = w'*$ for some $w' \in \{*, k\}^{i-1}$.

Proposition B.5. *Let $i \in \mathbb{N}^+$ be a positive integer, and let $w' \in \{*, k\}^{i-1}$. It holds that $\Sigma_i^P[w'*, 1] = \Sigma_i^P[w'*, 2] = \cdots = \Sigma_i^P[w'*, \text{SAT}] = \Sigma_i^P[w'*, P]$.*

Proof. The proof is entirely analogous to the proof of Theorem 6.2.

Similarly to the case for $\Sigma_2^P[k*]$, we will drop t from the notation $\Sigma_i^P[w, t]$ for each $i > 0$ and each $w \in \{*, k\}$ such that $w = w'*$ for some $w' \in \{*, k\}^{i-1}$.

Convention B.6. *Let $i \in \mathbb{N}^+$ be a positive integer, and let $w' \in \{*, k\}^{i-1}$. We use $\Sigma_i^P[w'*]$ to denote the class $\Sigma_i^P[w'*, 1] = \Sigma_i^P[w'*, P]$.*

We also straightforwardly get the following inclusions.

Observation B.7. *Let $i, i_1, i_2 \in \mathbb{N}$, let $w \in \{*, k\}^i$, $w_1 \in \{*, k\}^{i_1}$ and $w_2 \in \{*, k\}^{i_2}$ be finite words over the symbols $*$ and k, and let $t \in \mathbb{N}^+ \cup \{\text{SAT}, P\}$. Then:*

- *if $t \in \mathbb{N}^+$, then $\Sigma_i^P[w, t] \subseteq \Sigma_i^P[w, t + 1]$;*
- *$\Sigma_i^P[w, t] \subseteq \Sigma_i^P[w, \text{SAT}]$;*
- *$\Sigma_i^P[w, \text{SAT}] \subseteq \Sigma_i^P[w, P]$;*
- *if i_1 is even, then $\Sigma_i^P[w, t] \subseteq \Sigma_{i_1+i+i_2}^P[w_1 w w_2, t]$; and*
- *if i_1 is odd, then $\Sigma_i^P[w, t] \subseteq \Pi_{i_1+i+i_2}^P[w_1 w w_2, t]$.*

These last two inclusions involve classes $\Sigma_i^P[w]$ and $\Sigma_i^P[w']$, where w is a particular type of subsequence of w'. We can generalize these inclusions. We do so by introducing the concept of parity-preserving subsequences. Let $w_1, w_2 \in \{*, k\}^*$ be two words. We say that w_1 is a *parity-preserving subsequence of w_2* if $w_1 = v_1 v_2 \ldots v_n$, for some symbols $v_1, \ldots, v_n \in \{*, k\}$, and $w_2 = w_0' v_1 w_1' v_2 \ldots w_{n-1}' v_n w_n'$ for some words $w_1', \ldots, w_n' \in \{*, k\}^*$, where for each $i \in [n-1]$ the word w_i' is of even length. We then get the following inclusions.

Observation B.8. *Let $i_1, i_2 \in \mathbb{N}$, let $w_1 \in \{*, k\}^{i_1}$ and $w_2 \in \{*, k\}^{i_2}$ be some finite words over the symbols $*$ and k, and let $t \in \mathbb{N}^+ \cup \{\text{SAT}, P\}$. If w_1 is a parity-preserving subsequence of w_2, then $\Sigma_{i_1}^P[w_1, t] \subseteq \Sigma_{i_2}^P[w_2, t]$.*

Observation B.9. *Let $i_1, i_2, i_3 \in \mathbb{N}$, where i_3 is odd, let $w_1 \in \{*, k\}^{i_1}$, $w_2 \in \{*, k\}^{i_2}$, and $w_3 \in \{*, k\}^{i_3}$ be some finite words over the symbols $*$ and k, and let $t \in \mathbb{N}^+ \cup \{\text{SAT}, P\}$. If w_1 is a parity-preserving subsequence of w_2, then $\Sigma_{i_1}^P[w_1, t] \subseteq \Pi_{i_3+i_2}^P[w_3 w_2, t]$.*

Next, we show that the class $\Sigma_i^P[w, t]$ is contained in $\Sigma_i^P[*^i]$, for any $i \in \mathbb{N}$ and any word $w \in \{*, k\}^i$.

Proposition B.10. *Let $i \in \mathbb{N}$, and let $w \in \{*, k\}^i$ be any finite word over the symbols $*$ and k. Then $\Sigma_i^P[w, P] \subseteq \Sigma_i^P[*^i]$.*

Proof (sketch). Similarly to our observation in the beginning of Sect. 6.4.1, for any quantifier block $Q_i X_i$, we can straightforwardly express the weight constraint for the corresponding truth assignments using a Boolean circuit C_i. Then, by appropriately using these additional Boolean circuits C_i, we can transform any instance of $\Sigma_i^P[w]$-WSAT into an equivalent instance of $\Sigma_i^P[*^n]$-WSAT.

This inclusion result can be generalized. We identify some conditions under which it holds that $\Sigma_i^P[w_1, t] \subseteq \Sigma_i^P[w_2, t]$, for any $i \in \mathbb{N}$, any $w_1, w_2 \in \{*, k\}^i$, and any $t \in \mathbb{N}^+ \cup \{\text{SAT}, P\}$. Namely, this is the case whenever w_2 can be obtained from w_1 by replacing occurrences of the symbol k to the symbol $*$. Formally, we say that for any $i \in \mathbb{N}$ and $w_1, w_2 \in \{*, k\}^i$—where $w_1 = v_1^1 v_2^1 \ldots v_i^1$ and $w_2 = v_1^2 v_2^2 \ldots v_i^2$—it is the case that w_2 *dominates* w_1 if for each $\ell \in [i]$ it holds that whenever $v_\ell^1 = *$, then also $v_\ell^2 = *$ (but not necessarily vice versa). If w_2 dominates w_1, then we get that $\Sigma_i^P[w_1, t] \subseteq \Sigma_i^P[w_2, t]$.

Proposition B.11. *Let $i \in \mathbb{N}$, let $w_1, w_2 \in \{*, k\}^i$ be words of length i over the symbols $*$ and k such that w_2 dominates w_1, and let $t \in \mathbb{N}^+ \cup \{\text{SAT}, P\}$. Then $\Sigma_i^P[w_1, t] \subseteq \Sigma_i^P[w_2, t]$.*

Proof. The proof is entirely analogous to the proof of Proposition B.10.

We finish this section by relating the classes $\Sigma_i^P[w, t]$ and $\Pi_i^P[w, t]$ to some parameterized complexity classes known from the literature. Firstly, we observe that each of the parameterized complexity classes $\Sigma_i^P[w, t]$ and $\Pi_i^P[w, t]$ is contained in the class para-PSPACE.

Observation B.12. *Let $i \in \mathbb{N}$, let $w \in \{*, k\}^i$ be a word of length i over the symbols $*$ and k, and let $t \in \mathbb{N}^+ \cup \{\text{SAT}, P\}$. Then $\Sigma_i^P[w, t] \subseteq$ para-PSPACE and $\Pi_i^P[w, t] \subseteq$ para-PSPACE.*

Several of the classes $\Sigma_i^P[w, t]$ and $\Pi_i^P[w, t]$ are also contained in the class XP.

Observation B.13. *Let $i \in \mathbb{N}$, and let $t \in \mathbb{N}^+ \cup \{\text{SAT}, P\}$. Then $\Sigma_i^P[k^i, t] \subseteq$ XP and $\Pi_i^P[k^i, t] \subseteq$ XP.*

More generally, some of the classes $\Sigma_i^P[w, t]$ and $\Pi_i^P[w, t]$ are contained in some of the classes $X\Sigma_j^P$ and $X\Pi_j^P$. Clearly, for each $i \in \mathbb{N}$, it holds that $\Sigma_i^P[w, t] \subseteq X\Sigma_i^P$ and $\Pi_i^P[w, t] \subseteq X\Pi_i^P$, for each $w \in \{*, k\}^i$ and each $t \in \mathbb{N}^+ \cup \{\text{SAT}, P\}$. However, in various cases, $\Sigma_i^P[w, t]$ and $\Pi_i^P[w, t]$ are contained in $X\Sigma_j^P$ or $X\Pi_j^P$ for $j < i$. For example, it holds that $\Sigma_3^P[*k*] \subseteq X\Sigma_1^P = \text{XNP}$. We capture several such inclusions in an inductive manner.

Observation B.14. *Let* $t \in \mathbb{N}^+ \cup \{SAT, P\}$. *It holds that:*

- $\Sigma_0^p[\epsilon, t] = \Pi_0^p[\epsilon, t] = FPT \subseteq XP = X\Sigma_0^p = X\Pi_0^p$.

Moreover, let $i \in \mathbb{N}^+$, *let* $w \in \{*, k\}^i$ *with* $w = v_1 \ldots v_i$, *let* $w' = v_2 \ldots v_i$, *let* $j \in [i-1]$, *and let* $K \in \{X\Sigma_j^p, X\Pi_j^p\}$.

- *If* $v_1 = k$, *then:*

  - *If* $\Pi_{i-1}^p[w', t] \subseteq K$, *then* $\Sigma_i^p[w, t] \subseteq K$.
  - *If* $\Sigma_{i-1}^p[w', t] \subseteq K$, *then* $\Pi_i^p[w, t] \subseteq K$.

- *If* $v_1 = *$, $j > 0$, *and* $\Pi_{i-1}^p[w', t] \subseteq X\Sigma_j^p$, *then* $\Sigma_i^p[w, t] \subseteq X\Sigma_j^p$.
- *If* $v_1 = *$ *and* $\Pi_{i-1}^p[w', t] \subseteq X\Pi_j^p$, *then* $\Sigma_i^p[w, t] \subseteq X\Sigma_{j+1}^p$.
- *If* $v_1 = *$ *and* $\Sigma_{i-1}^p[w', t] \subseteq X\Sigma_j^p$, *then* $\Pi_i^p[w, t] \subseteq X\Pi_{j+1}^p$.
- *If* $v_1 = *$, $j > 0$, *and* $\Sigma_{i-1}^p[w', t] \subseteq X\Pi_j^p$, *then* $\Pi_i^p[w, t] \subseteq X\Pi_j^p$.

B.3   Separation Results

Finally, we identify some conditions under which we can rule out inclusions between the classes $\Sigma_i^p[w, t]$ and $\Pi_i^p[w, t]$ (under various complexity-theoretic assumptions). The conditions for which we established inclusions and the conditions for which we can rule out inclusions are not exhaustive. We will also briefly discuss some of the cases that neither satisfy the conditions that we identify in this section nor the conditions that we identified in Sect. B.2. For such cases the relation between the classes $\Sigma_i^p[w, t]$ and $\Pi_i^p[w, t]$ remains open.

Consider arbitrary $i_1, i_2 \in \mathbb{N}$ and arbitrary words $w_1 \in \{*, k\}^{i_1}$ and $w_2 \in \{*, k\}^{i_2}$ over the symbols $*$ and k. We consider various cases: either $i_1 > i_2, i_1 = i_2$, or $i_1 < i_2$. We discuss these cases one after another.

B.3.1   The Case for $i_1 > i_2$

In the first case, where $i_1 > i_2$, we can establish a conditional result that $\Sigma_{i_1}^p[w_1, t] \not\subseteq \Sigma_{i_2}^p[w_2, t]$.

Proposition B.15. *Let* $i_1, i_2 \in \mathbb{N}$ *such that* $i_1 > i_2$, *and let* $w_1 \in \{*, k\}^{i_1}$ *be any word of length* i_1 *over the symbols* $*$ *and* k. *Then* $\Sigma_{i_1}^p[w, 1] \not\subseteq \Sigma_{i_2}^p[*^{i_2}]$, *unless there is a subexponential-time reduction from* $QSAT_{i_1}$ (3CNF $\cup$ 3DNF) *to* $QSAT_{i_2}$. *Similarly,* $\Sigma_{i_1}^p[w, 1] \not\subseteq \Pi_{i_2}^p[*^{i_2}]$, *unless there is a subexponential-time reduction from* $QSAT_{i_1}$ (3CNF $\cup$ 3DNF) *to* co-$QSAT_{i_2}$.

Proof. We know that $A[i_1] \subseteq \Sigma_{i_1}^p[w, 1]$, $\Sigma_{i_2}^p[*^{i_2}] =$ para-$\Sigma_{i_2}^p$, and $\Pi_{i_2}^p[*^{i_2}] =$ para-$\Pi_{i_2}^p$. Therefore, in case $i_1 \geq 2$, the result follows from Proposition 14.4. In case $i_1 < 2$, the result follows from known results [45–47].

Corollary B.16. *Let $i_1, i_2 \in \mathbb{N}$ such that $i_1 > i_2$, and let $w_1 \in \{*, k\}^{i_1}$ and $w_2 \in \{*, k\}^{i_2}$ be any words over the symbols $*$ and k. Then $\Sigma_{i_1}^{\mathrm{p}}[w, 1] \not\subseteq \Sigma_{i_2}^{\mathrm{p}}[w_2, \mathrm{P}]$, unless there is a subexponential-time reduction from $\mathrm{QSAT}_{i_1}(\mathrm{3CNF} \cup \mathrm{3DNF})$ to QSAT_{i_2}. Similarly, $\Sigma_{i_1}^{\mathrm{p}}[w, 1] \not\subseteq \Pi_{i_2}^{\mathrm{p}}[w_2, \mathrm{P}]$, unless there is a subexponential-time reduction from $\mathrm{QSAT}_{i_1}(\mathrm{3CNF} \cup \mathrm{3DNF})$ to co-QSAT_{i_2}.*

For the special case where $w_1 = *^{i_1}$, we can establish non-inclusion even under a weaker assumption.

Observation B.17. *Let $i_1, i_2 \in \mathbb{N}$ such that $i_1 > i_2$. Then $\Sigma_{i_1}^{\mathrm{p}}[*^{i_1}] \not\subseteq \Sigma_{i_2}^{\mathrm{p}}[*^{i_2}]$, and $\Sigma_{i_1}^{\mathrm{p}}[*^{i_1}] \not\subseteq \Pi_{i_2}^{\mathrm{p}}[*^{i_2}]$, unless the PH collapses.*

Corollary B.18. *Let $i_1, i_2 \in \mathbb{N}$ such that $i_1 > i_2$, and let $w_2 \in \{*, k\}^{i_2}$ be any word of length i_2 over the symbols $*$ and k. Then $\Sigma_{i_1}^{\mathrm{p}}[*^{i_1}] \not\subseteq \Sigma_{i_2}^{\mathrm{p}}[w_2, \mathrm{P}]$, and $\Sigma_{i_1}^{\mathrm{p}}[*^{i_1}] \not\subseteq \Pi_{i_2}^{\mathrm{p}}[w_2, \mathrm{P}]$, unless the PH collapses.*

B.3.2 The Case for $i_1 = i_2$

Next, we turn to the case where $i_1 = i_2 = i$. In the particular case where $i = 2$, we managed to obtain inclusion or (conditional) separation results in all cases. The techniques that we used in this case turn out not to suffice for a similarly nice result in the case for $i > 2$. We give several (conditional) separation results for the general case where $i_1 = i_2$, and we discuss some of the cases where it remains open what the relation is between the classes $\Sigma_i^{\mathrm{p}}[w, t]$.

In Sect. B.2, we showed that if a word $w_2 \in \{*, k\}^i$ dominates another word $w_1 \in \{*, k\}^i$, then $\Sigma_i^{\mathrm{p}}[w_1, t] \subseteq \Sigma_i^{\mathrm{p}}[w_2, t]$, for any $t \in \mathbb{N}^+ \cup \{\mathrm{SAT}, \mathrm{P}\}$. Conversely, however, if w_1 dominates w_2—and $w_1 \neq w_2$—we get that $\Sigma_i^{\mathrm{p}}[w_1, t] \not\subseteq \Sigma_i^{\mathrm{p}}[w_2, t]$, unless the PH collapses.

Proposition B.19. *Let $i \in \mathbb{N}$, let $w_1, w_2 \in \{*, k\}^i$ such that $w_1 \neq w_2$ and w_1 dominates w_2, and let $t \in \mathbb{N}^+ \cup \{\mathrm{SAT}, \mathrm{P}\}$. Then $\Sigma_i^{\mathrm{p}}[w_1, t] \not\subseteq \Sigma_i^{\mathrm{p}}[w_2, t]$, unless the PH collapses.*

Proof. Because $w_1 \neq w_2$ and w_1 dominates w_2, we know that $i > 0$ and there exists some $j_1, j_2 \in \mathbb{N}$ with $i \geq j_1 > j_2$ such that one of the following holds:

(i) para-$\Sigma_{j_1}^{\mathrm{p}} \subseteq \Sigma_i^{\mathrm{p}}[w_1, t]$ and $\Sigma_i^{\mathrm{p}}[w_2, t] \subseteq \mathrm{X}\Sigma_{j_2}^{\mathrm{p}}$,
(ii) para-$\Sigma_{j_1}^{\mathrm{p}} \subseteq \Sigma_i^{\mathrm{p}}[w_1, t]$ and $\Sigma_i^{\mathrm{p}}[w_2, t] \subseteq \mathrm{X}\Pi_{j_2}^{\mathrm{p}}$,
(iii) para-$\Pi_{j_1}^{\mathrm{p}} \subseteq \Sigma_i^{\mathrm{p}}[w_1, t]$ and $\Sigma_i^{\mathrm{p}}[w_2, t] \subseteq \mathrm{X}\Sigma_{j_2}^{\mathrm{p}}$, or
(iv) para-$\Pi_{j_1}^{\mathrm{p}} \subseteq \Sigma_i^{\mathrm{p}}[w_1, t]$ and $\Sigma_i^{\mathrm{p}}[w_2, t] \subseteq \mathrm{X}\Pi_{j_2}^{\mathrm{p}}$.

In each of these cases, $\Sigma_i^{\mathrm{p}}[w_1, t] \subseteq \Sigma_i^{\mathrm{p}}[w_2, t]$ implies a collapse of the PH.

In various cases, for words w_1 and w_2 for which neither w_1 dominates w_2 nor w_2 dominates w_1, we can establish that $\Sigma_i^{\mathrm{p}}[w_1, t] \not\subseteq \Sigma_i^{\mathrm{p}}[w_2, t]$ (assuming that the PH does not collapse). Whenever there exist $j_1, j_2 \in \mathbb{N}$ with $j_1 > j_2$ such that para-$\Pi_{j_1}^p \subseteq \Sigma_i^p[w_1, t]$ and $\Sigma_i^p[w_2, t] \subseteq X\Sigma_{j_2}^p$, for instance, we know that $\Sigma_i^p[w_1, t] \not\subseteq \Sigma_i^p[w_2, t]$, unless the PH collapses. To take a particular example, we get that $\Sigma_3^p[k**] \not\subseteq \Sigma_3^p[*k*]$, unless the PH collapses, because para-$\Pi_2^p \subseteq \Sigma_3^p[k**]$ and $\Sigma_3^p[*k*] \subseteq X\Sigma_1^p$.

For many other words w_1 and w_2 for which neither w_1 dominates w_2 nor w_2 dominates w_1, it remains open whether $\Sigma_i^p[w_1, t] \subseteq \Sigma_i^p[w_2, t]$. An example of this is the case where $w_1 = kk***$ and $w_2 = **kk*$—that is, it is an open problem whether $\Sigma_5^p[kk***] \subseteq \Sigma_5^p[**kk*]$. Similarly, it remains open whether $\Sigma_5^p[**kk*] \subseteq \Sigma_5^p[kk***]$.

Next, we turn to the relation between $\Sigma_i^p[w_1, t]$ and $\Pi_i^p[w_2, t]$. We can establish the following (conditional) result that $\Sigma_i^p[w, t] \subseteq \Pi_i^p[w, t]$.

Proposition B.20. *Let $i \in \mathbb{N}$, let $w_1, w_2 \in \{*, k\}^i$, and let $t \in \mathbb{N}^+ \cup \{SAT, P\}$. Then $\Sigma_i^p[w_1, t] \not\subseteq \Pi_i^p[w_2, t]$, unless there exists a subexponential-time reduction from $QSAT_i(3CNF \cup 3DNF)$ to co-$QSAT_i$.*

Proof. We know that $A[i] \subseteq \Sigma_i^p[w_1, t]$ and $\Pi_i^p[w_2, t] \subseteq$ para-Π_i^p. Therefore by Proposition 14.4, we know that $\Sigma_i^p[w_1, t] \not\subseteq \Pi_i^p[w_2, t]$ unless there exists a subexponential-time reduction from $QSAT_i(3CNF \cup 3DNF)$ to co-$QSAT_i$.

In the particular case where $w_1 = w_2 = w \neq k^i$, we can even show that $\Sigma_i^p[w, t] \not\subseteq \Pi_i^p[w, t]$ under a weaker assumption (namely, under the assumption that the PH does not collapse).

Proposition B.21. *Let $i \in \mathbb{N}$, let $w \in \{*, k\}^i$ such that $w \neq k^i$, and let $t \in \mathbb{N}^+ \cup \{SAT, P\}$. Then $\Sigma_i^p[w, t] \not\subseteq \Pi_i^p[w, t]$, unless the PH collapses.*

Proof. We know that there must be some $j \in [i]$ with $j > 0$ such that one of the following holds:

(i) para-$\Sigma_j^p \subseteq \Sigma_i^p[w, t]$ and $\Pi_i^p[w, t] \subseteq X\Pi_j^p$; or
(ii) para-$\Pi_j^p \subseteq \Sigma_i^p[w, t]$ and $\Pi_i^p[w, t] \subseteq X\Sigma_j^p$.

In both of these cases $\Sigma_i^p[w, t] \not\subseteq \Pi_i^p[w, t]$ implies that the PH collapses.

B.3.3 The Case for $i_1 < i_2$

Finally, we consider the case where $i_1 < i_2$. In Sect. B.2, we have already seen that whenever a word $w_1 \in \{*, k\}^{i_1}$ is a parity-preserving subsequence of a word $w_2 \in \{*, k\}^{i_2}$, or whenever w_2 dominates w_1, then $\Sigma_{i_1}^p[w_1, t] \subseteq \Sigma_{i_2}^p[w_2, t]$, for any $t \in \mathbb{N}^+ \cup \{SAT, P\}$. Of course, the inclusion relations obtained on the basis of these two properties also compose. That is, whenever w_1 is dominated by a third word

$w_3 \in \{*, k\}^{i_1}$, and w_3 is a parity-preserving subsequence of w_2, then also $\Sigma_{i_1}^P[w_1, t] \subseteq \Sigma_{i_2}^P[w_2, t]$. For example, $\Sigma_2^P[kk, P] \subseteq \Sigma_3^P[k**]$, because the word $w_3 = k*$ dominates kk and is a parity-preserving subsequence of $k**$.

For various cases, we can establish (conditional) separation results that state that $\Sigma_{i_1}^P[w_1, t] \not\subseteq \Sigma_{i_2}^P[w_2, t]$. For instance, whenever there exist $j_1, j_2 \in \mathbb{N}$ with $j_1 > j_2$ such that para-$\Pi_{j_1}^P \subseteq \Sigma_{i_1}^P[w_1, t]$ and $\Sigma_{i_2}^P[w_2, t] \subseteq X\Pi_{j_2}^P$, we can conclude that $\Sigma_{i_1}^P[w_1, t] \not\subseteq \Sigma_{i_2}^P[w_2, t]$, unless the PH collapses. An example of this is the result that $\Sigma_3^P[k**] \not\subseteq \Sigma_4^P[k*k*]$, unless the PH collapses—namely, we have that para-$\Pi_2^P \subseteq \Sigma_3^P[k**]$ and $\Sigma_4^P[k*k*] \subseteq X\Pi_1^P$. The conditional separation results that we can obtain using this type of argument are summarized in the following result. (This result also captures similar separation results for the case where $i_1 = i_2$, that we discussed in Sect. B.3.2.)

Observation B.22. *Let* K_1, K_2 *be two parameterized complexity classes, e.g.,* $K_1, K_2 \in \{ \Sigma_i^P[w, t], \Pi_i^P[w, t] : i \in \mathbb{N}, w \in \{*, k\}^i, t \in \mathbb{N}^+ \cup \{SAT, P\} \}$. *Moreover, let* $j_1, j_2 \in \mathbb{N}$ *with* $j_1 > j_2$.

- *If* para-$\Sigma_{j_2}^P \subseteq K_1$ *and* $K_2 \subseteq X\Pi_{j_2}^P$, *then* $K_1 \not\subseteq K_2$, *unless the PH collapses.*
- *If* para-$\Pi_{j_2}^P \subseteq K_1$ *and* $K_2 \subseteq X\Sigma_{j_2}^P$, *then* $K_1 \not\subseteq K_2$, *unless the PH collapses.*
- *If* para-$\Sigma_{j_1}^P \subseteq K_1$ *and* $K_2 \subseteq X\Sigma_{j_2}^P$, *then* $K_1 \not\subseteq K_2$, *unless the PH collapses.*
- *If* para-$\Pi_{j_1}^P \subseteq K_1$ *and* $K_2 \subseteq X\Pi_{j_2}^P$, *then* $K_1 \not\subseteq K_2$, *unless the PH collapses.*

There are also cases where neither of the above two conditions applies (assuming that the PH does not collapse)—that is, cases where inclusion cannot be established on the basis of dominating words and parity-preserving subsequences, and where the separation results of Observation B.22 do not apply. For these cases, it remains open what the relation is between the classes $\Sigma_i^P[w, t]$ and $\Pi_i^P[w, t]$. For example, it is an open problem whether $\Sigma_3^P[*k*] \subseteq \Sigma_4^P[k**k, P]$.

References

1. Aaronson, S.: Quantum Computing Since Democritus. Cambridge University Press, New York (2013). (cit. on p. 85)
2. Abrahamson, K.A., Downey, R.G., Fellows, M.R.: Fixed-parameter tractability and completeness. IV. On completeness for W[P] and PSPACE analogues. Ann. Pure Appl. Logic **73**(3), 235–276 (1995). (cit. on p. 39)
3. Abramsky, S., Gottlob, G., Kolaitis, P.G.: Robust constraint satisfaction and local hidden variables in quantum mechanics. In: Rossi, F. (ed.) Proceedings of the 23rd International Joint Conference on Artificial Intelligence (IJCAI 2013). AAAI Press/IJCAI (2013). (cit. on p. 174)
4. Adams, D.: The Hitchhiker's Guide to the Galaxy. Pan Books, London (1979). (cit. on pp. 137, 375)
5. Ajtai, M., Fagin, R., Stockmeyer, L.J.: The closure of monadic NP. J. Comput. Syst. Sci. **60**(3), 660–716 (2000). (cit. on pp. 261–263)
6. Allender, E.W., Rubinstein, R.S.: P-printable sets. SIAM J. Comput. **17**(6), 1193–1202 (1988). (cit. on p. 311)
7. Arora, S., Barak, B.: Computational Complexity - A Modern Approach. Cambridge University Press, New York (2009). (cit. on pp. 16, 21, 194)
8. Atserias, A., Oliva, S.: Bounded-width QBF is PSPACE-complete. In: Portier, N., Wilke, T. (eds.) Proceedings of the 30th International Symposium on Theoretical Aspects of Computer Science (STACS 2013), Kiel, Germany, 27 February–2 March 2013. LIPIcs, vol. 20, pp. 44–54. Schloss Dagstuhl - Leibniz-Zentrum fuer Informatik (2013). (cit. on p. 57)
9. Ayari, A., Basin, D.: QUBOS: deciding quantified boolean logic using propositional satisfiability solvers. In: Aagaard, M.D., O'Leary, J.W. (eds.) FMCAD 2002. LNCS, vol. 2517, pp. 187–201. Springer, Heidelberg (2002). https://doi.org/10.1007/3-540-36126-X_12. (cit. on p. 52)
10. Bäckström, C., Nebel, B.: Complexity results for SAS+ planning. Comput. Intell. **11**, 625–656 (1995). (cit. on p. 251)
11. Baier, C., Katoen, J.P.: Principles of Model Checking. The MIT Press, Cambridge (2008). (cit. on pp. 60, 181, 187, 202)
12. Baral, C., Kreinovich, V., Trejo, R.: Computational complexity of planning and approximate planning in the presence of incompleteness. Artif. Intell. **122**(1–2), 241–267 (2000). (cit. on p. 253)
13. Barrett, C.W., Sebastiani, R., Seshia, S.A., Tinelli, C.: Satisfiability modulo theories. In: Biere, A., Heule, M., van Maaren, H., Walsh, T. (eds.) Handbook of Satisfiability, vol. 185, pp. 825–885. IOS Press, Amsterdam (2009). (cit. on p. 50)
14. Beigel, R.: Bounded queries to SAT and the Boolean hierarchy. Theoret. Comput. Sci. **84**(2), 199–223 (1991). (cit. on p. 144)

15. Belov, A., Lynce, I., Marques-Silva, J.: Towards efficient MUS extraction. AI Commun. **25**(2), 97–116 (2012). (cit. on p. 137)
16. Ben-Eliyahu, R., Dechter, R.: Propositional semantics for disjunctive logic programs. Ann. Math. Artif. Intell. **12**(1), 53–87 (1994). (cit. on p. 162)
17. Benedetti, M., Bernardini, S.: Incremental compilation-to-SAT procedures. In: Hoos, H.H., Mitchell, D.G. (eds.) SAT 2004. LNCS, vol. 3542, pp. 46–58. Springer, Heidelberg (2005). https://doi.org/10.1007/11527695_4. (cit. on p. 140)
18. Biere, A.: Bounded model checking. In: Biere, A., Heule, M., van Maaren, H., Walsh, T. (eds.) Handbook of Satisfiability, Frontiers in Artificial Intelligence and Applications, vol. 185, pp. 457–481. IOS Press, Amsterdam (2009). (cit. on pp. 49, 181, 192)
19. Biere, A.: Resolve and expand. In: Hoos, H.H., Mitchell, D.G. (eds.) SAT 2004. LNCS, vol. 3542, pp. 59–70. Springer, Heidelberg (2005). https://doi.org/10.1007/11527695_5. (cit. on p. 52)
20. Biere, A., Cimatti, A., Clarke, E.M., Strichman, O., Zhu, Y.: Bounded model checking. In: Advances in Computers, vol. 58, pp. 117–148 (2003). (cit. on pp. 49, 181)
21. Biere, A., Cimatti, A., Clarke, E., Zhu, Y.: Symbolic model checking without BDDs. In: Cleaveland, W.R. (ed.) TACAS 1999. LNCS, vol. 1579, pp. 193–207. Springer, Heidelberg (1999). https://doi.org/10.1007/3-540-49059-0_14. (cit. on pp. 49, 181, 192)
22. Biere, A., Heule, M., van Maaren, H., Walsh, T. (eds.): Handbook of Satisfiability, Frontiers in Artificial Intelligence and Applications, vol. 185. IOS Press, Amsterdam (2009). (cit. on pp. 3, 45)
23. Bodlaender, H.L.: A linear-time algorithm for finding tree-decompositions of small treewidth. SIAM J. Comput. **25**(6), 1305–1317 (1996). (cit. on pp. 53, 54)
24. Bodlaender, H.L.: A partial k-arboretum of graphs with bounded treewidth. Theoret. Comput. Sci. **209**(1–2), 1–45 (1998). (cit. on p. 54)
25. Bodlaender, H.L.: Dynamic programming on graphs with bounded treewidth. In: Lepistö, T., Salomaa, A. (eds.) ICALP 1988. LNCS, vol. 317, pp. 105–118. Springer, Heidelberg (1988). https://doi.org/10.1007/3-540-19488-6_110. (cit. on p. 56)
26. Bodlaender, H.L.: Fixed-parameter tractability of treewidth and pathwidth. In: Bodlaender, H.L., Downey, R., Fomin, F.V., Marx, D. (eds.) The Multivariate Algorithmic Revolution and Beyond. LNCS, vol. 7370, pp. 196–227. Springer, Heidelberg (2012). https://doi.org/10.1007/978-3-642-30891-8_12. (cit. on pp. 35, 54)
27. Bodlaender, H.L., Downey, R., Fomin, F.V., Marx, D. (eds.): The Multivariate Algorithmic Revolution and Beyond. LNCS, vol. 7370. Springer, Heidelberg (2012). https://doi.org/10.1007/978-3-642-30891-8. (cit. on pp. 3, 45)
28. Bodlaender, H.L., Downey, R.G., Fellows, M.R., Hermelin, D.: On problems without polynomial kernels. J. Comput. Syst. Sci. **75**(8), 423–434 (2009). (cit. on pp. 335, 343)
29. Bodlaender, H.L., Thomassé, S., Yeo, A.: Kernel bounds for disjoint cycles and disjoint paths. Theoret. Comput. Sci. **412**, 4570–4578 (2011). (cit. on p. 346)
30. Brandt, F., Conitzer, V., Endriss, U., Lang, J., Procaccia, A.: Handbook of Computational Social Choice. Cambridge University Press, Cambridge (2016). (cit. on pp. 221, 222)
31. Brewka, G., Eiter, T., Truszczynski, M.: Answer set programming at a glance. Commun. ACM **54**(12), 92–103 (2011). (cit. on pp. 51, 59, 72)
32. Bundala, D., Ouaknine, J., Worrell, J.: On the magnitude of completeness thresholds in bounded model checking. In: Proceedings of the 27th Annual IEEE Symposium on Logic in Computer Science (LICS 2012), pp. 155–164. IEEE Computer Society (2012). (cit. on p. 184)
33. Buss, S.R., Hay, L.: On truth-table reducibility to SAT. Inf. Comput. **91**(1), 86–102 (1991). (cit. on pp. 141, 153)
34. Cadoli, M., Donini, F.M., Liberatore, P., Schaerf, M.: Preprocessing of intractable problems. Inf. Comput. **176**(2), 89–120 (2002). (cit. on pp. 286, 301–305, 308, 309)
35. Cai, J., Gundermann, T., Hartmanis, J., Hemachandra, L.A., Sewelson, V., Wagner, K.W., Wechsung, G.: The Boolean hierarchy I: structural properties. SIAM J. Comput. **17**(6), 1232–1252 (1988). (cit. on pp. 30, 139, 143–145, 231)

36. Cai, L., Chen, J., Downey, R.G., Fellows, M.R.: Advice classes of parameterized tractability. Ann. Pure Appl. Logic **84**(1), 119–138 (1997). (cit. on pp. 130, 289)
37. Cai, L., Juedes, D.: On the existence of subexponential parameterized algorithms. J. Comput. Syst. Sci. **67**(4), 789–807 (2003). (cit. on p. 272)
38. Cesati, M.: Compendium of parameterized problems (2006). http://bravo.ce.uniroma2.it/home/cesati/research/compendium.pdf, (cit. on p. 355)
39. Cesati, M., Ianni, M.D.: Computation models for parameterized complexity. Math. Logic Q. **43**, 179–202 (1997). (cit. on p. 130)
40. Chandra, A.K., Kozen, D.C., Stockmeyer, L.J.: Alternation. J. ACM **28**(1), 114–133 (1981). (cit. on p. 29)
41. Chang, R., Kadin, J.: The Boolean hierarchy and the polynomial hierarchy: a closer connection. SIAM J. Comput. **25**, 169–178 (1993). (cit. on pp. 30, 139, 147, 232)
42. Chang, R., Kadin, J.: The Boolean hierarchy and the polynomial hierarchy: a closer connection. SIAM J. Comput. **25**(2), 340–354 (1996). (cit. on p. 137)
43. Chen, H.: Parameter compilation. CoRR abs/1503.00260 (2015). (cit. on p. 301)
44. Chen, H.: Parameterized compilability. In: Proceedings of the 19th International Joint Conference on Artificial Intelligence (IJCAI 2005) (2005). (cit. on pp. 285–287, 301, 303, 304, 307, 308)
45. Chen, J., Chor, B., Fellows, M., Huang, X., Juedes, D., Kanj, I.A., Xia, G.: Tight lower bounds for certain parameterized NP-hard problems. Inf. Comput. **201**(2), 216–231 (2005). (cit. on pp. 272, 382)
46. Chen, J., Huang, X., Kanj, I.A., Xia, G.: Strong computational lower bounds via parameterized complexity. J. Comput. Syst. Sci. **72**(8), 1346–1367 (2006). (cit. on pp. 272, 382)
47. Chen, J., Kanj, I.A.: Parameterized complexity and subexponential-time computability. In: Bodlaender, H.L., Downey, R., Fomin, F.V., Marx, D. (eds.) The Multivariate Algorithmic Revolution and Beyond. LNCS, vol. 7370, pp. 162–195. Springer, Heidelberg (2012). https://doi.org/10.1007/978-3-642-30891-8_11. (cit. on pp. 272, 382)
48. Chen, Z.Z., Toda, S.: The complexity of selecting maximal solutions. Inf. Comput. **119**, 231–239 (1995). (cit. on pp. 30, 153)
49. Clarke, E., Grumberg, O., Jha, S., Lu, Y., Veith, H.: Progress on the state explosion problem in model checking. In: Wilhelm, R. (ed.) Informatics. LNCS, vol. 2000, pp. 176–194. Springer, Heidelberg (2001). https://doi.org/10.1007/3-540-44577-3_12. (cit. on p. 181)
50. Clarke, E.M., Emerson, E.A., Sifakis, J.: Model checking: algorithmic verification and debugging. Commun. ACM **52**(11), 74–84 (2009). (cit. on p. 181)
51. Clarke, E.M., Grumberg, O., Peled, D.A.: Model Checking. MIT Press, Cambridge (1999). (cit. on p. 181)
52. Clarke, E., Kroening, D., Ouaknine, J., Strichman, O.: Completeness and complexity of bounded model checking. In: Steffen, B., Levi, G. (eds.) VMCAI 2004. LNCS, vol. 2937, pp. 85–96. Springer, Heidelberg (2004). https://doi.org/10.1007/978-3-540-24622-0_9. (cit. on pp. 49, 181, 184, 192)
53. Clarke, E.M., Wing, J.M.: Formal methods: state of the art and future directions. ACM Comput. Surv. (CSUR) **28**(4), 626–643 (1996). (cit. on p. 181)
54. Cook, S.A.: The complexity of theorem-proving procedures. In: Proceedings 3rd Annual Symposium on Theory of Computing, Shaker Heights, Ohio, pp. 151–158 (1971). (cit. on pp. 24, 25, 48, 128, 150, 292)
55. Cormen, T.H., Leiserson, C.E., Rivest, R.L., Stein, C.: Introduction to Algorithms, vol. 6. MIT Press, Cambridge (2001). (cit. on p. 51)
56. Courcelle, B.: The monadic second-order logic of graphs. I. Recognizable sets of finite graphs. Inf. Comput. **85**, 12–75 (1990). (cit. on p. 234)
57. Cygan, M., et al.: Parameterized Algorithms. Springer, Cham (2015). https://doi.org/10.1007/978-3-319-21275-3. (cit. on pp. 3, 33)
58. Darwiche, A., Marquis, P.: A knowledge compilation map. J. Artif. Intell. Res. **17**, 229–264 (2002). (cit. on pp. 301, 309)

59. Dasgupta, S., Papadimitriou, C.H., Vazirani, U.: Coping with NP-completeness. In: Algorithms. McGraw-Hill Inc. (2008). Chap. 9, (cit. on p. 2)

60. Demri, S., Laroussinie, F., Schnoebelen, P.: A parametric analysis of the state-explosion problem in model checking. J. Comput. Syst. Sci. **72**(4), 547–575 (2006). (cit. on pp. 181, 183)

61. Dietrich, F., List, C.: Judgment aggregation by quota rules: majority voting generalized. J. Theor. Polit. **19**(4), 391–424 (2007). (cit. on p. 237)

62. Dietrich, F., List, C.: Judgment aggregation without full rationality. Soc. Choice Welfare **31**(1), 15–39 (2008). (cit. on p. 222)

63. Downey, R.G., Fellows, M.R.: Parameterized complexity: a framework for systematically confronting computational intractability. In: Contemporary Trends in Discrete Mathematics: From DIMACS and DIMATIA to the Future. AMS-DIMACS, vol. 49, pp. 49–99. American Mathematical Society (1999). (cit. on p. 2)

64. Downey, R.: A basic parameterized complexity primer. In: Bodlaender, H.L., Downey, R., Fomin, F.V., Marx, D. (eds.) The Multivariate Algorithmic Revolution and Beyond. LNCS, vol. 7370, pp. 91–128. Springer, Heidelberg (2012). https://doi.org/10.1007/978-3-642-30891-8_9. (cit. on p. 33)

65. Downey, R.G., Fellows, M.R.: Fixed-parameter tractability and completeness. II. On completeness for $W[1]$. Theor. Comput. Sci. **141**(1–2), 109–131 (1995). (cit. on p. 116)

66. Downey, R.G., Fellows, M.R.: Fundamentals of Parameterized Complexity. TCS. Springer, London (2013). https://doi.org/10.1007/978-1-4471-5559-1. ISBN: 978-1-4471-5558-4, 978-1-4471-5559-1, (cit. on pp. 3, 21, 33, 45, 87, 116)

67. Downey, R.G., Fellows, M.R., Stege, U.: Parameterized complexity: a framework for systematically confronting computational intractability. In: Contemporary Trends in Discrete Mathematics: From DIMACS and DIMATIA to the Future. AMS-DIMACS, vol. 49, pp. 49–99. American Mathematical Society (1999). (cit. on pp. 1, 3, 33, 41, 45, 87, 116, 118, 285, 289)

68. Downey, R.G., Fellows, M.R., Stege, U.: Computational tractability: the view from mars. Bull. Eur. Assoc. Theor. Comput. Sci. **69**, 73–97 (1999). (cit. on p. 2)

69. Du, D.Z., Ko, K.I.: Theory of Computational Complexity. Wiley, Hoboken (2011). (cit. on p. 21)

70. Dvořák, W., Järvisalo, M., Wallner, J.P., Woltran, S.: Complexity-sensitive decision procedures for abstract argumentation. Artif. Intell. **206**, 53–78 (2014). (cit. on p. 137)

71. Eiter, T., Gottlob, G.: On the computational cost of disjunctive logic programming: propositional case. Ann. Math. Artif. Intell. **15**(3–4), 289–323 (1995). (cit. on pp. 59, 73, 75, 78, 90, 91, 122, 162)

72. Eiter, T., Gottlob, G.: The complexity of logic-based abduction. J. ACM **42**(1), 3–42 (1995). (cit. on pp. 58, 162)

73. Endriss, U.: Judgment aggregation. In: Brandt, F., Conitzer, V., Endriss, U., Lang, J., Procaccia, A. (eds.) Handbook of Computational Social Choice. Cambridge University Press, Cambridge (2016). (cit. on p. 222)

74. Endriss, U., Grandi, U., de HaanDede Haan, R., Lang, J.: Succinctness of languages for judgment aggregation. In: Proceedings of the 15th International Conference on the Principles of Knowledge Representation and Reasoning (KR 2016), Cape Town, South Africa, 25–29 April 2016, pp. 176–186. AAAI Press (2016). http://www.ac.tuwien.ac.at/files/tr/ac-tr-16-006.pdf, (cit. on p. 224)

75. Endriss, U., Grandi, U., Porello, D.: Complexity of judgment aggregation. J. Artif. Intell. Res. **45**, 481–514 (2012). (cit. on pp. 61, 222, 225, 228, 235, 236, 239)

76. Endriss, U., DE HaanDede Haan, R.: Complexity of the winner determination problem in judgment aggregation: Kemeny, Slater, Tideman, Young. In: Weiss, G., Yolum, P., Bordini, R.H., Elkind, E. (eds.) Proceedings of the 14th International Conference on Autonomous Agents and Multiagent Systems (AAMAS 2015), pp. 117–125. IFAAMAS/ACM (2015). https://www.ac.tuwien.ac.at/files/tr/ac-tr-15-010.pdf, (cit. on p. 245)

77. Endriss, U., de HaanDede Haan, R., Szeider, S.: Parameterized complexity results for agenda safety in judgment aggregation. In: Proceedings of the 5th International Workshop on Computational Social Choice (COMSOC 2014). Carnegie Mellon University (2014). https://www.ac.tuwien.ac.at/files/tr/ac-tr-16-005.pdf, (cit. on pp. 18, 135, 157, 250)

78. Endriss, U., de HaanDede Haan, R., Szeider, S.: Parameterized complexity results for agenda safety in judgment aggregation. In: Weiss, G., Yolum, P., Bordini, R.H., Elkind, E. (eds.) Proceedings of the 14th International Conference on Autonomous Agents and Multiagent Systems (AAMAS 2015), pp. 127–136. IFAAMAS/ACM (2015). https://www.ac.tuwien.ac.at/files/tr/ac-tr-16-005.pdf, (cit. on pp. 18, 135, 157, 250)

79. Fages, F.: Consistency of Clark's completion and existence of stable models. Methods Logic Comput. Sci. 1(1), 51–60 (1994). (cit. on p. 162)

80. Fellows, M.R., Hermelin, D., Rosamond, F.A., Vialette, S.: On the parameterized complexity of multiple-interval graph problems. Theoret. Comput. Sci. 410(1), 53–61 (2009). (cit. on pp. 38, 306)

81. Fichte, J.K., Truszczyński, M., Woltran, S.: Dual-normal logic programs-the forgotten class. Theory Pract. Log. Program. 15(4–5), 495–510 (2015). (cit. on pp. 72, 82)

82. Fichte, J.K., Szeider, S.: Backdoors to normality for disjunctive logic programs. In: Proceedings of the 27th AAAI Conference on Artificial Intelligence (AAAI 2013), pp. 320–327. AAAI Press (2013). (cit. on pp. 3, 59, 75, 162, 362)

83. Fisher, M.: Temporal representation and reasoning. In: Handbook of Knowledge Representation, Foundations of Artificial Intelligence, vol. 3, pp. 513–550. Elsevier (2008). (cit. on p. 181)

84. Flum, J., Grohe, M.: Describing parameterized complexity classes. Inf. Comput. 187(2), 291–319 (2003). (cit. on pp. 37, 41, 55, 62, 77, 102, 130, 132, 138, 262, 263)

85. Flum, J., Grohe, M.: Parameterized Complexity Theory. TTCSAES. Springer, Heidelberg (2006). https://doi.org/10.1007/3-540-29953-X. (cit. on pp. 3, 22, 28, 33, 40, 45, 60, 65, 88, 95, 119, 130, 181, 183, 187, 195, 234, 272, 285, 289, 338, 341)

86. Garey, M.R., Johnson, D.R.: Computers and Intractability. W. H. Freeman and Company, New York, San Francisco (1979). (cit. on pp. 2, 25, 48)

87. Gasarch, W.I.: Sigact news complexity theory column 36. SIGACT News 33, 34–47 (2002). (cit. on p. 25)

88. Gaspers, S., Szeider, S.: Backdoors to satisfaction. In: Bodlaender, H.L., Downey, R., Fomin, F.V., Marx, D. (eds.) The Multivariate Algorithmic Revolution and Beyond. LNCS, vol. 7370, pp. 287–317. Springer, Heidelberg (2012). https://doi.org/10.1007/978-3-642-30891-8_15. (cit. on pp. 35, 58)

89. Gebser, M., Kaufmann, B., Neumann, A., Schaub, T.: Conflict-driven answer set solving. In: Proceedings of the 20th International Joint Conference on Artificial Intelligence (IJCAI 2007), pp. 386–392. MIT Press (2007). (cit. on p. 162)

90. Gelfond, M.: Answer sets. In: van Harmelen, F., Lifschitz, V., Porter, B. (eds.) Handbook of Knowledge Representation. Elsevier Science, San Diego (2007). (cit. on pp. 51, 59, 72)

91. Gelfond, M., Lifschitz, V.: Classical negation in logic programs and disjunctive databases. New Gener. Comput. 9(3/4), 365–386 (1991). (cit. on pp. 59, 72)

92. Giunchiglia, E., Lierler, Y., Maratea, M.: Answer set programming based on propositional satisfiability. J. Autom. Reasoning 36, 345–377 (2006). (cit. on p. 162)

93. Goldreich, O.: Computational Complexity: A Conceptual Perspective. Cambridge University Press, Cambridge (2008). (cit. on p. 21)

94. Goldreich, O.: P, NP, and NP-Completeness: The Basics of Complexity Theory. Cambridge University Press, Cambridge (2010). (cit. on p. 21)

95. Göller, S.: The fixed-parameter tractability of model checking concurrent systems. In: Proceedings of the 22nd EACSL Annual Conference on Computer Science Logic (CSL 2013). LIPIcs, vol. 23, pp. 332–347 (2013). (cit. on pp. 181, 183)

96. Gomes, C.P., Kautz, H., Sabharwal, A., Selman, B.: Satisfiability solvers. In: van Harmele, F., Lifschitz, V., Porter, B. (eds.) Handbook of Knowledge Representation, Foundations of Artificial Intelligence, vol. 3, pp. 89–134. Elsevier, Oxford (2008). (cit. on pp. 45, 47)

97. Gomes, C.P., Selman, B., Kautz, H., et al.: Boosting combinatorial search through randomization. In: 15th National Conference on Artificial Intelligence and 10th Innovative Applications of Artificial Intelligence Conference (AAAI 1998/IAAI 1998), pp. 431–437 (1998). (cit. on p. 49)

98. Gottlob, G.: On minimal constraint networks. Artif. Intell. **191–192**, 42–60 (2012). (cit. on p. 174)

99. Gottlob, G., Fermüller, C.G.: Removing redundancy from a clause. Artif. Intell. **61**(2), 263–289 (1993). (cit. on pp. 156, 342)

100. Gottlob, G., Pichler, R., Wei, F.: Bounded treewidth as a key to tractability of knowledge representation and reasoning. Artif. Intell. **174**(1), 105–132 (2010). (cit. on p. 54)

101. Gottlob, G., Scarcello, F., Sideri, M.: Fixed-parameter complexity in AI and nonmonotonic reasoning. Artif. Intell. **138**(1–2), 55–86 (2002). (cit. on p. 88)

102. Grandi, U.: Binary aggregation with integrity constraints. Ph.D. thesis, University of Amsterdam (2012). (cit. on pp. 222, 223, 239)

103. Grandi, U., Endriss, U.: Lifting integrity constraints in binary aggregation. Artif. Intell. **199–200**, 45–66 (2013). (cit. on pp. 222, 223)

104. Grossi, D., Pigozzi, G.: Judgment Aggregation: A Primer. Morgan & Claypool Publishers, Los Altos (2014). (cit. on p. 222)

105. de Haan, R.: An overview of non-uniform parameterized complexity. Technical report, TR15-130, Electronic Colloquium on Computational Complexity (ECCC) (2015). (cit. on pp. 18, 135, 332)

106. de Haan, R.: Parameterized complexity results for the Kemeny rule in judgment aggregation. In: Proceedings of the 6th International Workshop on Computational Social Choice (COMSOC 2016). University of Toulouse (2016). (cit. on pp. 18, 250)

107. de Haan, R.: Parameterized complexity results for the Kemeny rule in judgment aggregation. In: Kaminka, G.A., et al. (eds.) Proceedings of the 22nd European Conference on Artificial Intelligence (ECAI 2016), Frontiers in Artificial Intelligence and Applications, vol. 285, pp. 1502–1510 (2016). (cit. on pp. 18, 250)

108. de Haan, R., Kanj, I.A., Szeider, S.: On the parameterized complexity of finding small unsatisfiable subsets of CNF formulas and CSP instances. ACM Trans. Comput. Log. **18**(3), 21:1–21:46 (2017). (cit. on pp. 18, 69)

109. de Haan, R., Kronegger, M., Pfandler, A.: Fixed-parameter tractable reductions to SAT for planning. In: Yang, Q., Wooldridge, M. (eds.) Proceedings of the 24th International Joint Conference on Artificial Intelligence (IJCAI 2015), Buenos Aires, Argentina, 25–31 July 2015, pp. 2897–2903 (2015). http://www.ac.tuwien.ac.at/files/tr/ac-tr-15-011.pdf, (cit. on pp. 18, 135, 157, 259)

110. de Haan, R., Pfandler, A., Rümmele, S., Szeider, S.: Backdoors to abduction (2013). (cit. on pp. 18, 179)

111. de Haan, R., Szeider, S.: Compendium of parameterized problems at higher levels of the Polynomial Hierarchy. Tech. report, TR14-143, Electronic Colloquium on Computational Complexity (ECCC) (2014). (cit. on pp. 18, 135, 157)

112. de Haan, R., Szeider, S.: Fixed-parameter tractable reductions to SAT. In: Sinz, C., Egly, U. (eds.) SAT 2014. LNCS, vol. 8561, pp. 85–102. Springer, Cham (2014). https://doi.org/10. 1007/978-3-319-09284-3_8. (cit. on pp. 18, 69, 135, 157, 218)

113. de Haan, R., Szeider, S.: Machine characterizations for parameterized complexity classes beyond para-NP. In: Italiano, G.F., Margaria-Steffen, T., Pokorný, J., Quisquater, J.-J., Wattenhofer, R. (eds.) SOFSEM 2015. LNCS, vol. 8939, pp. 217–229. Springer, Heidelberg (2015). https://doi.org/10.1007/978-3-662-46078-8_18. (cit. on pp. 18, 135, 268)

114. de Haan, R., Szeider, S.: Parameterized complexity classes beyond para-NP. J. Comput. Syst. Sci. **87**, 16–57 (2017). (cit. on pp. 18, 135, 283)

115. de Haan, R., Szeider, S.: Parameterized complexity results for symbolic model checking of temporal logics. In: Proceedings of the 15th International Conference on the Principles of Knowledge Representation and Reasoning (KR 2016), Cape Town, South Africa, 25–29 April 2016, pp. 453–462. AAAI Press (2016). http://www.ac.tuwien.ac.at/files/tr/ac-tr-15-002.pdf, (cit. on pp. 18, 204)

116. de Haan, R., Szeider, S.: The parameterized complexity of reasoning problems beyond NP. In: Baral, C., De Giacomo, G., Eiter, T. (eds.) Proceedings of the 14th International Conference on the Principles of Knowledge Representation and Reasoning (KR 2014), Vienna, Austria, 20–24 July 2014. AAAI Press (2014). Full version available as [117], (cit. on pp. 18, 83, 135, 179)

117. de Haan, R., Szeider, S.: The parameterized complexity of reasoning problems beyond NP. Technical report, arXiv.org/abs/1312.1672 (2014). (cit. on pp. 18, 83, 135, 179, 268, 283, 393)

118. Hartmanis, J., Immerman, N., Sewelson, V.: Sparse sets in NP-P: EXPTIME versus NEXP-TIME. Inf. Control **65**(2), 158–181 (1985). (cit. on p. 315)

119. Hartmanis, J.: New developments in structural complexity theory. Theoret. Comput. Sci. **71**(1), 79–93 (1990). (cit. on p. 137)

120. Hartmanis, J., Stearns, R.E.: On the computational complexity of algorithms. Am. Math. Soc. **117**, 285–306 (1965). (cit. on p. 294)

121. Hemachandra, L.A.: The strong exponential hierarchy collapses. J. Comput. Syst. Sci. **39**(3), 299–322 (1989). (cit. on pp. 141, 153)

122. Hooker, J.N.: Solving the incremental satisfiability problem. J. Logic Program. **15**(1&2), 177–186 (1993). (cit. on p. 140)

123. Hopcroft, J.E., Motwani, R., Ullman, J.D.: Introduction to Automata Theory, Languages, and Computation, 2nd edn. Addison-Wesley Series in Computer Science, pp. I–XIV, 1–521. Addison-Wesley-Longman (2001). ISBN: 978-0-201-44124-6, (cit. on p. 104)

124. Janhunen, T., Niemelä, I., Seipel, D., Simons, P., You, J.H.: Unfolding partiality and disjunctions in stable model semantics. ACM Trans. Comput. Log. **7**(1), 1–37 (2006). (cit. on p. 162)

125. Jenner, B., Torán, J.: Computing functions with parallel queries to NP. Theoret. Comput. Sci. **141**(1), 175–193 (1995). (cit. on p. 157)

126. Kadin, J.: The polynomial time hierarchy collapses if the Boolean hierarchy collapses. SIAM J. Comput. **17**(6), 1263–1282 (1988). (cit. on pp. 30, 139, 147, 232)

127. Karp, R.M., Lipton, R.J.: Some connections between nonuniform and uniform complexity classes. In: Proceedings of the 12th Annual ACM Symposium on Theory of Computing (STOC 1980), pp. 302–309. Association for Computing Machinery, New York (1980). https://doi.org/10.1145/800141.804678, (cit. on pp. 32, 285, 300)

128. Karp, R.M., Lipton, R.J.: Turing machines that take advice. L'enseignement mathématique **28**(2), 191–209 (1982). (cit. on pp. 32, 285, 300)

129. Katebi, H., Sakallah, K.A., Marques-Silva, J.P.: Empirical study of the anatomy of modern sat solvers. In: Sakallah, K.A., Simon, L. (eds.) SAT 2011. LNCS, vol. 6695, pp. 343–356. Springer, Heidelberg (2011). https://doi.org/10.1007/978-3-642-21581-0_27. (cit. on p. 47)

130. Kautz, H., Selman, B.: Pushing the envelope: planning, propositional logic, and stochastic search. In: Proceedings of the 13th AAAI Conference on Artificial Intelligence (AAAI 1996), pp. 1194–1201. AAAI Press (1996). (cit. on p. 49)

131. Kautz, H.A., Selman, B.: Planning as satisfiability. In: Proceedings of the 10th European Conference on Artificial Intelligence (ECAI 1992), pp. 359–363 (1992). (cit. on p. 49)

132. Kloks, T.: Treewidth: Computations and Approximations. Springer, Heidelberg (1994). https://doi.org/10.1007/BFb0045375. (cit. on p. 54)

133. Köbler, J., Schöning, U., Wagner, K.W.: The difference and truth-table hierarchies for NP. RAIRO - Theor. Inf. Appl. **21**(4), 419–435 (1987). (cit. on pp. 141, 144, 153)

134. Köbler, J., Thierauf, T.: Complexity classes with advice. In: Proceedings of the 5th Annual Structure in Complexity Theory Conference, Universitat Politècnica de Catalunya, Barcelona, Spain, 8–11 July 1990, pp. 305–315. IEEE Computer Society (1990). (cit. on p. 153)

135. Kolaitis, P.G., Vardi, M.Y.: Conjunctive-query containment and constraint satisfaction. J. Comput. Syst. Sci. **61**(2), 302–332 (2000). Special issue on the Seventeenth ACM SIGACT-SIGMOD-SIGART Symposium on Principles of Database Systems (Seattle, WA, 1998), (cit. on p. 57)

136. Kozen, D.: Theory of Computation. Springer, London (2006). https://doi.org/10.1007/1-84628-477-5. (cit. on p. 21)
137. Krajicek, J.: Bounded Arithmetic, Propositional Logic and Complexity Theory. Cambridge University Press, New York (1995). (cit. on pp. 154, 156)
138. Krentel, M.W.: The complexity of optimization problems. J. Comput. Syst. Sci. **36**(3), 490–509 (1988). (cit. on pp. 30, 137)
139. Kroening, D., Ouaknine, J., Strichman, O., Wahl, T., Worrell, J.: Linear completeness thresholds for bounded model checking. In: Gopalakrishnan, G., Qadeer, S. (eds.) CAV 2011. LNCS, vol. 6806, pp. 557–572. Springer, Heidelberg (2011). https://doi.org/10.1007/978-3-642-22110-1_44. (cit. on pp. 60, 184, 186, 192, 197, 204)
140. Kronegger, M., Pfandler, A., Pichler, R.: Parameterized complexity of optimal planning: a detailed map. In: Proceedings of the 23rd International Joint Conference on Artificial Intelligence (IJCAI 2013), pp. 954–961. AAAI Press (2013). (cit. on p. 253)
141. Kupferman, O., Vardi, M.Y., Wolper, P.: An automata-theoretic approach to branching-time model checking. J. ACM **47**(2), 312–360 (2000). (cit. on p. 188)
142. Lang, J., Slavkovik, M.: How hard is it to compute majority-preserving judgment aggregation rules? In: Proceedings of the 21st European Conference on Artificial Intelligence (ECAI 2014). IOS Press (2014). (cit. on pp. 222, 239)
143. Lee, J., Lifschitz, V.: Loop formulas for disjunctive logic programs. In: Palamidessi, C. (ed.) ICLP 2003. LNCS, vol. 2916, pp. 451–465. Springer, Heidelberg (2003). https://doi.org/10.1007/978-3-540-24599-5_31. (cit. on p. 162)
144. Levin, L.: Universal sequential search problems. Probl. Inf. Transm. **9**(3), 265–266 (1973). (cit. on pp. 24, 25, 48, 128, 150, 292)
145. Liberatore, P.: Redundancy in logic I: CNF propositional formulae. Artif. Intell. **163**(2), 203–232 (2005). (cit. on p. 64)
146. Lifschitz, V., Razborov, A.: Why are there so many loop formulas? ACM Trans. Comput. Log. **7**(2), 261–268 (2006). (cit. on p. 162)
147. Lin, F., Zhao, X.: On odd and even cycles in normal logic programs. In: Cohn, A.G. (ed.) Proceedings of the 19th National Conference on Artifical Intelligence (AAAI 2004), pp. 80–85. AAAI Press (2004). (cit. on p. 162)
148. List, C.: The theory of judgment aggregation: an introductory review. Synthese **187**(1), 179–207 (2012). (cit. on pp. 222, 237)
149. List, C., Puppe, C.: Judgment aggregation: a survey. In: Anand, P., Pattanaik, P., Puppe, C. (eds.) Handbook of Rational and Social Choice. Oxford University Press, Oxford (2009). (cit. on p. 222)
150. Lokshtanov, D., Marx, D., Saurabh, S.: Lower bounds based on the exponential time hypothesis. Bull. Eur. Assoc. Theor. Comput. Sci. **105**, 41–72 (2011). (cit. on p. 272)
151. Lück, M., Meier, A., Schindler, I.: Parameterized complexity of CTL. In: Dediu, A.-H., Formenti, E., Martín-Vide, C., Truthe, B. (eds.) LATA 2015. LNCS, vol. 8977, pp. 549–560. Springer, Cham (2015). https://doi.org/10.1007/978-3-319-15579-1_43. (cit. on pp. 181, 183)
152. Mahaney, S.R.: Sparse complete sets for NP: solution of a conjecture of Berman and Hartmanis. In: Proceedings of the 21st Annual IEEE Symposium on Foundations of Computer Science (FOCS 1980), pp. 54–60. IEEE Computer Society (1980). (cit. on p. 315)
153. Malik, S., Zhang, L.: Boolean satisfiability from theoretical hardness to practical success. Commun. ACM **52**(8), 76–82 (2009). (cit. on p. 45)
154. Marek, V.W., Truszczynski, M.: Stable models and an alternative logic programming paradigm. In: Apt, K.R., Marek, V.W., Truszczynski, M., Warren, D.S. (eds.) The Logic Programming Paradigm: A 25-Year Perspective, pp. 169–181. Springer, Heidelberg (1999). https://doi.org/10.1007/978-3-642-60085-2_17. (cit. on pp. 51, 59, 72)
155. Marek, W., Truszczyński, M.: Autoepistemic logic. J. ACM **38**(3), 588–619 (1991). (cit. on p. 78)
156. Marques-Silva, J., Janota, M., Belov, A.: Minimal sets over monotone predicates in Boolean formulae. In: Sharygina, N., Veith, H. (eds.) CAV 2013. LNCS, vol. 8044, pp. 592–607. Springer, Heidelberg (2013). https://doi.org/10.1007/978-3-642-39799-8_39. (cit. on p. 137)

157. Marques-Silva, J., Sakallah, K.: GRASP - a new search algorithm for satisfiability. In: Proceedings of the 1996 International Conference on Computer-Aided Design (ICCAD 1996), San Jose, CA, USA, 10–14 November 1996, pp. 220–227. ACM, IEEE (1996). (cit. on p. 49)
158. Marques-Silva, J., Sakallah, K.: GRASP: a search algorithm for propositional satisfiability. IEEE Trans. Comput. **48**(5), 506–521 (1999). (cit. on p. 49)
159. Marques-Silva, J.P., Lynce, I., Malik, S.: Conflict-driven clause learning SAT solvers. In: Biere, A., Heule, M., van Maaren, H., Walsh, T. (eds.) Handbook of Satisfiability, pp. 131–153. IOS Press, Amsterdam (2009). (cit. on p. 45)
160. Marquis, P.: Compile! In: Bonet, B., Koenig, S. (eds.) Proceedings of the 29th AAAI Conference on Artificial Intelligence (AAAI 2015), pp. 4112–4118. AAAI Press (2015). (cit. on pp. 301, 309)
161. Meier, A.: On the complexity of modal logic variants and their fragments. Ph.D. thesis, Gottfried Wilhelm Leibniz Universität Hannover (2011). (cit. on pp. 181, 183)
162. Mengel, S.: Private communication (2015). (cit. on p. 315)
163. Meyer, A.R., Stockmeyer, L.J.: The equivalence problem for regular expressions with squaring requires exponential space. In: Proceedings of the 1972 Symposium on Switching and Automata Theory (SWAT 1972), pp. 125–129. IEEE Computer Society (1972). (cit. on p. 26)
164. Moskewicz, M.W., Madigan, C.F., Zhao, Y., Zhang, L., Malik, S.: Chaff: engineering an efficient sat solver. In: Proceedings of the 38th Design Automation Conference (DAC 2001), Las Vegas, NV, USA, 18–22 June 2001, pp. 530–535. ACM (2001). (cit. on p. 49)
165. Nehring, K., Puppe, C.: The structure of strategy-proof social choice - Part I: general characterization and possibility results on median spaces. J. Econ. Theor. **135**(1), 269–305 (2007). (cit. on pp. 61, 225)
166. Niedermeier, R.: Invitation to Fixed-Parameter Algorithms. Oxford Lecture Series in Mathematics and its Applications. Oxford University Press, Oxford (2006). (cit. on pp. 3, 33, 36, 45)
167. Nishimura, N., Ragde, P., Szeider, S.: Detecting backdoor sets with respect to Horn and binary clauses. In: Proceedings of the 7th International Conference on Theory and Applications of Satisfiability Testing (SAT 2004), Vancouver, BC, Canada, 10–13 May 2004, pp. 96–103 (2004). (cit. on p. 58)
168. Nordh, G., Zanuttini, B.: What makes propositional abduction tractable. Artif. Intell. **172**(10), 1245–1284 (2008). (cit. on pp. 58, 162)
169. Pan, G., Vardi, M.Y.: Fixed-parameter hierarchies inside PSPACE. In: Proceedings of the 21st IEEE Symposium on Logic in Computer Science (LICS 2006), Seattle, WA, USA, 12–15 August 2006, pp. 27–36. IEEE Computer Society (2006). (cit. on p. 57)
170. Papadimitriou, C.H.: Computational Complexity. Addison-Wesley, Reading (1994). (cit. on pp. 16, 21, 26)
171. Papadimitriou, C.H., Zachos, S.K.: Two remarks on the power of counting. In: Cremers, A.B., Kriegel, H.-P. (eds.) GI-TCS 1983. LNCS, vol. 145, pp. 269–275. Springer, Heidelberg (1982). https://doi.org/10.1007/BFb0036487. (cit. on p. 139)
172. Pednault, E.P.D.: ADL: exploring the middle ground between STRIPS and the situation calculus. In: Proceedings of the 1st International Conference on Principles of Knowledge Representation and Reasoning (KR 1989), pp. 324–332 (1989). (cit. on p. 252)
173. Pfandler, A., Rümmele, S., Szeider, S.: Backdoors to abduction. In: Rossi, F. (ed.) Proceedings of the 23rd International Joint Conference on Artificial Intelligence (IJCAI 2013). AAAI Press/IJCAI (2013). (cit. on pp. 3, 58, 163, 363)
174. Rintanen, J.: Asymptotically optimal encodings of conformant planning in QBF. In: Holte, R.C., Howe, A.E. (eds.) Proceedings of the 22nd AAAI Conference on Artificial Intelligence (AAAI 2007), pp. 1045–1050. AAAI Press (2007). (cit. on p. 255)
175. Rossi, F., van Beek, P., Walsh, T. (eds.): Handbook of Constraint Programming. Elsevier, New York (2006). (cit. on p. 50)
176. Sakallah, K.A., Marques-Silva, J.: Anatomy and empirical evaluation of modern SAT solvers. Bull. Eur. Assoc. Theor. Comput. Sci. **103**, 96–121 (2011). (cit. on p. 45)

177. Samer, M., Szeider, S.: Constraint satisfaction with bounded treewidth revisited. J. Comput. Syst. Sci. **76**(2), 103–114 (2010). (cit. on p. 57)

178. Samulowitz, H., Davies, J., Bacchus, F.: Preprocessing QBF. In: Benhamou, F. (ed.) CP 2006. LNCS, vol. 4204, pp. 514–529. Springer, Heidelberg (2006). https://doi.org/10.1007/11889205_37. (cit. on p. 27)

179. Schaefer, M., Umans, C.: Completeness in the polynomial-time hierarchy: a compendium. SIGACT News **33**(3), 32–49 (2002). (cit. on pp. 15, 355, 375)

180. Selman, B., Kautz, H.A.: Knowledge compilation and theory approximation. J. ACM **43**, 193–224 (1996). (cit. on p. 301)

181. Selman, B., Levesque, H.J.: Abductive and default reasoning: a computational core. In: Proceedings of the 8th National Conference on Artificial Intelligence (AAAI 1990), Boston, Massachusetts, 29 July–3 August 1990, pp. 343–348 (1990). (cit. on pp. 58, 162)

182. Siekmann, J., Wrightson, G. (eds.): Automation of Reasoning. Classical Papers on Computer Science 1967–1970, vol. 2. Springer, Heidelberg (1983). https://doi.org/10.1007/978-3-642-81955-1. (cit. on p. 396)

183. Sipser, M.: Introduction to the Theory of Computation. Cengage Learning, Boston (2012). (cit. on p. 21)

184. Spakowski, H.: Completeness for parallel access to NP and counting class separations. Ph.D. thesis, Heinrich-Heine-Universität Düsseldorf (2005). (cit. on pp. 30, 147, 154)

185. Stockmeyer, L.J.: The polynomial-time hierarchy. Theoret. Comput. Sci. **3**(1), 1–22 (1976). (cit. on pp. 26, 27)

186. Stockmeyer, L.J., Meyer, A.R.: Word problems requiring exponential time. In: Proceedings of the 5th Annual ACM Symposium on Theory of Computing (STOC 1973), pp. 1–9. ACM (1973). (cit. on p. 28)

187. Truszczyński, M.: Trichotomy and dichotomy results on the complexity of reasoning with disjunctive logic programs. Theory Pract. Log. Program. **11**(6), 881–904 (2011). (cit. on p. 78)

188. Tseitin, G.S.: Complexity of a derivation in the propositional calculus. Zap. Nauchn. Sem. Leningrad Otd. Mat. Inst. Akad. Nauk SSSR **8**, 23–41 (1968). English translation reprinted in [182], (cit. on pp. 55, 171, 216, 218, 226, 244, 247, 322)

189. Turing, A.M.: On computable numbers, with an application to the entscheidungsproblem. Proc. Lond. Math. Soc. **42**, 230–265 (1936). (cit. on p. 23)

190. Umans, C.: Approximability and completeness in the polynomial hierarchy. Ph.D. thesis, University of California, Berkeley (2000). (cit. on pp. 206, 209)

191. Vardi, M.Y.: Boolean satisfiability: theory and engineering. Commun. ACM **57**(3), 5 (2014). (cit. on pp. 45, 47)

192. Vardi, M.Y.: The moral hazard of complexity-theoretic assumptions. Commun. ACM **59**(2), 5 (2016). (cit. on p. 16)

193. Wagner, K.W.: Bounded query classes. SIAM J. Comput. **19**(5), 833–846 (1990). (cit. on pp. 30, 137)

194. Wechsung, G.: On the Boolean closure of NP. In: Budach, L. (ed.) FCT 1985. LNCS, vol. 199, pp. 485–493. Springer, Heidelberg (1985). https://doi.org/10.1007/BFb0028832. (cit. on p. 144)

195. Whittemore, J., Kim, J., Sakallah, K.A.: SATIRE: a new incremental satisfiability engine. In: Proceedings of the 38th Design Automation Conference (DAC 2001), Las Vegas, NV, USA, 18–22 June 2001, pp. 542–545. ACM (2001). (cit. on p. 140)

196. Williams, H.P.: Logic and Integer Programming. Springer, Boston (2009). (cit. on p. 51)

197. Williams, R., Gomes, C., Selman, B.: Backdoors to typical case complexity. In: Gottlob, G., Walsh, T. (eds.) Proceedings of the 18th International Joint Conference on Artificial Intelligence (IJCAI 2003), pp. 1173–1178. Morgan Kaufmann, San Francisco (2003). (cit. on p. 58)

198. Wrathall, C.: Complete sets and the polynomial-time hierarchy. Theoret. Comput. Sci. **3**(1), 23–33 (1976). (cit. on pp. 26, 27)

Index of Parameterized Problems

© Springer-Verlag GmbH Germany, part of Springer Nature 2019
R. de Haan: Parameterized Complexity in the Polynomial Hierarchy, LNCS 11880,
https://doi.org/10.1007/978-3-662-60670-4

Printed in the United States
By Bookmasters